Marine Glycoconjugates

Marine Glycoconjugates

Trends and Perspectives

Special Issue Editors

Vladimir I. Kalinin
Valentin A. Stonik
Natalia V. Ivanchina

MDPI • Basel • Beijing • Wuhan • Barcelona • Belgrade • Manchester • Tokyo • Cluj • Tianjin

Special Issue Editors
Vladimir I. Kalinin
G.B. Elyakov Pacific Institute of Bioorganic Chemistry
Russia

Valentin A. Stonik
G.B. Elyakov Pacific Institute of Bioorganic Chemistry
Russia

Natalia V. Ivanchina
G.B. Elyakov Pacific Institute of Bioorganic Chemistry
Russia

Editorial Office
MDPI
St. Alban-Anlage 66
4052 Basel, Switzerland

This is a reprint of articles from the Special Issue published online in the open access journal *Marine Drugs* (ISSN 1660-3397) (available at: https://www.mdpi.com/journal/marinedrugs/special_issues/Marine_Glycoconjugates_Trends_Perspectives).

For citation purposes, cite each article independently as indicated on the article page online and as indicated below:

LastName, A.A.; LastName, B.B.; LastName, C.C. Article Title. *Journal Name* **Year**, *Article Number*, Page Range.

ISBN 978-3-03928-558-7 (Pbk)
ISBN 978-3-03928-559-4 (PDF)

Cover image courtesy of Vladimir B. Krasokhin.

Contents

About the Special Issue Editors . . . . **vii**

Preface to "Marine Glycoconjugates" . . . . **ix**

Vladimir I. Kalinin, Valentin A. Stonik and Natalia V. Ivanchina
Marine Glycoconjugates: Trends and Perspectives
Reprinted from: *Mar. Drugs* **2020**, *18*, 120, doi:10.3390/md18020120 . . . . **1**

Valentin A. Stonik and Inna V. Stonik
Sterol and Sphingoid Glycoconjugates from Microalgae
Reprinted from: *Mar. Drugs* **2018**, *16*, 514, doi:10.3390/md16120514 . . . . **7**

Christian Galasso, Genoveffa Nuzzo, Christophe Brunet, Adrianna Ianora, Angela Sardo, Angelo Fontana and Clementina Sansone
The Marine Dinoflagellate *Alexandrium minutum* Activates a Mitophagic Pathway in Human Lung Cancer Cells
Reprinted from: *Mar. Drugs* **2018**, *16*, 502, doi:10.3390/md16120502 . . . . **27**

Eduard Kostetsky, Natalia Chopenko, Maria Barkina, Peter Velansky and Nina Sanina
Fatty Acid Composition and Thermotropic Behavior of Glycolipids and Other Membrane Lipids of *Ulva lactuca* (Chlorophyta) Inhabiting Different Climatic Zones
Reprinted from: *Mar. Drugs* **2018**, *16*, 494, doi:10.3390/md16120494 . . . . **39**

Elena Catanzaro, Cinzia Calcabrini, Anupam Bishayee and Carmela Fimognari
Antitumor Potential of Marine and Freshwater Lectins
Reprinted from: *Mar. Drugs* **2020**, *18*, 11, doi:10.3390/md18010011 . . . . **53**

Tao Wu, Yulin Xiang, Tingting Liu, Xue Wang, Xiaoyuan Ren, Ting Ye and Gongchu Li
Oncolytic Vaccinia Virus Expressing *Aphrocallistes vastus* Lectin as a Cancer Therapeutic Agent
Reprinted from: *Mar. Drugs* **2019**, *17*, 363, doi:10.3390/md17060363 . . . . **93**

Imtiaj Hasan, Marco Gerdol, Yuki Fujii and Yasuhiro Ozeki
Functional Characterization of OXYL, A SghC1qDC LacNAc-specific Lectin from The Crinoid Feather Star *Anneissia Japonica*
Reprinted from: *Mar. Drugs* **2019**, *17*, 136, doi:10.3390/md17020136 . . . . **105**

Timofey V. Malyarenko, Olesya S. Malyarenko, Alla A. Kicha, Natalia V. Ivanchina, Anatoly I. Kalinovsky, Pavel S. Dmitrenok, Svetlana P. Ermakova and Valentin A. Stonik
In Vitro Anticancer and Proapoptotic Activities of Steroidal Glycosides from the Starfish *Anthenea aspera*
Reprinted from: *Mar. Drugs* **2018**, *16*, 420, doi:10.3390/md16110420 . . . . **131**

Roman S. Popov, Natalia V. Ivanchina, Alla A. Kicha, Timofey V. Malyarenko, Boris B. Grebnev, Valentin A. Stonik and Pavel S. Dmitrenok
The Distribution of Asterosaponins, Polyhydroxysteroids and Related Glycosides in Different Body Components of the Far Eastern Starfish *Lethasterias fusca*
Reprinted from: *Mar. Drugs* **2019**, *17*, 523, doi:10.3390/md17090523 . . . . **145**

Alexandra S. Silchenko, Anatoly I. Kalinovsky, Sergey A. Avilov, Vladimir I. Kalinin, Pelageya V. Andrijaschenko, Pavel S. Dmitrenok, Roman S. Popov, Ekaterina A. Chingizova, Svetlana P. Ermakova and Olesya S. Malyarenko
Structures and Bioactivities of Six New Triterpene Glycosides, Psolusosides E, F, G, H, H_1, and I and the Corrected Structure of Psolusoside B from the Sea Cucumber *Psolus fabricii*
Reprinted from: *Mar. Drugs* **2019**, *17*, 358, doi:10.3390/md17060358 . . . . **159**

Alexandra S. Silchenko, Anatoly I. Kalinovsky, Sergey A. Avilov, Vladimir I. Kalinin, Pelageya V. Andrijaschenko, Pavel S. Dmitrenok, Roman S. Popov and Ekaterina A. Chingizova
Structures and Bioactivities of Psolusosides B_1, B_2, J, K, L, M, N, O, P, and Q from the Sea Cucumber *Psolus fabricii*. The First Finding of Tetrasulfated Marine Low Molecular Weight Metabolites
Reprinted from: *Mar. Drugs* **2019**, *17*, 631, doi:10.3390/md17110631 . . . . **183**

Yunmei Chen, Yuanhong Wang, Shuang Yang, Mingming Yu, Tingfu Jiang and Zhihua Lv
Glycosaminoglycan from *Apostichopus japonicus* Improves Glucose Metabolism in the Liver of Insulin Resistant Mice
Reprinted from: *Mar. Drugs* **2020**, *18*, 1, doi:10.3390/md18010001 . . . . **211**

Olesya I. Zhuravleva, Alexandr S. Antonov, Galina K. Oleinikova, Yuliya V. Khudyakova, Roman S. Popov, Vladimir A. Denisenko, Evgeny A. Pislyagin, Ekaterina A. Chingizova and Shamil Sh. Afiyatullov
Virescenosides from the Holothurian-Associated Fungus *Acremonium striatisporum* Kmm 4401
Reprinted from: *Mar. Drugs* **2019**, *17*, 616, doi:10.3390/md17110616 . . . . **225**

Katarzyna Dworaczek, Dominika Drzewiecka, Agnieszka Pękala-Safińska and Anna Turska-Szewczuk
Structural and Serological Studies of the O6-Related Antigen of *Aeromonas veronii* bv. *sobria* Strain K557 Isolated from *Cyprinus carpio* on a Polish Fish Farm, which Contains L-perosamine (4-amino-4,6-dideoxy-L-mannose), a Unique Sugar Characteristic for *Aeromonas* Serogroup O6
Reprinted from: *Mar. Drugs* **2019**, *17*, 399, doi:10.3390/md17070399 . . . . **241**

About the Special Issue Editors

Vladimir I. Kalinin PhD., Dr. Sc. was born in Vladivostok, Soviet Union, now the Russian Federation, in 1957. He graduated from Far Eastern State University, Chemical Department (Vladivostok) in 1979. From 1979 to date, he has been part of the G.B. Elyakov Pacific Institute of Bioorganic Chemistry of the Far East Branch of the Russian Academy of Sciences. From 1979 to 1981, he was a research intern, from 1981 to 1987 a junior scientist, from 1987 to 1989 a scientist, from 1989 to 1998 a senior scientist, from 1998 to today, a leading scientist. He has a PhD in chemistry (1989) and a Dr. Sc. in biology (biochemistry) (1998). His scientific interests include structure, biological activities, taxonomical distribution, chemotaxonomy significance and the evolution of sea cucumber triterpene glycosides. The author and co-author of more than 80 chapters and scientific articles, in particular international scientific journals referred by SCOPUS and WOS such as Marine Drugs, Molecules, Tetrahedron, Journal of Natural Products (Lloydia), Natural Product Research, Natural Product Communications, Carbohydrate Research, Journal of Theoretical Biology, Toxicon, Biological Systematics and Ecology etc. He is Associate Editor of the Natural Product Communications, Member of Editorial Board of Marine Drugs. Guest Editor of the SI of the Marine Drugs "Marine Glycoconjugates, Trends and Perspectives"; "Echinoderms Metabolites: Structure, Functions and Biomedical Perspectives".

Valentin A. Stonik PhD, Dr. Sc., Professor was born in Vladivostok, Russia on December 4, 1942. He graduated from the Department of Chemistry of the Far Eastern State University (Vladivostok) in 1965. PhD degree (Candidat of Science) in organic chemistry in 1969; the title of dissertation is "Synthesis of hydroacrydines and relative compounds". From 1970, he has been part of the G.B. Elyakov Pacific Institute of Bioorganic Chemistry, Far East Division of the Russian Academy of Sciences. From 1970, he was part of the G.B. Elyakov Pacific Institute of Bioorganic Chemistry, Far East Division of the Russian Academy of Sciences. He became head of the Laboratory of Biosynthesis from 1976, Head of the Laboratory of Chemistry of Marine Natural Products from 1985, Deputy Director of the Institute from 1990 to 2002 and Director from 2002. He is the scientific Advisor of the Institute from 2018. D.Sc. degree in bioorganic chemistry, chemistry of natural and physiologically active compounds in 1988; title of dissertation "Biphyllic physiologically active compounds from echinoderms and sponges. Structure and properties"; Russian Academy of Sciences M.M. Shemyakin Prize Winner (1995), Corresponding Member of the Russian Academy of Sciences (1997), Full Member of the Russian Academy of Sciences (Academician) (2000). Scientific interests: alkaloids, unusual lipids, isoprenoids, polyhydroxysteroids, sterol glycosides of polyhydroxysteroids, steroidal oligoglycosides, monosaccharides. Author and co-author of more than 400 scientific articles in Russian and international journals; 3 monographs; 21 patents. Member of the Editorial Boards of the Journals: Marine Drugs, Natural Product Communications, Natural Product Letters, Bioorganic Chemistry, and others. Guest Editor in SI in Marine Drugs: Marine Glycoconjugates: Trends and Perspectives; Selected Papers from the 3rd International Symposium on Life Science; Carbohydrate-Containing Marine Compounds of Mixed Biogenesis.

Natalia V. Ivanchina PhD. Born in Vladivostok, Russia on August 31, 1971. Graduated from the Department of Chemistry of the Far Eastern State University (Vladivostok) in 1993. PhD degree (Candidat of Science) in bioorganic chemistry in 2000; title of dissertation "Isolation and structure elucidation of starfish polyhydroxysteroids and polyhydroxysteroidal glycosides". From 1996 in G.B. Elyakov Pacific Institute of Bioorganic Chemistry, Far Eastern Branch of the Russian Academy of Sciences. Junior Scientist from 1996, Research Scientist from 2000, Senior Researcher from 2005, Head of the Laboratory of the Chemistry of Marine Natural Products from 2018. Winner of G.B. Elyakov Far Eastern Branch of the Russian Academy of Sciences Prize for work in the field of organic and bioorganic chemistry (2018). Scientific interests: starfish polyhydroxysteroids and polyhydroxysteroidal glycosides, structures, biosynthesis, biological activities, metabolomics. Author and co-author of more than 80 scientific articles in Russian and international journals. Member of the Editorial Board of the Marine Drugs. Guest Editor in SI in Marine Drugs: Marine Glycoconjugates: Trends and Perspectives; Carbohydrate-Containing Marine Compounds of Mixed Biogenesis.

Preface to "Marine Glycoconjugates"

Glycoconjugates, biomolecules in which carbohydrate moieties are attached by a covalent bond to any aglycone, play a very significant role in biological systems. Glycoproteins, peptidoglycans, lipopolyshaccharides and other biopolymer glycoconjugates are responsible for cellular interactions, including cell–cell recognition and the binding of cells to intercellular matrix, as well as carrying out other signal, antigenic and transport functions, participating in the formation of receptors and other important membrane and blood constituents. Low molecular weight glycoconjugates, such as triterpene and steroidal glycosides, glycolipids, phenolic glycosides are also well known as molecules having important internal and exterior functions for their organism producers such as protection against predators, competition for space, food resources, etc. Diverse glycoconjugates provide a very wide spectrum of biological activities, including defensive, cytotoxic, antitumor, immunomodulatory, and antioxidant properties. Water environment requires high solubility for signal or anti-predatory exometabolites of marine organism. All these peculiarities explain the very wide diversity of glycoconjugates of marine origin. The main goal of this Special Issue "Marine Glycoconjugates: Trends and Perspectives" is to provide a convenient platform for discussion of all possible scientific aspects concerning low molecular weight and biopolymer glycoconjugates of marine origin, their isolation and chemical structures, taxonomical distribution, methods of analysis, biological activities, biosynthesis and evolution, biological roles and biomedical perspectives.

Vladimir I. Kalinin, Valentin A. Stonik, Natalia V. Ivanchina

Special Issue Editors

marine drugs

Editorial

Marine Glycoconjugates: Trends and Perspectives

Vladimir I. Kalinin *, Valentin A. Stonik and Natalia V. Ivanchina

G.B. Elyakov Pacific Institute of Bioorganic Chemistry, Far Eastern Branch of the Russian Academy of Sciences, Pr. 100-letya Vladivostoka 159, 690022 Vladivostok, Russia; stonik@piboc.dvo.ru (V.A.S.); ivanchina@piboc.dvo.ru (N.V.I.)

* Correspondence: kalininv@piboc.dvo.ru; Tel.: +7-914-705-08-45

Received: 31 January 2020; Accepted: 11 February 2020; Published: 18 February 2020

Glycoconjugates play significant roles in biological systems and are used in medicine, for example as vaccines. Glycoproteins, peptidoglycans, lipopolysaccharides, and other biopolymer glycoconjugates are responsible for cellular interactions, including cell–cell recognition and the binding of cells to the intercellular matrix. These molecules perform signal, antigenic and transport functions, and participate in the formation of receptors and other important membrane and blood constituents. Due to the negative charges of some sulfated glycoconjugates and the binding of water, they are critical for maintaining the physical status of connective tissue [1]. Low molecular weight glycoconjugates, such as triterpene and steroidal glycosides [2–4], glycolipids, are also well known as molecules playing important internal and exterior roles. Diverse glycoconjugates show a very wide spectrum of biological activities, including defensive, cytotoxic, antitumor, immunomodulatory, and antioxidant properties. The water environment requires high solubility for signal or anti-predatory exometabolites of marine organisms, and carbohydrate moieties provide this property. All these peculiarities explain the very wide diversity of glycoconjugates of marine origin, including those presented in the current Special Issue.

Sterol and sphingolipid glycoconjugates, which are widespread, but insufficiently studied metabolites of microalgae, were discussed by Stonik and Stonik in the review article. Glycosylated sterols play important biological roles in microalgae and show different beneficial properties useful for medicine and healthy food. Dietary sterols and their glycoconjugates of microalgae enter into marine invertebrates through food chains and may be converted into 7(8)-unsaturated sterols and their derivatives, such as polyhydroxylated sterols and, probably, glycosides of starfishes and sea cucumbers. The knowledge of microalgal glycosphingolipids still remains poor, despite intensive investigations. Some of them are important for their interactions with pathogens and may induce apoptosis in microalgae. They also participate in the termination of microalgal blooms [5].

In their experimental article, Galasso et al. have discussed the obtaining and properties of a water-soluble bioactive fraction isolated from the toxic dinoflagellate *Alexandrium minutem*. This substance is a glycoprotein with a molecular weight above 20 kDa. It demonstrates specific antiproliferative activity (IC_{50} = 0.4 µg/mL) against the A549 cell line (the human lung adenocarcinoma cells). Moreover, the glycoconjugate did not reveal a cytotoxicity against human normal lung fibroblasts (WI38), but induced cell death, triggered by mitochondrial autophagy (mitophagy) in tumor cells. No mitophagic events were activated by it in normal WI38 cells [6].

Kostetsky et al. from the Far East Federal University (Vladivostok, Russia) have compared the fatty acid composition and thermal transitions of membrane lipids from green macroalgae *Ulva lactuca*, collected in the Sea of Japan and the Adriatic Sea. The adaptation to a warmer climatic zone was accompanied by a significant decrease in the ratio between unsaturated and saturated fatty acids (UFA/SFA) in membrane lipids. The decreasing ratio n-3/n-6 polyunsaturated fatty acids (PUFAs) was found in extra-plastidial lipids and only in the major glycolipid, non-lamellar monogalactosyldiacylglycerol. The opposite thermotropic behavior of non-lamellar and lamellar glycolipids can contribute to the maintenance of the highly dynamic structure of thylakoid membranes [7].

The role of marine and freshwater lectins, compounds specifically recognizing carbohydrate ligands, as anticancer agents, has been discussed by Italian and American scientists in the review "Antitumor Potential of Marine and Freshwater Lectins," of Catanazaro et al. The co-authors have concluded that lectins from aquatic organism, demonstrating a great variety of inhibitory effects against tumor cells and triggering apoptosis and other forms of cell death, promise a bright future for such lectins in anticancer research. Some these lectins are able to enhance the antineoplastic activity of common antitumor medications. In the majority of cases, the lectins can distinguish normal and transformed cells and even different types of tumor cells due to their ability to recognize glycosylated patterns of various cell types. The lectins tested on animals revealed optimal antitumor activities at negligible toxicity [8].

In their experimental work, Chinese scientists (Tao Wu et al.) from Zhejiang Sci-Tech University, Hangzhou, have reported that a gene encoding marine sponge *Aphrocallistes vastus* C-type lectin (AVL) was inserted into an oncolytic vaccinia virus vector (oncoVV) to form a recombinant virus oncoVV-AVL. In vivo experiments showed that oncoVV-AVL induces a significant antitumor effect in colorectal cancer and liver cancer mouse models. These findings open the possibility of using the virus, containing marine lectin AVL, in oncolytic viral therapies [9].

One more lectin, OXYL, a type-2 LacNAc-binding 14 kDa lectin belonging to the C1qDC family, was isolated and characterized from the feather star *Anneissia japonica* (Echinodermata, Crinoidea) by an international team (Hasan, et al.) led by Professor Yasuhiro Ozeki (Japan). The structural studies of this lectin, carried out by Edman degradation, revealed its N-terminal region, including the first 40 amino acids of the mature protein. Its functional characteristics have been studied in detail. The lectin function relates to immunity in the organism-producer. It caused a strong aggregation of bacterial cells, but did not act on their growth. The lectin also displayed remarkable LacNAc recognition-dependent anti-biofilm activity. The authors supposed that, due to its novel primary structure and unique activities, this lectin may be estimated as a living fossil, revealing the structural and functional diversification of metazoan lectins [10].

New low-molecular-weight glycoconjugated compounds of echinoderms were isolated from different species and studied by scientists from G.B. Elyakov Pacific Institute of Bioorganic Chemistry, PIBOC (Russian Federation). For example, three new steroidal glycosides, anthenosides V–X, along with seven previously known anthenosides, E, G, J, K, S_1, S_4, and S_6, have been isolated from the extract of tropical starfish *Anthenea aspera* by Malyarenko et al. It is of interest that anthenoside V contains a rare 5α-cholest-8(14)-en-3α,7β,16α-trihydroxysteroidal nucleus. All the investigated compounds at nontoxic concentrations inhibit the colony formation of human melanoma RPMI-7951, breast cancer T-47D, and colorectal carcinoma HT-29 cell lines. The mixture of anthenosides J and K possesses significant anticancer activity and induces apoptosis of HT-29 cells. The mechanism of the proapoptotic action of this mixture was found to be associated with the regulation of anti- and proapoptotic protein expression, followed by the activation of initiator and effector caspases [11].

Metabolomic studies concerning the localization of polar steroids, including their glycoconjugates, in different body components of the Far Eastern starfish *Lethasterias fusca*, have been carried out using nanoflow liquid chromatography/mass spectrometry with captive spray ionization (nLC/CSI–QTOF–MS) by Popov et al. The assumptions concerning the digestive function of polyhydroxysteroids and their derivatives and the protective role of asterosaponins in starfish were in good agreement with the obtained results. The highest level of polar steroids and their glycoconjugates was found in the stomach. Asterosaponins were found in all other body components; the main share of free polyhydroxysteroids and related glycosides were located in the pyloric caeca [12].

Seven new sulfated triterpene glycosides, psolusosides B, E, F, G, H, H_1, and I, and the earlier-known psolusoside A and colochiroside D, have been isolated by Sichenko et al. from the sea cucumber *Psolus fabricii*, collected in the Sea of Okhotsk. The structure of psolusoside B has been revised using modern 2D-NMR and HR-ESIMS procedures and re-established as a disulfated tetraoside. The structures of other glycosides were elucidated in the same manner. Cytotoxic activities of psolusosides E, F, G, H, H_1, and Iagainst the mouse ascite Ehrlich carcinoma cells, erythrocytes, neuroblastoma Neuro 2A and normal epithelial JB-6 cells were quite different, while the hemolytic action of the tested compounds was higher than their cytotoxicity against other cells, particularly against the Ehrlich ascites carcinoma. Psolusoside G was not cytotoxic against normal JB-6 cells, but revealed high activity against Neuro 2A cells. The cytotoxic activity against human colorectal adenocarcinoma HT-29 cells and the influence on the colony formation and growth of HT-29 cells by psolusosides B, E, F, H, H_1, and I, along with psolusoside A as a control, were examined. The highest inhibitory activities were shown for psolusosides E and F [13].

Additionally, 10 new di-, tri- and tetrasulfated triterpene glycosides, psolusosides B_1, B_2, J, K, L, M, N, O, P, and Q, have been also isolated from the sea cucumber *Psolus fabricii* by the same team. The cytotoxic activities of these psolusosides on several mouse cell lines, including Ehrlich ascites carcinoma cells, neuroblastoma Neuro 2A, normal epithelial JB-6 cells, and erythrocytes, were quite different depending both on the structural peculiarities of these glycosides and the type of cells. The most interesting finding was that psolusosides P and Q contain four sulfate groups in their carbohydrate chains. [14]. The presence of four sulfate groups is extremely rare in low molecular weight metabolites. However, examples of substances having four and even more sulfate groups were earlier reported in the studies on another class of metabolites, namely on polysulfated sterol dimers, hamigerols A and B, from the Mediterranean sponge *Hamigera hamigera* [15]. Moreover, axinelloside A, an unprecedented, highly sulfated lipooligosaccharide from the marine sponge *Axinella infundibula*, contains 19 sufate groups attached to 12 sugars [16]. Nevertheless, the finding of tetrasulfated derivative is also unique, particularly in relation to echinoderm metabolites.

Yunmei Chen et al. from Gingdao Ocean University of China have discussed an in vivo affect of glycosaminoglycan AHG from the edible sea cucumber *Apostichopus japonicus* on hyperglycemia in the liver of insulin-resistant mice. The obtained results demonstrated that AHG supplementation significantly decreased body weight, level of blood glucose and content of serum insulin in a dose-dependent manner in the mice. The protein levels and gene expression of gluconeogenesis rate-limiting enzymes G6Pase and PEPCK were significantly decreased. Although the total expression of IRS1, Akt, and AMPK in the insulin-resistant liver was not affected by AHG supplementation, the phosphorylations of IRS1, Akt, and AMPK were clearly increased by AHG treatment. The authors of this article concluded that AHG could be a promising candidate for the development of an antihyperglycemic agent [17].

Several reports in this Issue concern the bioactive compounds of microorganisms. Zhuravleva et al. from PIBOC (Vladivostok, Russian Federation) have described in their article 10 new diterpene glycosides, virescenosides Z9–Z18 isolated from a marine strain of the fungus *Acremonium striatisporum* KMM 4401 associated with the sea cucumber *Eupentacta fraudatrix*. Glycosides of this class bear rare monosaccharide units such as altrose. Moreover, virescenosides Z12–Z16 are monosides containing unique methyl esters of altruronic acid. The carbohydrate moiety of virescenoside Z18 was found to be a methylester of mannuronic acid. The effects of some glycosides and aglycons on urease activity and the regulation of ROS and NO production in murine macrophages, stimulated with lipopolysaccharide (LPC), were also studied [18].

A lipopolysaccharide (LPS), including the O-specific polysaccharide (O-antigen) of *Aeromonas veronii* bv. *sobria* strain K557, serogroup O6, isolated from the carp *Cyprinus carpio* during an outbreak of motile aeromonad septicemia (MAS) on a Polish fish farm, has been immunochemically studied by Dworaczek et al. Freshwater *C. carpio* is a common inhabitant of the bays and lagoons of Southern part of the Baltic Sea. The O-polysaccharide was obtained by acid hydrolysis of the LPS and studied by chemical transformations and spectral methods including ^{1}H and ^{13}C NMR spectroscopy. The O-antigen comprises two O-polysaccharides, both containing a unique sugar, 4-amino-4,6-dideoxy-l-mannose (N-acetyl-l-perosamine, l-Rhap4NAc). Western blotting and an enzyme-linked immunosorbent assay

(ELISA) revealed that the cross-reactivity between the LPS of *A. veronii* bv. *Sobria* K557 and the *A. hydrophila* JCM3968 O6 antiserum, and vice versa, is caused by the occurrence of common disaccharides, whereas an additional →4)-α-D-GalpNAc-associated epitope defines the specificity of the O6 reference antiserum. This investigation provides additional knowledge of the immunospecificity of *Aeromonas* bacteria and seems to be significant for epidemiological studies and for finding the routes of transmission of pathogenicity [19].

The diversity of marine glycoconjugate sources and their chemical structures, described in this Special Issue, is impressive. Two articles concern microalgal metabolites such as steroid and sphingoid glycoconjugates, and a glycoprotein from a sea cucumber with interesting biological activities, respectively [5,6]. One article discusses the fatty acid composition and thermotropic behavior of glycolipids and other membrane lipids of green macrophyte *Ulva lactuca* [7]. Three articles cover lectin subjects [8–10]. One review article analyzes the results and perspectives of marine and freshwater lectins' application in experimental oncology and the therapy of oncological diseases; another article describes the use of a sponge lectin in the construction of a recombinant virus. The third article concerns the function of the immunity of a lectin in producing this compound crinoid. Two articles concern steroid glycosides from starfish [11,12], and two others concern triterpene glycosides from sea cucumbers [13,14]. One article describes the effect of a glycosaminoglycan from the sea cucumber *Apostichopus japonicus* on hyperglycemia in the liver of insulin-resistant mice [17]. One article concerns the isolation of 10 new triterpene glycosides from a fungus associated with a sea cucumber [18]. The article by Dworaczek et al. characterizes the O-specific polysaccharide (O-antigen) of a bacterial pathogen of common carp by chemical and immunochemical methods [19]. In total, the Special Issue comprises 13 articles, including two reviews. It is interesting that six articles are about glycoconjugates from echinoderms and one article concerns the glycoconjugates of a microorganism associated with an echinoderm, i.e., the subject of more than half of the articles, is linked with echinoderms that reveals significant the biomedical potential of this group of marine invertebrates.

In conclusion, the editors are very appreciative to all authors that contributed their excellent works to our Special Issue and wish them new and exciting discoveries.

References

1. Wiederschain, G.Y. Glycobiology: Progress, problems, and perspectives. *Biochemistry (Moscow)* **2013**, *78*, 679–696. [CrossRef] [PubMed]
2. Stonik, V.A.; Ivanchina, N.V.; Kicha, A.A. New polar steroids from starfish. *Nat. Prod. Commun.* **2008**, *3*, 1587–1610. [CrossRef]
3. Kalinin, V.I.; Ivanchina, N.V.; Krasokhin, V.B.; Makarieva, T.N.; Stonik, V.A. Glycosides from marine sponges (Porifera, Demospongiae): Structures, taxonomical distribution, biological activities and biological roles. *Mar. Drugs* **2012**, *10*, 1671–1710. [CrossRef] [PubMed]
4. Kalinin, V.I.; Silchenko, A.S.; Avilov, S.A. Taxonomic significance and ecological role of triterpene glycosides from holothurians. *Biol. Bull.* **2016**, *43*, 532–540. [CrossRef]
5. Stonik, V.A.; Stonik, I.V. Sterol and sphingoid glycoconjugates from microalgae. *Mar. Drugs* **2018**, *16*, 514. [CrossRef] [PubMed]
6. Galasso, C.; Nuzzo, G.; Brunet, C.; Ianora, A.; Sardo, A.; Fontana, A.; Sansone, C. The marine dinoflagellate Alexandrium minutum activates a mitophagic pathway in human lung cancer cells. *Mar. Drugs* **2018**, *16*, 502. [CrossRef] [PubMed]
7. Kostetsky, E.; Chopenko, N.; Barkina, M.; Velansky, P.; Sanina, N. Fatty acid composition and thermotropic behavior of glycolipids and other membrane lipids of Ulva lactuca (Chlorophyta) inhabiting different climatic zones. *Mar. Drugs* **2018**, *16*, 494. [CrossRef] [PubMed]
8. Catanzaro, E.; Calcabrini, C.; Bishayee, A.; Fimognari, C. Antitumor potential of marine and freshwater lectins. *Mar. Drugs* **2020**, *18*, 11. [CrossRef] [PubMed]
9. Wu, T.; Xiang, Y.; Liu, T.; Wang, X.; Ren, X.; Ye, T.; Li, G. Oncolytic vaccinia virus expressing Aphrocallistes vastus lectin as a cancer therapeutic agent. *Mar. Drugs* **2019**, *17*, 363. [CrossRef] [PubMed]

10. Hasan, I.; Gerdol, M.; Fujii, Y.; Ozeki, Y. Functional characterization of OXYL, a SghC1qDC LacNAc-specific lectin from the crinoid feather star Anneissia japonica. *Mar. Drugs* **2019**, *17*, 136. [CrossRef] [PubMed]
11. Malyarenko, T.V.; Malyarenko, O.S.; Kicha, A.A.; Ivanchina, N.V.; Kalinovsky, A.I.; Dmitrenok, P.S.; Ermakova, S.P.; Stonik, V.A. In vitro anticancer and proapoptotic activities of steroidal glycosides from the starfish Anthenea aspera. *Mar. Drugs* **2018**, *16*, 420. [CrossRef] [PubMed]
12. Popov, R.S.; Ivanchina, N.V.; Kicha, A.A.; Malyarenko, T.V.; Grebnev, B.B.; Stonik, V.A.; Dmitrenok, P.S. The distribution of asterosaponins, polyhydroxysteroids and related glycosides in different body components of the Far Eastern starfish Lethasterias fusca. *Mar. Drugs* **2019**, *17*, 523. [CrossRef] [PubMed]
13. Silchenko, A.S.; Kalinovsky, A.I.; Avilov, S.A.; Kalinin, V.I.; Andrijaschenko, P.V.; Dmitrenok, P.S.; Popov, R.S.; Chingizova, E.A.; Ermakova, S.P.; Malyarenko, O.S. Structures and bioactivities of six new triterpene glycosides, psolusosides E, F, G, H, H1, and I and the corrected structure of psolusoside B from the sea cucumber Psolus fabricii. *Mar. Drugs* **2019**, *17*, 358. [CrossRef] [PubMed]
14. Silchenko, A.S.; Kalinovsky, A.I.; Avilov, S.A.; Kalinin, V.I.; Andrijaschenko, P.V.; Dmitrenok, P.S.; Popov, R.S.; Chingizova, E.A. Structures and bioactivities of psolusosides B1, B2, J, K, L, M, N, O, P, and Q from the sea cucumber Psolus fabricii. The first finding of tetrasulfated marine low molecular weight metabolites. *Mar. Drugs* **2019**, *17*, 631. [CrossRef] [PubMed]
15. Cheng, J.-F.; Lee, J.-S.; Sun, F.; Jares-Erijman, E.A.; Cross, S.; Rinehart, K.L. Hamigerols A and B, unprecedented polysulfate sterol dimers from the Mediterranean sponge Hamigera hamigera. *J. Nat. Prod.* **2007**, *70*, 1195–1199. [CrossRef]
16. Warabi, K.; Hamada, T.; Nakao, Y.; Matsunaga, S.; Hirota, H.; van Soest, R.W.M.; Fusetani, N. Axinelloside A, an unprecedented highly sulfated lipopolysaccharide inhibiting telomerase, from the marine sponge, Axinella infundibula. *J. Am. Chem. Soc.* **2005**, *127*, 13262–13270. [CrossRef]
17. Chen, Y.; Wang, Y.; Yang, S.; Yu, M.; Jiang, T.; Lv, Z. Glycosaminoglycan from Apostichopus japonicus improves glucose metabolism in the liver of insulin resistant mice. *Mar. Drugs* **2020**, *18*, 1. [CrossRef] [PubMed]
18. Zhuravleva, O.I.; Antonov, A.S.; Oleinikova, G.K.; Khudyakova, Y.V.; Popov, R.S.; Denisenko, V.A.; Pislyagin, E.A.; Chingizova, E.A.; Afiyatullov, S.S. Virescenosides from the holothurian-associated fungus Acremonium striatisporum KMM 4401. *Mar. Drugs* **2019**, *17*, 616. [CrossRef] [PubMed]
19. Dworaczek, K.; Drzewiecka, D.; Pekala-Safinska, A.; Turska-Szewczuk, A. Structural and serological studies of the O6-related antigen of Aeromonas veronii bv. sobria Strain K557 isolated from Cyprinus carpio on a Polish fish farm, which contains l-perosamine (4-amino-4,6-dideoxy-l-mannose), a unique sugar characteristic for Aeromonas Serogroup O6. *Mar. Drugs* **2019**, *17*, 399.

marine drugs

Review

Sterol and Sphingoid Glycoconjugates from Microalgae

Valentin A. Stonik [1] and Inna V. Stonik [2,*]

[1] G.B. Elyakov Pacific Institute of Bioorganic Chemistry, Far Eastern Branch, Russian Academy of Sciences, Pr. 100-let Vladivostoku 159, 690022 Vladivostok, Russia; stonik@piboc.dvo.ru
[2] National Scientific Center of Marine Biology, Far Eastern Branch, Russian Academy of Sciences, Palchevskogo Str, 17, 690041 Vladivostok, Russia
* Correspondence: innast2004@mail.ru; Tel.: + 7-423-231-7107

Received: 28 October 2018; Accepted: 14 December 2018; Published: 17 December 2018

Abstract: Microalgae are well known as primary producers in the hydrosphere. As sources of natural products, microalgae are attracting major attention due to the potential of their practical applications as valuable food constituents, raw material for biofuels, drug candidates, and components of drug delivery systems. This paper presents a short review of a low-molecular-weight steroid and sphingolipid glycoconjugates, with an analysis of the literature on their structures, functions, and bioactivities. The discussed data on sterols and the corresponding glycoconjugates not only demonstrate their structural diversity and properties, but also allow for a better understanding of steroid biogenesis in some echinoderms, mollusks, and other invertebrates which receive these substances from food and possibly from their microalgal symbionts. In another part of this review, the structures and biological functions of sphingolipid glycoconjugates are discussed. Their role in limiting microalgal blooms as a result of viral infections is emphasized.

Keywords: microalgae; sterol glycoconjugates; glycosylceramides; structures; biological activities; functions

1. Introduction

The search for new marine-derived molecules in previously insufficiently studied groups of marine organisms is one of the most important stages of drug discovery. Metabolites of microalgae are a promising source of such compounds. Microalgae (microphytes) are unicellular or colonial eukaryotic, mainly autotrophic organisms occurring in fresh, brackish, and seawaters from the surface layers down to the bottom sediments [1–3]. Diatoms (kingdom Chromista, phylum Ochrophyta) are an abundant and widely-distributed component of the marine phytoplankton, including over 8000 species [2]. A characteristic feature of diatoms is the porous frustule (cell wall) with two overlapping valves (epitheca and hypotheca) composed of silica. Some species of diatoms can produce a variety of natural products, including low-molecular-weight compounds [4]. The phylum Ochrophyta includes, along with diatoms, some other groups of marine, freshwater, and soil heterokontophytes (Eustigmatophyceae, Chrysophyceae, Chrysomerophyceae, Dictyochophyceae, Pelagophyceae, Raphidophyceae, Xanthophyceae, and others). Eustigmatophytes, coccoid microalgae inhabiting mainly freshwater habitats and soil, and so-called golden microalgae, belonging to the class Chrysophyceae, as well as species belonging to the related class Sinurophyceae with marine genera *Mallomonas* and *Synura*, sometimes produce toxins and other bioactive compounds [5]. Dinoflagellates (phylum Dinophyta, kingdom Chromista) represent another important component of phytoplankton, including about 2300 marine, freshwater, or parasitic species [2]. Dinophytes are abundant in warmer seas and in temperate areas during the warm seasons. They possess two dissimilar flagella and a cell cover composed of several membranes, containing polysaccharide plates (in armored dinoflagellates) [6]. Some species of diatoms and other heterokontophytes, as well as many dinoflagellates, can cause bloom events and produce dangerous toxins [6,7].

There are a series of other phyla of microalgae belonging to the kingdom Chromista. For example, the phylum Haptophyta (97 species) is an algal group with several classes, including very abundant in open ocean waters coccolithophorids (Coccolithophyceae) [1], which have an exoskeleton of calcareous plates (coccoliths). Coccolithophorids comprise the bloom-forming species belonging to the genera *Emiliania, Phaeocystis,* and *Chrysochromulina*, as well as the *Isochrysis* species, which are used in the oyster and shrimp aquaculture as food for larvae. Pavlovophyceae is another class of haptophytes whose metabolites have been extensively studied recently [8]. Cryptophytes (phylum Cryptophyta, kingdom Chromista) is a small group of marine and freshwater microalgae (about 200 species) which is abundant in brackish-water habitats and oligotrophic lakes. Some species live symbiotically inside ciliates, providing the host with photosynthates [3].

Glaucophytes (phylum Glaucophyta, approximately 15 species) are a very small group of rare freshwater microalgae, which, together with red and green algae, are combined into the kingdom Plantae [1]. Unicellular species of green (Chlorophyta, a total of ca. 8000 species) and red algae (Rhodophyta, ca. 7000 species) may be also referred to as microalgae. Euglenophytes (phylum Euglenozoa, ca. 2000 species) belong to the kingdom Protozoa [1,2], consisting mostly of unicellular motile microalgae. This algal group is abundant in freshwater habitats: puddles, ditches, ponds, streams, lakes, and rivers, particularly in waters exposed to pollutants or decaying organic matter [3].

Generally, most microalgae are capable of performing photosynthesis, and globally produce about a half of all organic substances and oxygen each year, although some groups are shown to be heterotrophs. All these lower plants lie at the base of food webs in the hydrosphere, providing the major part of matter and energy for higher trophic levels. Metabolites of microalgae play important roles in the lives of their producers, but also greatly influence other living organisms in the hydrosphere.

Glycoconjugates and other conjugated forms are known as bioactive metabolites and signal compounds. Recently, new information appeared on the interesting biological properties of microalgal glycoconjugates. The aim of this first review in this scientific field is to summarize the current data concerning glycoconjugate chemical forms of sterols and sphingolipids from microalgae, which contain hydrophobic fragments of different biosynthetic origins with attached polar groups such as sugars. The review covers the period from 1978 to 2017.

2. Sterols and Sterol Glycoconjugates. Structural Diversity and Taxonomic Distribution

2.1. Free Sterols

Sterols of microalgae, as well as their fatty acids, were better examined in comparison with any other metabolites, but not in all the taxonomic groups of microalgae, and therefore, sterols in some taxa remain largely unstudied to date. Microalgal sterols are well known as the main source of steroidal materials of many other marine organisms, which obtain these compounds through their diet. The structural diversity of microalgal sterols and their biological significance were thoroughly reviewed [9–13]. One of the recognized experts in this scientific field, Volkman, recently wrote that "the published information on sterol compositions of microalgae continues to increase, but it is still far from comprehensive" [13].

Usually, the structural diversity of sterols is illustrated using the formulae of their tetracyclic nuclei differing from each other mainly in their unsaturation degree and positions of double bonds. These core fragments bear variable side chains. More than a dozen different tetracyclic cores and many variants of side chains were found in microalgal sterols, as shown in Figure 1. Significant variability in the corresponding fractions and their main constituents was established by comparing sterols of different taxonomic groups. Moreover, it was shown that many taxa contain unusual and rare sterols. A number of sterols of microalgae have not been found in terrestrial higher plants.

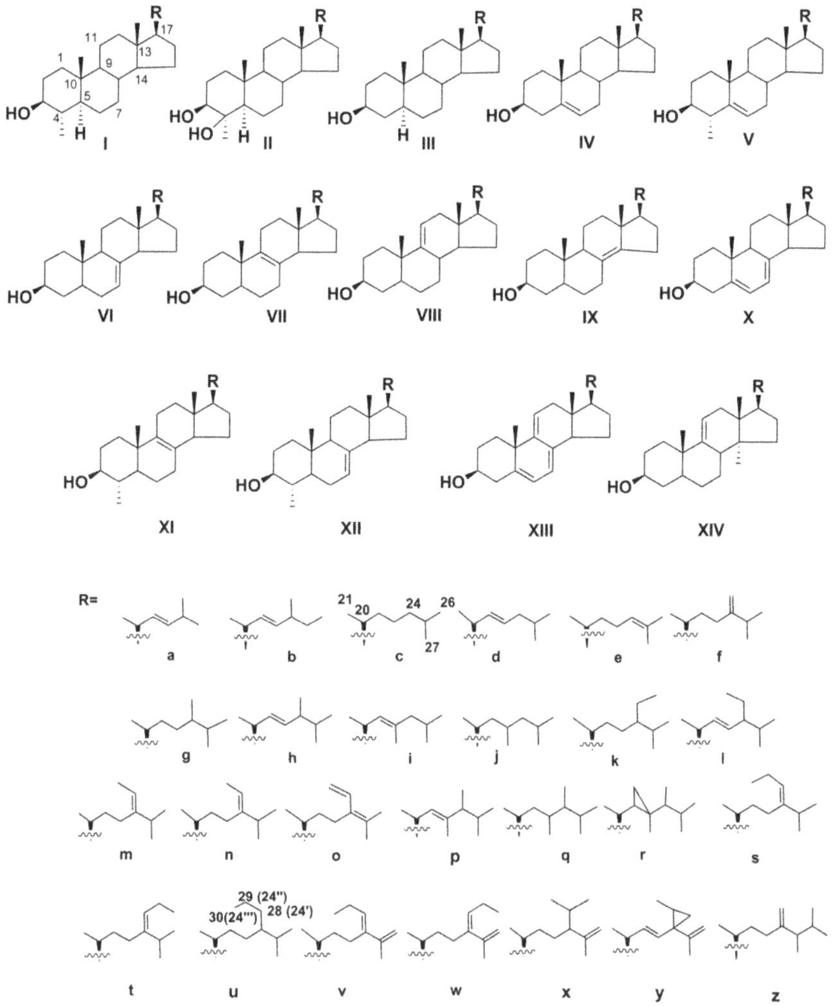

Figure 1. Sterols from microalgae.

For example, dinophytes are characterized by a great variety of unusual sterols, and often contain a saturated 4α-methyl-5α-cholestane type core, as in dinosterol (24*R*-Ip) and dinostanol (23*R*,24*R*-Iq). In Pelagophyceae, extremely rare C30 sterols such as 24-propylidenecholesterols (IVs,t) and other highly-alkylated in side chain compounds were found. Some other exotic sterols of microalgal origin are mentioned in Table 1, which shows the taxonomic distribution of sterols.

Frequently, the taxonomic distribution of sterols is analyzed on the class level, because the compositions of larger taxa, such as phyla, are still the subject of discussion and revisions by taxonomists. The structural diversity of sterols is great (Figure 1 and Table 1), and many of these compounds are discovered each year.

Table 1. Distribution of sterols in some taxa of microalgae.

Taxa	Sterols
	Ochrophyta
Bacillariophyceae	$C_{28}\Delta^{5,24(28)}$ (IVf), 24*R*-$C_{28}\Delta^{5,22}$ (epibrassicasterol IVh) [10–13], $C_{27}\Delta^{5}$ (IVc), 27-nor-$C_{28}\Delta^{5,22}$ (ocellasterol IVb), $C_{29}\Delta^{5,22}$ (IVl [11,13], $C_{29}\Delta^{5}$ (IVk) [14], previously unknown $C_{28}\Delta^{7,22}$ (VIi), $C_{27}\Delta^{5,22}$ (IVd) [13], $C_{29}\Delta^{22}$ (IIIp), $C_{30}\Delta^{22}$ (dinosterol Ip), cyclopropane sterols (Ir, gorgosterol IVr [15,16], $C_{28}\Delta^{7,22}$ (VIh), $C_{28}\Delta^{8(9)}$ (VIIg), $C_{29}\Delta^{24,28}$ (IVo and IIIo) [13].
Eustigmatophyceae	$C_{28}\Delta^{5}$ (IVg), $C_{29}\Delta^{5,24(28)}$ (IVn), $C_{29}\Delta^{5,24(28)}$ (IVm), $C_{29}\Delta^{5}$ (IVk), $C_{28}\Delta^{5,24(28)}$ (IVf), $C_{27}\Delta^{5}$ (IVc) [12,13].
Pelagophyceae	$C_{30}\Delta^{5}$ (IVu), $C_{30}\Delta^{5,24(28)}$ (IVs,t), rare $C_{30}\Delta^{5,24(28),25(26)}$ (IVv,w), trace C_{30} sterols (IVx-z) [13,17]
Chrysophyceae	24*S*- and 24*R*-$C_{29}\Delta^{5,22}$ (IVl), 24*S*-$C_{28}\Delta^{5,22}$ (IVh), $C_{28}\Delta^{5,7,22}$ (Xh), 24*R*- and 24*S*-$C_{29}\Delta^{5}$ (IVk), $C_{29}\Delta^{5,7,22}$ (Xl), $C_{29}\Delta^{5,24(28)}$ (IVm,n), $C_{29}\Delta^{7,24(28)}$ (Xm),
Synurophyceae	$C_{27}\Delta_{5}$ (IVc), $C_{29}\Delta^{5}$ (IVk) [13]
Chrysomerophyceae	$C_{29}\Delta^{5}$ (IVk), $C_{28}\Delta^{5,24(28)}$ (IVf), $C_{27}\Delta^{5}$ (IVc), $C_{29}\Delta^{5,22}$ (IVl), $C_{29}\Delta^{5,24(28)}$ (IVn), $C_{28}\Delta^{5,22}$ (IVh) [13]
Xanthophyceae	$C_{27}\Delta_{5}$ (IVc), $C_{29}\Delta^{5}$ (IVk) [13]
Dictyophyceae	$C_{28}\Delta^{5,24(28)}$ (IVf), $C_{28}\Delta^{5,22}$ (Vd) [13]
Rhaphidophyceae	24*S*- and 24*R*-$C_{29}\Delta^{5}$ (IVk), $C_{27}\Delta^{5}$ (IVc), $C_{27}\Delta^{8(9)}$ (VIIc), $C_{29}\Delta^{5,22}$ (IVl), 27-nor-$C_{27}\Delta^{5,22}$ (IVb), $C_{29}\Delta^{0}$ (IIIk) [13,18]
	Dinophyta
Dinophyceae	$C_{30}\Delta^{22}$ (Ip dinosterol), $C_{29}\Delta^{0}$ (Iq), $C_{29}\Delta^{0}$ (1g), $C_{29}\Delta^{24(28)}$(If), $C_{28}\Delta^{22}$ (IIIh), $C_{29}\Delta^{22}$ (IIIp), $C_{28}\Delta^{8(14)}$ (IXc), $C_{28}\Delta^{8(14),24(28)}$ (amphisterol IXf), $C_{28}\Delta^{8(14),22}$ (IXh), $C_{27}\Delta^{8(14),22}$ (IXb), $C_{28}\Delta^{8(14),22}$ (IXi) and others [13,19,20]
	Cryptophyta
Cryptophyceae	$C_{28}\Delta^{5,22}$ (IVh), $C_{27}\Delta^{5}$ (IVc), $C_{29}\Delta^{5,22}$ (IVl) [13,18]
	Haptophyta
Coccolithophyceae	$C_{28}\Delta^{5,22}$ (*epi*brassicasterol IVh), $C_{29}\Delta^{5,22}$ (stigmasterol IVl), $C_{27}\Delta_{5}$ (IVc), $C_{28}\Delta^{5,24(28)}$ (IVf) [13]
Pavlovophyceae	$C_{29}\Delta^{5,22}$ (IVl), $C_{27}\Delta^{5}$ (IVc), $C_{29}\Delta^{22}$ (IIIl), $C_{30}\Delta^{22}$ (Il), pavlovols IIg, IIk,IId, minor Ig,h [13,21,22]
	Euglenophyta
Euglenophyceae	$C_{28}\Delta^{5,7,22}$ (ergosterol Xh), $C_{29}\Delta^{8(9)}$ (XIg), $C_{28}\Delta^{5,7,24(28)}$ (Xf), $C_{29}\Delta^{5,7}$ (Xk), $C_{27}\Delta^{5}$ (IVc), $C_{29}\Delta^{5,22}$ (IVl), $C_{28}\Delta^{5}$ (IVg), $C_{28}\Delta^{5,22}$ (IVh), $C_{27}\Delta^{0}$ (IIIc), 23-unsaturated $C_{29}\Delta^{5,7}$ (Xk) [13]
	Glaucophyta
Glaucophyceae	$C_{28}\Delta^{5,24(28)}$ (IVf), $C_{29}\Delta^{5,22}$ (IVl), $C_{29}\Delta^{5}$ (IVk) [13,23]
	Cercozoa
Chlorarachniophyceae	$C_{28}\Delta^{5,22}$ (IVh), $C_{29}\Delta^{5,22}$ (IVl) [24]
	Rhodophyta
Porphyridiophyceae	$C_{28}\Delta^{5,7,22}$ (ergosterol Xh), $C_{27}\Delta^{5,22}$ (IVd), $C_{28}\Delta^{8,22}$ (XId), $C_{29}\Delta^{8,22}$ (XIh), $C_{28}\Delta^{8}$ (VIIg), $C_{28}\Delta^{8}$ (XIc) [13]
Stylonematophyceae	$C_{28}\Delta^{5,22}$ (IVh), $C_{27}\Delta^{5}$ (IVc), $C_{28}\Delta^{7}$ (XIIc), $C_{28}\Delta^{7,22}$ (XIId) [13]
	Chlorophyta
Prasinophyceae	$C_{29}\Delta^{5,24(28)}$ (IVm), $C_{29}\Delta^{5,24(28)}$ (IVn) [24], $C_{28}\Delta^{5,24(28)}$ (IVf), $C_{29}{}^{,5,7,22}$ (Xl), rare $C_{28}\Delta^{5,7,9(11),22}$ (XIIIh), $C_{29}\Delta^{5,7,9(11),22}$ (XIIIl) [10,13]
Chlorophyceae	$C_{27}\Delta^{5}$ (IVc), $C_{28}\Delta^{5}$ (IVg), $C_{29}\Delta^{5,22}$ (IVl) $C_{29}\Delta^{7,22}$ (VId), $C_{28}\Delta^{7}$ (VIg) [13,25,26]
Trebouxiophyceae	$C_{27}\Delta^{5}$ (IVc), 24S-$C_{29}\Delta^{5}$ (clionasterol IVk), $C_{29}\Delta^{5,22}$ (poriferasterol IVl), $C_{28}\Delta^{5}$ (IVg), $C_{29}\Delta^{5,7,22}$ (7-dehydroporiferasterol Xl), $C_{29}\Delta^{7,22}$ (chondrillasterol VIl), $C_{28}\Delta^{8}$ (VIIg), $C_{28}\Delta^{8,22}$ (VIIh), $C_{28}\Delta^{5,7,22}$ (ergosterol Xh), unusual $\Delta^{9(11)}$-sterols: $C_{28}\Delta^{9(11)}$ (VIIIg), $C_{29}\Delta^{9(11)}$ (XIVg), $C_{28}\Delta^{55,7,9(11),22}$ (XIIIh), $C_{29}\Delta^{5,7,9(11),22}$ (XIIIl) [13]
Chlorodendrophyceae	$C_{28}\Delta^{5,24(28)}$ (IVf), $C_{28}\Delta^{5}$ (IVg), $C_{27}\Delta^{5}$ (IVc), $C_{27}\Delta^{5,22}$ (IVd), $C_{27}\Delta^{5,24}$ (IVe) [13] $C_{29}\Delta^{5,22}$ (stigmasterol IVl), $C_{28}\Delta^{5}$ (IVg) [13]

Generally, the diversity of numerous microalgal free sterols was found to be higher in comparison with that of higher plants. Sterol compositions of microalgae were discussed in numerous works, which considered the chemotaxonomic importance of these compounds, their application as markers for marine and terrigenous organic matter, their significance for marine life, and their influence on the development of their producers in wild conditions and cultures [9–13,20,26,27].

There is no doubt that microalgal sterols make a great contribution to steroid metabolism in marine animals, being an important food component of echinoderms, mollusks, and other invertebrates, and penetrating via food chains into fish and marine mammals. Another almost unstudied biological signification of these metabolites is associated with their potential participation in symbiotic relationships, because, as is well known, a tremendous number of marine organisms, particularly invertebrates, host microbial symbionts. Such membrane constituents as sterols in both micro- and macro-organisms might have co-evolved in these symbiotic consortiums.

2.2. Structural Diversity and Analysis of Steryl Glycoconjugates

Sterols are present in microalgae in not only free, but also in conjugated forms. Actually, free sterols are frequently, if not always, accompanied in these microorganisms by conjugated forms such as steryl glycosides (SG), acyl steryl glycosides, esters (bearing fatty acid residues), and sulfates. Glycosylated forms of sterols in both higher and lower plants mainly include two types of natural compounds: glycosides of sterols and their acylated-by-fatty-acids derivatives. In these acylated glycosides, acylation proceeds, as considered, predominantly to the C-6 position of a sugar moiety (in majority cases glucose, but other monosaccharides may also be present), as shown below in the example of β-sitosterol glycoconjugates which are widely distributed in higher plants and are found in some microalgae (Figure 2).

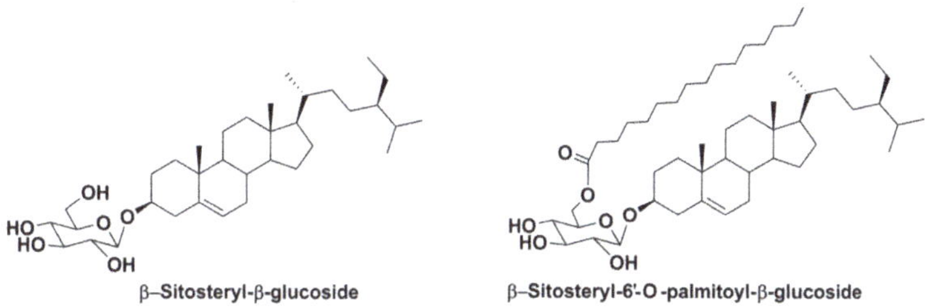

Figure 2. An example of glycoconjugated sterols from higher plants.

Steryl glycoconjugates are known for their bioactivity (see Section 2.3 of this review). These metabolites have been found many times in nature, being isolated from different higher and lower plants, such as olives, soybeans, potatoes, and algae. Moreover, they are also abundant in many species of fungi, and have been detected in bacteria and animals. The concentration of these metabolites is commonly (but not always) almost one order of magnitude lower than that of free sterols. That is why these compounds were not detected in algae for a long time, and their absence was even considered a characteristic difference between higher plants and algae [28].

Frequently, information about these biochemical forms of sterols is lost in many studies on microalgal species, because researchers either hydrolyze extracts before isolation in order to determine the total sterol composition, or isolate only fractions of free sterols. In many papers, the authors established only the presence of a glycoconjugated fraction, without specifying its sterol profile, and provided the data on total sterol composition of all the sterol forms. For example, in a study on the heterocontophyte *Nannochloropsis* sp. (Eustigmatophyceae) and the fungoid organism *Schizochytrium limacinum* (superphylum Heteroconta, Thraustochytriaceae, Labyrinthulomycetes), which are promising for biofuel production, Wang and Wang [29] found fractions of steryl glucosides in both species, but did not establish their sterol profiles. They reported only the total sterol composition with main constituents IVc and IVm from *Nannochloropsis* and an unusual polyunsaturated sterol from *Schizochytrium*. However, later, the presence of this sterol was not confirmed [30].

As a rule, steryl glycosides from algae differ from those of higher plants in aglycone moieties. It was reported that isofucosteryl glucoside and clionasteryl glucoside are present in the macrophyte green algae *Ulva gigantea* and *Cladophora rupestris*, while fucosteryl glucoside and desmosteryl glucoside are characteristic metabolites of brown and red macrophytes *Fucus vesiculosus* and *Rhodymenia palmata*, respectively [31]. It was suggested that the sterol profile of steryl glycosides reflects the free sterol profile of the studied species, as confirmed by studies on terrestrial plants [32]. Moreover, first studies on microalgal steryl glycoconjugates from *Porphyridium* sp. also confirmed this suggestion [32].

However, subsequent investigations showed that it is not always true as regards microalgae. In fact, the microalgae *Tetraselmis chui* (Phylum Chlorophyta, Chlorodendrophyceae), *Nannochloropsis salina* (Phylum Ochrophyta, Eustigmatophyceae), and *Skeletonema costatum* (Phylum Ochrophyta, Bacillariophyceae), which are extensively used in mariculture, differ from each other in their free sterol compositions. The first of these species contained mainly 24-methylenecholesterol (IVf) and campesterol (24*R*-IVg) in free sterol fractions, the second, cholesterol (IVc) and its 22(23)-dehydroderivative (IVd); and *S. costatum*, at least four main sterols, including cholesterol (IVc), dehydrocholesterol (IVd), stigmasterol (IVl), and brassicasterol (24*S*-IVh). However, cholesterol was found as the only common fraction of free and all conjugated forms in these microalgae, although it is a minor constituent of free sterols in some of these species [33]. These results showed that the glycosylation may occur specifically in the studied microalgae, and that glycosylation leads to the formation of a variety of known and previously-unknown compounds. Thus, main glycoconjugated forms, when compared with free sterol fractions, may contain minor sterols as aglycones.

Similar results were obtained by comparing free and conjugated sterols of *Pavlova lutheri* (Phylum Haptophyta, Pavlovophyceae). Uncommon 4α-methyl-24β-ethylcholest-22(23)-en-3β-ol (III) was identified as the dominant sterol constituent along with eight common desmethylsterols, including poriferasterol (24*R*-IVl) and cholesterol. In contrast to free sterol fraction, this dominant sterol was not found as aglycone in steryl glycosides, while glucosylated cholesterol and so-called pavlovols (3,4-dihydroxy-4α-methylsterol derivatives such as II c,d,f,g, Figure 1) were indicated as the main sterol constituents in glycosylated forms. Moreover, neither II, nor pavlovols were identified in acetylated glycoconjugates with acylated cholesteryl glucoside as a main constituent. This shows that the acylation also occurs specifically, and cholesteryl glucoside is probably acetylated more effectively in comparison with other steryl glucosides in this species [34].

Later, a more detailed comparison of sterol compositions of free sterols, esterified sterol forms, steryl glucosides, and acylated steryl glucosides was carried out by the same group of researchers on seven unicellular algae [35]. The total sterol content varied from 91 μg/g dry weight in the diatom *Skeletonema costatum* to 1354 μg/g in *Isochrysis* aff. *galbana* (phylum Haptophyta, Isochrysidaceae). The latter species contained almost the same amount of conjugated and non-conjugated forms as *P. lutheri*, another haptophyte microalga studied earlier. It was of particular interest that free sterols were dominant not in all the studied species, but only in four of them: centric diatom *S. costatum*, haptophyte *Isochrysis* aff. *galbana*, chlorophyte *Tetraselmis suecica*, and pennate diatom *Haslea ostrearia*. At the same time, the eustigmatophyte *Nannochloropsis oculata* and centric diatom *Thalassiosira pseudonana* contained most of their sterols in esterified form, while the centric diatom *Chaetoceros calcitrans* had about 60% of its sterols in glycoconjugated form (as a sum of steryl glucosides and acylated steryl glucosides). Only *C. calcitrans* contained the same sterol (cholesterol) as the dominant constituent in conjugated and non-conjugated forms. Other studied microalgae had more or less different sterol compositions in free and glycoconjugated forms. For example, the diatom *H. ostrearia* contained mainly unusual 23,23-dimethylcholest-5-en-3β-ol in its glycoconjugated form.

In the pennate diatom *Phaeodactylum tricornutum*, a significant portion of sterols was found in glycosylated form with glucoside of 24-methylcholesta-5,22-dien-3β-ol as the main constituent. Levels and compositions of free and conjugated sterol forms in microalgae depend on culture conditions. The content of SG was 100-fold lower in *P. tricornutum* when this microalga was grown at 23 °C instead of 13 °C [36].

Generally, glycoconjugated sterols in microalgae (Figure 3) are more poorly-studied compared to those in higher plants. However, even the limited available data on these metabolites in microalgae clearly show that the structural diversity of their steryl moieties is greater than that in higher plants (Figure 3).

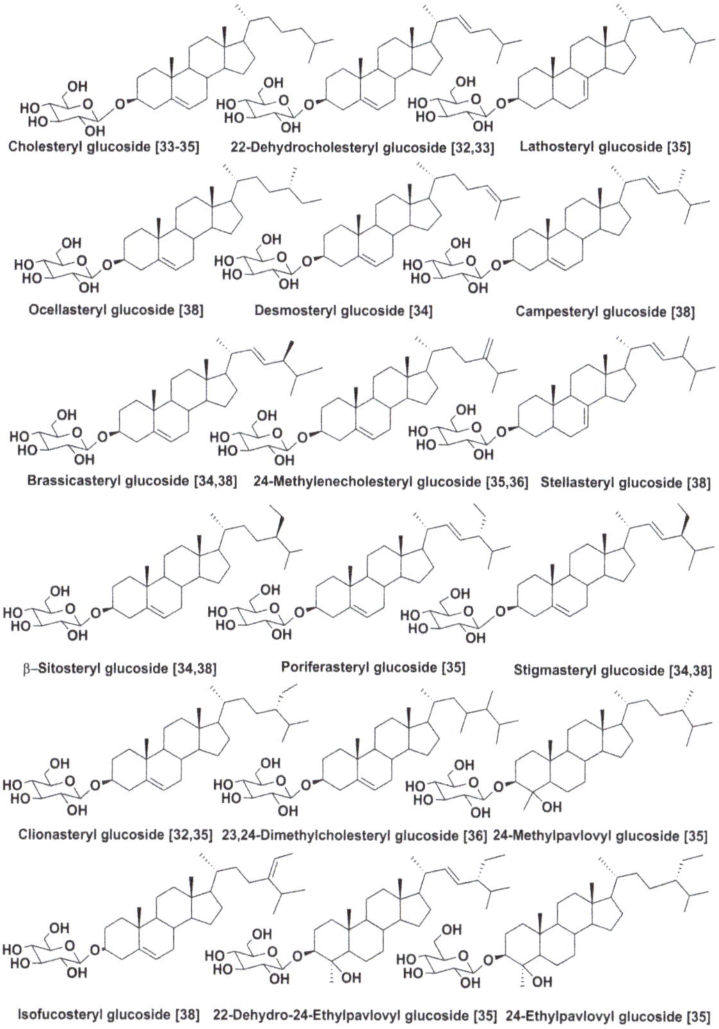

Figure 3. Steryl glucosides from microalgae.

Glycosides, including glucosides, glucoronides, galactosides, ribosides, and even compounds containing several monosaccharide residues, were found among steryl glycoconjugates of different origin [37]. In spite of this fact, all the thus far indentified glycoconjugates from microalgae, including those studied with application of mass-spectrometry, were found to contain only hexoses, probably D-glucose residues.

The main difficulty of a structural study on this class of natural compounds in microalgae is their low content and often the small amount of initial biomass, which could be used for their isolation and identification. The common approach to their identification includes the isolation of total glycosylated fraction using thin-layer or column chromatography on silica gel and application of NMR spectroscopy, that indicates the low-field shift of anomeric CH group signals and the appearance of characteristic signals of sugar moiety in comparison with spectra of free sterols. However, NMR spectroscopy has not yet been used in studies on microalgal steryl glycosides out of their low levels in biological sources. As a rule, acid and alkaline hydrolyses were carried out to obtain free sterols, and thereafter, to study them by GC or GC-mass-spectrometry (GS-MS) methods.

The progress in this scientific field was stimulated by the development and application of more sensitive chromatographic and mass-spectrometric techniques, which made possible the identification and quantitative analysis of respective natural compounds in microgram quantities in some unstudied species. Also, new attempts have been made to find available sources of the respective glycosidases and esterases as biochemical tools for mild transformation of sterol glycoconjugates to free sterols. The additional interest in analysis of steryl glycoside content arose due to difficulties in obtaining biodiesel, associated with the steryl glycosides responsible for poor filterability of this product [38,39].

Solid-phase extraction followed by gas chromatography was proposed as a rapid and convenient method of analysis of steryl glycosides in foods and dietary supplements. For example, steryl glucosides, extracted with a hexane-diethyl ether (1:1, v/v) mixture after alkaline saponification of lipids, were isolated by solid-phase extraction, derivatized, and quantitively analyzed as trimethylsilyl ethers by capillary gas chromatography (GC) with a 5% diphenyl-95% dimethylpolysiloxane column [38].

A similar proposed variant of analysis involved the use of GS–MS of pretreated silylated samples, followed by single ion monitoring at 147, 204, 217 m/z, which are specific ions for the silylated sugar moiety [39].

Mass spectrometry combined with liquid chromatography also opens new possibilities and promises a rapid progress in identification of new variants of steryl glycosides [40].

Recently, a group of Chinese scientists used gas chromatography–triple quadrupole mass spectrometry to characterize steryl glycosides as their trimethyl silyl derivatives in eight microalgae [41]. As a result, not only nine compounds were identified, including several previously unknown glycosides, but also, characteristic fragmentations in the corresponding mass spectra were established. This method proved to be a powerful tool for finding new glycoconjugated sterols and promising new biological sources of highly valuable natural product substances in this class. The fragmentation pathways included the cleavage of a glycosidic bond that made it possible to identify ion peaks of both monosaccharide and sterol moieties in the studied conjugates (Figure 4).

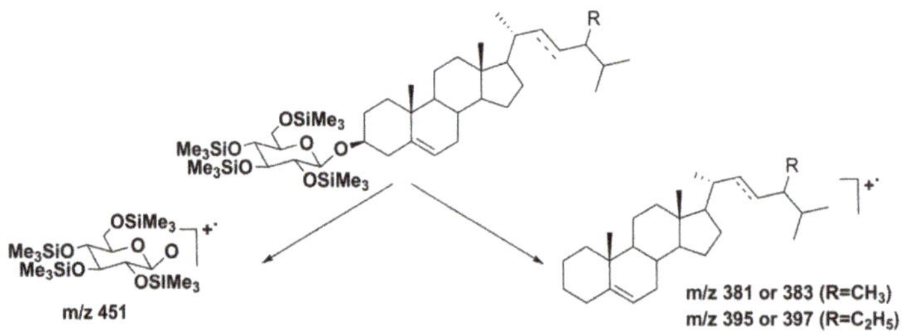

Figure 4. Fragmentation pathways of steryl glucosides in EI MS with the formation of either sugar or sterol cationic radical species.

Unfortunately, the application of solely chromatography-mass-spectrometry does not allow researchers to precisely identify the monosaccharide (or monosaccharides) present in glycoconjugates, nor to establish the type of glycoside bonds between sugars in di- or triglycosylated compounds, when these glycosides will be found. The additional use of NMR spectroscopy (including different 2D NMR techniques such as COSY, ROESY, HMBC and HSQC) or chemical transformations such as hydrolysis of permethylated derivative, followed by analysis of obtained sugar derivatives, is required in such cases.

2.3. Biological Activities and Biological Functions

The most studied compounds of this class, e.g., β-sitosteryl and campesteryl glycosides from edible terrestrial higher plants and other biological sources, have attracted attention by their influence on immunological processes. An injection with sitosteryl glucoside increased the survival rate of mice infected by *Candida albicans*. A mixture of β-sitosterol and sitosteryl glycosides, so-called sitosterolin, increased the production IL-2 and IFN-γ cytokines by murine helper T cells stimulated by phytohemagglutinin. The immunomodulatory and protective effects of this preparation were used in treating patients suffering from allergic diseases. Daily administration of sitosterolin improved the health of patients with pulmonary tuberculosis and helped their recovery [42,43]. The fact that sitosterol and campesteryl glucosides are also produced by some microalgae is of particular interest (Figure 3).

Sterylglycosides with structural peculiarities in sugar moiety also showed noteworthy immunological properties. For example, the gram-negative bacterium *Helicobacter pylori*, colonizing the stomach and causing gastric diseases, is auxotrophic for cholesterol and extracts this lipid from plasma membranes of epithelial cells of the host stomach. This bacterium converts cholesterol into cholesteryl 3α-glucoside. Since the incorporation of cholesterol promotes immune responses of the host, *Helicobacter pylori* converts cholesterol to cholesteryl glucoside and then to the corresponding cholesteryl 6′-*O*-acyl glucoside, thus evading the immune surveillance. In another experiment, the analogous steryl β-galactoside caused a rise of antibodies against a *Borrelia burgdorferi* infection [43,44].

It was shown that phytosteryl glucosides, separated from the crude soybean lecithin and solubilized in purified soybean oil, influenced the cholesterol adsorption. When plasma cholesterol of patients after consumption of a test breakfast with pudding containing lecithin and cholesterol was analyzed, it was highest after the placebo test. However, the LDL cholesterol absorption in eleven patients 4 and 5 days later was reduced by almost 40% after the addition of phytosterol glucosides to their diet [45]. It was shown that acylated steryl glycosides also reduce cholesterol absorption in mice as efficiently as phytosteryl esters. Cleavage of the glycosidic bond in steryl glycosides is not required for their biological activity. Moreover, these bonds are not cleaved in animals which consume food containing these compounds [46].

The nutritional value of food supplements derived from chlorophytes *Chlorella*, *Scenedesmus*, *Nannochloropsis*, and *Dunaliella* is well known [47]. Generally, phytosterol-containing products, including phytosterols as well as their glucosides and acylated glucosides (phytosterolins), have been commercialized as nutriceuticals or pharmaceuticals capable of lowering the blood cholesterol level and preventing the onset of cardiovascular disorders. The administration of these agents and a proper diet, enriched in fruits and vegetables, improves immune status and may help in treating various health problems related to chronic immune-mediated abnormalities. Microalgae can be considered as a promising source of these valuable compounds. The sterol content of such microalgal species as the haptophyte *P. lutheri*, chlorophytes *Tetraselmis* sp. and *Nannochloropsis* sp. was reported to range from 0.4–2.6% of dry weight. Taking into consideration their fast-growing characteristics, the annual production of fatty acid- and sterol-enriched oil from microalgae may vary from 19,000 to 57,000 L, which is greater than that obtained from terrestrial higher plants [48].

The great nutritional value of microalgae is first of all associated with the accumulation of polyunsaturated fatty acids in their lipids. The study on lipid classes of the haptophyte *P. lutheri* showed that acylated steryl glucosides in this species contain mainly C16:1(n-7), C14:0, and C16:0 fatty acids, but with a much higher level of docosahexaenic acid than that of eicosapentaenic acid in minor fatty acids [49]. Thus, acylated steryl glycosides of food supplements from microalgae also provide some contribution not only to the total phytosterol content, but also to the pool of useful polyunsaturated acids, characteristic of microalgae.

Steryl glycosides are also involved in cellular stress response. The level of sterol glycosylation is graduated in microalgae exposed to heat shock [36].

Steryl glycosides of unusual structures, possessing by other biological properties, can be also found in microalgae. For example, astasin, an unusual cytotoxic modified steryl glycoside from the colorless euglenophyte *Astasia longa*, consisting of ergosterol, xylopyranose and oxalic acid (Figure 5), has become an unexpectedly valuable find [50]. This compound inhibits the growth of human lymphoma HL-60 cells.

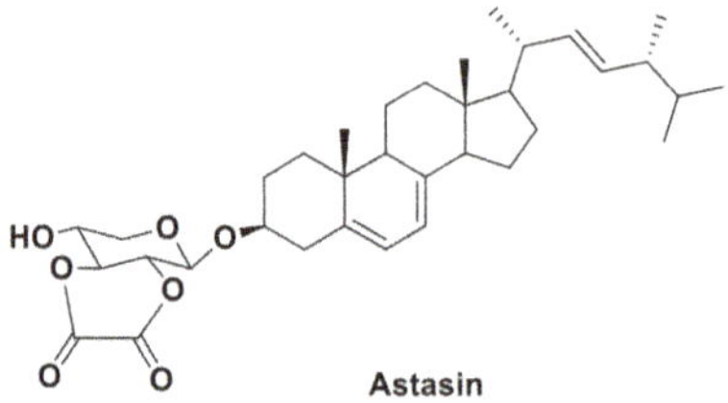

Figure 5. Unusual steryl glycoside astasin.

Biological functions of glycosylated sterols are most probably associated with the alteration of biophysical properties of cell membranes by these compounds. It was established that cholesteryl glycosides were much less effective in comparison with cholesterol in ordering the hydrocarbon chain region in the sphigolipid bilayer [51]. Steryl glycosides, which are amongst the main plant membrane components, were supposed to regulate the action of hormonal and environmental signals, thus providing the organization and biophysical properties of these membranes [52,53]. The consequences of variations in the proportions of free and glycosylated sterols in membranes are still not quite clear; however, the absence or shortage of steryl glycosides leads to dramatic dysfunctions in their producers [52]. Nevertheless, to the best of our knowledge, the biological functions of sterylglycosides on a molecular level and their effects on the growth and development in microalgae remain insufficiently studied so far.

Recently, it was found that an additional biological function of steryl glycosides possibly consists of their participation in biosynthesis of polysaccharides. In plants, cellulose synthase initiates the glucan synthesis using sitosterol-β-D-glucoside as primer. β-Sitosterol cellodextrins are formed from sitosterol-β-D-glucoside and uridine-5′-diphosphate glucose under conditions favorable for cellulose synthesis [54]. However, it is unclear whether this function is implemented in lower plants such as microalgae.

2.4. *Biosynthesis of Glycosylated Sterols*

The biosynthesis of steryl glycosides in plants, including some microalgae, occurs with the participation of sterol glycosyltransferases and uridine diphosphate glucose (UDP) as a sugar donor; this was shown using the example of the parasitic chlorophyte *Protothecha zopfii* [55]. As suggested, 4-methyl and 4,4-dimethyl sterols are poor substrates for these glycosylating enzymes [56]. Nevertheless, several glycosides of 4α-methylated sterols were detected in microalgae, which contradicts this suggestion (see Figure 3).

The membrane-bound sterol glycosytransferases (UDP-sugar:sterol glycopyranosyltransferases) belong to the family 1 of glycosyltransferases (UGT-superfamily) and hold an important position in plant metabolism [57,58]. The recent cloning of sterol glycosyltransferase genes from higher plants, algae, fungi, and bacteria was used to analyze the steryl glycoside functions. The corresponding full-length amino acid sequences of diatoms *Phaeodactylum tricornutum* and *Thalassiosira pseudonana*, as well as the sequences of the chlorophyte *Chlamydomonas reinhardtii*, were compared with those of related enzymes, and were functionally identified as sterol β-glycosyltransferases, deduced from genomic DNA. It was shown that the down regulation of sterol glycoside biosynthesis in the higher plant *Arabidopsis thaliana* causes dysfunctions in seed development [52].

Sterol glycoconjugates, acylated at C-6 of sugar portion, found in many plants including microalgae, are biosynthesized with the participation of another group of enzymes, namely steryl glycoside:acyltransferases [59]. A scheme of biosynthesis of steryl glycoconjugates, using glycosylated forms of sitosterol as an example, is given in Figure 6.

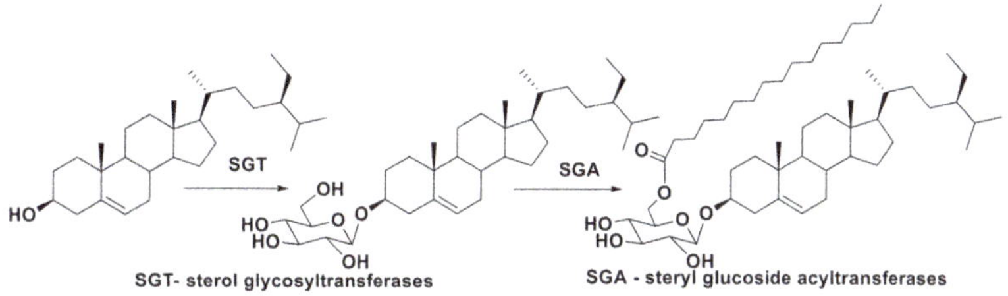

Figure 6. Biosynthesis of sterol glycoconjugates.

3. Sphingoid Glycoconjugates

Sphingolipids, found in almost all animals, plants, and fungi, as well as in some prokaryotic organisms and viruses, are an important part of the corresponding lipidomes. These compounds perform important structural and intracellular functions and participate in extracellular signaling. Along with sterols, they form the specialized microdomains in plasma membranes which are involved in a great variety of cellular processes, and are known as "rafts" [60].

Glycosphingolipids (Figure 7), as a class of sphingolipids, differ from non-glycosylated forms (ceramides) by a greater structural diversity: in addition to amine-containing lipid backbones, consisting of so-called sphingoid bases (sphingosine, sphinganine, phytosphinganine, and others) and fatty acid residues, they may contain a variety monosaccharide or oligosaccharide moieties. Moreover, these metabolites are sometimes sulfated and rarely phosphorylated [61].

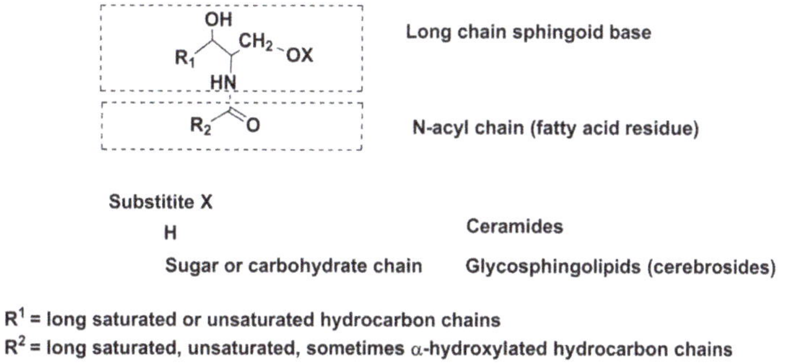

Figure 7. General structures of some sphingolipids (ceramides and glycosphingolipids).

The structural diversity of glycosphingolipids from mammals, higher plants, and fungi has been extensively studied for many years. A significant difference has been shown to exist between these metabolites from different taxa, particularly in their ceramide backbones. However, lower plants remain poorly studied in this respect to date. Many of marine animal species ingest microalgal metabolites obtained via the food chain. Moreover, there are many marine invertebrates containing microalgal symbionts. These host-symbiont consortiums produce a variety of highly bioactive natural compounds such as glycosphingolipids, biosynthesizing most probably with the participation of these symbiotic microorganisms. For example, the obligate symbiotic relationship between cnidarians such as corals and some dinoflagellates of the family Symbiodiniaceae is well known [62]. A comparison of their sphingolipids with those of whole symbiotic complexes would be of interest to better understand the peculiarities of the biosynthesis of these compounds in the plant and animal kingdoms.

3.1. Structural Diversity

Glycosylceramides (GlyCer) were detected in a large number of microalgae species, but in most of these cases, their structures were not fully characterized. As a rule, the determination of positions and configurations of double bonds, as well as the exact identification of sugars, were not performed. The first structural study on glycosylceramides from marine microalgae belonging to the class Prasinophyceae was carried out by Japanese scientists, who isolated two unusual glycolipids from *Tetraselmis* sp. (Figure 8). The chemical structures of these compounds were determined by NMR spectroscopy and GC-MS. Their sphingosine bases have (4*E*,8*E*)-sphinga-4,8-dienic structures (d18:2) and resemble the related lipids from terrestrial higher plants, while fatty acid moieties consist of 2-hydroxy-Δ^3-unsaturated fatty acid residues (h18:1 and h24:1). Δ^3(*E*)-unsaturation appears to be earlier found in some fungal glycosylceramides, but all these metabolites from fungi contain a characteristic C9-methyl branching in sphingoid moieties. Moreover, such long fatty acyl chains as C24 are rarely identified in fungal cerebrosides. Thus, the isolated metabolites are representatives of previously unknown structural series [63].

Three new glycosylceramides named isogalbamides A–C were isolated from the microalga *Isochrysis galbana* (Haptophyta, class Coccolithophyceae) using silica gel column chromatography and reversed phase HPLC, and studied by NMR and MS/MS methods. These first galactose-containing microalgal glycoconjugates have unprecedented tetraunsaturated sphigoid bases with structures of monosaccharide, N-acyl chains and bases, confirmed by COSY, HSQC, and HMBS NMR spectra. Isogalbamide A contains methyl branching at C-9 in the sphingoid base, like some fungal glycosylceramides, but closely-related isogalbamides C and D bear normal chains in these moieties. Isogalbamides showed moderate activity as inhibitors of the production of pro-inflammatory cytokine TNF-α in lipopolysaccharide-stimulated human THP-1 macrophages [64]. The procedures of extraction, separation, and spectroscopic analysis provided in this paper will be significant for forthcoming studies on microalgal glycosylceramides.

The identification of GlyCer with dC18:2 long chain base and 2-keto-3-deoxynonic acid (Kdn) as monosaccharide, as well the corresponding compound with methylated long-chain base, acylated by C22:0, C22:1, C22:2 and C-22:3 fatty acids, from the *Emiliana huxleyi* (phylum Haptophyta, Coccolithophyceae) has become another unexpectedly valuable finding (Figure 8) [65].

The diatom *Skeletonema costatum* was also recently examined for this type of important membrane component. A separation of lipids, extracted from this alga, using reverse-phase liquid chromatography, provided a fraction, which was subsequently studied by tandem mass spectrometry with collision–induced dissociation of the lithiated adducts using electrospray ionization quadrupole time-of-flight mass analyzer. As result, many types of novel glycosphingolipids were characterized from three strains of this alga, although their exact structures were not completely established. One of these strains (SCXMB02) contained disaccharides type of the corresponding compounds with 12 variants of mono- and diunsaturated C18 bases. N-acyl groups were primarily saturated or monoenic with 16C and 20 to 24C. The polar head groups were disaccharides consisting of heptose and hexose. Another strain (SKPXS0711) contained 13 related disaccharide ceramides, including those with triunsaturated long-chain bases.

The third strain (SKSPXs0807zjj) had novel glycosphingolipids of another type with trisaccharide moiety (heptose-hexose-hexose). This strain was considered as another species of the genus *Skeletonema* [66,67]. It is of particular interest that no trace of hexosylceramides and lactosylceramides, which are characteristic of higher plants, was found in the studied strains of *Skeletonema*. N-acyl groups of these glycosphingolipids were non-hydroxylated fatty acids, in contrast with the corresponding metabolites of higher plants. Thus, these studies demonstrated a great diversity of novel sphingoid glycoconjugates in diatoms.

Moreover, glycosphingolipid compositions of 17 microalgal strains belonging to species of the phyla Ochrophyta (2 strains of *Skeletonema costatum* and one strain of each *Skeletonema* sp., *S. tropicum*, *Thalassiosira pseudonana*, *Stephanodiscus* sp., *Conticribra weissflogii*, *Ceratoneis closterium*,

Ceratoneis sp., *Phaeodactylum tricornutum*, and *Amphora* sp.), Dinophyta (*Alexandrium minitum*, *Prorocentrum donghaiense*, and *Karlodinium veneficum*), and Haptophyta (*Isochrysis galbana*, *I. zhanjiangensis*, and *Pleurochrysis carterae*) have been also studied. A total of more than 40 variants of glycosylated ceramides have been detected in these strains [68]. As a result, the studied microalgae showed a great diversity of glyconjugated compounds (Table 2).

It has been found that the content of these compounds depends on the conditions of their cultivation. In the diatom *T. pseudonana* grown in the phosphorus-limited conditions, the level of diglycosylceramides increased by up to ten-fold. Glycosphingolipids in this species were identified as new representatives of diglycosylated compounds with the main $(Gly)_2$Cer d18:3/24:0 constituent [69]. Some results of glycosphingolipid identification and their taxonomic distribution are given in Table 2.

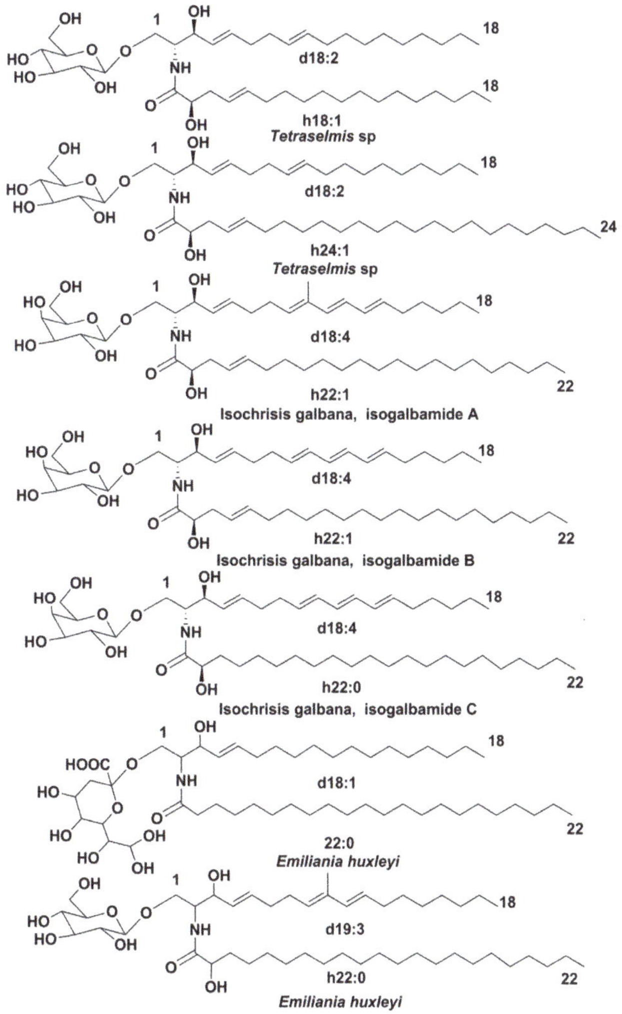

Figure 8. Glycosylceramides from the microalgae *Tetraselmis* sp., *Isochrisis galbana*, and *Emiliania huxleyi*. Sphingoid base is designated with the chain length and number of double bonds; prefix 'd ' is used to designate dihydroxylated bases. Fatty acids residues are designated with the chain length and number of double bonds; prefix 'h' is used for hydroxylated fatty acids.

Table 2. Distribution of some structural types of glycosphingolipids in major taxa of marine microalgae.

Glycosphingolipids	Taxa		
	Ochrophyta	Dinophyta	Haptophyta
Ceramide moiety	d18:0/16:0; d18:1/16:0; d18:2/16:0; d18:1/22:0; d18:2/22:0; d18:3/22:0; d18:2/22:1; d18:3/23:0; d18:1/24:0; d18:2/24:0; d18:3/24:0; 18:2/24:1; d18:3/24:1; d18:2/24:2 d18:1/26:0;d18:2/26:0 [66,67]; d18:2/14:0; d18:3/14:0 [68]	d18:3/16:0; d18:4/16:0; d18:3/16:1; d18:4/16:1; d19:3/16:0; 19:4/16:1; d19:3/h18:1; 19:3/h19:1; d19:3/h24:1; d19:4/h24:1 [68]	d18:$2^{4,8}$/h18:1^4; d18:$1^{4,8}$/h24:1^4 [63]; d18:$4^{4,8,10,12}$/h22:1^4; 9-methyl-d18:$4^{4,8,10,12}$/h22:1^4; d18:$4^{4,8,10,12}$/h22:1^4; d18:$1^{4,8,10,12}$ /h22:0 [64]; h18:1^4/h22:0; C19:$3^{4,8,10}$/d22:0 [65]; d18:0/h22:0; d18:0/ h22:1; d18:0/h22:2, d19:2/h22:0; d19:2/h22:1; d19:2/h22:2; d18:3/h23:2 and others [68]
Glycosyl moiety	Monosaccharide Disaccharide Trisaccharide	Monosaccharide	Glucose [63], galactose [64], glucose and sialic acid [65]

The obtained results showed the structural diversity of these metabolites in microalgae: the identified glycosphingolipids contained various long chain bases, a set of different non-hydroxylated and α-hydroxylated fatty acids, and from one to three monosaccharide units in their carbohydrate moieties. Despite the fact that the exact structures of glycosphingolipids have been determined only in a few cases, it may be expected that many novel structural variants of these glycoconjugates should be discovered in the near future.

3.2. Biosynthesis, Biological Activities and Biological Roles of Microalgal Glycosphingolipids

Structures and taxonomic distribution of glycosphingolipids suggest that biosynthesis of these metabolites proceeds in general terms as in other plants. However, in different micoalgal taxa, there are numerous peculiarities and deviations that will undoubtedly be the subject of further research. The main pathway of glycosylceramide biosynthesis is shown in the Figure 9 [70]. However, not one compound, but whole sets of metabolites can be formed at each its stage depending from organism producer. For example, sphinganine may be transformed into 4-hydroxysphinganine as a result of 4-hydroxylation in some plants and marine organisms, and long chain bases may be formed as a result of sphinganine degradation followed by other transformations. Ceramide synthesis in some cases is accompanied by hydroxylation of fatty acid residues, giving a pool of different ceramides. Their glycosylation is sometimes not limited by the attachment of one monosaccharide unit, and so on.

Glycosylceramides, as ubiquitous glycosphingolipids occurring in plants, fungi, and animals, proved to be major components of the membranes in most eukaryotic cells. There is little information on the functions of individual representatives of these membrane constituents in both higher and lower plants. It was shown that glycolipids in membrane regulate some intracellular processes. The GlyCer content of the plant plasmatic membrane was observed to decrease in stress conditions such as, for example, cold acclimation. In these cases, the compositions of glycosphingolipids in microalgae are also changed [71].

It was established that GlcCer participate in cell-to-cell interactions [69]. These compounds may facilitate binding of pathogens to cells of the host, as with e.g., the human pathogen *Helicobacter pylori* binds monoglucosylceramides [72].

An important recent discovery is the participation of glycosphingolipids in the regulation of cell death caused by a viral infection in marine phytoplankton. The basis for this study was created by the complete genome sequencing of recently-discovered *Coccolithovirus*, the giant double-stranded DNA virus encoding approximately 600 proteins. Among the genes of this virus, a variety of previously unknown genes was found, including those involved in biosynthesis of sphingolipids [73].

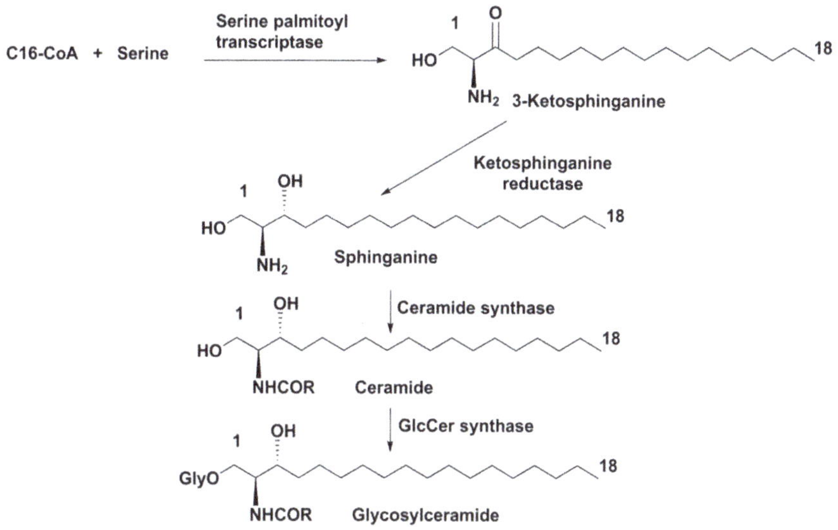

Figure 9. The main pathway of glycosylceramide biosynthesis.

The virus infects the coccolithophorid *E. huxleyi*, a cosmopolitan microalga, accounting for approximately a third of the global marine $CaCO_3$ production due to its calcite skeleton and emission of dimethyl sulfide, which has an influence on cloud formation. The extensive annual rapid development into short-term large populations (blooms) of this species in the North Atlantic are terminated by the viral infection resulting in the lysis of microalgal cells. The genome of this type of viruses penetrates into host cells and changes the metabolism of glycosphingolipids during infection. Particularly, the virus EhV86 encodes the serine palmitoyl transferase (SPT), which induces a metabolic switch in sphingolipid biosynthesis in infected microalga, modulating the algal physiology. Viral glycosphingolipids, when interacting with cells of the microalga, suppressed the cell growth, elevated in vivo caspase activities, and caused mortality of the cells, inducing their apoptosis [74]. Composed of unique hydroxylated sphingoid base (t17:0), viral glycosphingolipids not only penetrate into microalgae, being essential for the infection process, but also participate in the virus assembly formation.

An abridged scheme of participation of the virus in sphingolipid biosynthesis during the viral infection in the coccolithophorid *E. huxleyi* is provided in Figure 10. The viral-encoded serine palmitoyl transferase (SPT), in comparison with this enzyme of the host, has a different substrate specificity, which allows C15-CoA instead of C16-CoA to produce the C17 sphinganine base instead of C18 sphinganine in the host cells. Further transformations, including glycosylation and hydroxylation, lead to a metabolic remodulation in sphingolipid biosynthesis. A viral SPT proved to be a key enzyme in rewiring the host ceramide synthesis [75].

The critical role of glycosphingolipids in the stimulation of programmed cell death of *E. huxleyi* was confirmed in the process of its natural bloom event. These metabolites also contribute to understanding the co-evolutionary "arm race" between populations of *E. huxleyi* and cocolithoviruses [76].

Some other conjugated lipids also were recognized as regulators of algal death at final stages of microalgal blooms. Actually, sterol sulfates were proved to be regulatory molecules of a cell death program in *Skeletonema marinoi*, a marine diatom bloom-forming species inhabiting temperate coastal waters. Intracellular level of sterol sulfates increases with cell ageing, that leads to an oxidative burst and the production of nitric oxide followed by apoptosis-like death of these microalgae [77].

An assumption that glycosphingolipids and sterol glycosides can be biogenetically related to each other is interesting. There is some evidence that glycosylceramides can be partly synthesized in plants through the interaction of ceramides with steryl glucosides as glucose donors [78–80]. However, this transformation has not yet been confirmed in microalgae.

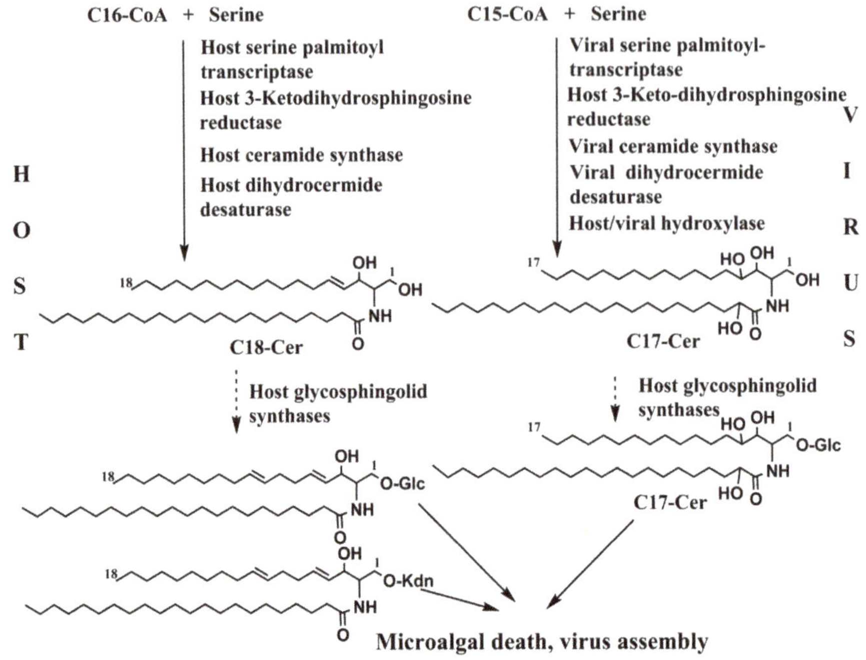

Figure 10. A simplified scheme of glycosylceramide biosynthetic pathways during interaction between the coccolithophorid *E. huxleyi* and its specific virus ExV.

4. Conclusions

Glycoconjugates of sterols and sphingolipids are widely distributed metabolites of microalgae, although they are still insufficiently studied on the structural level in many taxa. Glycosylated sterols play important functional roles in their producers and exhibit different bioactivities, making them promising agents for applications in medicine and as supplementary components to healthy food. Sterols and their conjugates enter marine invertebrates from microalgae through the food chain and may be transformed into other steroid derivatives such as 7(8)-unsaturated compounds in many starfishes and sea cucumbers, or polyhydroxylated sterols and glycosides in starfishes and some other marine invertebrates. The corresponding metabolites from invertebrates often have the same structural peculiarities of their side chains as those of microalgal sterols and their conjugates [81,82], although the microalgal source of some invertebrate sterols, for example, containing a side chain (a) (Figure 1), has not yet been confirmed.

Glycosphingolipids have been intensively studied in animals for decades, but knowledge of these metabolites in microalgae still remains poor. Recently microbial metabolites of this class have been successfully studied using modern variants of mass-spectrometry and chromatography.

It was found that glycosphingolipids and sterol conjugates are regulators of membrane functions. Some of them are important for the interactions of their producers with pathogens, and may induce apoptosis in microalgae. Also, these compounds participate in the termination of microalgal blooms.

Author Contributions: Authors contributed equally to the work. V.A.S. writing and discussion of the review parts concerning structural diversity, I.V.S. writing and discussion of the review parts concerning taxonomy and biological properties of glycoconjugates.

Funding: This work was supported by the Grant No. 17-04-00034 from the Russian Foundation for Basic Research.

Acknowledgments: The study was carried out on the equipment of the Collective Facilities Center, «The Far Eastern Center for Structural Molecular Research (NMR/MS) of PIBOC FEB RAS».

Conflicts of Interest: The authors declare no conflict of interest.

References

1. Borowitzka, M. Systematics, taxonomy and species names: Do they matter? In *The Physiology of Microalgae*; Developments in Applied Phycology; Borowitzka, M.A., Beardall, J., Raven, J.A., Eds.; Springer International Publishing: Dordrecht, the Netherlands, 2016; pp. 655–681, ISBN 978-3-319-24945-2 or 978-3-319-24943-8.
2. Guiry, M.D. How many species of algae are there? *J. Phycol.* **2012**, *48*, 1057–1063. [CrossRef] [PubMed]
3. Lee, R.E. *Phycology*, 4th ed.; Cambridge University Press: New York, NY, USA, 2008; ISBN 978-0-511-38669-5.
4. Stonik, V.; Stonik, I. Low-molecular weight metabolites from diatoms: Structures, biological roles and biosynthesis. *Mar. Drugs* **2015**, *13*, 3672–3709. [CrossRef] [PubMed]
5. Kristiansen, J.; Škaloud, P. Chrysophyta. In *Handbook of the Protists*; Archibald, J., Simpson, A., Slamovits, C., Eds.; Springer: Cham, Switzerland, 2017; pp. 331–366. ISBN 978-3-319-28147-6.
6. Steindinger, K.A.; Jangen, K. Dinoflagellates. In *Identifying Marine Diatoms and Dinoflagellates*; Thomas, C.R., Ed.; Academic Press: San-Diego, CA, USA, 1997; pp. 387–584. ISBN 0-12-693018-X.
7. Hasle, G.; Syvertsen, E. Diatoms. In *Identifying Marine Diatoms and Dinoflagellates*; Thomas, C.R., Ed.; Academic Press: San-Diego, CA, USA, 1997; pp. 5–385. ISBN 0-12-693018-X.
8. Sirakov, I.; Velichkova, K.; Stoyanova, S.; Staykov, Y. The importance of microalgae for aquaculture industry. Review. *Int. J. Fish. Aquat. Stud.* **2015**, *2*, 81–84.
9. Volkman, J.K. A review of sterol markers for marine and terrigenous organic matter. *Org. Geochem.* **1986**, *9*, 83–99. [CrossRef]
10. Volkman, J.K. Sterols in microorganisms. *Appl. Microbiol. Biotechnol.* **2003**, *60*, 495–506. [CrossRef] [PubMed]
11. Rampen, S.W.; Abbas, B.A.; Shouten, S.; Sinninghe Dumste, J.J.S. A comprehensive study of sterols in marine diatoms (Bacillariophyta): Implications for their use as tracers for diatom productivity. *Limnol. Oceanogr.* **2010**, *55*, 91–105. [CrossRef]
12. Volkman, J.K. Sterols and other triterpenoids: Source specificity and evolution of biosynthetic pathways. *Org. Geochem.* **2005**, *36*, 139–159. [CrossRef]
13. Volkman, J.K. Sterols in microalgae. In *The Physiology of Microalgae*; Developments in Applied Phycology, Borowitzka, M.A., Beardall, J., Raven, J.A., Eds.; Springer International Publishing: Dordrecht, the Netherlands, 2016; pp. 485–505, ISBN 978-3-319-24945-2 or 978-3-319-24943-8.
14. Stonik, I.V.; Kapustina, I.I.; Aizdaicher, N.A.; Svetashev, V.I. Sterols and fatty acids from *Attheya* planktonic diatoms. *Chem. Nat. Compd.* **2017**, *53*, 422–425. [CrossRef]
15. Giner, J.L.; Wikfors, G.H. "Dinoflagellate sterols" in marine diatoms. *Phytochemistry* **2011**, *72*, 1896–1901. [CrossRef]
16. Rampen, S.W.; Volkman, J.K.; Hur, S.B.; Abbas, D.A.; Shouten, S.; Jameson, I.D.; Holdsworth, D.G.; Bae, J.H.; Sinninghe Dumste, J.J.S. Occurrence of gorgosterol in diatoms of the genus *Delphineis*. *Org. Geochem.* **2009**, *40*, 144–147. [CrossRef]
17. Giner, L.L.; Zhao, H.; Boyer, G.L.; Satchell, M.F.; Anderson, R.A. Sterol chemotaxonomy of marine Pelagophyte algae. *Chem. Biodivers.* **2009**, *6*, 1111–1130. [CrossRef] [PubMed]
18. Dunstan, G.A.; Brown, M.R.; Volkman, J.K. Cryptophyceae and Raphidophyceae; chemotaxonomy, phylogeny, and application. *Phytochemistry* **2005**, *66*, 2557–2570. [CrossRef] [PubMed]
19. Leblond, J.D.; Lasiter, A.D.; Li, C.; Logares, R.; Rengefors, K.; Evens, T.J. A data mining approach to dinoflagellate clustering according to sterol composition: Correlations with evolutionary history. *Int. J. Data Min. Bioinform.* **2010**, *4*, 431–451. [CrossRef] [PubMed]
20. Giner, J.L.; Ceballos, H.; Tang, Y.Z.; Gobler, C.J. Sterols and fatty acids of the harmful dinoflagellate *Cochlodinium polykrikoides*. *Chem. Biodivers.* **2016**, *13*, 249–252. [CrossRef]
21. Volkman, J.K.; Kearney, P.; Jeffrey, S.W. A new source of 4-methyl sterols and 5a(H)-stanols in sediments: Premnesiophyte microalgae of the genus *Pavlova*. *Org. Geochem.* **1990**, *15*, 489–497. [CrossRef]
22. Volkman, J.K.; Farmer, C.L.; Barrent, S.M.; Sikes, E.L. Unusual dihydroxysterols as chemotaxonomic markers for microalgae from the order *Pavlovales (Haptophyceae)*. *J. Phycol.* **1997**, *33*, 1016–1023. [CrossRef]

23. Leblond, J.D.; Timofte, H.I.; Roche, S.A.; Porter, N.M. Sterols of glaucocystophytes. *Phycol. Res.* **2011**, *59*, 129–134. [CrossRef]
24. Leblond, J.D.; Dahmen, J.L.; Seipelt, R.L.; Elrod-Erickson, M.J.; Kincaid, R.; Howard, J.C.; Evans, T.J.; Chapman, P.J. Lipid composition of chlorarachniophytes (Chlorarachniophyceae) from genera *Bigelowiella, Gymnochlora,* and *Lotharella*. *J. Phycol.* **2005**, *41*, 311–321. [CrossRef]
25. Bilbao, P.G.S.; Damani, C.; Salvador, G.A.; Leonardi, P. *Haematococcus pluvalis* as a source of fatty acids and phytosterols potential nutritional and biological applications. *J. Appl. Phycol.* **2016**, *28*, 3283–3294. [CrossRef]
26. Martin-Creuzberg, D.; Merkel, P. Sterols in freshwater microalgae: Potential applications for zooplankton nutrition. *J. Plankt. Res.* **2016**, *38*, 865–877. [CrossRef]
27. Gladu, P.K.; Paterson, G.W.; Wikfors, G.W.; Chitwood, D.J.; Lusby, W.R. The occurrence of brassicasterol and epibrassicasterol in the Chromophycota. *Comp. Biochem. Physiol.* **1990**, *97B*, 491–494. [CrossRef]
28. Lewin, R.A. Biochemical taxonomy. In *Algal Physiology and Biochemistry (Botanical Monographs)*; Stewart, W.D.P., Ed.; Blackwell Sci.: Oxford, UK, 1974; pp. 1–32. ISBN 978-0632091003.
29. Wang, G.; Wang, T. Characterization of lipid components in two microalgae for biofuel application. *J. Am. Oil Chem. Soc.* **2011**, *89*, 135–143. [CrossRef]
30. Yao, L.; Gerde, J.A.; Lee, S.-L.; Wang, T.; Harrata, K.A. Microalgae lipid characreization. *J. Agric. Food Chem.* **2015**, *63*, 1773–1787. [CrossRef] [PubMed]
31. Duperon, R.M.; Thiersault, M.; Duperon, P. Occurrence of steryl glycosides and acylated sreyl glycosides in some marine algae. *Phytochemistry* **1983**, *22*, 535–538. [CrossRef]
32. Nyström, L.; Schär, A.; Lampi, A.M. Steryl glycosides and acetylated steryl glycosides in plant foods reflect unique sterol patterns. *Eur. J. Lipid Sci. Technol.* **2012**, *114*, 656–669. [CrossRef]
33. Mohammady, N.G. Total, free and conjugated sterolic forms in three microalgae used in mariculture. *Z. Naturforsch. C* **2004**, *59*, 9–10. [CrossRef]
34. Vernon, B.; Dauguet, J.-C.; Billard, C. Sterolic biomarkers in marine phytoplankton. I. Free and conjugated sterols of *Pavlova lutheri* (Haptophyta). *Eur. J. Phycol.* **1996**, *31*, 211–215. [CrossRef]
35. Vernon, B.; Dauguet, J.-C.; Billard, C. Sterolic biomarkers in marine phytoplankton. I. Free and conjugated sterols of seven species used in mariculture (Haptophyta). *Eur. J. Phycol.* **1998**, *34*, 273–279. [CrossRef]
36. Veron, B.; Billard, C.; Dauguet, J.C.; Hartmann, M.-A. Sterol composition of *Phaeodactylum tricornutum* as influenced by growth temperature and light spectral quality. *Lipids* **1996**, *31*, 989–994. [CrossRef]
37. Kovganko, N.V.; Kashkan, Z.N. Sterol glycosides and acylglycosides. *Chem. Nat. Compd.* **1999**, *35*, 479–497. [CrossRef]
38. Phillips, K.M.; Ruggio, D.M.; Ashraf-Khorassani, M. Analysis of steryl glucosides in foods and dietary supplement by solid-phase extraction and gas chromatography. *J. Food Lipids* **2005**, *12*, 124–140. [CrossRef]
39. Pieber, B.; Schnober, S.; Goebl, C.; Mittelbach, M. Novel sensitive determination of steryl glycosides in biodiesel by gas chromatography—Mass spectrometry. *J. Chromatogr. A* **2010**, *1217*, 6555–6561. [CrossRef]
40. Oppliger, S.R.; Mungen, L.H.; Nystrom, L. Rapid and highly accurate detection of steryl glycosides by ultraperformance liquid chromatography-quadropole time-of-flight mass spectrometry (UPLC-Q-TOF MS). *J. Agric. Food Chem.* **2014**, *62*, 9410–9419. [CrossRef] [PubMed]
41. Yu, S.; Zhang, Y.; Ran, Y.; Lai, W.; Ran, Z.; Xu, J.; Zhou, C.; Yan, X. Characterization of steryl glycosides in marine microalgae by gas chromatography-triple guadropole mass spectrometry (GC-QQQ-MS). *J. Sci. Food Agric.* **2018**, *98*, 1574–1583. [CrossRef] [PubMed]
42. Bouic, P.J. The role of phytosterols and phytosterolines in immune modulation: A review of the past 10 years. *Curr. Opin. Clin. Nutr. Metab. Care* **2001**, *4*, 471–475. [CrossRef] [PubMed]
43. Shimamura, M. Immunological functions of steryl glycosides. *Arch. Immunol. Ther. Exp. (Warsz)* **2012**, *60*, 351–359. [CrossRef]
44. Lee, J.H.; Lee, J.Y.; Park, J.H.; Jung, H.S.; Kim, J.S.; Kang, S.S.; Kim, Y.S.; Han, Y. Immunoregulatory activity by daucosterol, a beta-sistosterol glycoside, induces protective Th1 immune response against disseminated Candidiasis in mice. *Vaccine* **2007**, *25*, 3834–3840. [CrossRef]
45. Lin, X.; Ma, L.; Racette, S.B.; Anderson Spearie, C.L.; Ostlund, R.E.J. Phytosterol glycosides reduce cholesterol absorption in humans. *Am. J. Physiol. Gastrointest. Liver Physiol.* **2009**, *296*, G931–G935. [CrossRef]
46. Lin, X.; Ma, L.; Moreau, R.A.; Ostlund, R.E.J. Glycosidic bond cleavage is not required for phyrosteryl glucoside-induced reduction of cholesterol absorption in mice. *Lipids* **2011**, *46*, 701–708. [CrossRef]

47. Kent, M.; Welladsen, H.M.; Mangott, A.; Li, Y. Nutritional evaluation of Australian microalgae as potential human health supplements. *PLoS ONE* **2015**, *10*, e0118985. [CrossRef]
48. Bouic, P.J.; Lamprecht, J.H. Plant sterols and sterolins: A review of their immune-modulating properties. *Altern. Med. Rev.* **1999**, *4*, 170–177. [PubMed]
49. Meireless, L.A.; Guedas, A.C.; Malcata, F.X. Lipid class composition of the microalga *Pavlova lutheri*: Eicosapentaenic and docosahexaenoic acids. *J. Agric. Food Chem.* **2003**, *51*, 2237–2241. [CrossRef] [PubMed]
50. Kaya, K.; Sano, T.; Shiraishi, F. Astasin, a novel cytotoxic carbohydrate–conjugated ergosterol from the colorless euglenoid, *Astasia longa*. *Biochim. Biophys. Acta* **1995**, *1255*, 201–204. [CrossRef]
51. Halling, K.K.; Ramstedt, B.; Slotte, J.P. Glycosylation induces shifts in the lateral distribution of cholesterol from ordered towards less ordered domains. *Biochim. Biophys. Acta* **2008**, *1778*, 1100–1111. [CrossRef] [PubMed]
52. Grille, S.; Zaslawski, A.; Thiele, S.; Plat, J.; Warnecke, D. The functions of steryl glycosides come to those who wait: Recent advances in plants, fungi, bacteria and animals. *Prog. Lipid Res.* **2010**, *49*, 262–288. [CrossRef]
53. Valitova, J.N.; Sulkarnayeva, A.G.; Minibayeva, F.V. Plant sterols: Diversity, biosynthesis, and physiological functions. *Biochemistry (Mosc.)* **2016**, *81*, 819–834. [CrossRef] [PubMed]
54. Peng, L.; Kawagoe, Y.; Hogan, P.; Delmer, D. Sitosterol-beta-glucoside as primer for cellulose synthesis in plants. *Science* **2002**, *5552*, 147–150. [CrossRef] [PubMed]
55. Hopp, H.; Romero, P.A.; Daleo, G.R.; Pont Lezica, R. Steryl glycoside biosynthesis in the alga *Prototeheca zopfii*. *Phytochemistry* **1978**, *17*, 1049–1052. [CrossRef]
56. Wojciechowski, Z.A. Biochermistry of phytosterol conjugates. In *Physiology and Biochemistry of Sterols*; Patterson, G.W., Ed.; AOCS Publishing: New York, NY, USA, 1992; pp. 361–395. ISBN 978-1439821831.
57. Chaturvedi, P.; Misra, P.; Tuli, R. Sterol glycosyltransferases—The enzymes that modify sterols. *Appl. Biochem. Biotechnol.* **2011**, *165*, 47–68. [CrossRef] [PubMed]
58. Ullman, P.; Ury, A.; Rimmele, D.; Benveniste, P.; Bouvier-Nave, P. UDP-glucose sterol β-D- glucosyltransferase, a plasma membrane-bound enzyme of plants: Enzymatic properties and lipid dependence. *Biochimie* **1993**, *75*, 713–723. [CrossRef]
59. Yoon, K.; Han, D.; Sommerfield, M.; Hu, Q. Phospholipid:diacylglycerol acyltransferase is a multifunctional enzyme involved in membrane lipid turnover and degradation while synthesizing triacylglycerol in the unicellular green microalga *Chlamydomonas reinhardtii*. *Plant Cell* **2012**, *24*, 3708–3724. [CrossRef]
60. Sonnino, S.; Prinetti, A. Membrane domains and the "lipid raft" concept. *Curr. Med. Chem.* **2013**, *20*, 4–21. [CrossRef]
61. Merril, A.H.J.; Wang, M.D.; Park, M.; Sullards, M.C. (Glyco)sphingolipidology: An amazing challenge and opportunity for systems biology. *Trends Biochem. Sci.* **2007**, *32*, 457–467. [CrossRef]
62. Davy, S.K.; Allemand, D.; Weis, V.M. Cell biology of cnidarian-dinoflagellate symbiosis. *Microbiol. Mol. Biol. Rev.* **2012**, *76*, 229–261. [CrossRef]
63. Arakaki, A.; Iwama, D.; Liang, D.; Murakami, N.; Ishikura, M.; Tanaka, T.; Matsunaga, T. Glycosylceramides from marine microalga *Tetraselmis* sp. *Phytochemistry* **2013**, *85*, 107–114. [CrossRef]
64. De los Reyes, C.; Ortega, M.J.; Rodriguez-Luna, A.; Talero, E.; Motilva, V.; Zubía, E. Molecular characterization and anti-onflammatory activity of galactosylglycerides and galactosylceramides from the microalga *Isochrysis galbana*. *J. Argric. Food Chem.* **2016**, *64*, 8783–8794. [CrossRef]
65. Fulton, J.M.; Fredericks, H.F.; Bidle, K.D.; Vardi, A.; DiTullio, R.; Van Mooy, A.D. Novel molecular determinants of viral susceptibility and resistance in the lipidome of *Emiliania huxleyi*. *Environ. Microbiol.* **2014**, *16*, 1137–1149. [CrossRef]
66. Zhao, F.; Xu, J.; Chen, J.; Yan, X.; Zhou, C.; Li, S.; Xu, X.; Ye, F. Sructural elucidation of two types of novel glycosphingolipids in three strains of *Skeletonema* by liquid chromatography coupled with mass spectrometry. *Rapid. Commun. Mass Spectrom.* **2013**, *27*, 1535–1547. [CrossRef]
67. Yan, X.J.; Chen, D.Y.; Xu, J.L.; Zhou, C.X. Profiles of photosynthetic glycerolipids in three strains of *Skeletonema* determined by UPLC-Q-TOF-MS. *J. Appl. Phycol.* **2010**, *23*, 271–282. [CrossRef]
68. Li, Y.; Lou, Y.; Mu, T.; Ke, A.; Ran, Z.; Xu, J.; Chen, J.; Zhou, C.; Yan, Q.; Xu, Q.; et al. Sphingoid in marine microalgae: Development and application of a mass spectrometric method for global structural characterization of ceramides in three major phyla. *Anal. Chim. Acta* **2017**, *986*, 82–94. [CrossRef]

69. Hunter, J.T.; Brandsma, J.; Dymond, M.K.; Koster, G.; Moore, C.M.; Postle, A.D.; Mills, R.A.; Attard, G.S. Lipidomics of *Thalassiosira pseudonana* under phosphorus stress reveal underlying phospholipid substitution dynamics and novel diglycosylceramide substitutes. *Appl. Environ. Microbiol.* **2018**. [CrossRef]
70. Sperling, P.; Henz, E. Plant sphingolipids: Structural diversity, biosynthesis, first genes and functions. *Biochim. Biophys. Acta* **2003**, *1632*, 1–15. [CrossRef]
71. Lynch, D.V.; Steponkus, P.L. Plasma membrane lipid alterations associated with cold acclimation of winter rye seedlings (*Secale* cereale L. cv Puma). *Plant Physiol.* **1987**, *83*, 761–767. [CrossRef]
72. Warntcke, D.; Heinz, E. Recently discovered functions of glycosylceramides in plants and fungi. *Cell. Mol. Life Sci.* **2003**, *60*, 919–941. [CrossRef]
73. Abul-Milh, M.; Foster, D.B.; Lingwood, C.A. In vitro binding of Helicobacter pylori to monohexosylceramides. *Glycoconj. J.* **2001**, *18*, 253–260. [CrossRef]
74. Wilson, W.H.; Schroeder, D.C.; Allen, M.J.; Holden, M.T.G.; Parkhill, J.; Barell, B.G.; Churcher, C.; Harnlin, N.; Mungall, K.; Norbertzak, H.; et al. Complete genome sequence and lytic phase transcription profile of a Coccolithovirus. *Science* **2005**, *309*, 1090–1092. [CrossRef]
75. Vardi, A.; Van Mooy, B.A.S.; Fredericks, H.F.; Ropendorf, K.J.; Ossolinski, J.E.; Yaramaty, L.; Bidle, K.D. Viral glycosphingolipids induce lytic infection and cell death in marine phytoplankton. *Science* **2009**, *326*, 861–865. [CrossRef]
76. Ziv, C.; Malitzky, S.; Othman, A.; Ben-Dor, S.; Wei, Y.; Zheng, S.; Aharoni, A.; Hornemann, T.; Vardi, A. Viral serine palmitoyltransferase induces metabolic switch in sphingolipid biosynthesis and is required for infection of a marine alga. *Proc. Natl. Acad. Sci. USA* **2016**, *113*, 1907–1916. [CrossRef]
77. Vardi, A.; Haramaty, L.; Van Mooy, B.A.S.; Fredericks, H.F.; Kimmance, S.A.; Larsen, A.; Bidle, K.D. Host-virus dynamics and subcellular control of cell fate in a natural coccolithophore population. *Proc. Natl. Acad. Sci. USA* **2012**, *109*, 19327–19332. [CrossRef]
78. Gallo, C.; d'Ippolito, G.; Nuzzo, G.; Sardo, A.; Fontana, A. Autoinhibitory sterol sulfates mediate programmed cell death in a bloom-forming marine diatom. *Nat. Commun.* **2017**, *8*, 1292. [CrossRef]
79. Lynch, D.V.; Criss, A.K.; Lehoczky, J.L.; Bui, V.T. Ceramide glucosylation in bean hypocotyl microsome: Evidence that steryl glucoside serves as glucose donor. *Arch. Biochem. Biophys.* **1997**, *340*, 3111–3316. [CrossRef]
80. Lynch, D.V.; Dunn, T.V. An introduction to plant sphingolipids and a review of recent advances in understanding their metabolism and function. *New Phytolog.* **2004**, *161*, 677–702. [CrossRef]
81. Stonik, V.A.; Ponomarenko, L.P.; Makarieva, T.N.; Boguslavsky, V.M.; Dmitrenok, A.S.; Fedrov, S.N.; Strobikin, S.A. Free sterol compositions from the sea cucumbers *Pseudostichopus trachus, Holothuria (Microtele) nobilis, Trochostoma orientale* and *Bathyplotes natans*. *Comp. Biochem. Physiol.* **1998**, *120B*, 337–347. [CrossRef]
82. Ivanchina, N.V.; Kicha, A.A.; Stonik, V.A. Steroid glycosides from marine organisms. *Steroids* **2011**, *76*, 425–454. [CrossRef]

marine drugs

Article

The Marine Dinoflagellate *Alexandrium minutum* Activates a Mitophagic Pathway in Human Lung Cancer Cells

Christian Galasso [1,*,†], Genoveffa Nuzzo [2,†], Christophe Brunet [1,*], Adrianna Ianora [1], Angela Sardo [2], Angelo Fontana [2] and Clementina Sansone [1]

1 Stazione Zoologica Anton Dohrn, Istituo Nazionale di Biologia, Ecologia e Biotecnologie Marine, Villa Comunale, 80121 Naples, Italy; adrianna.ianora@szn.it (A.I.); clementina.sansone@szn.it (C.S.)
2 Bio-Organic Chemistry Unit, Institute of Biomolecular Chemistry-CNR, Via Campi Flegrei 34, Pozzuoli, 80078 Naples, Italy; nuzzo.genoveffa@icb.cnr.it (G.N.); angela.sardo@icb.cnr.it (A.S.); a.fontana@icb.cnr.it (A.F.)
* Correspondence: christian.galasso@szn.it (C.G.); christophe.brunet@szn.it (C.B.); Tel.: +39-081-5833221 (C.G.); Tel.: +39-081-5833603 (C.B.)
† These authors contributed equally to this work.

Received: 6 November 2018; Accepted: 10 December 2018; Published: 12 December 2018

Abstract: Marine dinoflagellates are a valuable source of bioactive molecules. Many species produce cytotoxic compounds and some of these compounds have also been investigated for their anticancer potential. Here, we report the first investigation of the toxic dinoflagellate *Alexandrium minutum* as source of water-soluble compounds with antiproliferative activity against human lung cancer cells. A multi-step enrichment of the phenol–water extract yielded a bioactive fraction with specific antiproliferative effect (IC_{50} = 0.4 $\mu g \cdot mL^{-1}$) against the human lung adenocarcinoma cells (A549 cell line). Preliminary characterization of this material suggested the presence of glycoprotein with molecular weight above 20 kDa. Interestingly, this fraction did not exhibit any cytotoxicity against human normal lung fibroblasts (WI38). Differential gene expression analysis in A549 cancer cells suggested that the active fraction induces specific cell death, triggered by mitochondrial autophagy (mitophagy). In agreement with the cell viability results, gene expression data also showed that no mitophagic event was activated in normal cells WI38.

Keywords: glycoprotein; mitophagy; marine antiproliferative compounds; *Alexandrium minutum*

1. Introduction

To date, many microalgae have been shown to inhibit proliferation and development of malignant cells. For example, the diatoms *Attheya longicornis*, *Chaetoceros socialis*, *Chaetoceros furcellatus*, *Skeletonema marinoi* and *Porosira glacialis* exhibited anti-cancer activity against melanoma A2058 cells [1]. An ethyl acetate crude extract from the diatom *Chaetoceros calcitrans* was able to induce apoptosis in breast cancer cells [2]. Aldehydes from *S. marinoi* induced apoptosis in colorectal Caco-2 tumor cells [3] and mes-c-myc A1 cell line [4], through an extrinsic apoptotic pathway [5]. Interestingly, the potential microalgal bioactivities can be modulated by culture conditions [6], highlighting that the synthesis of secondary metabolites responsible for such biological effects is an adaptive response to environmental cues. These molecules are probably synthetized to protect themselves and/or to reinforce responses to environmental stimuli, through activation of specific molecular pathways [7]. In addition, carotenoids from marine sources have been reported for their anti-proliferative effects [8]. From a study on the green alga *Dunaliella tertiolecta*, the carotenoid violaxanthin showed a potent antiproliferative activity on MCF-7 breast cancer cells and induced biochemical changes typical of early apoptosis [9]. Another carotenoid, peridinin, less known because uniquely present in some dinoflagellate species, induces apoptosis in DLD-1 human colon cancer cells [10].

Dinoflagellates represent a promising marine source, with more than 2000 species, covering a large biodiversity of ecological strategies. They can be free living, symbiotic, parasitic or mixotrophic. Most of them are known for the potent neurotoxins that they produce causing ecosystem and human health problems [11]. Human diseases associated with exposure to marine dinoflagellate toxins include paralytic, diarrheic, neurotoxic and ciguatera fish poisoning [12]. The toxins inducing these pathologies are chemically diverse and include macrolides, cyclic polyethers, spirolides and purine alkaloids [13]. In previous studies, some compounds produced by dinoflagellates showed interesting and specific in vitro antiproliferative effects against various cancer cell lines [14–16]. Species of the genus *Alexandrium* present peculiar features from this point of view [17]. Indeed, *Alexandrium pseudogonyaulax* excretes antimicrobial and antifungal substances such as goniodomin-A [18], which is also able to inhibit angiogenesis [19]. *Alexandrium ostenfeldii* produces the cyclic imine toxin 13-desmethyl spirolide C [20], a polyketide recently discovered for its anti-Alzheimer's activity, being able to cross the blood–brain barrier in mice targeting nicotinic receptors [21].

These results obtained from *Alexandrium* species give strong impulse to screen this promising group of marine dinoflagellates, analyzing the chemical diversity of their secondary metabolites and their potential bioactivities and applications for human health.

We investigated the biological activity of extracts from *Alexandrium minutum*, known for toxin production affecting bivalves (e.g., oysters) [22]. Through a bioassay-guided fractionation, we isolated from a polar extract of this species a bioactive fraction with a specific antiproliferative effect on the human lung adenocarcinoma cells (A549), without affecting cell viability in human normal lung fibroblasts (Figure 3). Chemical analysis suggested that the hydrophilic high molecular weight molecule likely responsible for the activity was a glycoprotein. We also discovered that the antiproliferative effect in A549 cancer cells of this molecule is linked to the mitochondrial autophagy cell death pathway. Autophagy is an evolutionarily conserved catabolic process, aiming to the maintenance of cellular homeostasis and correcting functioning of intracellular organelles [23]. Mitochondrial are fundamental intracellular organelles responsible of the regulation of cellular homeostasis and cell death [24]; thus, the removal of damaged mitochondria is critical for maintaining proper cellular functions and viability. Mitochondria with damages or dysfunction are removed through a specific autophagic process called "mitophagy". Mitophagy has been described in yeast as a process mediated by autophagy-related 32 gene (Atg32) and in mammals by NIP3-like protein X (NIX; also known as BNIP3L (BCL2 Interacting Protein 3 Like)) [25]. Mitophagy is regulated in many metazoan cell types by Parkin and PTEN-induced putative kinase protein 1 (PINK1), while mutations in these genes are linked to Parkinson's disease [26–28].

To our knowledge, this is the first report of cell death activation triggered by specific mitophagic pathway induced by marine dinoflagellate derived compound, without affecting normal cells.

2. Results

2.1. Chemical Analyses

Extraction of *A. minutum* biomass (5.7 g wet weight) with TRI reagent® gave 350 mg water soluble extract. Fractionation of this material by HR-X column led to four enriched fractions that were tested against human lung adenocarcinoma cells A549 (see Supplementary Materials, Figure S1). Fraction 1B (5.3 mg), eluted with ACN/H_2O 7:3, was the only one that showed cytotoxic effect, with an IC_{50} of 1.3 $\mu g \cdot mL^{-1}$. This fraction was also tested on a panel of human cells. Fraction 1B exhibited stronger cytotoxic effect on A549, with respect to human colorectal adenocarcinoma cells (HT29) and human prostate cancer cells (PC3). Moreover, the same fraction did not show cytotoxic effect on human normal lung fibroblasts (WI38) (see Supplementary Materials, Figure S4). Diffusion NMR experiments are used to determine the size of macromolecules and aggregates according to their diffusion coefficients in solution. The spectra of the active Fraction 1B confirmed the presence of a family of macromolecules with different molecular weights. On the other hand, ^{1}H NMR experiment of this fraction was characterized by signals between 3 and 5 ppm that were in agreement with a predominance of carbohydrates (Figure 1). After further fractionation by sequential ultrafiltration over membranes with cut-off of 3 kDa and 10 kDa, the activity was retained in the fraction above 10 kDa (Fraction 3B 0.7 mg). Significantly, this fraction (IC_{50} of 0.4 μg·mL^{-1}) was four times more potent than the parent Fraction 1B. Further ultrafiltration over exclusion membranes led an enrichment of the cytotoxic above 30 kDa but the activity was also present in the filtrate.

Electrophoresis gel corroborated the co-presence of three major proteins in the active Fraction 3B of *A. minutum*. According to the intensity of the band stained by silver nitrate, two of these proteins accounted for minor components with a molecular weight higher than 50 kDa even if the main band was around 20 kDa. Analysis by colorimetric phenol–sulforic acid method [29] and Bradford assay [30] indicated that almost 96% of this sample was composed of carbohydrates and only 4% of protein. Hydrolysis of this material by acid treatment with a solution 2 M of Trifluoroacetic acid (TFA) led to a complete loss of the activity against A549 cells (data not shown). Analysis of the hydrolyzed sugars by high-performance anion-exchange chromatography (HPAEC) supported a large presence of D-galactose and D-glucose (63% and 37%, respectively) (see Supplementary Materials, Figure S2).

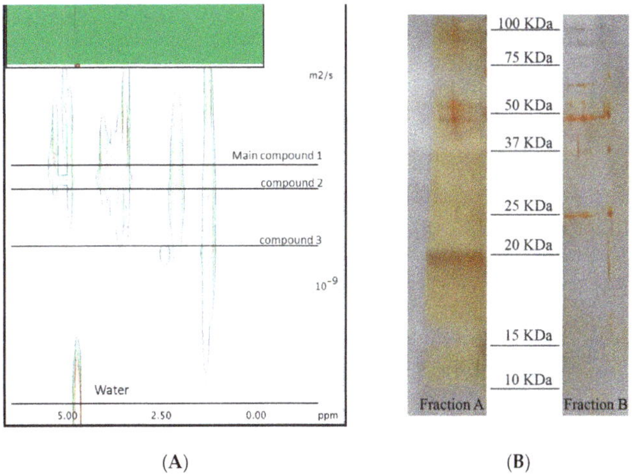

Figure 1. (**A**) 2D-Diffusion Ordered Spectroscopy (DOSY) spectra recorded in D_2O at 600 MHz of Fraction 1B; and (**B**) Electrophoresis gel of Fractions 3B (active sample) and 4B (deglycosylated Fraction 3B sample).

2.2. *Bioassay and Mechanism of Action*

The activity of the *A. minutum* fractions was tested on human lung adenocarcinoma cells. Fraction 1B exhibited a strong cytotoxicity on A549 cells, with an IC_{50} = 1.3 μg·mL^{-1}. The following steps of fractionation still enhanced the activity, also lowering the IC_{50}. In particular, Fraction 2B presented an IC_{50} = 0.8 μg·mL^{-1} (Figure S5), while Fraction 3B reached an IC_{50} = 0.4 μg·mL^{-1} (Figure 2A). Fraction 3A did not significantly affect cell viability for all concentrations tested (Figure 2A and Figure S5). Interestingly, Fraction 3B lowered the A549 cell viability in a dose-dependent manner and did not exhibited cytotoxicity on human normal lung fibroblasts (WI38) (Figure 2B).

To establish the cell death signaling pathway induced by the active Fraction 3B, gene expression of both cell lines treated with 0.4 μg·mL^{-1} of Fraction 3B for 2 h was analyzed. The exposure time of 2 h was selected after having verified that cell death pathways were already expressed and activated after this time.

Control housekeeping genes for real-time qPCR were actin-beta (ACTB), beta-2-microglobulin (B2M), hypoxanthine phosphoribosyltransferase (HPRT1) and large ribosomal protein P0 (RPLP0). Figure 3 shows the relative expression ratios of the analyzed genes with respect to controls. Only expression values greater than a two-fold difference with respect to the controls were considered significant. The 2 h-treatment on A549 cells with Fraction 3B (Figure 3A) induced a strong up-regulation of the autophagy-related protein 12/gene (Atg12, 12.2-fold change) with a consequent increase in ATPase H+ Transporting V1 Subunit G2 (ATP6V1G2, 35.5-fold change) and BCL2/Adenovirus E1B 19kDa Interacting Protein 3 (BNIP3, 2.4-fold change). Two genes involved in mitophagy were significantly up-regulated: PTEN induced putative kinase 1 (PINK1, 3.6-fold change) and Parkin gene (11-fold change). Moreover, an up-regulation of the mitofusin-1/2 gene (MFN-1/2, 6.6-fold change) and the Voltage Dependent Anion Channel gene (VDAC, 5.9-fold change) was revealed. Activation of these genes caused a strong up-regulation of the Neighbor of BRCA1 gene 1 (NBR1, 12.6-fold change) and NIX gene also known as BCL2/Adenovirus E1B 19 kDa Interacting Protein 3-Like (5.9-fold change). The autophagy receptor Sequestosome 1 (SQSTMS1, 13.3-fold change) was strongly up-regulated indicating the degradation of intracellular vesicles.

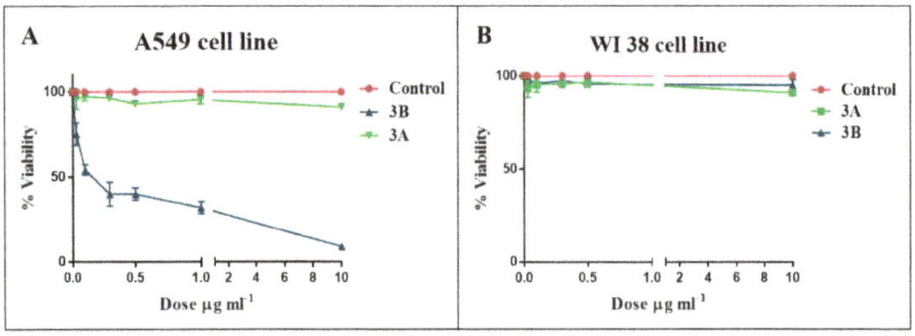

Figure 2. Effect of Fractions 3A (<10 KDa) and 3B (>10 KDa) on cell viability of human lung adenocarcinoma cells of (**A**) A549 and human normal lung fibroblasts and (**B**) WI38. Values are reported as mean ± S.D. compared to controls (100% viability) of three independent experiments. Concentrations tested were 0.1, 1 and 10 $\mu g \cdot mL^{-1}$ for 48 h.

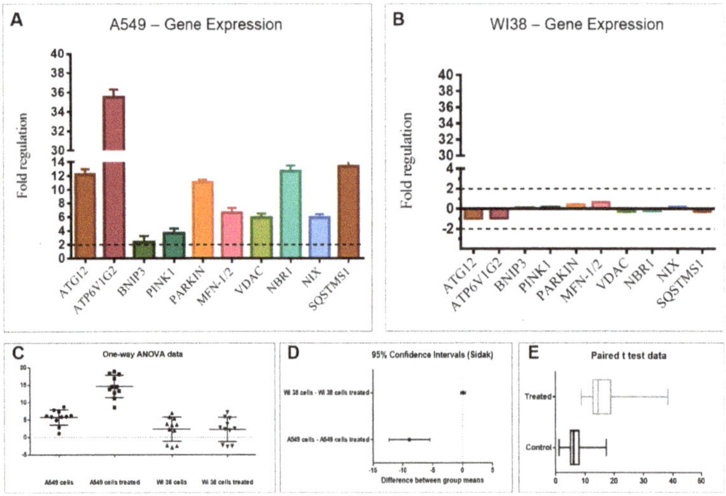

Figure 3. Effect of Fraction 3B on the expression levels of target genes in: human lung adenocarcinoma cells (A549) (**A**); and human normal lung fibroblasts (WI38) (**B**). All experiments were performed with RNA extracted from three different biological replicates and error bars represent ±S.D. Statistical analyses on the results obtained in (**A**,**B**); (**C**) One-way ANOVA); (**D**) Sidak; and (**E**) student's t-test.

Conversely, the same analysis carried out on the WI38 cells treated with Fraction 3B did not show any significant variation. Mitophagic genes in normal cells were not activated by the treatment, confirming that the active fraction did not induce cell death (Figure 3B).

Statistical analysis (ANOVA, *t*-student and Sidàk test; Figure 3C–E) performed on the gene expression data validated the significant differences of the biological activity of Fraction 3B on the A549 cells vs. WI38 cells.

3. Discussion

In this study, we investigated the cytotoxic activity of the dinoflagellate *Alexandrium minutum* specific against human lung adenocarcinoma cells A549. In particular, we use an innovative method of extraction by TRI Reagent® that led to enrichment of cytotoxic macromolecules. Fractionation by hydrophobic column (Chromabond® HR-X) followed by sequential steps of exclusion membranes increased the activity from IC_{50} = 1.3 μg·mL^{-1} in the extract to IC_{50} 0.4 μg·mL^{-1} in the fraction retained above 10 kDa (Fraction 3B). Chemical analysis of this material is consistent with the presence of one or more molecules with a molecular weight at least of 20 kDa and composed by almost 96% of carbohydrates and only 4% of protein. Differential gene expression analysis of A549 cells after treatment with the active Fraction 3B suggested an up-regulation of the genes involved in the mitochondrial autophagy (mitophagy) cell death pathway.

First, the gene ATG12 belonging to the autophagy-related gene family and involved in the first steps of formation and elongation of autophagosomes is activated. Then, ATP6V1G2, a downstream factor acting after ATG genes and responsible of the autophagosome degradation, is up-regulated [31]. Other genes, i.e., BNIP3 and NIX, are up-regulated by the active fraction treatment. These genes are probably involved in the connection between autophagy and cell death, even though the molecular mechanisms of BNIP3/NIX genes and connection with cell death are not well understood. Indeed, some studies report the activation of necrotic cell death by BNIP3/NIX genes, without the involvement of apaf-1, caspase 9 or 3 and cytochrome [32,33].

Confirmation of the mitophagic gene activation by Fraction 3B comes from up-regulation of Pink and Parkin genes. PINK1 is considered an upstream regulator of Parkin function, since the recruitment of Parkin to impaired mitochondria requires PINK1 expression and its kinase activity [34–37]. The change in membrane potential is the first signal for the mitophagic pathway through PINK1/parkin cascade [38,39]. PINK1, together with parkin protein, is responsible for the Mfn1/2 ubiquitination, playing a key role in the autophagic degradation of dysfunctional mitochondria (mitophagy) [40]. During the elimination of defective mitochondria, PINK1 mediates translocation of Parkin from the cytosol to mitochondria by an unknown mechanism. Sun et al. [41] reported that the three most abundant interacting proteins were the voltage-dependent anion channels (VDACs), pore-forming proteins in the outer mitochondrial membrane demonstrating their role as mitochondrial docking sites to recruit Parkin from the cytosol to defective mitochondria. Fraction 3B also up-regulated the gene encoding for VDAC in A549 cells. Moreover, the up-regulation of NBR1, a functional homolog of SQSTM1 (Sequestosome 1) with a role in PARK2-mediated mitophagy, indicates mitochondrial degradation [42].

These data predict that an extensive mitophagy may induce mitochondrial dysfunction and degradation (such as reduced mitochondrial membrane potential) with a consequential irreversible cell death [43]. Autophagy, as well as other peculiar death pathways, such as pyroptosis [44], are desirable targets for cancer therapy, since a new trend in anticancer research has arisen focusing on the discovery of new natural drugs that induce specific programmed cell death and ensure an optimal release of immunostimulatory signals [45], able to act as co-adjuvant stimuli for the reinforcement of the anti-cancer effect.

Until now, no natural products from marine microalgae are reported as inducers of mitophagic cell death in human tumor cells, although other studies show that compounds extracted from marine organisms are able to fight cancer cells, activating this specific mechanism. For instance, Kahalalide

F (KF) (C75H124N14O16), a natural depsipeptides isolated from the Hawaiian herbivorous marine mollusk *Elysia rufescens*, activates mitochondrial autophagy in vitro in human prostate cancer cells [46].

This study is the first report that a water soluble high molecular weight compound isolated from the marine dinoflagellate, *Alexandrium minutum*, already known for toxins production, is able to up-regulate genes involved in a specific cell death in human lung adenocarcinoma cells through mitophagy activation, without affecting normal cell viability (human normal lung fibroblasts WI38). This mitophagic gene expression described in this study is likely attributed to a glycopeptide. The isolated fraction is much more active compared to the total extract, thus confirming the enrichment of the active fraction along with the chromatographic purification. Macromolecules based on polysaccharide and protein structure have already been described in other microalgae, including dinoflagellate [47]. These compounds have been associated to allelopathic activity by outcompeting other photoautotrophic species through growth inhibition. A large non-proteinaceous toxin with cytotoxic activity has been also reported from the dinoflagellate *Alexandrium tamarense* [48]. Although we have no direct proof that the cytotoxic molecule(s) of *A. minutum* belong to the same family, the indirect chemical evidence points out to this direction.

Further chemical studies are needed to better characterize the glycopeptide, a promising molecule with application for human health. Moreover, new chemical information about this molecule will help in elucidating the pathway for its synthesis, a crucial requirement for its biotechnological production. The latter might be pursued chemically or biologically, with for instance the improvement of its synthesis in *A. minutum* in modulating the growth-environmental conditions.

4. Materials and Methods

4.1. General Experimental Procedures

NMR spectra were recorded on a Bruker Avance DRX 600 equipped with a cryoprobe operating at 600 MHz for proton in D_2O solution containing 1 mM sodium 3-trimethylsilyl [2,2,3,3-D_4] propionate as a chemical shift reference for ^1H spectra. Water was deionized and purified (Milli-Q, Millipore, Germany), whereas TRI Reagent® and all solvents were purchased from Sigma (Aldrich, Milan, Italy). Chromabond® HR-X resin was obtained from Macherey-Nagel GmbH (Düren, Germany). Ultrafiltration was carried out by Vivaspin 500 centrifugal concentrators (Sigma-Aldrich) with 3, 10 and 30 kDa molecular weight cut-off membrane of polyethersulfone (PES). Electrophoresis (SDS-PAGE) was performed by a mini gel apparatus BioRAD (Milan, Italy). To estimate the molecular mass, Precision Plus Protein™ Dual Color Standards (BioRad, Hercules, CA, USA) with 10 recombinant proteins of precise molecular weights (10–250 kD) were used.

4.2. Biological Material

Alexandrium minutum, isolated from the Gulf of Naples, was cultured in K medium [49], prepared with filtered (0.22 μm) natural sterile seawater, maintained at room temperature (22 ± 1 °C), under a 12:12 light/dark regime at 100 μmol $m^{-2}\cdot s^{-1}$, in a 10 L sterilized polycarbonate carboy. The culture was harvested at the end of the stationary phase by centrifugation in a swing-out Allegra X12R (Beckman Coulter Inc., Palo Alto, CA, USA) at 2300× *g*, 4 °C, for 10 min. Cell growth was estimated by daily cells counts using a Bürker counting chamber (Merck, Leuven, Belgium) (depth 0.100 mm) under an inverted microscope. The harvested cell pellet (wet weight 5.7 g) was stored at −80 °C until analysis.

4.3. Extraction of *A. minutum*

According to manufacturer's instructions, cells of *A. minutum* (5.7 g) were lysed in 45.6 mL TRI Reagent ®, which is a homogeneous water mixture of guanidine thiocyanate and phenol, for extraction of RNA, DNA and protein. The homogenate was centrifuged at 12,000× *g* for 10 min at 4 °C and the supernatant was added to 9.12 mL of $CHCl_3$. The sample was vortexed and, after 15 min at room temperature, was centrifuged at 12,000× *g* for 15 min at 4 °C to separate the mixture into three phases including a red organic phase, an interphase and a colorless upper aqueous phase. The aqueous phase was transferred to a fresh tube and 31.9 mL EtOH was added to the sample. The suspension was centrifuged at 2000× *g* for 5 min at 4 °C and the supernatant was diluted with 114 mL acetone. After centrifugation at 12,000× *g* for 10 min at 4 °C, the pellet was suspended with MilliQ water and the resulting suspension was further centrifuged at 10,000× *g* for 10 min at 4 °C. The clear supernatant was recovered and lyophilized to give 350 mg of dry material.

4.4. Fractionation of the Hydrophilic Extract of *A. minutum*

The lyophilized extract of *A. minutum* was fractionated by HR-X ® column according to a modified protocol of the method previously reported for the isolation of marine natural products [50]. Briefly, 3.8 g of Chromabond ® HR-X resin was loaded into a glass column with methanol and equilibrated with MilliQ water. The dry extract of *A. minutum* (350 mg) was suspended in a minimal volume of MilliQ water and loaded onto the column. Fractionation was achieved by a four-step elution (see Supplementary Materials, Figure S1) with 140 mL H_2O (Fraction 1A, 340 mg), 110 mL ACN/H_2O 7:3 (Fraction 1B, 5.3 mg), 85 mL acetonitrile (Fraction 1C, 0.8 mg), and 85 mL dichloromethane/MeOH 9:1 (Fraction 1D, 0.5 mg).

4.5. Ultrafiltration of the Active Fractions

Fraction 1B was dissolved in 2 mL of water and subjected to two sequential steps of ultracentrifugation on Vivaspin filters with cut-off of 3 and 10 kDa at 5000× *g* for 10 min at 24 °C. The fraction retained above 10 kDa was lyophilized to give 0.7 mg of the active Fraction 3B. Further ultrafiltration of this material with membrane of 30 kDa cut-off did not produce any useful separation of the activity.

4.6. Sizing of the Active Macromolecule by Diffusion NMR Analysis

The active Fraction 1B (5.3 mg) was dissolved in 700 μL of D_2O with sodium 3-trimethylsilyl [2,2,3,3-D_4] propionate for calibration. Diffusion edited ^{1}H NMR spectra (1D-DOSY) were acquired with the pulse sequence using stimulated echo and 1 spoil gradient strength, 16 scans, big delta 0.1 s and little delta 2 ms and a gradient (gpz6) between 1% and 100%. The fitting of the diffusion dimension in the 2D-DOSY spectra was achieved using AU-program dosy to calculate gradient-diff ramp processing in Bruker TopSpin 3.5 [51].

4.7. Hydrolysis of the Active Fraction

The active Fraction 3B (0.3 mg) was dissolved in 200 μL of water and reacted with an equal volume of 2 M TFA at 100 °C for 4 h. The reaction mixture was dried under vacuum and the residue was dissolved in 1 mL of water. This solution was loaded on a Vivaspin membrane with 3 KDa molecular weight cut-off and centrifuged at 5000× *g* for 10 min at 24 °C. The filtrate was then analyzed by high-performance anion-exchange chromatography (HPAEC) using a Dionex LC30 (Dionex, Sunnyvale, CA, USA) for the analysis of sugars.

4.8. SDS-Polyacrylamide Gel Electrophoresis (SDS-PAGE)

Fractionation obtained by cut-off partitions, as described above, were analyzed by electrophoresis (SDS-PAGE) on 12% gels (1 mm) according to the method of Laemmli [52], loading 10 μg of each samples. After electrophoresis, the gels were detected by silver nitrate. Precision Plus Protein Dual Color Standards (BioRad) were used for molecular weight estimation.

4.9. Treatment of Human Cells

The human lung adenocarcinoma cell line A549 was purchased from the American Type Culture Collection (ATCC® CCL185™) and grown in DMEM-F12 (Dulbecco's modified Eagle's medium) supplemented with 10% fetal bovine serum (FBS), 100 units mL^{-1} penicillin and 100 μg·mL^{-1} streptomycin in a 5% CO_2 atmosphere at 37 °C. Human normal lung fibroblasts WI38 were grown in MEM (Eagle's minimal essential medium) supplemented with 10% fetal bovine serum (FBS), 100 units mL^{-1} penicillin, 100 μg·mL^{-1} streptomycin, 2 mM of L-glutamine and non-essential amino acids (NEAA, 2 mM) in a 5% CO_2 atmosphere at 37 °C. A549 and WI38 cells (2×10^3 cells·well^{-1}) were seeded in a 96-well plates and kept overnight for attachment. Chemical fractions were dissolved in dimethyl sulfoxide (DMSO) and used for the treatment of cells. The final concentration of DMSO used was 1% (*v*/*v*) for each treatment. Eighty percent confluent cells were treated in triplicate with fractions at 0.1, 1 and 10 μg mL^{-1} for 24 and 48 h in complete cell medium. Control cells were incubated with complete cell medium with 1% of DMSO.

4.10. Cell Viability

The antiproliferative effect of chemical fractions on cell viability was evaluated using the 3-(4,5-Dimethylthiazol-2-yl)-2,5-diphenyl tetrazolium bromide (MTT) assay (Applichem A2231) according to Gerlier et al. [53]. A549 and WI38 cells, after treatment with chemical fractions, were incubated with 10 μL (5 mg mL^{-1}) of MTT for 3 h at 37 °C in a 5% CO_2 atmosphere. Isopropanol (100 μL) was used to stop the incubation time and to solubilize purple crystals formed in each well by only viable cells. The absorbance was recorded on a microplate reader at a wavelength of 570 nm (Multiskan FC, THERMO SCIENTIFIC, Waltham, MA, USA). The effect of the fractions at different concentrations was reported as percent of cell viability calculated as the ratio between mean absorbance of each treatment and mean absorbance of control cells (A549 and WI38 cell lines treated with only 1% of DMSO).

4.11. RNA Extraction and Real-Time PCR

A549 (2×10^6) and WI38 (2×10^6) cells used for RNA extraction and analysis were seeded in Petri dishes (100 mm diameter) and kept overnight for attachment. After 2 h of treatment with 0.4 μg·mL^{-1} with Fraction 3B, cells were washed directly in the Petri dish by adding Phosphate-Buffered Saline (PBS) and rocking gently. Cells grown in complete medium without any treatment constituted experimental control. Cells were lysed in the Petri dish by adding 1 mL of Trisure Reagent (Bioline, Galgagnano, Lodi, Italy cat. BIO-38033). RNA was isolated according to the manufacturer's protocol. RNA concentration and purity were assessed using the nanophotomer NanodroP (Euroclone, Pero, Milan, Italy). About 200 ng RNA was subjected to reverse transcription reaction using the RT^2 first strand kit (Qiagen, Hilden, Germany, cat.330401) according to the manufacturer's instructions. The qRT-PCR analysis was performed in triplicate using the RT^2 Profiler PCR Array kit (Qiagen, cat.330231) to analyze the expression of genes involved in cell death signaling pathways. Plates were run on a ViiA7 (Applied Biosystems, Foster City, CA, USA, 384-well blocks) using a Standard Fast PCR Cycling protocol with 10 μl reaction volumes. Cycling conditions used were: 1 cycle initiation at 95 °C for 10 min followed by amplification for 40 cycles at 95 °C for 15 s and 60 °C for 1 min. Amplification data were collected via ViiA 7 RUO Software (Applied Biosystems). The cycle threshold (Ct)-values were analyzed with PCR array data analysis online software (GeneGlobe Data Analysis Center http://pcrdataanalysis.sabiosciences.com/pcr/arrayanalysis.php, Qiagen). Real time data were expressed as fold expression, describing the changes in gene expression between treated cells and untreated cells (control).

4.12. Statistical Analysis

One-way ANOVA was used for the assessment of variance within the control and treated groups and between these two experimental groups. Šidák method was applied to gene expression data to counteract the problem of multiple comparisons and to control the familywise error rate, pinpointing confidence intervals. *T*-test analysis determined statistical difference between means of the two experimental groups. Gene expression data were analyzed by PCR array data analysis online software (http://pcrdataanalysis.sabiosciences.com/pcr/arrayanalysis.php, Qiagen®). Only gene expression values greater than a 2.0-fold difference with respect to the controls were considered significant.

Supplementary Materials: The following are available online at http://www.mdpi.com/1660-3397/16/12/502/s1, Figure S1: Bioassay-guided fractionation of *A. minutum* on lung adenocarcinoma cells A549., Figure S2: High-performance anion-exchange chromatography (HPAEC) of hydrolyzed sugars. Figure S3: Whole electrophoresis gel. Figure S4: Effect of the fraction 1B on cell viability of human normal lung fibroblasts (WI38), human prostate cancer cells (PC3), human colorectal adenocarcinoma cells (HT29) and human lung adenocarcinoma cells (A549). Figure S5: Effect of the fractions 2A and 2B on cell viability of human lung adenocarcinoma cells (A549).

Author Contributions: C.G., G.N., A.F. and C.S. conceived and designed the experiments. C.S. and C.G. took part and contributed equally in the experiments with cell culture and real-time-PCR. G.N. and A.F. performed the chemical extractions and analysis. A.S. cultured the microalga. C.G., G.N., C.B., A.F., C.S. analyzed the data. A.I. and A.F. contributed reagents/materials/analysis tools. C.G., G.N., C.B., A.I., A.S., A.F. and C.S. wrote the paper.

Funding: Stazione Zoologica Anton Dohrn flagship project MARCAN and The European Marine Biological Research Infrastructure Cluster (EMBRIC) EU project funded this work.

Acknowledgments: We thank Massimo Perna for their technical assistance and support.

Conflicts of Interest: The authors declare no conflict of interest.

References

1. Martínez Andrade, K.A.; Lauritano, C.; Romano, G.; Ianora, A. Marine Microalgae with Anti-Cancer Properties. *Mar. Drugs* **2018**, *16*, 165. [CrossRef] [PubMed]
2. Goh, S.H.; Alitheen, N.B.; Yusoff, F.M.; Yap, S.K.; Loh, S.P. Crude ethyl acetate extract of marine microalga, *Chaetoceros calcitrans*, induces Apoptosis in MDA-MB-231 breast cancer cells. *Pharmacogn. Mag.* **2014**, *10*, 1–8. [PubMed]
3. Miralto, A.; Barone, G.; Romano, G.; Poulet, S.A.; Ianora, A.; Russo, G.L.; Buttino, I.; Mazzarella, G.; Laabir, M.; Cabrini, M.; et al. The insidious effect of diatoms on copepod reproduction. *Nature* **1999**, *402*, 173–176. [CrossRef]
4. Ianora, A.; Miralto, A.; Poulet, S.A.; Carotenuto, Y.; Buttino, I.; Romano, G.; Casotti, R.; Pohnert, G.; Wichard, T.; Colucci-D'Amato, L.; et al. Aldehyde suppression of copepod recruitment in blooms of a ubiquitous planktonic diatom. *Nature* **2004**, *429*, 403–407. [CrossRef] [PubMed]
5. Sansone, C.; Braca, A.; Ercolesi, E.; Romano, G.; Palumbo, A.; Casotti, R.; Francone, M.; Ianora, A. Diatom-derived polyunsaturated aldehydes activate cell death in human cancer cell lines but not normal cells. *PLoS ONE* **2014**, *9*. [CrossRef] [PubMed]
6. Ingebrigtsen, R.A.; Hansen, E.; Andersen, J.H.; Eilertsen, H.C. Light and temperature effects on bioactivity in diatoms. *J. Appl. Phycol.* **2016**, *28*, 939–950. [CrossRef] [PubMed]
7. Barra, L.; Chandrasekaran, R.; Corato, F.; Brunet, C. The Challenge of Ecophysiological Biodiversity for Biotechnological Applications of Marine Microalgae. *Mar. Drugs* **2014**, *12*, 1641–1675. [CrossRef] [PubMed]
8. Galasso, C.; Corinaldesi, C.; Sansone, C. Carotenoids from Marine Organisms: Biological Functions and Industrial Applications. *Antioxidants* **2017**, *6*, 96. [CrossRef]
9. Pasquet, V.; Morisset, P.; Ihammouine, S.; Chepied, A.; Aumailley, L.; Berard, J.B.; Serive, B.; Kaas, R.; Lanneluc, I.; Thiery, V.; et al. Antiproliferative activity of violaxanthin isolated from bioguided fractionation of *Dunaliella tertiolecta* extracts. *Mar. Drugs* **2011**, *878*, 819–831. [CrossRef]
10. Sugawara, T.; Yamashita, K.; Sakai, S.; Asai, A.; Nagao, A.; Shiraishi, T.; Imai, I.; Hirata, T. Induction of apoptosis in DLD-1 human colon cancer cells by peridinin isolated from the dinoflagellate, *Heterocapsa Triquetra*. *Biosci. Biotechnol. Biochem.* **2007**, *71*, 1069–1072. [CrossRef]
11. Berdalet, E.; Fleming, L.E.; Gowen, R.; Davidson, K.; Hess, P.; Backer, L.C.; Moore, S.K.; Hoagland, P.; Enevoldsen, H. Marine harmful algal blooms, human health and wellbeing: Challenges and opportunities in the 21st century. *J. Mar. Biol. Assoc. UK* **2016**, *96*, 61–91. [CrossRef] [PubMed]
12. Camacho, F.G.; Rodríguez, J.G.; Mirón, A.S.; García, M.C.; Belarbi, E.H.; Chisti, Y.; Grima, E.M. Biotechnological significance of toxic marine dinoflagellates. *Biotechnol. Adv.* **2007**, *25*, 176–194. [CrossRef] [PubMed]
13. Kellmann, R.; Stüken, A.; Orr, R.J.S.; Svendsen, H.M.; Jakobsen, K.S. Biosynthesis and Molecular Genetics of Polyketides in Marine Dinoflagellates. *Mar. Drugs* **2010**, *8*, 1011–1048. [CrossRef] [PubMed]

14. Yaakob, Z.; Ali, E.; Zainal, A.; Mohamad, M.; Takriff, M.S. An overview: Biomolecules from microalgae for animal feed and aquaculture. *J. Biol. Res. (Thessalon.)* **2014**, *21*, 6. [CrossRef] [PubMed]
15. Kobayashi, J.; Kubota, T. Bioactive Macrolides and Polyketides from Marine Dinoflagellates of the Genus *Amphidinium*. *J. Nat. Prod.* **2007**, *70*, 451–460. [CrossRef] [PubMed]
16. Bhakuni, D.S.; Rawat, D.S. *Bioactive Marine Natural Products*; Springer: New York, NY, USA, 2005.
17. Sansone, C.; Nuzzo, G.; Galasso, C.; Casotti, R.; Fontana, A.; Romano, G.; Ianora, A. The Marine Dinoflagellate *Alexandrium andersoni* Induces Cell Death in Lung and Colorectal Tumor Cell Lines. *Mar. Biotechnol.* **2018**, *20*, 343–352. [CrossRef] [PubMed]
18. Brosnahan, M.L.; Ralston, D.K.; Fischer, A.D.; Solow, A.R.; Anderson, D.M. Bloom termination of the toxic dinoflagellate *Alexandrium catenella*: Vertical migration behavior, sediment infiltration, and benthic cyst yield. *Limnol. Oceanogr.* **2017**, *62*, 2829–2849. [CrossRef]
19. Assunção, J.; Guedes, A.; Malcata, F. Biotechnological and Pharmacological Applications of Biotoxins and Other Bioactive Molecules from Dinoflagellates. *Mar. Drugs* **2017**, *15*, 393. [CrossRef] [PubMed]
20. Hu, T.; Burton, I.W.; Cembella, A.D.; Curtis, J.M.; Quilliam, M.A.; Walter, J.A.; Wright, J.L. Characterization of Spirolides A, C, and 13-Desmethyl C, New Marine Toxins Isolated from Toxic Plankton and Contaminated Shellfish. *J. Nat. Prod.* **2001**, *64*, 308–312. [CrossRef]
21. Alonso, E.; Otero, P.; Vale, C.; Alfonso, A.; Antelo, A.; Giménez-Llort, L.; Chabaud, L.; Guillou, C.; Botana, L.M. Benefit of 13-desmethyl Spirolide C Treatment in Triple Transgenic Mouse Model of Alzheimer Disease: Beta-Amyloid and Neuronal Markers Improvement. *Curr. Alzheimer Res.* **2013**, *10*, 279–289. [CrossRef]
22. Castrec, J.; Soudant, P.; Payton, L.; Tran, D.; Miner, P.; Lambert, C.; Le Goïc, N.; Huvet, A.; Quillien, V.; Boullot, F.; et al. Bioactive extracellular compounds produced by the dinoflagellate *Alexandrium minutum* are highly detrimental for oysters. *Aquat. Toxicol.* **2018**, *199*, 188–198. [CrossRef] [PubMed]
23. Lv, S.; Wang, X.; Zhang, N.; Sun, M.; Qi, W.; Li, Y.; Yang, Q. Autophagy facilitates the development of resistance to the tumor necrosis factor superfamily member TRAIL in breast cancer. *Int. J. Oncol.* **2015**, *46*, 1286–1294. [CrossRef] [PubMed]
24. Osellame, L.D.; Blacker, T.S.; Duchen, M.R. Cellular and molecular mechanisms of mitochondrial function. *Best Pract. Res. Clin. Endocrinol. Metab.* **2012**, *26*, 711–723. [CrossRef] [PubMed]
25. Glick, D.; Barth, S.; Macleod, K.F. Autophagy: Cellular and molecular mechanisms. *J. Pathol.* **2010**, *221*, 3–12. [CrossRef] [PubMed]
26. Kundu, M.; Lindsten, T.; Yang, C.Y.; Wu, J.; Zhao, F.; Zhang, J.; Selak, M.A.; Ney, P.A.; Thompson, C.B. ULK1 plays a critical role in the autophagic clearance of mitochondria and ribosomes during reticulocyte maturation. *Blood* **2008**, *112*, 1493–1502. [CrossRef] [PubMed]
27. Mortensen, M.; Ferguson, D.J.; Edelmann, M.; Kessler, B.; Morten, K.J.; Komatsu, M.; Simon, A.K. Loss of autophagy in erythroid cells leads to defective removal of mitochondria and severe anemia in vivo. *Proc. Natl. Acad. Sci. USA* **2010**, *107*, 832–837. [CrossRef] [PubMed]
28. Aerbajinai, W.; Giattina, M.; Lee, Y.T.; Raffeld, M.; Miller, J.L. The proapoptotic factor Nix is coexpressed with Bcl-xL during terminal erythroid differentiation. *Blood* **2003**, *102*, 712–717. [CrossRef] [PubMed]
29. Saha, S.K.; Brewer, C.F. Determination of the concentrations of oligosaccharides, complex type carbohydrates, and glycoproteins using the phenol-sulfuric acid method. *Carbohydr. Res.* **1994**, *17*, 157–167. [CrossRef]
30. Bradford, M.M. A Rapid and Sensitive Method for the Quantitation of Microgram Quantities of Protein Utilizing the Principle of Protein-Dye Binding. *Anal. Biochem.* **1976**, *72*, 248–254. [CrossRef]
31. Sopariwala, D.H.; Yadav, V.; Badin, P.M.; Likhite, N.; Sheth, M.; Lorca, S.; Vila, I.K.; Kim, E.R.; Tong, Q.; Song, M.S.; et al. Long-term PGC1β overexpression leads to apoptosis, autophagy and muscle wasting. *Sci. Rep.* **2017**, *7*, 10237. [CrossRef]
32. Vande Velde, C.; Cizeau, J.; Dubik, D.; Alimonti, J.; Brown, T.; Israels, S.; Hakem, R.; Greenberg, A.H. BNIP3 and genetic control of necrosis-like cell death through the mitochondrial permeability transition pore. *Mol. Cell. Biol.* **2000**, *20*, 5454–5468. [CrossRef] [PubMed]
33. Zhang, J.; Ney, P.A. Role of BNIP3 and NIX in cell death, autophagy, and mitophagy. *Cell Death Differ.* **2009**, *16*, 939–946. [CrossRef] [PubMed]
34. Geisler, S.; Holmström, K.M.; Treis, A.; Skujat, D.; Weber, S.S.; Fiesel, F.C.; Kahle, P.J.; Springer, W. The PINK1/Parkin-mediated mitophagy is compromised by PD-associated mutations. *Autophagy* **2010**, *6*, 871–878. [CrossRef] [PubMed]

35. Matsuda, N.; Sato, S.; Shiba, K.; Okatsu, K.; Saisho, K.; Gautier, C.A.; Sou, Y.S.; Saiki, S.; Kawajiri, S.; Sato, F.; et al. PINK1 stabilized by mitochondrial depolarization recruits Parkin to damaged mitochondria and activates latent Parkin for mitophagy. *J. Cell Biol.* **2010**, *189*, 211–221. [CrossRef] [PubMed]
36. Narendra, D.P.; Jin, S.M.; Tanaka, A.; Suen, D.F.; Gautier, C.A.; Shen, J.; Cookson, M.R.; Youle, R.J. PINK1 is selectively stabilized on impaired mitochondria to activate Parkin. *PLoS Biol.* **2010**, *8*. [CrossRef] [PubMed]
37. Vives-Bauza, C.; Zhou, C.; Huang, Y.; Cui, M.; de Vries, R.L.; Kim, J.; May, J.; Tocilescu, M.A.; Liu, W.; Ko, H.S.; et al. PINK1-dependent recruitment of Parkin to mitochondria in mitophagy. *Proc. Natl. Acad. Sci. USA* **2010**, *107*, 378–383. [CrossRef] [PubMed]
38. Chen, H.; Detmer, S.A.; Ewald, A.J.; Griffin, E.E.; Fraser, S.E.; Chan, D.C. Mitofusins Mfn1 and Mfn2 coordinately regulate mitochondrial fusion and are essential for embryonic development. *J. Cell Biol.* **2003**, *160*, 189–200. [CrossRef] [PubMed]
39. Gegg, M.E.; Cooper, J.M.; Chau, K.Y.; Rojo, M.; Schapira, A.H.; Taanman, J.W. Mitofusin 1 and mitofusin 2 are ubiquitinated in a PINK1/parkin-dependent manner upon induction of mitophagy. *Hum. Mol. Genet.* **2010**, *19*, 4861–4870. [CrossRef] [PubMed]
40. Gegg, M.E.; Schapira, A.H.V. PINK1-parkin-dependent mitophagy involves ubiquitination of mitofusins 1 and 2: Implications for Parkinson disease pathogenesis. *Autophagy* **2011**, *7*, 243–245. [CrossRef]
41. Sun, Y.; Vashisht, A.A.; Tchieu, J.; Wohlschlegel, J.A.; Dreier, L. Voltage-dependent anion channels (VDACs) recruit Parkin to defective mitochondria to promote mitochondrial autophagy. *J. Biol. Chem.* **2012**, *287*, 40652–40660. [CrossRef]
42. Shi, J.; Fung, G.; Deng, H.; Zhang, J.; Fiesel, F.C.; Springer, W.; Li, X.; Luo, H. NBR1 is dispensable for PARK2-mediated mitophagy regardless of the presence or absence of SQSTM1. *Cell Death Dis.* **2015**, *6*, e1943. [CrossRef] [PubMed]
43. Gargini, R.; García-Escudero, V.; Izquierdo, M. Therapy mediated by mitophagy abrogates tumor progression. *Autophagy* **2011**, *7*, 466–476. [CrossRef] [PubMed]
44. Sannino, F.; Sansone, C.; Galasso, C.; Kildgaard, S.; Tedesco, P.; Fani, R.; Marino, G.; de Pascale, D.; Ianora, A.; Parrilli, E.; et al. *Pseudoalteromonas haloplanktis* TAC125 produces 4-hydroxybenzoic acid that induces pyroptosis in human A459 lung adenocarcinoma cells. *Sci. Rep.* **2018**, *7*, 41215. [CrossRef] [PubMed]
45. Pietrcola, F.; Bravo-San Pedro, J.M.; Galluzzi, L.; Kroemer, G. Autophagy in natural and therapy-driven anticancer immunosurveillance. *Autophagy* **2017**, *13*, 2163–2170. [CrossRef] [PubMed]
46. Faircloth, G.T.; Grant, W.; Smith, B.; Supko, J.G.; Brown, A.; Geldof, A.; Jimeno, J. Preclinical development of Kahalalide F, a new marine compound selected for clinical studies. *Proc. Am. Assoc. Cancer Res.* **2000**, *41*, 600.
47. Yamasaki, Y.; Shikata, T.; Nukata, A.; Ichiki, S.; Nagasoe, S.; Matsubara, T.; Shimasaki, Y.; Nakao, M.; Yamaguchi, K.; Oshima, Y.; et al. Extracellular polysaccharide-protein complexes of a harmful alga mediate the allelopathic control it exerts within the phytoplankton community. *ISME J.* **2009**, *3*, 808–817. [CrossRef] [PubMed]
48. Ma, H.; Krock, B.; Tillmann, U.; Cembella, A. Preliminary characterization of extracellular allelochemicals of the toxic marine dinoflagellate *Alexandrium tamarense* using a *Rhodomonas salina* bioassay. *Mar. Drugs* **2009**, *7*, 497–522. [CrossRef] [PubMed]
49. Keller, D.M.; Selvin, C.R.; Claus, W.; Guillard, R.L.R.J. Media for the culture of oceanic ultraphytoplankton. *J. Phycol.* **1987**, *23*, 633–638. [CrossRef]
50. Cutignano, A.; Nuzzo, G.; Ianora, A.; Luongo, E.; Romano, G.; Gallo, C.; Sansone, C.; Aprea, S.; Mancini, F.; D'Oro, U.; et al. Development and application of a novel SPE-method for bioassay guided fractionation of marine extracts. *Mar. Drugs* **2015**, *13*, 5736–5749. [CrossRef]
51. Viel, S.; Capitani, D.; Mannina, L.; Segre, A. Diffusion-Ordered NMR Spectroscopy: A Versatile Tool for the Molecular Weight Determination of Uncharged Polysaccharides. *Biomacromolecules* **2003**, *4*, 1843–1847. [CrossRef]
52. Laemmli, U.K. Cleavage of Structural Proteins during the Assembly of the Head of Bacteriophage T4. *Nature* **1970**, *227*, 680–685. [CrossRef] [PubMed]
53. Gerlier, D.; Thomasset, N. Use of MTT colorimetric assay to measure cell activation. *J. Immunol. Methods* **1986**, *94*, 57–63. [CrossRef]

marine drugs

MDPI

Article

Fatty Acid Composition and Thermotropic Behavior of Glycolipids and Other Membrane Lipids of *Ulva lactuca* (Chlorophyta) Inhabiting Different Climatic Zones

Eduard Kostetsky [1], Natalia Chopenko [1], Maria Barkina [1], Peter Velansky [1,2] and Nina Sanina [1,*]

1 Department of Biochemistry, Microbiology and Biotechnology, School of Natural Sciences, Far Eastern Federal University, Vladivostok 690091, Russia; kostetskiy.yeya@dvfu.ru (E.K.); natali_1389@mail.ru (N.C.); marybarkin@yandex.ru (M.B.); velanskiy.pv@dvfu.ru (P.V.)

2 National Scientific Center of Marine Biology, Far Eastern Branch of Russian Academy of Sciences, Vladivostok 690041, Russia

* Correspondence: sanina.nm@dvfu.ru; Tel.: +7-423-265-2429

Received: 13 November 2018; Accepted: 5 December 2018; Published: 7 December 2018

Abstract: Increasing global temperatures are expected to increase the risk of extinction of various species due to acceleration in the pace of shifting climate zones. Nevertheless, there is no information on the physicochemical properties of membrane lipids that enable the adaptation of the algae to different climatic zones. The present work aimed to compare fatty acid composition and thermal transitions of membrane lipids from green macroalgae *Ulva lactuca* harvested in the Sea of Japan and the Adriatic Sea in summer. *U. lactuca* inhabiting the Adriatic Sea had bleached parts of thalli which were completely devoid of chloroplast glycolipids. The adaptation to a warmer climatic zone was also accompanied by a significant decrease in the ratio between unsaturated and saturated fatty acids (UFA/SFA) of membrane lipids, especially in bleached thalli. Hence, bleaching of algae is probably associated with the significant decrease of the UFA/SFA ratio in glycolipids. The decreasing ratio of n-3/n-6 polyunsaturated fatty acids (PUFAs) was observed in extra-plastidial lipids and only in the major glycolipid, non-lamellar monogalactosyldiacylglycerol. The opposite thermotropic behavior of non-lamellar and lamellar glycolipids can contribute to maintenance of the highly dynamic structure of thylakoid membranes of algae in response to the increasing temperatures of climatic zones.

Keywords: glycolipids; phospholipids; betaine lipid; fatty acids; differential scanning calorimetry; thermal adaptation

1. Introduction

Over the last few decades, the temperature of seawaters has experienced changes that affect the Earth's ecological state on a global scale. Increasing global temperature is expected to increase the risk of extinction of various species in the future due to acceleration in the pace of shifting climate zones [1].

The primary interface between the environment and the organism's cells are the cell membranes, which are structurally based on the lipid matrix. The maintenance of the liquid crystalline state of the lipid matrix is necessary for the optimal functioning of biological membranes at different changes in the ambient temperature. Therefore, the primary effects of temperature, which is the most powerful environmental factor, are related to the compensatory molecular mechanisms directed to maintain the liquid crystalline state of the membrane lipid matrix [2,3]. This mechanism, which is called homeoviscous adaptation, is mainly realized due to rearrangements in the fatty acid composition of membrane lipids [4,5]. The efficiency of homeoviscous adaptation could be estimated by the lipid thermal transitions from a crystalline (gel) state to a liquid crystalline one. Our earlier studies on the fatty acid composition and phase transitions of the individual polar lipids from taxonomically different marine macrophytes inhabiting the Sea of Japan showed that the decrease in the ratio between unsaturated (UFA) and saturated fatty acids (SFA) was accompanied by the decrease in the ratio between n-3 and n-6 polyunsaturated fatty acids (PUFAs) during acclimatization from winter to summer. Despite the larger changes of these ratios in photosynthetic lipids (glycolipids monogalactosyldiacylglycerol (MGDG), digalactosyldiacylglycerol (DGDG), sulfoquinovosyldiacylglycerol (SQDG) and phospholipid phosphatidyldiacylglycerol (PG)), peak maximum temperatures (T_{max}) of their thermal transitions did not increase in contrast to the respective adaptive changes of non-photosynthetic phospholipids (phosphatidylcholine (PC)), (phosphatidylethanolamine) (PE)) and betaine lipid 1,2-diacylglycero-O-4′-(*N,N,N*-tri-methyl)-homoserine (DGTS) [6].

Extreme temperatures strongly affect the distribution of plants in the world. *Ulva lactuca* (Ulvales; Chlorophyta), widely known as *U. fenestrata*, is the most widespread edible green algae in the seas of all climatic zones. Nevertheless, there is no information on what features in physicochemical properties of membrane lipids facilitate the adaptation of the algae to the conditions of different climatic zones.

Moreover, abnormally high seawater temperatures could result in the partial or even total bleaching thalli of algae due to the loss of chlorophyll and other photosynthetic pigments that is accompanied by growth inhibition of plants [7,8]. However, the underlying mechanism and the role of the lipid matrix in these processes remain unknown.

Therefore, the aim of present work was to clarify the differences between the fatty acid composition and the thermotropic behavior of membrane lipids from *U. lactuca* adjusted to the conditions of different climatic zones – the Adriatic Sea (Mediterranean subtropical climatic zone) and the Sea of Japan (moderate climatic zone) as well as between green and bleached thalli of *U. lactuca* inhabiting the Adriatic Sea. The major membrane lipid of *U. lactuca* and other marine macrophytes MGDG is an important component of tubular immunostimulating complexes (TI-complexes) [9] that affect the conformation and immunogenicity of protein antigens [10,11]. Therefore, the results of the present work are also needed to modulate the lipid surrounding of subunit antigens incorporated in TI-complexes and to enhance their adjuvant effect.

2. Results and Discussion

To characterize differences in physicochemical properties of membrane lipids from *U. lactuca* adapted to the conditions of different climatic zones, glycolipids (MGDG, DGDG, SQDG) and the major phospholipids (PG and PE) [12] were isolated from algae harvested in the Adriatic Sea and the Sea of Japan in August, when the average seawater temperatures reach 26 °C and 22 °C, respectively. Betaine lipid DGTS which substitutes PC in green algae [13,14] was also isolated from the algae.

Our earlier study has shown that the percentage of glycolipids, phospholipids and DGTS in *U. lactuca* harvested in the Sea of Japan in summer is 52%, 17% and 15% of total lipids without triacylglycerols, respectively. MGDG is the dominant glycolipid (37% of total lipids without triacylglycerols), whereas the content of DGDG and SQDG is much less (10% and 5% of total lipids without triacylglycerols, respectively). Major phospholipids of *U. lactuca* are PG and PE (6% and 3% of total lipids without triacylglycerols, respectively). Lipid composition of *U. lactuca* is not highly dependent on season [15].

U. lactuca harvested in the Adriatic Sea had green and bleached parts of thalli (Figure 1). The last ones seem to appear due to the high-temperature stress, because the seawater temperature of 26 °C and higher greatly inhibits the growth of *U. lactuca* [16]. This phenomenon is called thermal bleaching [7] because of the loss of chlorophyll and other pigments. The analysis of lipid composition has shown that bleached parts of the algal thalli were completely devoid of chloroplast-specific glycolipids, which are essential not only for the formation of lipid bilayers of chloroplast membranes, but also for embedding photosynthetic complexes in thylakoid membranes and as integral components of these complexes [17].

It was confirmed earlier that deficient synthesis of chloroplast-specific lipids in Arabidopsis mutants really results in a decrease of chlorophyll content, defects in chloroplast ultrastructure and reduced photosynthetic activity [18].

Figure 1. Green and bleached parts of thalli of *Ulva lactuca* harvested in the Adriatic Sea.

However, thermal bleaching parts of algae thalli contained extra-plastidial PE and DGTS as well PG, which not only plays a crucial role in the structure and function of photosynthetic complexes in thylakoid membranes but is also found in extra-plastidial membranes [19]. As such, cells of the bleaching parts were completely devoid of chloroplasts but retained extra-plastidial membranes and the ability to adjust them under conditions of heat stress, taking into account the adequate rearrangements in the fatty acid composition of PE, PG and DGTS (Tables 1 and 2).

Table 1. The fatty acid composition of phospholipids from green and bleached thalli of *Ulva lactuca* harvested in the Adriatic Sea and from green thalli of the same algae harvested in the Sea of Japan (% of total fatty acids).

Fatty Acids	Phosphatidylglycerol			Phosphatidylethanolamine		
	The Sea of Japan	The Adriatic Sea		The Sea of Japan	The Adriatic Sea	
	Green Thalli		Bleached Thalli	Green Thalli		Bleached Thalli
16:0	35.5 ± 1.2	45.2 ± 0.8	56.2 ± 1.4	22.0 ± 0.6	29.7 ± 0.7	56.2 ± 1.3
16:1 (n-7)	0.9 ± 0.1	2.6 ± 0.2	2.1 ± 0.1	5.4 ± 0.3	4.4 ± 0.2	2.2 ± 0.1
16:3 (n-3)	8.6 ± 0.3	0.6 ± 0.1	0.2 ± 0.1	0.4 ± 0.1	0.5 ± 0.1	tr.
16:4 (n-3)	2.1 ± 0.2	0.8 ± 0.1	0.4 ± 0.1	0.7 ± 0.1	6.7 ± 0.3	0.9 ± 0.1
18:0	0.3 ± 0.1	0.5 ± 0.1	1.5 ± 0.1	4.6 ± 0.3	2.0 ± 0.1	2.8 ± 0.2
18:1 (n-9)	0.6 ± 0.1	2.3 ± 0.2	2.8 ± 0.2	4.2 ± 0.2	3.4 ± 0.2	4.7 ± 0.2
18:1 (n-7)	17.0 ± 0.7	19.6 ± 0.5	16.9 ± 0.3	10.5 ± 0.8	14.3 ± 0.5	18.7 ± 0.3
18:2 (n-6)	2.3 ± 0.2	2.5 ± 0.1	1.4 ± 0.1	2.7 ± 0.2	5.8 ± 0.4	2.1 ± 0.1
18:3 (n-3)	28.1 ± 0.8	14.1 ± 0.3	3.9 ± 0.1	26.2 ± 0.5	9.7 ± 0.3	2.5 ± 0.1
18:4 (n-3)	1.9 ± 0.1	6.1 ± 0.2	2.0 ± 0.1	1.4 ± 0.1	9.8 ± 0.2	1.5 ± 0.1
22:6 (n-3)	tr.	1.3 ± 0.1	0.9 ± 0.1	2.2 ± 0.1	4.5 ± 0.1	0.3 ± 0.1
SFA	36.4 ± 1.0	46.5 ± 0.7	60.7 ± 0.5	30.0 ± 0.5	34.6 ± 0.5	61.5 ± 0.7
MUFA	19.3 ± 0.4	25.5 ± 0.6	28.0 ± 0.3	24.0 ± 0.4	26.6 ± 0.6	29.2 ± 0.4
PUFA	44.3 ± 0.3	28.0 ± 0.4	11.3 ± 0.3	46.0 ± 0.7	38.8 ± 0.4	9.3 ± 0.3
n-3/n-6 PUFA	16.1 ± 0.3	6.5 ± 0.2	3.1 ± 0.1	9.9 ± 0.2	5.0 ± 0.2	1.9 ± 0.1
UI	152 ± 2	119 ± 1	66 ± 1	173 ± 2	183 ± 2	58 ± 1
UFA/SFA	1.7 ± 0.1	1.1 ± 0.2	0.6 ± 0.1	2.3 ± 0.1	1.9 ± 0.1	0.6 ± 0.1

SFA, UFA, MUFA, PUFA—saturated, unsaturated, monounsaturated, polyunsaturated fatty acids, respectively; UI—unsaturation index. Fatty acids with content below 2% are excluded but considered in calculations of total values. tr.—traces (content less than 0.1%). Mean values ± standard error of triplicate determination.

Table 2. The fatty acid composition of betaine lipid 1,2-diacylglycero-O-4′-(*N*,*N*,*N*-tri-methyl)-homoserine (DGTS) from green and bleached thalli of *Ulva lactuca* harvested in the Adriatic Sea and from green thalli of the algae harvested in the Sea of Japan (% of total fatty acids).

Fatty Acids	The Sea of Japan	The Adriatic Sea	
	Green Thalli		Bleached Thalli
14:0	0.4 ± 0.1	1.2 ± 0.1	2.3 ± 0.1
16:0	23.0 ± 0.3	31.0 ± 0.2	38.3 ± 0.3
16:1 (n-9)	0.4 ± 0.1	1.0 ± 0.1	4.3 ± 0.2
16:1 (n-7)	1.5 ± 0.1	5.8 ± 2	5.3 ± 0.2
18:0	0.8 ± 0.1	0.7 ± 0.1	0.7 ± 0.1
18:1 (n-9)	1.8 ± 0.1	1.8 ± 0.1	1.8 ± 0.1
18:1 (n-7)	9.9 ± 0.2	9.8 ± 0.2	9.8 ± 0.2
18:2 (n-6)	7.1 ± 0.2	2.2 ± 0.1	2.2 ± 0.1
18:3 (n-6)	4.5 ± 0.1	2.3 ± 0.1	2.3 ± 0.1
18:3 (n-3)	9.7 ± 0.1	6.6 ± 0.2	6.6 ± 0.2
18:4 (n-3)	28.2 ± 0.1	21.2 ± 0.2	21.2 ± 0.2
20:5 (n-3)	0.6 ± 0.1	3.0 ± 0.1	3.0 ± 0.1
22:5 (n-3)	7.1 ± 0.2	8.6 ± 0.2	8.6 ± 0.2
SFA	24.2 ± 0.2	34.2 ± 0.2	42.6 ± 0.2
MUFA	15.6 ± 0.1	18.0 ± 0.2	25.3 ± 0.2
PUFA	60.2 ± 0.3	47.8 ± 0.1	32.1 ± 0.2
n-3/n-6 PUFA	3.8 ± 0.1	5.8 ± 0.1	2.9 ± 0.1
UI	203 ± 2	195 ± 1	127 ± 1
UFA/SFA	3.1 ± 0.1	1.9 ± 0.1	1.3 ± 0.1

SFA, UFA, MUFA, PUFA—saturated, unsaturated, monounsaturated, polyunsaturated fatty acids, respectively; UI—unsaturation index. Fatty acids with content below 2% are excluded but considered in calculations of total values. Mean values ± standard error of triplicate determination.

2.1. The Fatty Acid Composition of Phospholipids

The fatty acid composition of major phospholipids isolated from *U. lactuca* is shown in Table 1. PG and PE from green thalli of algae inhabiting the Sea of Japan and the Adriatic Sea comprised the major fatty acid 16:0 with lesser amounts of 18:3 (n-3) and 18:1 (n-7). The first two fatty acids seem to play the main role in the adaptation of algae to different climatic zones, judging by the significant differences in their content in PG and PE from algae harvested in the Sea of Japan and the Adriatic Sea. However, the most profound changes occurred in bleached parts of algae from the Adriatic Sea. The content of PUFA 18:3 (n-3) was greatly reduced, whereas the level of 16:0 increased in the following line of the PG and PE sources: green algae from the Sea of Japan → green algae from the Adriatic Sea → bleached algae from the Adriatic Sea.

The percentage of 18:1 (n-7) in PE also increased in the same order. Especially pronounced changes occurred in total parameters of the fatty acid composition. The adaptation of *U. lactuca* to the higher temperature environment was accompanied by the essential increasing content of SFAs and monounsaturated fatty acids (MUFAs), as well as by decreasing percentage of PUFAs in PG and PE. Both phospholipids from bleached thalli were characterized by the prevailing content of SFAs and the lowest level of PUFAs and unsaturation index (UI). Simultaneously, the ratio of n-3/n-6 PUFAs consequently decreased, probably because of the increased demand for more potent mediators derived from n-6 PUFAs [20–22] in a warmer environment [6]. Then, changes in total parameters of the fatty acid composition observed in phospholipids from *U. lactuca* are mostly similar to earlier data on season acclimatization of marine macrophytes [6].

2.2. The Fatty Acid Composition of Betaine Lipid 1,2-diacylglycero-O-4′-(N,N,N-tri-methyl)-homoserine (DGTS)

The fatty acid composition of betaine lipid DGTS (Table 2) was somewhat different from that of phospholipids from *U. lactuca* (Table 1). It comprised the same major fatty acids: 16:0, 18:1 (n-7),

18:3 (n-3). However, the dominant major UFA was 18:4 (n-3), which percentage was much lower in phospholipids. In addition, the percentage of 22:5 (n-3), which was absent in phospholipids, was considerable in DGTS. Similarly, to phospholipids, the level of SFAs (mostly 16:0) increased and the content of PUFAs (mostly 18:4 (n-3) and 18:3 (n-3)) decreased in the line of DGTS sources: green algae from the Sea of Japan → green algae from the Adriatic Sea → bleached algae from the Adriatic Sea. Unlike phospholipids, the level of MUFAs also increased in this line, although the dominant MUFA 18:1 (n-7) did not participate in the adaption process. Dissimilarity in the fatty acid compositions of DGTS and phospholipids was marked in other algae [23].

The mentioned changes were accompanied by a decrease in UI and ratios of UFA/SFA. However, the ratio of n-3/n-6 PUFAs decreased only in the bleached parts of the algae in comparison with green thalli.

2.3. *The Fatty Acid Composition of Glycolipids*

Glycolipids are the main components of photosynthetic membranes and associated with the functioning of the photosynthetic apparatus, whereas phospholipids, except PG, are mainly accumulated in non-photosynthetic membranes and differ from glycolipids by biosynthetic pathway [24]. Nevertheless, trends observed in the fatty acid composition of phospholipids and DGTS (Tables 1 and 2) also remained in glycolipids from *U. lactuca* (Table 3), which mainly contained the same fatty acids (16:0, 16:1 (n-7), 18:1 (n-9), 18:1 (n-7), 18:2 (n-6), 18:3 (n-3) and 18:4 (n-3)) as well 16:3 (n-3) and 16:4 (n-3) found in phospholipids.

Table 3. The fatty acid composition of glycolipids from green parts of thalli of *Ulva lactuca* harvested in the Adriatic Sea and the Sea of Japan (% of total fatty acids).

Fatty Acids	MGDG		DGDG		SQDG	
	The Sea of Japan	The Adriatic Sea	The Sea of Japan	The Adriatic Sea	The Sea of Japan	The Adriatic Sea
16:0	2.8 ± 0.1	6.4 ± 0.3	22.2 ± 0.3	28.8 ± 0.3	56.7 ± 0.4	53.4 ± 0.2
16:1 (n-7)	0.8 ± 0.1	2.4 ± 0.2	0.7 ± 0.1	3.9 ± 0.2	1.5 ± 0.1	1.5 ± 0.1
16:2 (n-6)	n.d.	n.d.	7.8 ± 0.2	2.8 ± 0.1	n.d.	n.d.
16:3 (n-3)	3.0 ± 0.2	1.5 ± 0.1	8.3 ± 0.2	2.8 ± 0.1	n.d.	tr.
16:4 (n-3)	37.8 ± 0.4	32.3 ± 0.4	5.6 ± 0.1	1.5 ± 0.1	n.d.	0.3 ± 0.1
18:1 (n-9)	1.0 ± 0.1	2.3 ± 0.1	1.6 ± 0.1	4.4 ± 0.1	0.2 ± 0.1	0.2 ± 0.1
18:1 (n-7)	4.7 ± 0.2	7.0 ± 0.5	5.9 ± 0.2	12.5 ± 0.2	13.4 ± 0.2	22.1 ± 0.3
16:0	2.8 ± 0.1	6.4 ± 0.3	22.2 ± 0.3	28.8 ± 0.3	56.7 ± 0.4	53.4 ± 0.2
18:2 (n-6)	1.5 ± 0.1	1.8 ± 0.1	11.8 ± 0.3	9.4 ± 0.2	3.5 ± 0.1	1.7 ± 0.1
18:3 (n-3)	18.5 ± 0.3	14.5 ± 0.1	30.4 ± 0.5	23.2 ± 0.2	23.6 ± 0.3	11.5 ± 0.2
18:4 (n-3)	25.7 ± 0.4	24.4 ± 0.4	3.5 ± 0.2	3.5 ± 0.1	1.6 ± 0.1	11.2 ± 0.2
18:2 (n-6)	1.5 ± 0.1	1.8 ± 0.1	11.8 ± 0.3	9.4 ± 0.2	3.5 ± 0.1	1.7 ± 0.1
SFA	3.8 ± 0.2	7.0 ± 0.3	23.6 ± 0.4	31.8 ± 0.3	56.7 ± 0.4	53.7 ± 0.3
MUFA	7.9 ± 0.2	13.2 ± 0.3	8.8 ± 0.2	21.6 ± 0.4	15.6 ± 0.3	23.0 ± 0.3
PUFA	88.3 ± 0.8	79.8 ± 0.9	67.6 ± 0.5	46.6 ± 0.7	27.7 ± 0.4	23.3 ± 0.3
n-3/n-6 PUFA	50.7 ± 0.3	35.6 ± 0.4	2.5 ± 0.2	2.7 ± 0.1	7.2 ± 0.2	13.8 ± 0.1
UI	334 ± 3	302 ± 2	203 ± 2	152 ± 1	89 ±1	109 ± 2
UFA/SFA	25.3 ± 0.4	13.3 ± 0.2	3.2 ± 0.1	2.1 ± 0.1	0.8 ± 0.1	0.9 ± 0.1

SFA, UFA, MUFA, PUFA—saturated, unsaturated, monounsaturated, polyunsaturated fatty acids, respectively; UI—unsaturation index. Fatty acids with content below 2% are excluded but considered in calculations of total values. n.d.—not detected, tr.—traces (content less than 0.1%). Mean values ± standard error of triplicate determination.

The greatest differences between glycolipids occurred in the composition of major fatty acids. The major fatty acids of MGDG were 16:4 (n-3), 18:4 (n-3), 18:3 (n-3), 18:1 (n-7), while DGDG mainly comprised 18:3 (n-3), 16:0, 18:2 (n-6) and 18:1 (n-7) in order of decreasing percentage. The major fatty acids of SQDG were similar to those of DGDG except for 18:2 (n-6), for which the percentage did not exceed 3.5%.

However, SQDG from *U. lactuca* contained the highest level of *cis*-vaccenic acid 18:1 (n-7), which is not characteristic for other marine macrophytes [6,12]. Also, this glycolipid was the most enriched in SFA 16:0 (more than 50% of total fatty acids) unlike MGDG where the percentage of 16:0 and SFAs as a whole was the lowest.

The adaption of *U. lactuca* to a warmer climatic zone was accompanied by the same changes in total parameters of the fatty acid composition in galactolipids, that was found in phospholipids and DGTS (Tables 1 and 2): the levels of SFAs and MUFAs increased, whereas other parameters (PUFAs, n-3/n-6 PUFAs, UI and UFA/SFA) decreased. This suggests that bleaching of *U. lactuca* has been accompanied by the strengthening of the trend found not only in extra-plastidial lipids, but also in chloroplast-specific glycolipids.

In SQDG, total parameters except MUFAs and PUFAs changed in the opposite way. The direction of changes in n-3/n-6 PUFAs of DGDG also did not coincide with the general trend.

The decline in the ratio of the UFA/SFA was greatest in MGDG, while SQDG had the smallest change. A decrease in the ratio of n-3/n-6 PUFAs was observed in MGDG only. However, this parameter increased or did not change in SQDG and DGDG, respectively. Perhaps the observed dissimilarity in the adaptive changes of the fatty acid composition may be the cause of different thermotropic behavior of glycolipids from *U. lactuca* harvested in the Adriatic Sea and the Sea of Japan (Figure 2).

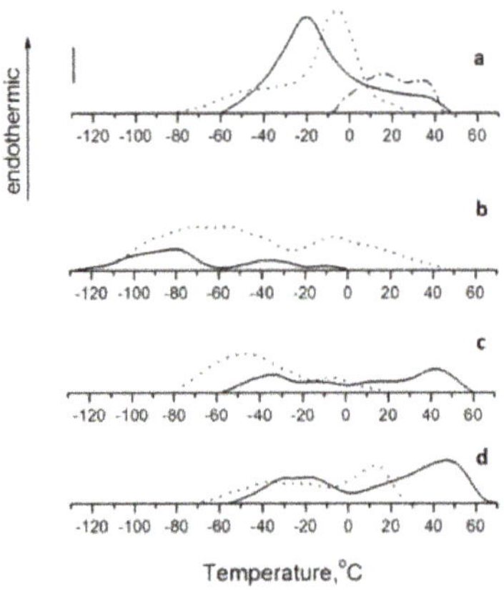

Figure 2. DSC thermograms of betaine lipid DGTS (**a**) and glycolipids MGDG (**b**), DGDG (**c**), SQDG (**d**) from the green thalli of *Ulva lactuca* inhabiting the Sea of Japan (solid curve), the green thalli of algae inhabiting the Adriatic Sea (dotted curve) and the bleached thalli of algae inhabiting the Adriatic Sea (dash-dotted curve). The vertical bar represents 0.5 mW. The scanning rate, 16 °C/min. The sample weight, 10 mg. Each sample was scanned at least three times.

2.4. Thermotropic Behavior and Molecular Species of Polar Lipids

Membrane lipid matrix of ectothermic organisms performs the role of thermosensor. The changes in the physical state of membrane lipids trigger compensatory adjustments in the fatty acid composition of membrane lipids directed to maintain the liquid crystalline state of the membrane lipid matrix, which is optimal for the functioning of living cells [3,25,26]. To assess the efficacy of adaptive rearrangements in the fatty acid composition of major polar lipids from *U. lactuca*, the thermotropic behavior of polar lipids was studied by differential scanning calorimetry (DSC). Calorimetry is the most appropriate method for a detection of the main lipid phase transition from crystalline to liquid crystalline state. Phase transitions of lipids occur due to trans-gauche rotational isomerization of methylene groups about the single C – C bonds along lipid acyl chains. When lipid is heated, the heat flow difference between a sample and a reference is scanned in some temperature range. This endothermic process is visualized by DSC as an integral curve (thermogram) of dependence between a heat capacity and temperature. The peak maximum temperature of thermal transition (T_{max}) is characteristic for a lipid sample of a definite chemical structure [27]. Additional peak(s) with lower heat capacity often occurred on thermograms of natural membrane lipids, which is probably due to their phase separation [28,29].

Thermograms and T_{max} of thermal transitions of extra-plastidial betaine lipid DGTS and chloroplast-specific glycolipids MGDG, DGDG and SQDG of *U. lactuca* inhabiting the Sea of Japan and the Adriatic Sea are shown in Figure 2 and Table 4, respectively.

Table 4. Peak maximum temperatures of thermal transitions (T_{max}, °C) of betaine lipid DGTS and glycolipids isolated from *Ulva lactuca* inhabiting the Sea of Japan and the Adriatic Sea.

Lipids		**T_{max}, °C** [1]
DGTS	Green thalli of *U. lactuca* inhabiting the Sea of Japan	−21
	Green thalli of *U. lactuca* inhabiting the Adriatic Sea	−6
	Bleached thalli of *U. lactuca* inhabiting the Adriatic Sea	15
MGDG	Green thalli of *U. lactuca* inhabiting the Sea of Japan	−78
	Green thalli of *U. lactuca* inhabiting the Adriatic Sea	−60
DGDG	Green thalli of *U. lactuca* inhabiting the Sea of Japan	41
	Green thalli of *U. lactuca* inhabiting the Adriatic Sea	−50
SQDG	Green thalli of *U. lactuca* inhabiting the Sea of Japan	50
	Green thalli of *U. lactuca* inhabiting the Adriatic Sea	12

[1] Standard deviations were less than 1 °C for three replicates.

Phase transition of DGTS from *U. lactuca* inhabiting the Sea of Japan was characterized by the cooperative endothermic peak at −21 °C, which was lower by 15 °C than T_{max} of DGTS from *U. lactuca* inhabiting the Adriatic Sea (Figure 2A). Such a classical adaptive change in the thermotropic behavior is usual for non-photosynthetic phospholipids, especially for PC [6], that confirms functional similarity of DGTS and PC in extra-plastidial membranes. The increase in T_{max} was accompanied by the 1.6-fold decrease in the ratio of UFA/SFA despite a little change in UI of the DGTS from *U. lactuca* inhabiting warmer climatic zone. Namely, the partial substitution (by 11.5%) of highly unsaturated molecular forms MUFA/PUFA and PUFA/PUFA for SFA/MUFA and SFA/PUFA (Table 5) contributed to the increase of T_{max}.

The further rise in T_{max} and the shift of the temperature range of the phase transition toward higher temperatures occurred at the thermal transition of DGTS from bleached thalli. The respective thermogram was located at the comparatively narrow temperature range between −8 °C and 44 °C and characterized by two poorly resolved peaks at 15 °C (T_{max}) and 35 °C. The cooperativity of phase transition significantly reduced likely due to the phase separation in this lipid sample. All changes in the thermotropic behavior of DGTS were accompanied by an increase in the share of SFA in its composition (Table 2).

Thermogram of MGDG from *U. lactuca*, harvested in the Sea of Japan lied in the temperature range between –128 °C and 0 °C. There were three peaks in this thermogram profile: the major peak at −78 °C (T_{max}) and additional peaks with lower heat capacity at −36 °C and −10 °C (Figure 2B). MGDG from *U. lactuca* inhabiting the Adriatic Sea revealed the similar thermotropic behavior, while peaks of phase transition occurred at essentially higher temperatures (major peak at about −60 °C and additional one at –6 °C). The high-temperature limit of phase transition was shifted to 44 °C.

Nevertheless, T_{max} of MGDG remained very low. This can be explained by the high content of UFAs (more than 90% of total fatty acids) (Table 3). Most of them are represented by PUFAs (about 80–90% of total fatty acids). In turn, MGDG is the major factor responsible for the functioning of photosynthetic protein complexes. Hence, the high unsaturation probably allows MGDG to adopt different shapes of co-existing membrane protein complexes and stabilize them [5]. Highly unsaturated non-bilayer MGDG can also facilitate membrane deformation and provide the curvature matching of different proteins to the thylakoid membranes [30]. The increase in saturation of MGDG likely decreases the curvature in thylakoid membranes and promotes the crowding out of membrane proteins from the lipid bilayer that may disrupt the architecture and the functioning of thylakoid membranes. Therefore, MGDG plays a crucial role in the structural flexibility of lipid–light-harvesting complex II (LHCII) macro-assemblies. This lipid mediates the dimerization of photosystem II (PSII), as well as the packaging of PSII and photosystem I (PSI) [17,31]. Also, MGDG is structurally involved in the cytochrome b6/f complex [32] and activates the CF0-CF1-ATPase in chloroplasts [33].

Table 5. The composition of molecular species of betaine lipid DGTS and glycolipids from *Ulva lactuca* inhabiting the Sea of Japan and the Adriatic Sea (% of the sum of molecular species).

Molecular Species	DGTS		MGDG		DGDG		SQDG	
	The Sea of Japan	The Adriatic Sea	The Sea of Japan	The Adriatic Sea	The Sea of Japan	The Adriatic Sea	The Sea of Japan	The Adriatic Sea
16:0/16:0	0.5 ± 0.1	0.3 ± 0.1	n.d.	tr.	0.3 ± 0.1	1.4 ± 0.1	14.2 ± 0.2	4.3 ± 0.2
14:0/18:1	n.d.	n.d.	tr.	0.5 ± 0.1	0.9 ± 0.1	8.3 ± 0.2	2.3 ± 0.1	4.7 ± 0.1
16:0/16:1	3.1 ± 0.1	9.8 ± 0.2	n.d.	n.d.	n.d.	n.d.	n.d.	n.d.
16:0/18:1	1.5 ± 0.1	2.9 ± 0.1	n.d.	0.2 ± 0.1	5.8 ± 0.2	6.7 ± 0.2	23.5 ± 0.3	38.1 ± 0.2
16:0/18:2	6.6 ± 0.2	4.8 ± 0.1	2.3 ± 0.1	1.2 ± 0.1	19.7 ± 0.2	16.2 ± 0.2	6.7 ± 0.2	2.9 ± 0.1
16:0/18:3	15.9 ± 0.1	15.4 ± 0.2	2.3 ± 0.1	3.6 ± 0.2	16.7 ± 0.2	21.1 ± 0.3	46.7 ± 0.2	22.9 ± 0.2
16:0/18:4	13.0 ± 0.1	18.1 ± 0.2	0.5 ± 0.1	1.0 ± 0.1	1.0 ± 0.1	2.9 ± 0.1	3.7 ± 0.1	21.5 ± 0.2
16:0/22:5	4.5 ± 0.2	5.9 ± 0.2	n.d.	n.d.	n.d.	n.d.	tr.	tr.
16:1/18:1	1.2 ± 0.1	1.8 ± 0.1	tr.	0.2 ± 0.1	0.4 ± 0.1	4.6 ± 0.2	tr.	0.3 ± 0.1
16:1/18:3	1.9 ± 0.1	0.6 ± 0.1	2.6 ± 0.2	3.5 ± 0.2	1.9 ± 0.1	6.2 ± 0.1	n.d.	tr.
16:1/16:4	tr.	0.2 ± 0.1	0.7 ± 0.1	3.7 ± 0.2	tr.	tr.	n.d.	n.d.
16:1/18:4	1.6 ± 0.1	3.1 ± 0.1	n.d.	0.8 ± 0.1	n.d.	tr.	n.d.	tr.
18:1/16:3	n.d	0.4 ± 0.1	1.9 ± 0.1	1.5 ± 0.1	2.8 ± 0.2	2.9 ± 0.1	tr.	tr.
18:1/18:3	6.9 ± 0.1	2.7 ± 0.1	0.6 ± 0.1	1.1 ± 0.1	2.3 ± 0.1	3.7 ± 0.2	0.2 ± 0.1	n.d.
18:1/16:4	0.9 ± 0.1	0.4 ± 0.1	7.4 ± 0.2	11.5 ± 0.2	0.6 ± 0.1	1.5 ± 0.1	n.d.	tr.
18:1/18:4	8.6 ± 0.2	6.2 ± 0.2	0.2 ± 0.1	0.9 ± 0.1	0.2 ± 0.1	1.2 ± 0.1	tr.	tr.
16:2/18:3	tr.	n.d.	4.6 ± 0.2	3.0 ± 0.1	10.4 ± 0.2	3.5 ± 0.2	tr.	n.d.
16:3/18:3	0.5 ± 0.1	n.d.	6.5 ± 0.2	4.0 ± 0.2	12.4 ± 0.2	2.1 ± 0.1	tr.	tr.
16:4/18:4	0.4 ± 0.1	0.2 ± 0.1	22.1 ± 0.2	27.8 ± 0.2	3.2 ± 0.1	1.1 ± 0.1	n.d.	n.d.
18:2/16:4	0.2 ± 0.1	n.d.	5.9 ± 0.2	3.6 ± 0.2	0.3 ± 0.1	tr.	n.d.	n.d.
18:3/18:3	1.6 ± 0.1	0.2 ± 0.1	1.6 ± 0.1	1.3 ± 0.1	3.7 ± 0.2	3.5 ± 0.2	tr.	n.d.
18:3/16:4	0.5 ± 0.1	tr.	30.1 ± 0.2	16.7 ± 0.2	5.4 ± 0.1	0.5 ± 0.1	n.d.	n.d.
18:3/18:4	4.3 ± 0.1	0.8 ± 0.1	1.2 ± 0.1	2.6 ± 0.1	0.6 ± 0.1	1.2 ± 0.1	n.d.	n.d.
20:5/22:5	3.8 ± 0.1	4.8 ± 0.1	n.d.	n.d.	n.d.	n.d.	n.d.	n.d.
SFA/SFA	0.5 ± 0.1	0.3 ± 0.1	tr.	tr.	0.9 ± 0.1	1.4 ± 0.1	14.2 ± 0.2	4.9 ± 0.2
SFA/MUFA	4.6 ± 0.1	12.7 ± 0.2	tr.	0.8 ± 0.1	6.8 ± 0.1	16.2 ± 0.2	26.2 ± 0.2	45.5 ± 0.2
SFA/PUFA	50.9 ± 0.2	54.3 ± 0.2	7.6 ± 0.2	7.6 ± 0.2	39.6 ± 0.2	44.1 ± 0.2	58.9 ± 0.2	49.0 ± 0.2
MUFA/MUFA	2.3 ± 0.1	2.7 ± 0.1	tr.	0.2 ± 0.1	0.6 ± 0.1	5.0 ± 0.1	tr.	0.3 ± 0.1
MUFA/PUFA	24.7 ± 0.2	17.9 ± 0.1	15.8 ± 0.3	24.9 ± 0.2	10.3 ± 0.2	18.5 ± 0.1	0.3 ± 0.1	0.3 ± 0.1
PUFA/PUFA	17.0 ± 0.1	12.1 ± 0.1	76.6 ± 0.2	66.5 ± 0.2	41.8 ± 0.2	14.8 ± 0.1	0.4 ± 0.1	tr.

Molecular species with content lower 2% are excluded, but considered in calculations of total values. SFA, MUFA, PUFA—saturated, monounsaturated, polyunsaturated fatty acid, respectively. n.d.—not detected, tr.—traces (content less than 0.1%). Mean values ± standard error of triplicate determination.

From the point of view of homeoviscous adaptation, the shift of T_{max} to the higher temperatures is the adequate compensatory adjustment, which was provided by the classical decrease in the unsaturation of fatty acids in MGDG. The same changes in the fatty acid composition as well as decreasing n-3/n-6 PUFAs of this glycolipid from *U. lactuca* and other marine macrophytes were observed at the season change from winter to summer [6]. As shown in Table 5, the following major molecular species of MGDGs likely contributed to the formation of the major peak on the respective thermograms: 18:3/16:4, 16:4/18:4, as well as 18:1/16:4, 16:3/18:3, 16:2/18:3 and 18:2/16:4 whose sum was about 70%. The positional distribution of fatty acids within MGDG from algae depends on the fatty acid chain length rather than on their unsaturation degree. The preferential occurrence of 18C and 16C fatty acid residues in *sn*-1 and *sn*-2 positions of the glycerol backbone of MGDG, respectively, is typical for green algae [34]. Higher T_{max} of MGDG from algae inhabiting warmer climatic zone was probably due to the two-fold lower content of major molecular species 18:3/16:4.

In spite of the lowering ratio of UFA/SFA, T_{max} of lamellar DGDG decreased, in contrast to non-lamellar MGDG (Figure 2B). An especially remarkable change occurred in the percentage of MUFAs in DGDG (by the 2.5-fold increase) which reduces phase transition temperature more effectively than PUFAs with three or more double bonds [35,36] (Table 3). MUFAs were situated in the following molecular species of DGDG: 14:0/18:1, 16:1/18:1and 16:1/18:3, whose total content was six-fold higher in DGDG from algae inhabiting the Adriatic Sea in comparison with DGDG from algae inhabiting the

Sea of Japan (Table 5). The total indicators (MUFA/PUFA and MUFA/MUFA) also demonstrated the major contribution of the DGDG molecular species containing MUFAs to the lowering of T_{max}. In turn, the sharp decrease in the level of highly unsaturated molecular species PUFA/PUFA probably does not effectively influence the thermotropic behavior of DGDG.

Adaptive changes in phase transition of SQDG was similar to that of DGDG (Figure 2C,D, respectively) in spite of essentially different polar groups and the fatty acid composition of these glycolipids. As such, a set of major molecular species of SQDG comprises 16:0/18:3, 16:0/18:1, 16:0/16:0, 16:0/18:4 (Table 5), while highly unsaturated molecular species MUFA/PUFA and PUFA/MUFA, whose high content was detected in DGDG, practically absent in SQDG. As a whole, SQDG is the most saturated glycolipid. Therefore, T_{max} of SQDG is the highest in comparison with T_{max} of other glycolipids from *U. lactuca* regardless of the temperature conditions of the habitat of the algae. The shift of T_{max} of SQDG from *U. lactuca* depending on climatic zones was not as sharp as it was found to be for DGDG (about 60 °C against about 90 °C, respectively) (Figure 2C,D, respectively), which was accompanied by a less pronounced increase in the level of MUFAs. The lower T_{max} of SQDG from *U. lactuca* inhabiting warmer climatic zone also correlated with a threefold lower level of the most high-melting molecular species SFA/SFA, as well as remarkably higher UI and UFA/SFA.

Homeoviscous adaptation is the compensatory mechanism that allows cell membranes to maintain the functionally optimal viscosity and, hence, the integrity of organelles at different environmental temperatures [5]. The infringement of the thylakoid membrane integrity initiates the bleaching of algae at elevated temperatures [7]. Our results on the fatty acid composition of PG, PE and DGTS of bleached parts of *U. lactuca* (Tables 1–3) allow us to propose that elevated ambient temperature also induces the significant increase of the fatty acid saturation in the major chloroplast glycolipids and the following destruction of thylakoid membranes, the release of glycolipids and chlorophyll [18].

Opposite changes in the thermotropic behavior of non-lamellar MGDG and lamellar DGDG and SCDG probably maintain the highly dynamic structure of the *U. lactuca* thylakoid membranes in response to an increase in temperature of climatic zones. Therefore, a highly dynamic flexibility of the thylakoid structure can be supported by the fine tuning of MGDG/DGDG ratio which influences the reversible transitions between non-lamellar and lamellar phases in thylakoid lipids [37]. As shown earlier [15], the MGDG/DGDG ratio increases at warm-acclimatization of *U. lactuca*. To compensate for the destabilizing effect of the MGDG elevated level at the adaptation of algae to warmer climatic zone, T_{max} of this glycolipid increased. Instead, the lowering of the DGDG share was accompanied by the decrease of its T_{max}.

On the other hand, the hydrogen bonds between the DGDG polar heads of adjacent bilayers result in thylakoid membrane stacking. However, electrostatic repulsion between positively charged molecules of SQDG hinders the stacking of thylakoid membranes, which additionally contributes to the dynamic structure of the thylakoid grana [37]. Probably, another way to maintain the highly dynamic structure of the thylakoid membranes is the decrease of the T_{max} and possibly the level [6] of SQDG from *U. lactuca* inhabiting the warmer climate zone.

The pronounced phase separation, which is characteristic for the thermotropic behavior of all glycolipids may provide a different lipid environment for PSI and PSII segregated into stroma and grana thylakoid membranes, respectively [38]. The higher mobility of protein complexes in stroma thylakoids indirectly confirms our assumption [39].

3. Materials and Methods

3.1. Plants

Marine macroalgae *U. lactuca* (Chlorophyta: Ulvales) was harvested in the Peter the Great Gulf of the Sea of Japan (moderate climatic zone) in August at seawater temperature of 24 °C, seawater salinity of 34.5‰, at a depth of about 1.5 m, and in the Adriatic Sea near the western coast of the Istrian peninsula (subtropical climatic zone) in August at seawater temperature of 27 °C, seawater salinity of

36‰, at a depth of about 1.5 m. The anatomical and morphological analysis of thalli of the algae was performed with the Axio Imager light microscope (Carl Zeiss, Oberkochen, Germany). In each case the investigated thalli were in a sterile state, without signs of reproduction [40]. Live algal biomass was collected per 500 g from each place. Adriatic algae had bleached parts of thalli, which were separated from green parts of thalli. Freshly harvested algae were thoroughly cleaned to remove epiphytes, small invertebrates and sand particles, and then heated for 2 minutes in boiling H_2O to inactivate enzymes.

3.2. Extraction and Isolation of Lipids

Total lipids were extracted from algae by the method of Bligh and Dyer [41]. Crude glyco- and phospholipids were isolated from total lipid extract by column chromatography on silica gel by elution with acetone, acetone/benzene/acetic acid/water (200:30:3:10, by vol.) and a gradient of chloroform and methanol, respectively [6]. Betaine lipid DGTS was eluted with chloroform/methanol/benzene/28% aqueous ammonia (65:30:10:6, by vol.). Then, these lipids were purified by preparative silica thin-layer chromatography (TLC) using acetone/benzene/acetic acid/water (200:30:3:10, by vol.) or chloroform/methanol/water (65:25:4, by vol.), respectively. The purity of lipids was checked by two-dimensional silica TLC [42,43].

3.3. Analysis of the FattyAcid Composition

The fatty acid composition of chromatographically pure lipids was studied by gas-liquid chromatography as described earlier [6]. Esterification of lipids was performed with acetylchloride/methanol (1:20) at 95 °C for 1 h. Fatty acid methyl esters were extracted with *n*-hexane and purified by TLC. Analysis of fatty acid methyl esters was performed by an Agilent 6898 gas chromatograph (Agilent Technologies, Santa Clara, USA), equipped with a flame-ionization detector, a silica capillary column (25 m × 0.25 mm) with Carbowax 20 M. The carrier gas was helium. Individual peaks of fatty acid methyl esters were identified by comparison of GC R_tS with those of authentic standards of fatty acid methyl esters and by equivalent chain length (ECL). Statistical analysis was carried out using the program Microsoft Excel. The results are presented as mean values ± standard error of triplicate determination.

3.4. Calorimetry

The thermotropic behavior of chromatographically pure lipids was studied by differential scanning calorimetry as described earlier [6]. Lipids solubilized in chloroform were introduced into standard aluminum pans. Vacuum dried samples of approximately 10 mg were sealed into pans and placed in a DSC-2M differential scanning calorimeter (Biopribor, Puschino, Russia). Samples were either heated or cooled at 16 °C min between −135 °C and 80 °C at a sensitivity of 5 mW. The position of the maximum of heat capacity vs. temperature plot was recorded as the phase transition temperature, T_{max}. The temperature range was calibrated by naphthalene, mercury and indium.

3.5. Analysis of the Molecular Species Composition

Analytical separation of molecular species of lipids was performed by high-performance liquid chromatography (HPLC) on chromatograph Shimadzu-LC20 with mass-detector LCMS-2010EV (Shimadzu Corp., Duisburg, Germany). It was used Ascentis C18 column (Supelco, Bellefonte, PA, USA), 25 cm × 2.1 mm, 5 μm particle size. The column was thermostated at a temperature of 45 °C. The flow rate was of 0.3 mL/min. Intervals with the constant eluent composition (5 mM aqueous solution of ammonium acetate/methanol/isopropanol, v/v) were following: 0 min—6:92:2, 30 min—6:79:15, 35–38 min—6:69:25 for MGDG, DGDG, SQDG; 0 min—6:92:2, 45 min—6:86:8, 50–65 min—6:54:40 for DGTS. Mass-detector's options: electrospray ionization, positive ion detection mode for MGDG, DGDG and DGTS, negative ion—for SQDG; nitrogen flow—1.5 L/min; voltage of capillary—4.5 kV for positive ionization, 3.5 kV for negative ionization; temperature of line of desolvation—250 °C; temperature of input interface—280 °C. The content of molecular species was determined by the peak areas on chromatograms of quasimolecular ions corresponding to each molecular species; *sn*-positions of acyls in lipid structure were not defined [44].

Statistical analysis was carried out using the program Microsoft Excel. The results are presented in the form of mean value ± standard error of triplicate determination.

4. Conclusions

The comparison of adaptive changes in the fatty acid composition and thermal transitions of polar lipids of *U. lactuca* allowed us to simulate the situation that is expected due to acceleration in the pace of shifting climate zones. Heat stress was shown to induce an intensified decline in the ratio of UFA/SFA in polar lipids of bleached parts of algal thalli, which is an adequate response from the point of view of the theory of homeoviscous adaptation. However, this process has a flip side. A significant increase in the share of SFA in chloroplast-specific non-bilayer MGDG likely decreases the curvature in thylakoid membranes and promotes the crowding out of different membrane proteins, participating in photosynthesis, from the lipid bilayer that may disrupt the architecture and the functioning of thylakoid membranes and chloroplasts as a whole. The opposite changes in the thermotropic behavior of non-lamellar MGDG and lamellar DGDG and SQDG, are probably directed to maintain a highly dynamic structure of thylakoid membranes of *U. lactuca* in response to increasing temperature of climatic zones. Despite the loss of glycolipids and, therefore, chloroplast membranes at heat stress, lipids of extra-plastidial membranes and other likely related functions exhibit thermotolerance, which can only prolong the life of *U. lactuca* and possibly of other algae in conditions of shifting climatic zones due to the increasing global temperature.

Author Contributions: conceptualization, E.K. and N.S.; methodology, N.S. and P.V.; software, M.B.; formal analysis, N.S., E.K., M.B., P.V. and N.C.; investigation, N.C., M.B. and P.V.; resources, E.K.; data curation, M.B.; writing, N.S., M.B., N.C. and E.K.; visualization, M.B.; supervision, N.S.; project administration, E.K.; funding acquisition, E.K. and N.S.

Funding: This research was funded by the Russian Science Foundation, grant number 15-15-00035-P and Ministry of Education and Science of the Russian Federation, the state assignment number 6.5736.2017/6.7.

Acknowledgments: Research described in this publication was supported in part concerned the harvesting, extraction, isolation, thermotropic behavior of lipids of *U. lactuca* by the Russian Science Foundation (project 15-15-00035-P). The part concerned with analysis of the fatty acid composition and molecular species of lipids from *U. lactuca* was supported by Ministry of Education and Science of the Russian Federation within the state assignment No.6.5736.2017/6.7.

Conflicts of Interest: The authors declare no conflict of interest.

References

1. Mahlstein, I.; Daniel, J.S.; Solomon, S. Pace of shifts in climate regions increases with global temperature. *Nature Clim. Chang.* **2013**, *3*, 739–743. [CrossRef]
2. Beney, L.; Gervais, P. Influence of the fluidity of the membrane on the response of microorganisms to environmental stresses. *Appl. Microbiol. Biotechnol.* **2001**, *57*, 34–42. [PubMed]
3. Török, Z.; Tsvetkova, N.M.; Balogh, G.; Horváth, I.; Nagy, E.; Pénzes, Z.; Hargitai, J.; Bensaude, O.; Csermely, P.; Crowe, J.H.; et al. Heat shock protein coinducers with no effect on protein denaturation specifically modulate the membrane lipid phase. *Proc. Natl. Acad. Sci. USA* **2003**, *100*, 3131–3136. [CrossRef] [PubMed]
4. Sinensky, M. Homeoviscous adaptation—A homeostatic process that regulates the viscosity of membrane lipids in *Escherichia coli*. *Proc. Natl. Acad. Sci. USA* **1974**, *71*, 522–525. [CrossRef] [PubMed]
5. Ernst, R.; Ejsing, C.S.; Antonny, B. Homeoviscous adaptation and the regulation of membrane lipids. *J. Mol. Biol.* **2016**, *428*, 4776–4791. [CrossRef] [PubMed]
6. Sanina, N.M.; Goncharova, S.N.; Kostetsky, E.Y. Seasonal change of fatty acid composition and thermotropic behavior of polar lipids marine macrophytes. *Phytochemistry* **2008**, *69*, 1517–1527. [CrossRef] [PubMed]

7. Tchernov, D.; Gorbunov, M.Y.; de Vargas, C.; Narayan Yadav, S.; Milligan, A.; Haggblom, M.; Falkowski, P. Membrane lipids of symbiotic algae are diagnostic of sensitivity to thermal bleaching in corals. *Proc. Natl. Acad. Sci. USA* **2004**, *101*, 13531–13535. [CrossRef] [PubMed]
8. De Silva, H.C.C.; Asaeda, T. Effects of heat stress on growth, photosynthetic pigments, oxidative damage and competitive capacity of three submerged macrophytes. *J. Plant Interact.* **2017**, *12*, 228–236. [CrossRef]
9. Kostetsky, E.Y.; Sanina, N.M.; Mazeika, A.N.; Tsybulsky, A.V.; Vorobyeva, N.S.; Shnyrov, V.L. Tubular immunostimulating complex based on cucumarioside A2-2 and monogalactosyldiacylglycerol from marine macrophytes. *J. Nanobiotechnol.* **2011**, *9*, 9–35. [CrossRef]
10. Sanina, N.M.; Kostetsky, E.Y.; Shnyrov, V.L.; Tsybulsky, A.V.; Novikova, O.D.; Portniagina, O.Y.; Vorobieva, N.S.; Mazeika, A.N.; Bogdanov, M.V. The influence of monogalactosyldiacylglycerols from different marine macrophytes on immunogenicity and conformation of protein antigen of tubular immunostimulating complex. *Biochimie* **2012**, *94*, 1048–1056. [CrossRef]
11. Sanina, N.; Davydova, L.; Chopenko, N.; Kostetsky, E.; Shnyrov, V. Modulation of immunogenicity and conformation of HA1 subunit of influenza A virus H1/N1 hemagglutinin in tubular immunostimulating complexes. *Int. J. Mol. Sci.* **2017**, *18*, E1895. [CrossRef] [PubMed]
12. Sanina, N.M.; Goncharova, S.N.; Kostetsky, E.Y. Fatty acid composition of individual polar lipid classes from marine macrophytes. *Phytochemistry* **2004**, *65*, 721–730. [CrossRef]
13. Dembitsky, V.M.; Rozentsvet, O.A. Diacylgliceryltrimethylhomoserines and phospholipids of some marine macrophytes. *Phytochemistry* **1989**, *28*, 3341–3343. [CrossRef]
14. Van Ginneken, V.; Gittenberger, A.; Rensing, M.; de Vries, E.; Peeters, E.T.H.M.; Verheij, E. Seaweed competition: Ulva sp. has the potential to produce the betaine lipid diacylglyceryl-O-4′-(*N,N,N*-trimethyl) homoserine (DGTS) in order to replace phosphatidylcholine (PC) under phosphate-limiting conditions in the P-limited Dutch Wadden Sea and outcompete an Aggressive non-indigenous *Gracilaria vermiculophylla* red drift algae out of this Unique Unesco world heritage coastal area. *Ocean & Fish Open Access J.* **2017**, *2*, 555596. [CrossRef]
15. Kostetsky, E.Y.; Goncharova, S.N.; Sanina, N.M.; Shnyrov, V.L. Season influence on lipid composition of marine macrophytes. *Bot. Mar.* **2004**, *47*, 134–139. [CrossRef]
16. Kalita, T.L.; Tytlianov, E.A. Effect of temperature and illumination on growth and reproduction of the green alga *Ulva fenestrata*. Russ. *J. Mar. Biol.* **2003**, *29*, 316–322. [CrossRef]
17. Kobayashi, K.; Endo, K.; Wada, H. Roles of lipids in photosynthesis. In *Lipids in Plant and Algae Development*; Subcell, B., Nakamura, Y., Li-Beisson, Y., Eds.; Springer International Publishing: Basel, Switzerland, 2016; Volume 86, pp. 21–50, ISBN 978-3-319-25979-6.
18. Wang, Z.; Benning, C. Chloroplast lipid synthesis and lipid trafficking through ER–plastid membrane contact sites. *Biochem. Soc. Trans.* **2012**, *40*, 457–463. [CrossRef]
19. Zhou, Y.; Hölzl, G.; vom Dorp, K.; Peisker, H.; Melzer, M.; Frentzen, M.; Dörmann, P. Identification and characterization of a plastidial phosphatidylglycerophosphate phosphatase in *Arabidopsis thaliana*. *Plant J.* **2017**, *89*, 221–234. [CrossRef] [PubMed]
20. Lauritzen, L.; Hansen, H.S.; Jorgesen, M.H.; Michaelsen, K.F. The essentiality of long chains n-3 fatty acids in relation to development and function of the brain and retina. *Prog. Lipid Res.* **2001**, *40*, 1–94. [CrossRef]
21. Bagga, D.; Wang, L.; Farias-Eisner, R.; Glaspy, J.A.; Reddy, S.T. Differential effects of prostaglandin derived from ω-6 and ω-3 polyunsaturated fatty acids on COX-2 expression and IL-6 secretion. *Proc. Natl. Acad. Sci. USA* **2003**, *100*, 1751–1756. [CrossRef] [PubMed]
22. Calder, P.C. n-3 Polyunsaturated fatty acids, inflammation, and inflammatory diseases. *Am. J. Clin. Nutr.* **2006**, *83*, 1505–1519. [CrossRef] [PubMed]
23. Weber, N.; Bergenthal, D.; Kokate, C.K.; Mangold, H.K. Biologically active ether lipids: incorporation of long-chain precursors into 1(3),2-diacylglycero-3(1)-O-4′-(*N,N,N*-trimethyl)homoserines and other lipids of *Chlorella fusca*. *J. Lipid Mediat.* **1989**, *1*, 37–48. [PubMed]
24. Kalisch, B.; Dörmann, P.; Hölzl, G. DGDG and glycolipids in plants and algae. In *Lipids in Plant and Algae Development*; Subcell, B., Nakamura, Y., Li-Beisson, Y., Eds.; Springer International Publishing: Basel, Switzerland, 2016; Volume 86, pp. 51–84, ISBN 978-3-319-25979-6.
25. Gombos, Z.; Wada, H.; Murata, N. The recovery of photosynthesis from low-temperature photoinhibition is accelerated by the unsaturation of membrane lipids: a mechanism of chilling tolerance. *Proc. Natl. Acad. Sci. USA* **1994**, *91*, 8787–8791. [CrossRef]

26. Vigh, L.; Nakamoto, H.; Landry, J.; Gomez-Munos, A.; Harwood, J.L.; Horvath, I. Membrane regulation of the stress response from prokaryotic models to mammalian cells. *Ann. N. Y. Acad. Sci.* **2007**, *1113*, 40–51. [CrossRef] [PubMed]
27. Huang, C.-H.; Li, S. Calorimetric and molecular mechanics studies of the thermotropic phase behavior of membrane phospholipids. *Biochim. Biophys. Acta* **1999**, *1422*, 273–307. [CrossRef]
28. Raison, J.K.; Wright, L.C. Thermal phase-transitions in the polar lipids of plant membranes—Their induction by desaturated phospholipids and their possible relation to chilling injury. *Biochim. Biophys. Acta* **1983**, *731*, 69–78. [CrossRef]
29. Welti, R.; Glaser, M. Lipid domains in model and biological membranes. *Chem. Phys. Lipids* **1994**, *73*, 121–137. [CrossRef]
30. Brown, M.F. Curvature forces in membrane lipid−protein interactions. *Biochemistry* **2012**, *51*, 9782–9795. [CrossRef]
31. Garab, G.; Ughy, B.; Goss, R. Role of MGDG and non-bilayer lipid phases in the structure and dynamics of chloroplast thylakoid membranes. In *Lipids in Plant and Algae Development*; Subcell, B., Nakamura, Y., Li-Beisson, Y., Eds.; Springer International Publishing: Basel, Switzerland, 2016; Volume 86, pp. 127–157, ISBN 978-3-319-25979-6.
32. Georgiev, G.A.; Ivanova, S.; Jordanova, A.; Tsanova, A.; Getov, V.; Dimitrov, M.; Lalchev, Z. Interaction of monogalactosyldiacylglycerol with cytochrome b6f complex in surface films. *Biochem. Biophys. Res. Commun.* **2012**, *419*, 648–651. [CrossRef]
33. Joyard, J.; Teyssier, E.; Miège, C.; Seigneurin-Berny, D.; Maréchal, E.; Block, M.; Dorne, A.-J.; Rolland, N.; Ajlani, G.; Douce, R. The biochemical machinery of plastid envelope membranes. *Plant Physiol.* **1998**, *118*, 715–723. [CrossRef]
34. Arao, T.; Yamada, M. Positional distribution of fatty acids in galactolipids of algae. *Phytochemistry* **1989**, *28*, 805–810. [CrossRef]
35. Keough, K.M.W.; Giffin, B.; Kariel, N. The influence of unsaturation on the phase transition temperatures of a series of heteroacid phosphatidylcholines containing twenty-carbon chains. *Biochim. Biophys. Acta* **1987**, *902*, 1–10. [CrossRef]
36. Kostetsky, E.Y.; Sanina, N.M.; Naumenko, N.V. The influence of fatty acid composition on the profile of the phase transition thermogram of phosphatidylcholine from holothurians *Cucumaria fraudatrix*. *Zh. Evol. Biochem. Physiol.* **1992**, *28*, 426–433.
37. Demé, B.; Cataye, C.; Block, M.A.; Maréchal, E.; Jouhet, J. Contribution of galactoglycerolipids to the 3-dimensional architecture of thylakoids. *FASEB J.* **2014**, *28*, 3373–3383. [CrossRef] [PubMed]
38. Austin, J.R.; Staehelin, A.L. Three-dimensional architecture of grana and stroma thylakoids of higher plants as determined by electron tomography. *Plant Physiol.* **2011**, *155*, 1601–1611. [CrossRef] [PubMed]
39. Kirchhoff, H.; Sharpe, R.M.; Herbstova, M.; Yarbrough, R.; Edwards, G.E. Differential mobility of pigment-protein complexes in granal and agranal thylakoid membranes of C_3 and C_4 plants. *Plant Physiol.* **2013**, *161*, 497–507. [CrossRef] [PubMed]
40. Vinogradova, K.L. *Manual of Algae of Far Eastern Seas of the USSR. Green Algae*; Nauka: Leningrad, Russia, 1979; pp. 46–110.
41. Bligh, E.G.; Dyer, W.I. A rapid method of total lipid extraction and purification. *Canad. J. Biochem. Physiol.* **1959**, *37*, 911–918. [CrossRef] [PubMed]
42. Vaskovsky, V.E.; Terekhova, T.A. HPTLC of phospholipids mixtures containing phosphatidylglycerol. *J. High Resolut. Chromatogr.* **1979**, *2*, 671–672. [CrossRef]
43. Vaskovsky, V.E.; Khotimchenko, S.V. HPTLC of polar lipids of algae and other plants. *J. Chromatogr.* **1982**, *5*, 635–636. [CrossRef]
44. Kostetsky, E.Y.; Velansky, P.V.; Sanina, N.M. Phase transition of phospholipids as criterion for assessing the capacity for thermal adaptation in fish. *Russ. J. Mar. Biol.* **2013**, *39*, 136–143. [CrossRef]

marine drugs

MDPI

Review

Antitumor Potential of Marine and Freshwater Lectins

Elena Catanzaro [1], Cinzia Calcabrini [1], Anupam Bishayee [2,*] and Carmela Fimognari [1,*]

1 Department for Life Quality Studies, Alma Mater Studiorum-Università di Bologna, Corso d'Augusto 237, 47921 Rimini, Italy; elena.catanzaro2@unibo.it (E.C.); cinzia.calcabrini@unibo.it (C.C.)

2 Lake Erie College of Osteopathic Medicine, Bradenton, FL 34211, USA

* Correspondence: abishayee@lecom.edu or abishayee@gmail.com (A.B.); carmela.fimognari@unibo.it (C.F.); Tel.: +1-941-782-5729 (A.B.); +39-0541-434658 (C.F.)

Received: 30 November 2019; Accepted: 18 December 2019; Published: 21 December 2019

Abstract: Often, even the most effective antineoplastic drugs currently used in clinic do not efficiently allow complete healing due to the related toxicity. The reason for the toxicity lies in the lack of selectivity for cancer cells of the vast majority of anticancer agents. Thus, the need for new potent anticancer compounds characterized by a better toxicological profile is compelling. Lectins belong to a particular class of non-immunogenic glycoproteins and have the characteristics to selectively bind specific sugar sequences on the surface of cells. This property is exploited to exclusively bind cancer cells and exert antitumor activity through the induction of different forms of regulated cell death and the inhibition of cancer cell proliferation. Thanks to the extraordinary biodiversity, marine environments represent a unique source of active natural compounds with anticancer potential. Several marine and freshwater organisms, ranging from the simplest alga to the most complex vertebrate, are amazingly enriched in these proteins. Remarkably, all studies gathered in this review show the impressive anticancer effect of each studied marine lectin combined with irrelevant toxicity in vitro and in vivo and pave the way to design clinical trials to assess the real antineoplastic potential of these promising proteins. It provides a concise and precise description of the experimental results, their interpretation as well as the experimental conclusions that can be drawn.

Keywords: marine lectins; cancer; cancer therapy; in vitro studies; in vivo studies; natural products

1. Introduction

Water covers 71% of the surface of our planet and hosts vibrant biodiversity. More than 2 million different species inhabit the seawater, many of which still await discovery [1]. Terrestrial natural compounds have always represented a source of biologically active molecules, while the marine environment is still a step behind. So far, the World Register of Marine Species checked and classified only 232,297 different species [2]. This number could be considered modest, but it mirrors how late oceans have started to be studied compared to land environments. Indeed, the spread of terrestrial natural history began during the 17th century, and escalated through the scientific travels of 18th and 19th centuries—including Charles Darwin's expeditions—until nowadays. On the contrary, it was not until the 20th century that ocean exploration started. It should be enough to underline that before the last century, technology was not sufficient to explore deep waters [3]. However, the number of discoveries is now exponentially increasing [4].

From the perspective of a pharmacologist, the marine biodiversity represents the ideal environment to quest for new substances with therapeutic potential. Indeed, since oceans give a home to many sessile or limited-mobility species, it is logical to assume that they synthesize metabolites to protect themselves and that those metabolites can be exploited for therapeutic uses [5]. Especially in anticancer research, the marine environment is an actual gold mine. By 2017, Food and Drug Administration of the United States approved seven marine-derived pharmaceutical drugs as anticancer agents, while many other compounds are currently being studied in different phases of clinical trials in oncology and hematology fields (4 in phase III, 6 in phase II, 8 in phase I) [6]. It has been estimated that 594,232 novel compounds of marine origin are waiting to be studied, assuming that between 55 and 214 could be approved as new anticancer drugs [4].

Nonetheless, if the natural world is so rich in bioactive molecules, why is the cure for cancer one of the hardest challenges scientists are facing for the last two centuries? First of all, it is incorrect to refer to cancer as a single disease. At least 200 different types of tumors exist. For instance, the organ where the tumor originates gives rise to specific neoplastic lesions; thus, breast cancer is different from lung cancer. Likewise, the type of cells from which cancer derives is crucial, and carcinomas derived from epithelial cells are different from lymphomas that derive from cells of the immune system. Furthermore, cancer is not quiescent, and the possibility to evolve and acquiring resistance to a full-blown effective therapy is very high [7]. Thus, it is pretty unlikely finding a single cure for all cancer types. Moreover, most of the anticancer agents used in clinic cause unacceptable toxicity that limits their compliance and effectiveness, hampering the positive outcome of cancer treatment. Since the cytotoxic effect of the antineoplastic drug also affects healthy cells, the primary cause of toxicity is the lack of selectivity towards tumors cells [8–10]. Thus, an interesting antitumor strategy would foresee a drug able to eradicate tumor cells distinguishing between them and healthy cells.

Therefore, the next question is: how do cancerous cells differ from normal ones? What all neoplastic cells have in common and that distinguish them from normal cells is the acquisition of different abilities. These abilities allow them to proliferate persistently, resist to cell death, circumvent growth restraint stimuli, reprogram energy metabolism, and escape the immune system recognition [11]. All the pathways involved in the acquisition of these capabilities make tumor cells unique and ideally can be exploited to obtain a therapy that selectively kills cancer cells. Targeting fast proliferating cells was one of the first strategies designed for this aim. However, some healthy tissues have a high physiological rate of proliferation and damage to normal cells cannot be avoided [12]. More, to survive in hostile environments, cancer cells deregulate cellular energetics. Indeed, reactive oxygen species (ROS) production is increased while the antioxidant machinery is often impaired. Thus, the basal oxidative stress level of cancer cells is higher than that of normal cells. By the consequence, enhancing ROS levels in cancer cells most likely provokes the selective eradication of these cells rather than normal cells [8,10,13]. In addition to oxidative stress, many other pathways, such as increased glycolysis (the Warburg effect), hypoxia and acidity have been targeted [9,11]. Nonetheless, hitting any one of these targets does not lead to a complete selectivity of action. Conversely, an effective strategy to identify and selectively hit cancer cells could be aiming at the phenotypic changes that come along with the malignancy transformation, such as the glycosylation pattern at the cell surface. In normal conditions, most proteins and lipids of eukaryotic and prokaryotic cells are coated by a sugar structure called glycocalyx. Glycocalyx plays a pivotal role in cell-cell recognition, communication, and intercellular adhesion. It is composed by mono-, oligo- or polysaccharides and changes in a predictable fashion throughout normal cell development. Glycosyltransferase enzymes, located in the Golgi apparatus, catalyze the construction of this carbohydrate structure. During malignant transformation, alterations in glycosyltransferases and their gene expression occur [14], and glycosylation profile of cells is irreversibly altered, conferring unique phenotypes to cancer cells [15]. For instance, protein fucosylation and sialylation, i.e., the addition of fucose moieties or sialic acid to proteins, is a common event in many cancer types [16]. The modifications of cell surface sugar patterns and in particular glycoproteins can be exploited to target cancer cells exclusively. Lectins are non-immunogenic glycoproteins that specifically bind unique carbohydrates residues on cell surfaces. Thus, they represent an attractive tool to detect and kill cancer cells [17].

Several studies have shown that plant and animal lectins exert different biological activities. These compounds inhibit bacterial [18] and fungal growth [19], and have immunomodulatory properties [20,21]. On cancer cell lines, they exhibit antiproliferative properties [20,22], promote apoptosis and autophagy, and block angiogenesis [20,22]. Moreover, their ability to bind to carbohydrates could make them useful for the detection of subtle modifications of carbohydrate composition, which appears during cancer transformation [23]. However, although anticancer activity of terrestrial lectins has been summarized in several reviews [20,24,25], to our knowledge no comprehensive report highlighting uniquely the antineoplastic activity of marine lectins has been written.

This review will explore the antitumor activity of lectins from both marine and freshwater organisms. A short overview of lectins will be followed by the presentation of the most studied lectins that showed overt anticancer potential. Diagnostic applications of lectins will not be discussed since it goes beyond the aim of this review.

2. Lectins

Lectins are expressed by animals, plants, fungi, and bacteria. They contain a domain for carbohydrate recognition that allows them to bind with carbohydrates without modification, i.e., they do not exhibit enzymatic activity on carbohydrate molecules. Since they reversibly and specifically bind carbohydrates, whether simple sugars or complex carbohydrates, lectins are capable of precipitating polysaccharides and glycoproteins or agglutinating cells [17].

Lectins can be classified using different criteria, including cellular localization, structural and evolutionary sequence similarities, taxonomic origin, carbohydrate recognition, function, and structure [26] (Figure 1).

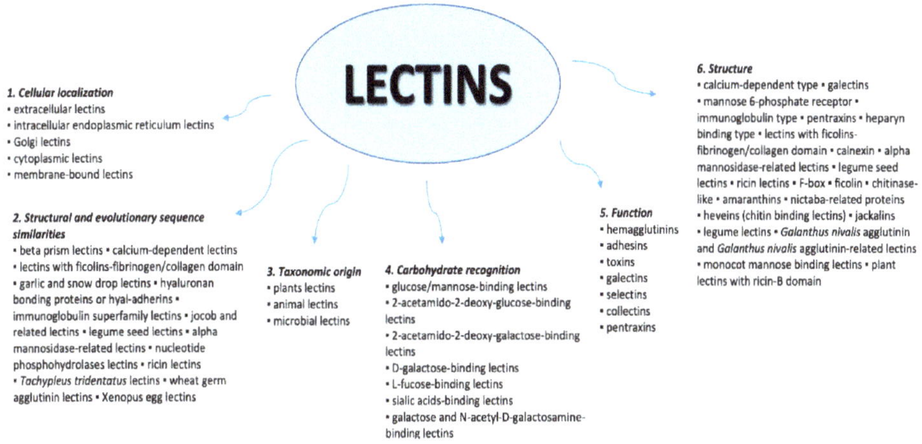

Figure 1. Classification of lectins.

Lectins play a pivotal role in cell-cell and host-pathogen interaction and communication, tissue development, sugar storage, and other mechanisms of cell survival and immune system stimulation. All endogenous lectins can be involved in both physiologic and pathologic processes [25]. Based on their characteristics, they can induce apoptosis and autophagy and inhibit angiogenesis, thus representing an interesting anticancer strategy [25]. Their mechanism of action consists of recognizing and reversibly binding specific carbohydrate moieties on cell surfaces. In particular, lectins bind to a specific sequence of monosaccharide moieties within glycosylated proteins, lipids, and glycans that are on cell surface. In humans, several types of cells express these proteins, from epithelial cells to antigen presenting ones and, depending on the type of lectin, they can be exposed on the cellular membrane or secreted into the extracellular matrix [27]. Each lectin usually has more than one binding site for the sugar units. As a consequence, they are able to bind different carbohydrates exposed on the surface of different cells, thus enabling cell-cell interaction [22]. Unique glycosylated-proteins expressed in tumor cells become the target of lectins that bind the characteristic glycan-chains. This process is the key to discriminate abnormal glycosylation moieties on the surface of cancer cell membranes [28]. Not all cancer types show the same glycosylation pattern, but they share common features. In particular, during the tumorigenesis, specific subsets of glycans on the cancer cell surface face different modifications, such as enrichment or decrease in their main components. The most recurrent changes on cell surface glycosylation patterns are enrichment in sialylation, branched-glycan structures, and the generation of the so-called "fucosylation core" [29,30]. For instance, the tetrasaccharide carbohydrate moiety sialyl-LewisX and sialyl-Tn antigen represent well-known markers of malignancies, since they are found in almost all cancer types [31]. Besides, O-linked and N-linked glycosylation create the most common branches on the tumor cell surface glycocalyx. For example, O-linked glycosylation is promoted by the entry of a GalNAc sequence, which can then be expanded with the addition of other different structures. Simultaneously, different blocks of sugars can be attached to the main branch through an amide group of an asparagine residue resulting in N-linked glycosylation. Among others, ovarian, skin, breast, and liver cancers show these types of modifications [30,32]. To complete the picture, core fucosylation, i.e., the enrichment in α1,6-fucose of the innermost GlcNAc residue of N-glycans, is a specific trait in many cancers, such as lung and breast cancer [29,33].

The recognition of glycosylated proteins on the outer cellular membrane represents only the first step necessary to trigger cancer cell death. After the binding with the carbohydrate moiety, lectins can be internalized, and there, in the intracellular environment, they trigger specific pathways that lead to cell death. Furthermore, taking these notions into consideration, exogenously induced expression of lectin inside the tumor cells could be another effective strategy to promote tumor eradication. For this reason, vaccinia viruses (VVs) have been exploited as replicating vectors harboring the genes needed for lectin expression and thus as an efficient way to transport lectins into cancer cells [34].

Marine and freshwater lectins show anticancer properties in vitro and in vivo thanks to the binding of cancer cell membranes, causing cytotoxicity, apoptosis, other forms of regulated cell death, and inhibition of tumor growth. A literature review was performed to identify publications on the antitumor potential of lectins isolated from water organisms such as invertebrates, vertebrates and tunicates, fish, and algae. The following paragraphs will present their activity as direct antitumor agents or through their delivery via VV.

3. Lectins from Marine and Freshwater Algae

Aquatic algae are a group of simple organisms containing chlorophyll. To sustain themselves, they almost only need light to carry out photosynthesis. They inhabit seas, rivers, lakes, soils, and grow on animal and plants as symbiotic partners. The complexity of these organisms ranges from the simplest unicellular to pluricellular entities that reorganize themselves to form simple tissues [28]. At least 44,000 species of algae have been identified and named [29], and among them 250 have been reported to contain lectins [3]. However, only a few of them have been adequately studied to investigate their antitumor potential (Table 1). For some of them, the only cytotoxic potential has been evaluated.

Table 1. Antineoplastic effects of lectins from algae based on in vitro and in vivo studies.

Origin	Lectin	Recognition Glycans	Cell Lines	In Vivo Model	Treatment Times and Doses	Cellular and Molecular Target	Ref.
Acanthophora spicifera	Crude lectins fraction		MCF-7		24 h (100 μg/mL)	% of cell growth inhibition in *A. spicifera*: 1.78 μg/mL (MCF-7); 4.27 μg/mL (HeLa)	[35]
Acrocystis nana			HeLa			% of cell growth inhibition in *A. nana*: 9.10 μg/mL (MCF-7); 47.68 (HeLa)	
Eucheuma serra	*Eucheuma serra* agglutinin (ESA)	Mannose	Colon26	Colon-26 cells injected in BALB/c mice	48 h (0–1000 μg/mL)	↓viability at concentrations > 8 μg/mL	[36]
					48 h (50 μg/mL)	% of AnnexinV$^+$/propidium iodide$^-$: 31.4%	
						↑caspase-3 activity	
					400 mg /200 mL PBS every 3 days up to 15 days (intravenously injection)	↓tumor volume	
						TUNEL-positive cells in tumor	
Eucheuma serra	*Eucheuma serra* agglutinin (ESA)	Mannose	OST		24 h (10–50 μg/mL)	Cell viability (50 μg/mL): 41.7 ± 12.3% (LM8); 54.7 ± 11.4% (OST)	[37]
			LM8		3–4 h (50 μg/mL)	AnnexinV$^+$/propidium iodide$^-$ (3 h): 68.2% (LM8); 74.8% (OST) AnnexinV$^+$/propidium iodide$^-$ (24 h): 24.1% (OST); 68.8% (LM8)	
					16 h (50 μg/mL)	OST: ↑caspase-3 activity (2.3-fold increase)	
	PEGylated vesicles with immobilized ESA (EPV)				24 h (1–5 μg/mL of ESA delivered by EPV)	Cell viability (1 μg/mL): ~50% (OST)	
Eucheuma serra	*Eucheuma serra* agglutinin (ESA)	Mannose	Colo201		72 h (0.05–150 μg/mL)	↓Viability at concentrations > 1.2 μg/mL (cancer cells)	[38]
			HeLa			No cytotoxicity at 10 μg/mL (MCF10-2A)	
			MCF-7				
					24 h (64 μg/mL)	DNA degradation (Colo201)	
			MCF10-2A		16 h (10.8 μg/mL)	↑caspase-3 activity (Colo201)	
Eucheuma serra	Span 80 vesicles containing immobilized ESA (EV) Span 80 vesicles containing DSPE-PEG2000 and immobilized ESA (EPV) Span 80 vesicles containing DSPE-PEG2000, immobilized ESA and entrapped ESA (EEPV)	Mannose	Colo201		0–24 h (54 μg/mL of ESA delivered by EV)	Cell viability (24 h): 17.2% (Colo201); no effect (MCF10-2A)	[39]
			MCF-7				
			Colon26		8 h (54 μg/mL of ESA delivered by EV)	DNA fragmentation in Colo201 and MCF-7	
			MCF10-2A				
				Colo201 cells transplanted in Balb/c-nu/nu mice	EPVs (containing 2.0 μg/mL of ESA) or EEPVs (containing 2.5 μg/mL of ESA) (0.01 mL/g b.w.) injected every 3 days up to 15 days	↓tumor volume (9th day): 51.1% (EEPV); 58.0% (EPV)	
					3 days after EPVs (0.01 mL/g b. w.) injection	TUNEL-positive cells around the blood vessels	

Table 1. *Cont.*

Origin	Lectin	Recognition Glycans	Cell Lines	In Vivo Model	Treatment Times and Doses	Cellular and Molecular Target	Ref.
Gloiocladia repens	Crude lectins fraction		MCF-7		24 h (100 μg/mL)	% of cell growth inhibition in G. repens: 14.19 μg/mL (HeLa); 28.54 μg/mL (MCF-7)	[35]
Helminthora divaricata			HeLa			% of cell growth inhibition in H. divaricata: HeLa cells: 3.63 μg/mL (HeLa); 12.25 μg/mL (MCF-7)	
Microcystis viridis	Recombinant *Microcystis viridis* lectin (R-MVL)	Mannose	HT-29		72 h (2–64 μg/mL)	IC50 [1]: 40.20 μg/mL (SCG-7904); 42.67 μg/mL (HepG2); 49.87 μg/mL (HT-29); 53.40 μg/mL (SKOV3)	[40]
			HepG2				
			SKOV3				
			SCG-7904				
Nitophylium punctatum	Crude lectins fraction		MCF-7		24 h (100 μg/mL)	% of cell growth inhibition: 2.97 (HeLa); 15.53 (MCF-7)	[35]
			HeLa				
Solieria filiformis	*Solieria filiformis* lectin (SfL) (mixture of isoforms 1 and 2)	Mannopentose	MCF-7		24 h (0–500 μg/mL)		[41]
			HDA				
					24 h (125 μg/mL)	AnnexinV⁺/propidium iodide⁻: 25.07%; AnnexinV⁺/propidium iodide⁺: 35.16%	
						↓*Bcl-2*; ↑*Bax*, ↑*caspase-3*, ↑caspase-*8*, ↑caspase-*9*	
Ulva pertusa	Adenovirus-*Ulva pertusa* lectin 1 (Ad-UPL1)	*N*-acetyl D-glucosamine	Huh7		48 h (100 MOI [2] Ad-UPL1 ± 10 μM U0126 [3])	Cell viability: ~50% (Ad-UPL1 + U0126, Huh7); ~90% (Ad-UPL1, BEL-7404 or Huh7)	[42]
			BEL-7404		48 h (50–100 MOI)	Huh-7: ↑pERK1/2, p-p38; ↓Akt	
						BEL-7404: ↑pERK1/2; ↓Akt	
						Huh-7: ↓Beclin-1; ↑LC3-II	
						BEL-7404: ↑Beclin-1; ↓LC3-II	

[1] Half maximal inhibitory concentration; [2] multiplicity of infections; [3] MEK 1/2 inhibitor.

The most characterized lectin of algae origin, the *Eucheuma serra* agglutinin (ESA), was extracted from the homonymous red macroalga. ESA is a mannose-binding lectin able to promote apoptosis on different cell lines and animal tumor models (Table 1). ESA amino acid structure is composed by four tandemly repeated motifs, each of them representing one binding site for mannose sequence. Specifically, each repeated motif binds specifically high mannose N-glycans with a minimum dimension of tetra- or penta-saccharide, such as Man(alpha1-3)Man(alpha1-6)Man(beta1-4)GlcNAc(beta1-4) GlcNAc [43].

ESA promotes cell death of many cancer cell lines, such as Colo201 (human colon adenocarcinoma), Colon26 (murine colon-carcinoma), HeLa (human cervix adenocarcinoma), MCF-7 (human breast adenocarcinoma), OST (human osteosarcoma), LM8 (murine osteosarcoma) [36–39]. In each of these cell lines, the mechanism of cell death is apoptosis, as demonstrated by DNA fragmentation, exposition of phosphatidylserine, and activation of caspase-3. The activity of ESA is tumor-type dependent and, comparing distinct studies, cervix adenocarcinoma and colon adenocarcinoma came back the most sensitive followed by osteosarcomas and breast cancer, which respond to a higher concentration of this lectin [36–38]. Since lectins, in general, have precise targets, this behavior probably reflects the different glycosylation pattern of different types of tumor. Certainly, the different glycosylation pattern between normal and cancer cells is behind the lack of ESA activity on non-transformed cells. Indeed, ESA did not affect the viability of normal fibroblasts and the non-tumorigenic epithelial MCF10-2A cell line [37,38]. Furthermore, their selectivity for tumor cells translates into lack of toxicity in vivo. For instance, ESA delayed the growth of Colon26 cells injected on BALB/c mice without affecting the body weight nor causing direct death, thus showing promising in vivo tolerability [36].

The ability of lectins to selectively target cancer cells can be exploited not merely to kill tumors, but also to deliver antitumor drugs on cancer cells. Furthermore, the antitumor activity of lectins could theoretically sum to that of the antitumor drug. With this aim, the development of a selective drug delivery system (DDS) was designed, and lipid vesicles resembling a microcapsule were tagged with ESA [38]. The microcapsules were made by sorbitan monooleate (Span80) with or without poly(ethylene glycol) (PEG) [37,39]. PEG was added in order to prolong the half-life of the vesicles compared to normal liposomes since PEGylation should decrease the reticuloendothelial uptake. First of all, it was demonstrated that both ESA-labelled DDSs target the exact same carbohydrate-sequence of free ESA, and that the drug transportation system does not abolish its cytotoxic activity, nor selectivity towards tumor cells [37,44]. ESA-immobilized lipid vesicles reached and bonded Colo201, HB4C5, OST, while no interaction was recorded with normal human fibroblasts and MCF10-2A [37–39] (Table 1). Indeed, all microcapsules directly exhibited pro-apoptotic activity on Colo201 and OST cells [37,39], while little effect was observed in normal MCF10-2A [39]. In vivo, the injection of the vesicles delayed tumor growth in nude mice bearing Colo201-derived tumors [39] (Table 1). No study directly compared the antitumor activity of the PEGylated vs. the not-PEGylated ESA-vesicles, while only in vivo bioavailability experiments were performed, showing no difference between the two systems [37,39]. In fact, after the injection of one or the other radioactive-labeled microcapsules in nude mice bearing Colo201 tumor cells, no difference in tumor accumulation nor uptake was recorded [39]. Finally, to understand if the DDS was able to transfer the encapsulated material into tumor cells, two different research groups [38,39] loaded them with the fluorophore fluorescein isothiocyanate (FITC) and treated cells with this complex. They demonstrated that FITC is internalized into tumor cells cytoplasm, both in vitro and in vivo, and showed the interesting potential of ESA-vesicles as a carrier of antitumor drugs as well as direct antineoplastic agents [38,39].

If ESA was not particularly active on MCF-7, two isolectins isolated from *Solieria filiformis* (Sfl) proved their ability to fight breast cancer (Table 1). Sfl-1 and Sfl-2 are mannopentose-binding lectins differing from each other for only 18% of the aminoacidic sequence. SfL-1 and SfL-2 consist of 268 amino acids in the form of four identical tandem-repeat protein domains. They both are constituted by two β-barrel-like domains made of five antiparallel β-strands, which are connected by a short peptide linker [41]. Despite the different biological activity of Sfls and ESA, both Sfl isoforms share at least 50% of their amino acid sequence with it [45] (Figure 2A). Thus, we can hypothesize that the similar structure allows them to both bind the same glycan structure, while the remaining part is responsible for the different cytotoxic activity.

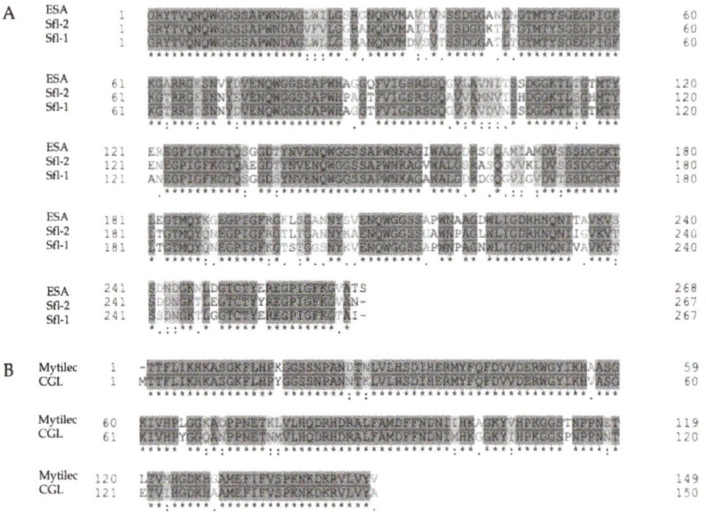

Figure 2. Comparison of amino acid sequence of ESA, Sfl-1 and Sfl-2 (**A**) and Mytilec and CGL (**B**). Data were analyzed for protein sequence similarity using BLAST technology, through the data base UniPROT [45]. "*" identical residues; ":" conserved substitution; "." Semi-conserved substitution; "_" gaps.

Both Sfls isoforms promote MCF-7 cell death to an equal extent, while they were not able to entirely kill primary human dermal fibroblasts. However, at a higher concentration than those promoting cell death, Sfls induce fibroblast proliferation [41]. Keeping these data in mind, the difference in the sensitivity of cancer and normal cells allows the identification of an interval of concentration toxic for the tumor ones. The mechanism of anticancer action of these lectins lies in the induction of apoptosis. Normally, apoptosis can be promoted by triggering the intrinsic and/or extrinsic pathway. The first one involves the B-cell lymphoma 2 (Bcl-2) protein family, mitochondrial permeability, and the creation of the apoptosome. The extrinsic pathway involves death receptors, death-inducing signaling complex and caspase activation. The two pathways share effector caspases. Due to congenital or acquired genetic alterations of cancer cells, one or, in the worst case, both pathways are often compromised [46]. Thus, the ability to activate both pathways maximizes the probability to promote programmed cell death successfully and minimizes the one to acquire resistance. Sfls upregulate the gene expression of the effector caspase-3 and trigger both the intrinsic [upregulation of the proapoptotic gene Bcl-2 associated X protein (Bax); downregulation of the anti-apoptotic gene Bcl-2; increased expression of caspase-9] and the extrinsic (increased expression of caspase-8) apoptotic pathways, qualifying themselves as very interesting anticancer agents. Further studies will be needed to understand if and how the 18% difference between the two isoforms affects their biological properties [41].

Ulva pertusa lectin 1 (UPL1) is an N-acetyl D-glucosamine-binding lectin that interacts with several intracellular pathways involved in proliferation and cell survival (Table 1). UPL1 primary structure does not share amino acid sequence similarity with any known plant or animal lectin. cDNA sequencing analysis showed that UPL1 is 1084 base-pair long and encodes for a premature protein of 203 amino acids. During post-translational modifications, 53 amino acids are lost and the N-terminal sequence of UPL1 starts at amino acid number 54 [47]. Intracellular delivery of UPL1 through a flag adenovirus does not exert an antiproliferative effect on the human papillomavirus-related endocervical adenocarcinoma BEL-7404, and hepatocyte-derived cellular carcinoma Huh7 cells. UPL1 mediates both the activation of the extracellular-signal-regulated kinase (ERK)-mitogen-activated protein kinase (MAPK), also known as MAPK/ERK kinase (MEK), signal and the phosphorylation of p38 MAPK. Since constitutive activation of ERK promotes cell survival and drug-resistance [47] and p38 MAPK can act as a survival factor, this could be the cause of UPL1 inability to stop cancer growth [42]. Indeed, the inhibition of each one of these pathways enhanced the antiproliferation activity of exogenous UPL1. Even if the cellular functions of tubulin in cancer is not clear [48], the ability of UPL1 to modulate MAPK pathways lies on its binding with β-tubulin. Probably, this critical protein is glycosylated in a specific manner providing recognition sites for the lectin. What is certain is that when the binding UPL1-tubulin was impeded, nor ERK-MEK or p38 MAPK were modulated, and UPL1 was able to kill cancer cells [42]. Moreover, only on Huh7, UPL1 triggered autophagy. If autophagy-associated cell death is considered a tumor suppressor characteristic, it can also enhance the survival of neoplastic cells exposed to metabolic stress and foster metastasis [49]. As a consequence, attention must be paid to assessing UPL1 real antitumor potential, alone or combined with survival signaling inhibitors, such as MAPK inhibitors [42].

Since lectins are such attractive pharmaceutical agents, the identification of new organisms able to produce them is compelling. However, the qualitative analysis of extracts could be time- and resource-consuming. Affinity electrophoresis or photometric assays are two tricky techniques used to carry out this task [48]. Additionally, an easy and fast method for understanding if a substance contains lectins is to evaluate its hemagglutination activity. Anam et al. [35] studied different crude extracts of red macroalgae (*Nitophylium punctatum, Acanthophora spicifera, Acrocystis nana, Helminthora divaricata* and *Gloiocladia repens*) and showed different degrees of hemagglutination activity. This activity, however, did not reflect the potency of their antitumor potential. All five crude extracts were able to kill MCF-7 and HeLa cells to a different extent (Table 1). They were all tested at a concentration of 100 μg/mL. *Acrocystis nana* extract was the second extract for hemagglutination activity, but the most potent in killing HeLa cells, inhibiting almost 50% of cell growth after 24 h. For MCF-7 cells, the extract more enriched in lectins (*Gloiocladia repens)* was the most cytotoxic with inhibition of almost 30% cell growth. All other extracts affected cell growth in an interval ranging between 2 and 15% [35]. These data, although not outstanding, may be deemed a starting point for the complete purification of the extracts in order to isolate the lectins. In general, consequential chromatographic steps are needed to purify lectins from crude extracts, such as affinity chromatography, ion-exchange chromatography, hydrophobic chromatography, and gel filtration chromatography [17]. In this case, the complete purification of the lectins from the crude extracts will clarify if lectins alone still have anticancer potential and whether that potential is higher than that of the extract. Anam et al. [35], who authored the study, suggest that extracts could contain impurities that dilute the samples and that purification should allow increasing the antitumor activity [35]. However, very often, extracts are more potent than the single molecules alone [49]. By the consequences, it will be interesting to see how these aspects pan out.

4. Lectins from Marine and Freshwater Invertebrates

Animals belonging to marine invertebrates have been exploited for their biological activity since the times of ancient Greece. Hippocrates and Galen, two pioneers of modern medicine, extensively narrated about dietary and pharmaceutical uses of shellfish, sponges or cephalopods, and their prescriptions containing marine invertebrates along with other ingredients have been found. Marine invertebrates consist of a large variety of organisms and have been categorized into over 30 different *phyla* [50]. They do not have an innate immune system nor develop an adaptive immune response against pathogens. Thus, as an immune defense they constitutively express small peptides that are induced upon danger is sensed. Lectins are among these proteins [51], and interestingly, they showed impressive antitumor activity both in vivo and in vitro (Table 2).

Table 2. Antineoplastic effects of lectins from invertebrates based on in vitro and in vivo studies.

Origin	Lectin	Recognition Glycans	Cell Lines	In Vivo Model	Treatment Times and Doses	Cellular and Molecular Target	Ref.
				Arthropoda and Mollusca			
Aplysia kurodai eggs	*Aplysia kurodai* egg lectin (AKL)	Galactose		Brine shrimp nauplii	24 h (2–32 μg/mL)	Mortality (32 μg/mL): 33.33% (PnL); 63.33% (AKL)	[52]
Crenomytilus grayanus	*Crenomytilus grayanus* lectin (CGL)	Galactose-lactosylceramide	MCF-7		24 h (50–200 μg/mL)	Cell viability: 33% at 200 μg/mL	[53]
Crenomytilus grayanus	*Crenomytilus grayanus* lectin (CGL)	Galactose-lactosylceramide	Raji		48 h (0–100 μg/mL)	IC50: 6.81 ± 0.83 μg/mL (Raji); unaffected (K562)	[54]
			K562		24 h (10 μg/mL CGL + 100 mM glucose, lactose, melibiose, raffinose or galactose)	Cell viability (Raji): ~55% (CGL + Lactose); ~70% (CGL + Glucose); ~100% (CGL + Melbiose, raffinose or galactose)	
					24 h (2.5–20 μg/mL)	↑% of cells in sub-G1 and G2/M phases and ↓G1 and S phases	
					24 h (2.5–20 μg/mL)	AnnexinV$^+$/propidium iodide (5 μg/mL): ~70%	
					24 h (2.5–10 μg/mL)	↑ caspase-9, caspase-3 and PARP cleavage	
Haliotis discus discus	*Oncolytic vaccinia* virus (oncoVV)-*Haliotis discus discus* sialic acid-binding lectin (HddSBL)	Sialic acid	C6	C6 tumor-bearing athymic BALB/c nude mice	60 days (10^7 pfu ^{1}/twice)	Mice survival: oncoVV-HddSBL > onco-VV	[55]
					15 days (10^7 pfu)	IL-2 secretion: oncoVV-HddSBL < onco-VV	
					24 h (5 MOI)	mRNA IL-2: oncoVV-HddSBL < onco-VV	
						NF-kB and AP-1 activity: oncoVV-HddSBL > onco-VV	
						IFIT2, IFIT3; DDX58: oncoVV-HddSBL < onco-VV	
					2–36 h (2 MOI)	OncoVV-HddSBL replication > onco-VV	
Haliotis discus discus	Adenovirus (Ad.FLAG)-*Haliotis discus discus* sialic acid binding lectin (Ad.FLAG-HddSBL)	Sialic acid	Hep3B		96 h (1–20 MOI)	Cell viability (20 MOI): ~40% (Hep3B); ~50% (A549); ~60% (H1299); ~80% (SW400)	[56]
			A549		48 h (20 MOI)	AnnexinV$^+$/propidium iodide (Hep3B): 19.8% (Ad.FLAG-HddSBL) vs 4.79% (Ad.FLAG)	
			H1299			↓Bcl-2;	
			SW480				
Haliotis discus discus	*Haliotis discus discus* sialic acid binding lectin (HddSBL) + coxsackie-adenovirus receptor (sCAR-HddSBL)	Sialic acid	K562/adr		48 h (5–30 MOI Ad-EGFP + 10 μg/mL sCAR-HddSBL)	Viral infection and replication: 13% sCAR-HddSBL vs 3.19% Ad-EGFP (K562/ADR); 48.6% sCAR-HddSBL vs 23.1% Ad-EGFP (U87MG)	[57]
			U87MG		96 h (8.2 MOI Ad-DIFBL + 10.6–31.8 μg/mL sCAR-HddSBL)	Cell viability (U87MG): ~40% (Ad-DIFBL); ~90% (31.8 μg/mL Ad-DIFBL-sCAR-HddSBL)	
					48 h (8.2 MOI Ad-DIFBL + 31.8 μg/mL sCAR-HddSBL)	U87MG: AnnexinV$^+$/propidium iodide$^-$ 10.2% (Ad-DIFBL-sCAR-DIFBL) vs 7.91% (Ad-DIFBL)	
					48 h (8.2 MOI Ad-DIFBL + 31.8 μg/mL sCAR-HddSBL)	U87MG: ↑pERK (Ad-DIFBL-sCAR-HddSBL); ↑E2F1 (sCAR-HddSBL; Ad-DIFBL-sCAR-HddSBL)	

Table 2. *Cont.*

Origin	Lectin	Recognition Glycans	Cell Lines	In Vivo Model	Treatment Times and Doses	Cellular and Molecular Target	Ref.
Ibacus novemdentatus	*N*-acetyl sugar-binding lectin (iNol)	*N*-acetylated glycan	MCF-7		48 h (0–100 μg/mL)	IC_{50}: 12.5 μg/mL (Caco2); 25 μg/mL (HeLa); 50 μg/mL (MCF-7); 100 μg/mL (TD47D)	[58]
			T47D		24 h (0–100 μg/mL)	HeLa: ↑caspase-9	
			HeLa		12–48h (100 μg/mL)	HeLa: ↑caspase-3 activity	
			Caco2		24 h (100 μg/mL)	HeLa: ↑DNA degradation; chromatin condensation	
Mytilus galloprovincialis	α-D-galactose-binding lectin (MytiLec)	Galactose-lactosylceramide	Raji		24 h (0-50 μg/mL)	Cell viability (50 μg/mL): ~40% (Raji); ~100% (K562)	[59]
			K562		24 h (20 μg/mL MytiLec + 100 mM Sucrose, Lactose or Melbiose)	Cell viability (Raji): ~40% (MytiLec + Sucrose or Lactose); ~100% (MytiLec + Melbiose)	
Mytilus galloprovincialis	α-D-galactose-binding lectin (MytiLec)	Galactose-lactosylceramide	Ramos		24 h (0.5–50 μg/mL)	Cell viability (50 μg/mL): ~45% (Raji); ~100% (K562)	[60]
			K562		12–24 h (0.5–50 μg/mL)	↑pMEK, pERK and p21; ↓CDK6, ↓cyclinD3	
					12–24 h (20 μg/mL)	↑pJNK, ↑pp38, ↑pERK	
					12–24 h (20 μg/mL)	↑caspase-9, ↑caspase-3, ↑TNFα	
					12–24 h (20 μg/mL + 10 μM U0126- pMEK inhibitor)	↑caspase-9, ↑caspase-3	
Perinereis nuntia	*Perinereis nuntia* lectin (PnL)	Galactose		Brine shrimp nauplii	24 h (2–32 μg/mL)	Mortality (32 μg/mL): 33.33%	[52]
Strongylocentrotus purpuratus	Adenovirus FLAG (Ad.FLAG)-*Strongylocentrotus purpuratus* rhamnose-binding lectin (SpRBL)	Rhamnose	Hep3B		48–96 h (1–20 MOI)	Cell viability (20MOI; 96 h; Ad.FLAG-SpRBL): ~20% (Hep3B); ~30% (BEL-7404, A549); ~40% (SW480)	[61]
			BEL-7404		48 h (20 MOI)	Hep3B: Annexin V^+/propidium iodide$^-$: 25.4% (Ad.FLAG-SpRBL) vs 1.35% (Ad.FLAG)	
			A549			Hep3B: = cleaved PARP; ↓Bcl-2, ↓XIAP	
			SW480		48 h (20 MOI)	Hep3B: ↓E2F-1	
Tachypleus tridentatus	*Oncolytic vaccinia* virus (oncoVV)-*Tachypleus tridentatus* Lectin (TTL)	Rhamnose	MHCC97-H	MHCC97-H tumor-bearing athymic BALB/c nude mice	44 days (10^7 pfu/twice)	↓tumor volume	[55]
					36 h (5 MOI)	OncoVV-TTL replication > onco VV	
			BEL-7404		24 h (5 MOI)	↑pERK (onco VV = oncoVV-TTL)	
						MAVS, IFI16, IFNβ: ↑ (oncoVV); = (onco VV-TTL)	
					(5 MOI ± U0126)	U0126 ↓ oncoVV-TTL replication	
				Chordata			
Didemnum ternatanum	*Didemnum ternatanum* lectin (DTL)	N-acetyl-D-glucosamine	HeLa in adhesion plates		72 h (2.5 μg/mL)	Cell proliferation: ~50%	[62]
			HeLa in soft agar		2 weeks (2.5 μg/mL)	Colony formation: 7 ± 1 (control); 25 ± 2 (DTL in agar); 60 ± 4 (DTL in plates and in agar)	

Table 2. *Cont.*

Origin	Lectin	Recognition Glycans	Cell Lines	In Vivo Model	Treatment Times and Doses	Cellular and Molecular Target	Ref.
				Porifera			
Aphrocallistes vastus	*Oncolytic vaccinia* virus (oncoVV)-*Aphrocallistes vastus* lectin (AVL) Adenovirus (Ad)-*Aphrocallistes vastus* lectin (AVL)	Galactose	HCT116	BEL-7404 or HCT116 tumor-bearing athymic BALB/c nude mice	48–72 h (10–20 MOI Ad-AVL)	Cell Viability (20 MOI Ad-AVL; 72h): ~40% (HCT116, U251); ~50% (HT-29, MHCC97-H, BEL-7404)	[63]
			U251		48-72 h (1-10 MOI Ad-AVL)	Cell Viability (5 MOI oncoVV-AVL; 72h): ~20% (HCT116); ~40% (U87, 4T1-LUC, BEL-7404)	
			BEL-7404		2–36 h (5 MOI)	OncoVV-AVL replication > onco VV	
			MHCC97-H		24 h (2 MOI)	AnnexinV$^+$/propidium iodide (HCT116): 6.49% (oncoVV-AVL) vs 1.26% (oncoVV)	
			HT-29		24 h (2 MOI) on HCT116	MDA5: no effect	
			4T1-LUC			↓caspase-3 (oncoVV-AVL); ↓caspase-8, ↓Bax (oncoVV, oncoVV-AVL)	
			U87			↓NIK, pNF-κB2 (oncoVV-AVL); ↑NIK (oncoVV)	
						↑pERK (onco VV = oncoVV-TTL)	
					24 h (5 MOI ± 10 µM U0126)	U0126 ↓ oncoVV-AVL replication	
					25 (BEL-7404) or 35 (HCT116) days (10^7 pfu)	↓tumor volume	
Cinachyrella apion	Lactose-Binding Lectin (CaL)	Lactose	HeLa		24-48 h (0.5 - 10 µg/mL)	Cell Viability (10 µg/mL; 24h): ~50% (HeLa); ~60% (PC3); ~75% (3T3)	[64]
			PC3		(10 - 20 µg/mL)	No cytotoxicity in peripheral blood cells	
			3T3		1 h (10 µg/mL)	No hemolysis in erythrocytes	
			Erythrocytes and peripheral blood cells		24 h (10 µg/mL)	Membrane blebbing and nuclear condensation	
					24 h (10 µg/mL ± 0.02 mM Z-VAD-FMK [2])	% of cells in S phase (HeLa): ~40% (control); ~ 50% (CaL + Z-VAD-FMK); 57.6% (CaL)	
						AnnexinV$^+$/propidium iodide (HeLa): 3.84% (control); 15.5% (CaL + Z-VAD-FMK); 23.2% (CaL)	
					6–24 h (10 µg/mL)	HeLa: ↑Bax, ↑pNF-κB (105 kDa), ↑JNK; =Bcl2, =pAKT; ↓pNFkB (50 kDa)	

Table 2. *Cont.*

Origin	Lectin	Recognition Glycans	Cell Lines	In Vivo Model	Treatment Times and Doses	Cellular and Molecular Target	Ref.
Cliona varians	*Cliona varians* lectin (CvL)	Galactose	Jurkat		72 h (1–150 μg/mL)	IC50: 70 μg/mL (K562); 100 μg/mL (Jurkat); no effect on lymphocytes	[65]
			K562		24 h (1–150 μg/mL)	No effect (B16, 786-O, PC3)	
			blood lymphocytes		72 h (70 μg/mL)	K562: ↑subG1 (28% CvL vs 14.1% control)	
			B16		72 h (50–70 μg/mL)	Apoptotic cells (K562; 70 μg/mL): 43% CvL vs 10% control	
			786-O		72 h (70 μg/mL)	K562: 25.3% (Annexin V-/propidium iodide+); 60.4% (Annexin V$^+$/propidium iodide$^+$);	
			PC3		72 h (50–70 μg/mL)	No increase in caspase-8, -9, and -3 activity	
					72 h (70 μg/mL)	Cathepsin B founded in cytoplasm and nucleus	
					2 h (5 μM E-64) + 72 h (50-80 μg/mL CvL)	Cell viability (80 μg/mL, K562): ~30% (CvL); ~100% (CvL + E-64)	
					72 h (50–70 μg/mL)	K562: ↑TNFR1, ↓NF-κB (p65 sub)	
						K562: ↑Bax, ↑Bcl-2	
						K562: ↑p21, ↓pRb	
Haliclona caerulea	Halilectin-3 (H3)	Mucin	MCF7		6-48 h (7.81–500 μg/mL)	Cell viability (250 μg/mL): 42% (MCF7); 75% (HDF); IC50: 100 μg/ml (MCF7)	[66]
			HDF		24 h (100 μg/mL)	↑% of cells in the G1 phase	
					24-48 h (100 μg/mL)	↑early apoptosis cells: 46% (24h); 55.4% (48h)	
					6-24 h (100 μg/mL)	24h: ↑caspase-3, ↑caspase-8, ↑caspase-9, ↑Bax, ↑TP53; ↓Bcl-2	
					8-24 h (100 μg/mL)	↑agglutination of MCF-7 cells; ↓cell adhesion	
					6 h (100 μg/mL)	↑LC3; ↓BECLIN-1	
						↑LC3II/LC3I	
						Autophagosoma vescicles	
Haliclona cratera	*Haliclona cratera* Lectin (HCL)	Galactose, *N*-Acetyl-D-galactosamine, Lactose	HeLa		48 h (0–40 μg/mL)	IC50: 9 μg/mL (HeLa); 11 μg/mL (FemX)	[67]
			FemX		72 h (0–15 μg/mL)	Lymphocytes: no toxicity	
			human T-lymphocytes		2 h (5 μg/mL PHA) + 72 h (0–15 μg/mL HCL)	Lymphocytes: 23% (PHA + HCL 15 μg/mL)	

Table 2. *Cont.*

Origin	Lectin	Recognition Glycans	Cell Lines	In Vivo Model	Treatment Times and Doses	Cellular and Molecular Target	Ref.
Halichondria okadai	18 kDa Lectin (HOL-18)	*N*-acetylhexosamine	Jurkat		24 h (1–25 µg/mL)	Cell Viability (25 µg/mL): ~30% (Jurkat); ~60% (K562)	[68]
			K562		24 h (25 µg/mL HOL-18 ± 50 mM D-GlcNAc [3], D-GalNAc [4] or Mannose)	Cell Viability: ~30% (Jurkat) - 50% (K562) (HOL-18 + Mannose); ~80% (HOL-18 +D-GalNAc); ~90% (HOL-18 + D-GlcNAc)	
Halichondria okadai	18 kDa Lectin (HOL-18)	*N*-acetylhexosamine	HeLa		48 h (6.25–100 µg/mL)	IC50: 40 µg/mL (HeLa); 52 µg/mL (MCF7); 63 µg/mL (T47D); no effect (Caco-2)	[69]
			MCF7		48 h (50 µg/mL HOL-18 + 20 mM Glucose, GlcNAc, Mannose, ManNAc [5])	Cell Viability: ~45% (HOL-18 ± Glucose or Mannose); ~75% (HOL-18 ± GlcNAC); ~90% (HOL-18 ± ManNAC)	
			T47D		48 h (6.25–100 µg/mL)	HeLa: ↑pERK, ↑caspase-3	
			Caco2				

[1] Plaque-forming unit; [2] carbobenzoxy-valyl-alanyl-aspartyl-[*O*-methyl]-fluoromethylketone; [3] *N*-acetyl D-glucosamine (D-GlcNAc); [4] *N*-acetyl D-galactosamine (D-GalNAc); [5] *N*-acetyl D-Mannosamine (D-ManNAc).

4.1. *Phylum Mollusca and Arthropoda*

The phylum Mollusca represents the second biggest *phylum* on earth. Approximately 90% of mollusks fall into the Gastropoda class, followed by Bivalves and Cephalopods [70]. Shelled mollusks are the most widely used in traditional medicine and together with the crustaceous (*phylum* Arthropoda) fall into the shellfish family [71]. Several studies revealed that the richness in hepato-pancreas mass, the characteristic open circulatory system, the filtering abilities and the shell arrangement make bivalves and crustaceans a remarkable source of molecules [72], such as the unique lectin proteins that can be exploited as therapeutic agents.

Crenomytilus grayanus lectin (CGL) is an intriguing lectin isolated from the homonymous bivalve belonging to the Mytilidae family. It consists of three highly similar tandem sequences of amino acids for a total of 150 residues. Secondary structure envisages a predominance of β-structure that sometimes alternates with α-ones [73]. CGL has a particular structure that makes it able to specifically recognize the globotriaosylceramide (Gb3) resulting in Gb3-dependent cytotoxicity. Gb3 is a globoside, and thus a non-protein cluster of differentiation composed by a galactose-lactosylceramide sequence [74]. It is overexpressed in several human tumors with intrinsic or acquired multidrug resistance [75]. CGL blocks cell proliferation and promotes cell death on Gb3-expressing tumor cells, such as Raji cells (Burkitt's lymphoma) and in a lesser extent MCF-7 (breast carcinoma) (Table 2). No effect was recorded on K562, a human myelogenous leukemia cell line not expressing Gb3 [53,54]. Furthermore, on Raji cells, the CGL cytotoxicity was completely hindered when Gb3 was enzymatically degraded [54]. On the same cell line, CGL promoted a G_2/M cell-cycle arrest that led to apoptosis, as demonstrated by phosphatidylserine externalization, activation of caspases-3 and caspase-9, and poly (ADP-ribose) polymerase (PARP) cleavage [54]. Even though further studies are needed to confirm the following assumption, the high expression of Gb3 on some cancer cells and the very specific action of CGL for these cells bodes well for a selective activity of this lectin towards neoplastic lesions and let us presume that it would bring to lack of toxicity in vivo.

Mytilus galloprovincialis, better known as Mediterranean mussel, contains a lectin, Mytilec, that shares a similar behavior with CGL. No wonder, they share 50% in amino acid sequence and bind the same glycan moiety [45] (Figure 2B; Table 2). Mytilec is incorporated into cells through the interaction with Gb3 and, thanks to that, it promotes cytotoxic effects. As CGL, it exhibits antitumor activity only on Gb3-expressing cell lines, such as Raji's and Ramos, another Burkitt's lymphoma cell line (Table 2), while no effect was induced on K562 [35,36]. The mechanism of action of Mytilec has been investigated in Ramos cells, where it promotes apoptosis and activates all MAPK pathways. MAPK system consists of three sequentially activated protein kinases that play a central role on different transduction pathways. These pathways are involved in processes, such as cell proliferation, differentiation, and cell death in eukaryotes [76,77]. Precise extracellular stimuli elicit the phosphorylation cascade and provoke the activation of a MAPK by the consecutive activation of a MAPK kinase kinase (MAPKKK) and a MAPK kinase (MAPKK). Briefly, after the stimulus is inferred, MAPKKK phosphorylates and triggers MAPKK, which, in turn, activates MAPK. Here, MAPK phosphorylates different substrates in the nucleus and cytoplasm, inducing variations in protein function and gene expression that deliver the proper biological response. Depending on which stimulus triggered the MAPK system in the first place and the intended biological effect, three leading families of MAPK can be activated. Growth factors and mitogens activate ERK, which controls cell survival, growth, differentiation and development; stress, inflammatory cytokines and, once more, growth factors prompt c-Jun *N*-terminal kinases (JNKs) and p38. These two latter MAPKs, in turn, deal with inflammatory response, apoptosis, cell growth, and differentiation [78]. Mytilec activates all these three pathways, but only in Gb3-expressing cell lines. The activation of the ERK pathway increased the levels of tumor necrosis factor-α (TNF-α) and the cell-cycle inhibitor p21, but it was excluded to be the cause of apoptosis induction. The role of JNK and p38 on the pro-apoptotic properties of Mytilic has still to be verified [60].

The gonads of the adult giant marine *Aplysia*, a sea slug belonging to Gastropoda class, contains the corresponding lectin: *Aplysia gonad* lectin (AGT). It is a Ca^{2+}-dependent lectin and is composed of two 32-33 kDa subunits [79]. Although the majority of studies about AGT describe its chemical characteristics, this galacturonic acid-galactose-lactose-binding lectin has been injected on mice and it was found able to abolish tumor appearance when tested at 30 μg and decrease lung tumor burden of 60% when tested at 5 μg, compared to untreated animals [79,80]. However, no information is available on treatment time and route of administration [80].

iNol is a very big lectin physiologically synthesized by the Arthropoda slipper lobster (*Ibacus novemdentatus*) probably to destroy pathogens. iNol consists of five subunits, which are composed by 70-, 40-, or 30-kDa polypeptides that are held together by disulfide bonds. Under physiological conditions, it has a polygonal structure. Despite the high molecular weight, this lectin is incorporated into mammalian cells, such as HeLa, through endocytosis thanks to the binding with N-acetylated glycan moieties on the cell surface. iNol kills HeLa, MCF-7, T47D (breast cancer) and Caco-2 (colon cancer) cells (Table 2). HeLa cells were the most sensitive to iNol activity, so they were used to understand its apoptotic potential. iNol promotes DNA fragmentation and activation of caspase-3 and caspase-9 only if the lectin is able to bind its carbohydrate-ligand, demonstrating that its antitumor potential is strictly linked to its lectin nature [58].

Chinese or Japanese horseshoe crab or tri-spine horseshoe crab are the common names of *Tachypleus tridentatus*, an Arthropoda resembling a crab, but more closely related to spiders and scorpions. Despite the not gracious description, *Tachypleus tridentatus* represents a source of several lectins, among which the rhamnose-binding Tachypleus tridentatus lectin (TTL). TTL characterization has not been completely clarified. Different studies agree on the fact that this lectin is a multimer, but it is not clear if it is composted by hexamers and octamers [81] or tetramers [82]. The antitumor activity of TTL was assessed in vivo, after its genomic insertion into an oncolytic VV (oncoVV-TTL). Balb/c nude mice were subcutaneously engrafted with MHCC97-H liver cancer cells and then treated with oncoVV-TTL. OncoVV-TTL was able to replicate inside tumor cells and significantly reduced tumor growth compared to onvoVV-only treated mice [55] (Table 2).

Another replication-deficient adenovirus was modified to encode the gene of *Haliotis discus discus* sialic acid-binding lectin (Ad.FLAG-HddSBL). HddSBL is an another Ca^{2+}-dependent binding lectin. It has 151 amino acid residues, but so far the glycosylation site has not been identified yet [83].

Ad.FLAG-HddSBL promoted cell death of hepatocellular carcinoma (Hep3B), lung cancer (A549 and H1299), and colorectal carcinoma (SW480) in a tumor-dependent manner, as the different sensitivity of these cell lines to this compound suggests [56]. Hep3B resulted in the most sensitive cell line to Ad.FLAG-HddSBL, followed by A549, H1299 and last SW480 (Table 2). Yang et al. [56], who authored the study, suggested that the different intracellular metabolism of sialic acids is the cause of the different activity of Ad.FLAG-HddSBL. On Hep3B, Ad.FLAG-HddSBL's mechanism of action has been investigated. It promotes a regulated form of cell death, since it triggered the mobilization of annexin-V to the outer cellular membrane and decreased the expression of the anti-apoptotic factors Bcl-2 and X-linked inhibitor of apoptosis protein (XIAP). Curiously, it did not affect PARP expression, and thereby it does not activate caspases at the analyzed time point [56]. Then, two possibilities arise: either it activates PARP at a different time point, or it promoted one of the so-called non-canonical cell deaths, which are caspase-independent forms of regulated cell death.

4.2. *Phylum Porifera*

The name Porifera means "pore bearer", and it is the most characteristic feature of sponges. Indeed, they are invertebrates sessile characterized by the lack of digestive, nervous, circulatory, and immune system. To get food, oxygen, and to reject all discards, they maintain a continuous water circulation in their body. Sponges do not have a physical barrier that protects them from predators, and the only way to survive is to produce secondary toxic metabolites. Furthermore, Porifera *phylum* is characterized by a unique biodiversity that results in a vast diversity of metabolites. Several studies have shown the chemopreventive and antitumor properties of many of them [5].

Two lectins originated from different species of the genus *Haliclona* showed an outstanding antitumor potential. *Haliclona cratera* lectin (HCL) was extracted, purified and tested for antitumor activity on human cervical and melanoma cell lines. It displayed cytotoxicity on both tumor models, which had almost the same sensitivity (half maximal inhibitory concentration 9 μg/mL vs. 11 μg/mL, respectively) [67] (Table 2). Another species, *Haliclona caerulea*, produces a peculiar lectin called halilectin-3 (H3). Its primary structure shows no similarity to any other animal lectin and consists of 251 amino acids, of which 145 arrange themselves into an α-chain and 106 into two β-chains. Alongside, quaternary structure has a heterotrimer conformation stabilized by disulfide bonds [84], which differs from the dimeric [67], trimeric [85], tetrameric [86] or multimeric [87] structure of all other lectin originated from sponges characterized so far. H3 is able to promote cancer cell death exploiting different mechanisms. On MCF-7 cells, H3 triggers both intrinsic and extrinsic pathways of apoptosis promoting an early upregulation of p53 and inhibits cell proliferation causing an accumulation of cells in the G1 phase. In addition to apoptosis, autophagy-associated cell death has been recorded after a few hours of MCF-7 treatment with H3. In particular, H3 upregulated the microtubule-associated protein 1A/1B-light chain 3 (LC3) expression, promoting the accumulation of LC3-II (Table 2). Contextually, autophagosome vesicles have been identified microscopically [66]. The authors of the study [40] hypothesize that both apoptosis and autophagy can be triggered thanks to Bcl-2 downregulation or p53 upregulation. To complete the picture, H3 has been found able to promote anoikis, a particular form of cell death that happens when cells are stimulated to detach from the extracellular matrix [66]. In normal conditions, cells need to adhere to the tissue where they grow since the essential growth factors and other survival signals are provided by the extracellular matrix and proximal cells [88]. H3 reduced the expression of integrin α6β1 and interacted with integrin α5β1, the fibronectin receptor, thus impairing MCF-7 adhesion and promoting cell death [66]. The involvement of apoptosis, autophagy, and anoikis in the cytotoxicity of H3 makes it a very promising antitumor lectin, able to fight cancer on several fronts and thus reduce drug resistance.

A lactose-binding lectin isolated from the sponge *Cinachyrella apion* (Cal) exhibited antitumor activity on human cervical (HeLa) and prostate adenocarcinoma (PC-3) cells, while milder cytotoxicity was recorded on non-transformed 3T3 mouse fibroblasts (Table 2). Cal primary structure consists of eight subunits of 15.5 kDa assembled by hydrophobic interactions and does not need Ca^{2+}, Mg^{2+}, and Mn^{2+} ions to bind sugars [87]. On HeLa, it promoted the typic phenotype changes of early apoptosis and modulated some of the pro-apoptotic proteins of the intrinsic pathway [64]. However, the inhibition of caspase activation did not significantly reduce its cytotoxicity. This suggests that apoptosis is not the leading mechanism through which Cal promotes cytotoxicity. Alternatively, the inhibition of the apoptotic pathway could activate a different pattern of regulated cell death [89]. Further studies are needed to untangle this knot.

Hol-18 is an *N*-acetylhexosamine-binding lectin isolated from the Japanese black sponge *Halicondria okadai*. It is a 72 kDa tetrameric lectin organized into four non-covalently bonded 18 kDa subunits [68]. Few studies demonstrated its newsworthy antitumor potential on Jurkat (acute T cell leukemia), K562 [68], HeLa, MCF-7 and T47D cells (Table 2). On Hela, MCF-7, and T47D, Hol-18 promoted apoptosis and activated the MAPK-ERK signal [69], showing once again how this pathway is often involved in marine lectin mechanism of action. No effect was recorded on Caco-2 cells since these cells were not able to internalize Hol-18 at all [69].

The recombinant oncoVV-AVL is the result of the insertion of the *Aphrocallistes vastus* lectin (AVL) into an oncolytic VV. AVL is a 34 kDa glycoprotein with a 24 kDa proteinaceous core *i.e.*, the lectin core. It counts 191 amino acids and is a Ca^{2+}-dependent lectin. It is characterized by vast hydrophobic sequences that allow it to bind the target cell membranes and facilitates the interaction with sugars [90]. The insertion of this lectin in the oncolityc VV allowed its exogenous expression into infected cells. It promoted cell death on several tumor cell types (Table 2), such colorectal cancer, glioma, and hepatocellular carcinoma cell lines. The mechanisms of cytotoxic activity of OncoVV-AVL have been investigated only on HCT116 colorectal carcinoma cells. It promoted regulated cell death but without the cleavage of caspase-8 nor the modulation of the pro-apoptotic Bax protein. Digging deeper into oncoVV-AVL mechanism of action, it has been demonstrated that ERK activation is necessary to allow virus replication. Since the virus replication is necessary to express the AVL lectin, it can be postulated that ERK is essential for oncoVV-AVL activity. OncoVV-AVL was tested also on Balb/c nude mice bearing BEL-7404- or HCT116-derived tumors. In both cases, oncoVV-AVL efficiently inhibited tumor growth [63] (Table 2).

Queiroz et al. [65] described the cytotoxic activity of the *Cliona varians* lectin (CVL) on different tumor cell lines and healthy cells. CVL is a Ca^{2+}-dependent glycoprotein. It is a tetramer of 114 kDa composed by different subunits of 28 kDa linked by disulphide bridges [91]. CVL killed K562 and Jurkat cells, while all solid tumor cell models (melanoma, renal carcinoma, and prostate tumor) and normal human peripheral blood lymphocytes were found to be insensitive to CVL (Table 2). On K562, it increased p21 and repressed the expression of pRb (retinoblastoma protein), suggesting the ability to block the cell cycle. More, CVL promoted morphological changes typical of apoptosis while cytofluorimetric analysis showed a scenario composed by the presence of both primary necrotic and late apoptotic cells. No caspase-3, caspase-8 or caspase-9 activation has been recorded, but CVL stimulated the translocation of cathepsins B (CTPB) from the vesicular compartment to the cytoplasm [65]. The role of the lysosomal cysteine protease CTPB in cell death is intriguing but not yet fully understood. CTPB modulates both necrosis and regulated cell-death processes in a tumor-type fashion. In some cell lines, it acts as an initiator of the intrinsic apoptotic pathway, modulates Bcl-2 family proteins such as Bax, Bcl-2, and Bid, and activates the caspases cascade [92,93]. In others, rather than activate effector caspases, it represents a caspase mediator, triggering some of the morphological changes of apoptosis [93]. For example, the release of CTPB enhances the release of cytochrome c and the caspase activation in hepatocytes treated with TNF [93], while on fibrosarcoma cells exposed to the same stimulus CTPB operates as a downstream mediator of the caspase apoptosis cascade [94]. CTPB is also involved in a caspase-independent form of regulated cell death, i.e., necroptosis. Necroptosis shares with necrosis the morphological features but, in contrast to the latter phenomenon, defined molecular pathways drive it. Necroptosis is triggered through the activation of death receptors such as tumor necrosis factor receptor 1 (TNFR1). TNFR1 promotes the formation of the so-called necrosome, a complex formed by receptor-interacting protein kinase 1 (RIP1)–RIP3–mixed lineage kinase domain-like protein (MLKL) [95]. The complex activates downstream events such as CTPB mobilization, which actively promotes programmed necrosis [96]. In CVL-mediated activity, the ablation of CTPB restored almost entirely cell viability, while TNFR1 was upregulated and the subunit p65 of the nuclear factor kB (Nf-kB) was downregulated. The authors of the study [65] suggest that this latter event could be the cause of the increase in both the pro- and anti-apoptotic Bax and Bcl-2 proteins. Even if the Bax/Bcl-2 protein ratio is not known, the authors still suggest that it could be the cause of what they call "caspase-independent apoptosis" [65]. Taken together, these data suggest the ability of CVL to promote a CTPB-mediated programmed form of cell death. Annexin V exposure—a hallmark of the early stages of regulated cell death—the upregulation of TNFR1, and the complete absence of caspase activation recall necroptosis, but further studies are needed to assess if apoptosis, regulated necrosis, or both events trigger CVL-mediated cell death.

4.3. Phylum Chordata

Chordata is another marine filter-feeding phylum. They have tubular openings through which water goes in and out. They are sessile and very often fixed to rocks or similar surfaces [97]. Tunicates, in particular, represent a source of didemins, such as aplidine and trabectin, molecules with anticancer, antiviral, anti-inflammatory, and immunosuppressive potential [98–100]. Furthermore, they contain lectins.

The most compelling tunicates' lectin is the N-acetyl-D-glucosamine-binding *Didemnum ternatanum* lectin (DTL). DTL is a homotrimer which activity is not dependent by Ca^{2+} and Mg^{2+}. It contains relatively high amounts of Gly, Ala, Asx, Glx Leu, Val residues and, as for the other lectin isolated from marine sponges, it is composed in part by carbohydrate (around 1.3%) [101]. Since one of the most critical functions of lectins is the cell-matrix interaction, the antitumor potential of DTL on HeLa cells has been studied in different microenvironment conditions. DTL showed different behavior on tumor cells depending on anchorage status (Table 2). In anchorage-independent conditions (cells cultivated in soft agar), DTL promotes cell growth, while in normal adhesion conditions (cells cultivated in adhesion plates) it inhibits cell proliferation and triggers cell differentiation in a way that promotes cell attachment [62]. Thus, cell microenvironment plays a key role in the DTL's activity. Taking this in mind, an interesting approach to better predict the anticancer potential of lectins could be using tumor 3D cultures built with specific bioreactors. Standard in vitro models do not capture the complex tumor biology and do not consider cell-to-cell and cell-to-matrix interactions. Perfusion-based bioreactor systems create heterogeneous cell populations, and an optimal physiological cell-cell and cell-extracellular matrix interactions, perfectly miming tumor microenvironment [102,103].

5. Lectins from Marine and Freshwater Vertebrates

5.1. Amphibians

Amphibians are ectothermic, tetrapod vertebrates that count several species worldwide, except Antarctica [104]. Behavioral as well as biochemical, physiological, and morphological adaptations make them able to survive in such different habitats. Driven by the knowledge that amphibians produce several metabolites mainly as an essential self-defense strategy, recently drug discovery dip into them. Many peptides with anticancer activities have been found in amphibian skin or oocyte cells and eggs (Table 3) [105].

Table 3. Antineoplastic effects of lectins from vertebrates based on in vitro and in vivo studies.

Origin	Lectin	Recognition Glycans	Cell Lines	In Vivo Model	Treatment Times and Doses	Cellular and Molecular Target	Ref.
					Amphibians		
Rana catesbeiana	Sialic acid-binding lectin (SBLc)	Sialic acid	P388		48 h (0.1–5 μM)	IC50: 0.3 μM (EDC-ED SBLc); 1.0 μM (EDC-GM SBLc); 1.5 μM (EDC-TA SBLc; SBLc)	[106]
	EDC-TA SBLc						
	EDC-GM SBLc						
	EDC-ED SBLc						
Rana catesbeiana	Sialic acid-binding lectin (SBLc)	Sialic acid	P388		48 h (0.1–5 μM)	GI50 [1] (P388): 1.56 (SBLj); 6.25 μM (SBLc)	[107]
Rana japonica	Sialic acid-binding lectin (SBLj)		L1210			GI1002 (L1210): 1.56 μM (SBLc and -j)	
				Sarcoma 180-bearing ddY mice	A single SBLc injection (2.5–10 mg/kg)	IC50: 5 mg/kg (Sarcoma 180-bearing mice) after 45 days	
				MepII-bearing ddI mice		IC50: 10 mg/kg (MepII-bearing mice) after 45 days	
				Sarcoma 180-bearing ddY mice	Continuous SBLc injection (0.5–2 mg/kg) for 10 days	IC50: <0.5 mg/kg (Sarcoma 180-bearing mice) after 45 days	
				MepII-bearing ddI mice		IC50: 0.5 mg/kg (MepII-bearing mice) after 45 days	
Rana catesbeiana	Sialic acid-binding lectin (SBLc)	Sialic acid	Jurkat		48 h (2 μM)	44% of cells in sub-G1 phase	[108]
					1–48 h (2 μM)	↑cleaved caspase-8, -9, -3	
					3–48 h (2 μM)	↑cleaved caspase-4, ↑Bip/GRP78	
Rana catesbeiana	Sialic acid-binding lectin (SBLc)	Sialic acid	P388		24 h (3 μM)	Cell viability: ~20% (P388); ~30% (K562); ~40% (HL60); ~80% (MCF-7); ~100% (Daudi; Raji; NHDF; NHEM; NHEK)	[109]
			K562				
			HL60		24 h (3, 20 μM)	No DNA fragmentation in Raji and NHDF cells	
			MCF-7			DNA fragmentation in P388 and K562 cells	
			Daudi		24 h (3 μM)	↑caspase-8, ↑caspase-3	
			Raji		1h (1 μM)	↑Hsp70 and Hsc70 on the cell membrane	
			NHDF				
			NHEM				
			NHEK				

Table 3. *Cont.*

Origin	Lectin	Recognition Glycans	Cell Lines	In Vivo Model	Treatment Times and Doses	Cellular and Molecular Target	Ref.
Rana catesbeiana	Sialic acid-binding lectin (SBLc)	Sialic acid	H28		48 h (0.2–20 µM) of treatment and 12 days of posttreatment	Colony formation (5 µM): <5% (H28); 20% (MESO-4); <70% (MESO-1)	[110]
			MESO-1			Colony formation (5 µM): >90% (Met-5A)	
			MESO-4		24–72 h (5 µM)	Annexin V+ (72h): ~5% (Met-5A); ~15% (MESO-1 and -4); ~50% (H28)	
			Met-5A		6–48 h (5 µM)	H28: ↑caspase-8, ↑caspase-9, ↑caspase-3	
						H28: ↑Bim, ↑Bik, ↑p-p38, ↑pJNK, ↑pERK	
					24 h (SBLc 5 µM ± TRAIL 2 ng/mL) on H28	↑cytotoxicity (~30%) vs single treatment (~70%)	
						↑ Annexin V+ (~50%) cells vs single treatment (~50%)	
						↑mitochondrial membrane depolarization vs single treatments	
						↑caspase-8, -9, -3 protein expression vs single treatment	
Rana catesbeiana	Sialic acid-binding lectin (SBLc)	Sialic acid	MCF-7		72 h (2 µM SBLc)	Cell viability: 25.5% (MDA-MB-231); 30.4% (MCF-7); 65.3% (SK-BR-3)	[111]
			SK-BR-3			↑p-p38	
			MDA-MB-231			↑ caspase-3/7 activity	
	SBLc mutant lacking RNase activity (H103A)				72 h (10 µM H103A)	Cell viability: 100% (MDA-MB-231)	
					72 h (2 µM H103A)	No effect on pp38, PARP expression	
						No effect on caspase-3/7	
Rana catesbeiana	Sialic acid-binding lectin (SBLc)	Sialic acid	H28		72 h (1–30 µM)	IC50: 0.46 µM (H28); 0.52 µM (H2452); 1.54 µM (MESO-4); 5.05 µM (MSTO); 5.51 µM (MESO-1); 52.22 (Met-5A)	[112]
			MESO-1		72 h (1 µM)	↑ % of cells in subG1-phase	
			MESO-4		72 h (1 µM)	↓cyclin A, ↓cyclin B1, ↓cyclin D1, ↓cyclin E, ↓p21, ↓pAkt	
			H2452		72 h (1 µM SBLc + 20 µM pemetrexed or 40 µM cisplatin)	CI3 (H28): 0.05 (SBLc + pemetrexed); 0.47 (SBLc + cisplatin)	
			Met-5A			Annexin V+/propidium iodide- (H28): no difference vs SBLc treatment	
						Caspase-3/7 activity (H28): no difference vs SBLc treatment	
						H28: ↑ % of cells in S- and subG1-phases (SBLc + pemetrexed); ↑ % of cells in S-, G2- and subG1-phases (SBLc + cisplatin)	
						H28: ↓cyclin B1, ↓p21, ↓pAkt (SBLc + pemetrexed or cisplatin)	

Table 3. *Cont.*

Origin	Lectin	Recognition Glycans	Cell Lines	In Vivo Model	Treatment Times and Doses	Cellular and Molecular Target	Ref.
Rana catesbeiana	Sialic acid-binding lectin (SBLc)	Sialic acid	NCI-H2452	H2452 or MSTO injected in BALB/C nu/nu Slc	24–72 h (H2452: 1 μM; MSTO: 0.4 μM)	Annexin V+ (72 h): 16.13% (H2452); 40.05% (MSTO)	[113]
			MSTO-211H		6–72 h (H2452: 5 μM; MSTO 2 μM)	↑nuclear fragmentation (72h): ~2.5-fold (H2452); ~4-fold (MSTO)	
					1–72 h (H2452: 5 μM; MSTO 2 μM)	↑activity and expression of caspase-9, -8, -3	
					72 h [(1 pM–1 μM SBLc) + (0.8 nM–800 μM pemetrexed)]	CI (H2452) < 1 at all combinations (SBLc + pemetrexed or SBLc + cisplatin)	
					72 h [(1 pM–1 μM SBLc) + (0.1 nM–100 μM cisplatin)]	CI (MSTO) < 1 up to 1 μM SBLc + 1.5 μM pemetrexed or 10 μM cisplatin	
					Pemetrexed (100 ng/kg) on days 1–5 and 15–19	↓tumor size after 47 (H2452) days of treatment	
					SBLc (2.5 mg/kg) 2/week for 4 weeks	↓tumor size after 36 (H2452) or 29 (MSTO) days of treatment	
Rana catesbeiana	Sialic acid-binding lectin (SBLc)	Sialic acid	ZR-75-1		72 h (1–20 μM)	Cell viability (20 μM): 40% (MDA-MB-468); 45% (MCF-7); 46% (SK-BR-3); 51% (BT-474); 52% (MDA-MB-231); 69% (ZR-75-1); 85% (MCF10A)	[114]
			BT474		72 h (1–10 μM) + 7–28 days in drug-free medium	↓cell number (except for MCF10A)	
			MCF-7		72 h (10 μM)	↓ number of colonies (except for MCF10A)	
			SK-BR-3		72–96 h (10 μM)	chromatin condensation and nuclear collapse (except in MCF10A)	
			MDA-MB-231			↑cleaved caspase-9 and PARP cleavage (except in MCF10A)	
			MDA-MB-468		72 h (10 μM)	↑pp38, ↑pJNK (ZR-75-1); ↑pp38, ↓JNK (MCF-7)	
			MCF10A			↓Bcl-2, ↓Bcl-xL, ↓Mcl-1 (MCF-7); ↓Bcl-2, ↑Bcl-xL; Mcl-1 (ZR-75-1)	
						↓ERα, ↓PgR, ↓HER2 (MCF-7); ↓ERα, ↓HER2 (ZR-75-1)	
						↓ErbB family in each cancer cells	
					3–24 h (10 μM)	Triple negative cells: ↑pp38 (MDA-MB-231 and -468); ↓EGFR/HER1, ↓pAKT (only in MDA-MB-231 cells)	
Rana catesbeiana	Sialic acid-binding lectin (SBLc)	Sialic acid	ZR-75-1		120 h (20 μg/mL)	Cell survival (72h): ~10% (MCF-7); ~30% (ZR-75-1)	[115]
			MCF-7		72 h (20 μg/mL)	MCF-7, ZR-75-1: ↑caspase-3 activity	
						MCF-7: extended pseudopodia, increase in phagocytic activity, cell debris	

Table 3. *Cont.*

Origin	Lectin	Recognition Glycans	Cell Lines	In Vivo Model	Treatment Times and Doses	Cellular and Molecular Target	Ref.
Rana catesbeiana	Sialic acid-binding lectin (SBLc)	Sialic acid	ZR-75-1		96 h (20 µg/mL)	Cell survival (120 h): <50% (MCF-7; ZR-75-1); >80% (MDA-MB-231; ZR-75-30)	[116]
			MCF-7		72 h (20 µg/mL)	↑ caspase-3 activity (MCF-7; ZR-75-1)	
			MDA-MB-231		96 h (0–40 µg/mL)	↓ER, ↓Bcl-2 (MCF-7)	
			ZR-75-30				
Rana catesbeiana	Sialic acid-binding lectin (SBLc)	Sialic acid	MCF-7		0–120 h (20 µg/mL)	Cell survival (120 h; SBLc): 13% (MCF-7); 31.3% (MCF-7/Bcl-xL)	[117]
Rana catesbeiana	Sialic acid-binding lectin (SBLc)	Sialic acid	Undifferentiated HL-60		120 h (2, 20 µg/mL)	Cell survival (120 h; 20 µM): 5.5% (undifferentiated)	[118]
			retinoic acid-differentiated HL-60		5–7–9 days of differentiation + 120 h (2, 20 µg/mL)	Cell viability (120 h; 20 µM; differentiated cells): 65% (5 days); 82.5% (7 and 9 days)	
					7 days of differentiation + 48, 96 h (20 µg/mL); 48, 96 h (20 µg/mL)	↑caspase-9, -3 and cleaved PARP (undifferentiated cells)	
						↑caspase-9 and -3 activity (undifferentiated cells)	
Rana catesbeiana	Sialic acid-binding lectin (SBLc)	Sialic acid	DBTRG		0–96 h (20 µg/mL)	Cell inhibition rate (96 h): ~10% (RG2); ~25% (GBM8401); ~45% (DBTRG; GBM8901)	[119]
			GBM8901		0–96 h (2–50 µg/mL)	Cell inhibition rate (50 µg/mL; 96 h): ~15% (RG2); ~40% (DBTRG); ~65% (GBM8901)	
			GBM8401		24, 72 h (50 µg/mL)	↑% of cells in sub-G1-phase (~30%; DBTRG)	
			RG2		72 h (50 µg/mL)	↑caspase-9 and -3 activity, not caspase-8 (DBTRG)	
				DBTRG cells subcutaneously injected in nude mice	a single injection (5 µg)	↓tumor size after 18 days of treatment	
Rana catesbeiana	Sialic acid-binding lectin (SBLc)	Sialic acid	HL60		120 h (20 µg/mL SBLc; 10 ng/mL IFN-γ)	Cell viability: SBLc + IFN-γ < SBLc (MCF-7; SK-Hep-1); SBLc + IFN-γ = SBLc (HL60)	[120]
			MCF-7		48, 96 h (20 µg/mL SBLc; 10 ng/mL IFN-γ)	HL60: ↑ caspase-3, -8, and -9 activity (SBLc + IFN-γ = SBLc)	
			SK-Hep-1			MCF-7: ↑caspase-7 activity (SBLc + IFN-γ > SBLc);	
						SK-Hep-1: caspase-3, -8, and 9 activity = control	
					48, 96 h (20 µg/mL SBLc + 10 ng/mL IFN-γ)	↑cleaved caspase-3 and PARP (HL60); ↑ cleaved caspase-7 and PARP (MCF-7)	

Table 3. *Cont.*

Origin	Lectin	Recognition Glycans	Cell Lines	In Vivo Model	Treatment Times and Doses	Cellular and Molecular Target	Ref.
Rana catesbeiana	Sialic acid-binding lectin (SBLc)	Sialic acid	SK-Hep-1		0–96 h (20 μM SBLc + 10 ng/mL TNF-α or -β)	Cell survival (96 h): ~40% (J5); ~50% (SK-Hep-1); ~90% (HepG2)	[121]
			J5		0–120 h (20 μM SBLc + 10 ng/mL IFN-γ)	Cell survival (120 h, SK-Hep-1): 13.3% (SBLc+IFN-γ) vs 64.7% (SBLc)	
			HepG2			Cell survival (120 h, J5): 27.8% (SBLc+IFN-γ) vs 76.8% (SBLc)	
			BHK21			Cell survival (120 h, HepG2): 64.2% (SBLc+IFN-γ) vs 93.9% (SBLc)	
						Cell survival (120 h, BHK21): 91.52% (SBLc+IFN-γ) vs 96.67% (SBLc)	
Rana catesbeiana	Tobacco-derived his-HR Recombinant hHscFv–RC-RNase protein	Sialic acid	SMMC7721		24 h (0.7–3.5 nM)	IC50: 2 nM (SMMC7721); 2.4 nM (HepG2); 4.8 nM (DV145)	[122]
			HepG2			Cell inhibition rate (3.5 nM): ~15% (HL-7702)	
			DV145				
			HL-7702				
Rana japonica	Sialic acid-binding lectin (SBLj)	Sialic acid	P388		48 h (0.1–5 μM)	GI501 (P388): 1.56 (SBLj)	[107]
			L1210			GI1001 (L1210): 1.56 (SBLj)	
					Fish		
Anguilla japonica	Adenovirus FLAG (Ad.FLAG) Anguilla japonica lectin 1 (AJL1)	β-galactoside	Hep3B		48–120 h (50–100 MOI)	Cell viability (100MOI; 96h): ~10% (SMMC-7721); ~20% (Hep3B, BEL-7404, QSG-7701); ~40% (Huh7); ~60% (A549)	[123]
			BEL-7404		48 h (50 MOI)	Hep3B: AnnexinV^{+}/propidium iodide^{-} 19.5% (Ad.FLAG-AJL1) vs 5.04% (Ad.FLAG)	
			Huh7			Hep3B: ↑cleaved PARP, Bcl-XL; ↓ procaspase-9, Bcl-2, XIAP	
			SMMC7721		48 h (50–100 MOI)	Hep3B: ↑PMRT5; ↓E2F-1	
			A549		48 h (50–100 MOI)	Hep3B: ↓ERK, ↓pERK, ↓p38	
			QSG-7701				
Aristichthys nobilis	Bighead carp gill lectin (GANL)		SMMC7721		24 h (0.5–64 μg/mL)	Cell inhibition rate (64 μg/mL): ~0% (SMMC7721, BGC803); ~20% (SKOV3, HepG2); ~-30% (LoVo); ~80% (HeLa)	[124]
			HepG2		72 h (0.5–64 μg/mL)	Cell survival (16 μg/mL): ~115% (splenocytes)	
			SKOV3				
			HeLa				
			BGC803				
			LoVo				
			BALB/c mice splenocytes				

Table 3. *Cont.*

Origin	Lectin	Recognition Glycans	Cell Lines	In Vivo Model	Treatment Times and Doses	Cellular and Molecular Target	Ref.
Dicentrarchus labrax	Adenovirus FLAG (Ad.FLAG)-*Dicentrarchus labrax* fucose-binding lectin (DIFBL)	Fucose	Hep3B		48–96 h (1–20 MOI)	Cell viability (20MOI; 96h; Ad.FLAG-DIFBL): ~30% (Hep3B); ~40% (BEL-7404, A549, SW480)	[61]
			BEL-7404		48 h (20 MOI)	Hep3B: Annexin V$^+$/propidium iodide$^-$ 21.5% (Ad.FLAG-DIFBL) vs 1.35% (Ad.FLAG)	
			A549			Hep3B: = cleaved PARP; ↓Bcl-2, ↓XIAP	
			SW480			Hep3B: ↓E2F-1	
Dicentrarchus labrax	*Dicentrarchus labrax* fucose-binding lectin (Ad-DIFBL) + coxsackie-adenovirus receptor (sCAR)-DIFBL	Fucose	K562/adr		48 h (5–30 MOI Ad-EGFP + 10 µg/mL sCAR-DIFBL)	Viral infection and replication: K562/ADR: 20% sCAR-DIFBL vs 3.19% Ad-EGFP; U87MG: 40.6% sCAR-DIFBL vs 23.1% Ad-EGFP	[57]
			U87MG		96 h (8.2 MOI Ad-DIFBL + 14-42 µg/mL sCAR-DIFBL)	Cell viability (U87MG): ~20% (42 µg/mL Ad-DIFBL-sCAR-DIFBL); ~50% (Ad-DIFBL)	
					48 h (8.2 MOI Ad-DIFBL + 42 µg/mL sCAR-DIFBL)	U87MG: AnnexinV$^+$/propidium iodide$^-$ 4.87% (Ad-DIFBL-sCAR-DIFBL) vs 7.91% (Ad-DIFBL)	
					48 h (8.2 MOI Ad-DIFBL + 42 µg/mL sCAR-DIFBL)	U87MG: ↑pERK (Ad-DIFBL-sCAR-DIFBL)	
Oncorhynchus keta	L-rhamnose-binding lectin (CSEL)	Rhamnose	Caco-2		24 h (1–100 µg/mL)	Cell viability: ~35% (Caco-2); no effects on DLD-1 or HCT-15	[125]
			DLD-1		24 h (1–100 µg/mL CSEL ± 0.1 M L-rhamnose or 2 µM PPMP)	Cell viability (Caco-2): ~35% (CSEL); ~90% (CSEL ± L-rhamnose or PPMP)	
			HCT-15		24 h (50–100 µg/mL)	DNA fragmentation in Caco-2	
						Annexin V$^+$/propidium iodide$^-$ (100 µg/mL; Caco-2): 32.8% (CSEL) vs 3.1% (control)	

Table 3. *Cont.*

Origin	Lectin	Recognition Glycans	Cell Lines	In Vivo Model	Treatment Times and Doses	Cellular and Molecular Target	Ref.
Oncorhynchus tshawytscha	Rhamnose-binding roe chinook salmon lectin (CSRL)	Rhamnose	*MCF-7*		24–48 h (3.9–250 μM)	IC50 (24–48 h): 93–45 μM (HepG2); 220–68 μM (MCF-7);	[126]
			Hep G2		48 h (68 μM)	WRL68: no effect	
			WRL68		24 h (0.625–20 μM)	NO production at 0.62 μM (at 20 μM CSRL)	
			Mouse peritoneal macrophages				
Silurus asotus	Rhamnose-binding lectin (SAL)	Rhamnose	*Raji*		5–30 min (2.5–10 μg/mL)	Raji (30 min; 10 μg/mL): ↑ AnnexinV+/propidium iodide- (16.7% SAL vs 4.31% control) and AnnexinV+/ propidium iodide+ (77.25% SAL vs 4.79% control)	[127]
			K562			Raji (30 min; 10 μg/mL): ↑ 30-fold shrunken cell population	
			K562/DXR			K562; K562/DXR: no effects	
					30 min (10 μg/mL SAL + 4 μM CsA)	Raji: ↓ necrotic cells (58.03% SAL + CsA vs 77.92% SAL)	
Silurus asotus	Rhamnose-binding lectin (SAL)	Rhamnose	Raji		24–120 h (0–100 μg/mL)	Cell viability: no effects	[128]
					24–48 h (100 μg/mL)	Cell proliferation: block at 50 μg/mL	
					72 h (100 μg/mL) + 48 h SAL-free medium	Cell proliferation: restored	
					24 h (100 μg/mL)	↑% of cells in G0/G1-phase (20%); ↓% of cells in S-phase (20%)	
					12–24 h (100 μg/mL SAL ± 20 mM Saccharide)	↓ CDK4, C-MYC (40%), CCND3 (30%); = CDK2; ↑ p21 (130%), p27 (70%); + saccharide reverts effects (except for p27)	
					12–24 h (100 μg/mL)	↓CDK4, ↓c-Myc, ↓cyclin D3, ↑ p21, ↑p27	
					0.5–24 h (100 μg/mL)	↑ GTP-Ras, pMEK, pERK; = pp38 and cJNK	
					(100 μg/mL) in A4GALT [4]siRNA Raji cells	↓pMEK, ↓pERK induced by SAL	
					2 h (10 μM U0126) ± 12 h (100 μg/mL SAL)	↓p21, ↓pERK induced by SAL (cell-cycle arrest depending on Ras-MEK-ERK pathway)	
						↑cell proliferation rate	

[1] Concentration that inhibits the growth of cells by 50%; [2] concentration that inhibits the growth of cells by 100%; [3] combination index; [4] A4GALT: α-1,4-galactosyltransferase.

A sialic acid-binding lectin (SBLc), better known as leczyme, is a multifunctional protein isolated from the oocytes of *Rana catesbeiana* and characterized by a dual nature: a lectin and RNAse activity. This nature is the key to its very interesting antitumor properties (Figure 3). This lectin is unique and not homologous to any other known protein. It is a single subunit of 111 residues of amino acids and does not contain any covalently bound carbohydrate. The amino terminus is a pyroglutamyl residue and all the half-cystinyl residues are in the form of oxidized disulfide bridges [129]. SBLs lectin-glycan binding has not been completely elucidated. Nitta et al. [130] showed that Lectin hemagglutination activity was inhibited blocking amino groups and in particular t-amino ones, but not tyrosine residues, letting presume that those residues are responsible for the binding with the sugar.

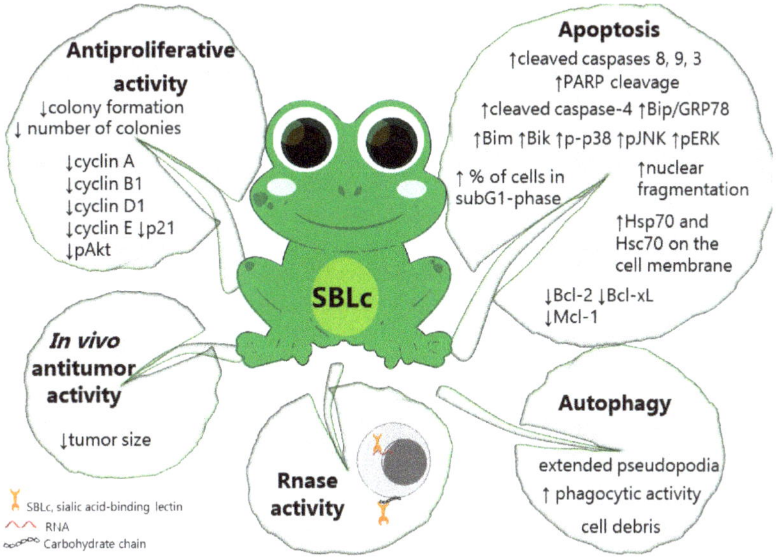

Figure 3. Anticancer mechanisms of SBLc, isolated from the oocytes of *Rana catesbeiana*.

The role of the lectin part of SBLc is to select cancer cells through the binding of specific sialic acid carbohydrates residues; then, specific receptors allow it to penetrate the cells, where the RNAse activity promotes the cleavage of RNA. Different structural changes on SBLc demonstrated that both intracellular incorporation and RNAse activity play a pivotal role for its anticancer activity. Indeed, increased incorporation with moderately increased RNAse activity is more effective than the increase in total RNAse activity [106,109]. For example, a less efficient internalization system does not affect or affects to a lesser extent the cytotoxicity of SBLc on estrogen negative cell lines, such as SK-BR3 [111,116]. On the other hand, a significant decrease in RNAse activity compromises SBLc's antitumor activity [111]. In other words, as Aristotle would suggest, *in medio stat virtus*.

When inside the cell, SBLc activates the intrinsic apoptotic pathway on more than 30 tumor cell lines representing leukemia [107–109,118,120,131,132], Burkitt's lymphoma [108,109], cervical [133], hepatocellular [120,121,133], and estrogen-positive and -negative breast carcinomas [113,114,116,117,120], glioblastoma [119], and malignant mesothelioma [110,113] (Table 3).

Just as in a jigsaw puzzle, many studies have tried to elucidate the proper molecular mechanism of action of SBLc. The pieces of the puzzles are a tumor-specific mechanism of action, the activation of the intrinsic apoptotic pathway with the involvement of heat shock protein 70 (HSP70) and the phosphorylation of p38 MAPK [108,110,116,120,134,135].

On MCF-7 estrogen receptor (ER)-positive breast cancer cells and H28 malignant mesothelioma cells, SBLc induced Bcl-2-mediated cell death [116]. On Jurkat cells, it promoted endoplasmic reticulum stress that may or may not be involved in its antitumor activity [134]. On p388 mouse lymphoma cells, HSP70 was linked to its ability to cause apoptosis. Even though the mechanism through which HSP70 assists SBLc cytotoxicity is not clear, it does not involve the binding between SBLc and cell membrane [109,135]. Finally, even if mitochondrial disruption is often part of SBLc mechanism of action, this lectin promoted a caspase-independent modality of cell death in SK-Hep1 hepatocellular carcinoma cells [120].

The activation of p38 MAPK pathway by SBLc has been recorded in leukemia, mesothelioma, and several breast cancer cells [108,111]. In these cell lines, the phosphorylation of p38 MAPK plays a key role in the pro-apoptotic activity of SBLc since the ablation of this pathway significantly decreases SBLc-mediated activation of caspase-3 and caspase-7 and cleavage of PARP and reduced cell death [111,114].

Special reference needs to be made to understand SBLc activity on breast cancer. The expression of ER, progesterone receptor (PR) and the human epidermal growth factor receptor 2 (HER2) represents the groundwork for the classification of this type of cancer and is useful for prognostic prediction and therapy selection. Usually, the worst prognosis refers to triple negative cancer (ER, PR and HER negative) and to those tumors overexpressing HER2 together with the epidermal growth factor (EGFR) [113]. Different studies about SBLc antitumor ability showed different results on the same tumor types. This is probably due to the different experimental procedures used in the studies, such as the presence of hormones, growth factor or fetal bovine serum in the medium. All studies agree on the fact that SBLc is very active on ER-positive cells, such as MCF-7 and ZR-75-1 [111,114,116], while contradictory data are published on ER-negative cell lines. For Tseng and colleagues [116], ERs are essential for cell binding of SBLc and thus its internalization. Accordingly, they reported a very mild activity of SBLc on MDA-MB-231 and ZR-75-30 ER-negative cell lines. On the contrary, Tatsuta et al. [114] and Kariya et al. [111] showed that both ER-negative (MDA-MB-231) and HER2-positive are sensitive to SBLc. Furthermore, Tatsuta et al. [114] showed a comprehensive screening of SBCc cytotoxic on several breast cancer cell lines with different phenotypes, such as ER-positive, PR-positive, HER2-positive, and triple-negative. As above mentioned, they demonstrated the involvement of p38 MAPK signal in its cytotoxic activity together with the modulation of the anti-apoptotic proteins of the Bcl-2 family on all cell lines. On MCF-7 and ZR-75-1, SBLc downregulated ERα, PR and HER2, while no modulation of EGFR was observed. This result is particularly interesting since hormonal therapies, and in particular receptor antagonism, represent efficient and usually well-tolerated treatments for breast cancer [136]. The cause of ER and PR downregulation could be linked to the activation of kinases, such as p38, drove by the ubiquitin proteasome system. Since SBLc promotes the phosphorylation of this kinase, Tatsua et al. [114] suggest that SBLc-promoted p38 phosphorylation can be linked to the destruction of hormone receptors. Likewise, the ubiquitin proteasome system may be involved in HER2 downregulation [114]. The hypothesis is based on the observation that SBLc modulates HSP70 [109,135]. Indeed, HSP70, together with HSP90, is responsible for the delivery of proteins, such as HER, to the ubiquitin-proteasome system in order to be degraded [114].

To avoid inflammation and promote dying-cell clearance, many types of specialized cells can phagocyte them. Cancer cells are not the first kind of professional phagocytes that come to mind; however, in rare situations, they are able to do that. It was already demonstrated that MCF-7 cells can play that role [137]. SBLc itself stimulated morphologic and phenotypic changes that allow viable MCF-7 cells phagocyting SBLc-triggered dying cells [115], promoting a very efficient auto-clearance. This unusual behavior represents an added value to the already promising anticancer activity of SBLc.

For many types of cancer, combination therapy is the best way to increase therapeutic efficacy, avoid drug resistance, and lessen the cytotoxicity of each single compound used in the therapy [138]. For example, malignant mesothelioma is treated with a combination of pemetrexed and cisplatin. SBLc alone was effective on cisplatin- and pemetrexate-resistant MESO-1 and MESO-4 cells (Table 3). Moreover, it synergistically increased the antitumor activity of pemetrexate in H28 [138] and MSTO cells [113] (Table 3) with an even stronger effect than that of pemetrexed plus cisplatin. The association of pemetrexed plus SBLs was particularly effective, because the first one blocks the proliferation of cells, which will be then easily killed thanks to the cytotoxic effect of the second one [113]. On the contrary, antagonistic effects were recorded when cisplatin was used together with SBLc. The antagonism is probably due to the opposite modulation of the tumor suppressor p21 by the two compounds. If cisplatin sharply increases p21 expression, SBLc antagonizes it, resulting in a lack of synergy [112]. In the case of SBLc plus TNF-related apoptosis-inducing ligand, a synergistic effect on H28 cells has been recorded, probably caused by an enhancement in the cleavage of the proapoptotic protein Bid that reinforces caspase activation [110]. SBLc also synergizes the antitumor cytotoxicity of interferon-γ (IFN-γ) in a cell-type-dependent way on different cell lines. The most marked effect occurred on MCF-7 cells. For the other tested cell lines, the degree of cell differentiation seems to play a role in this synergism: less degree of cell differentiation, more synergistic activity of SBLc plus IFN-γ. Then, an increasing trend of potency accompanied the well-differentiated HepG2 cells, the intermediate-differentiated J5 cells, and the poorly differentiated SK-hep1 cells, respectively. No evidence of synergism has been reported on HL-60 cells [120]. Considering that IFN-γ promotes differentiation of HL-60, this explanation perfectly fits with the abovementioned remarks [120]. Even if the vast majority of studies ascribe the SBLc's selectivity to its lectin nature (i.e., to its ability to bind tumor cells because of their membrane glycosylation pattern), a correlation with the degree of differentiation of target cells and the ability of SBLc to recognize them is probably involved here as well. Indeed, the antitumor activity of SBLc decreases with the increase of differentiation of both hepatoma [121] and leukemia cells [118]. SBLc does not affect any of the several non-transformed cell lines tested that are all characterized by a high degree of differentiation: fibroblasts [109,120,121,133,139], epidermal melanocytes [109], keratinocytes [109], and mesothelial cells [110].

The effective and safe profile of SBLc reported above was also observed in animal models. Already in 1994, Nitta et al. [107] demonstrated the ability of SBLc to counteract sarcoma 180 cells inoculated on mice (Table 3) with doses at least 150 times lower than the lethal ones. More, SBLc suppressed mesothelioma (H2452 or MSTO cells) growth in xenografted nude mice, without causing toxicity nor body weight changes [113]. For the record, pemetrexed, which is one of the drugs of choice currently used for the cure of mesothelioma, failed to eradicate murine MSTO-xenografted tumors on the same experimental conditions [113]. In addition, SBLc effectively inhibited the growth of glioblastoma tumors in nude mice once again without causing any side effect [119].

5.2. Fish

The economic value of fish is not entirely due to the food industry. On the contrary, many pieces of research are focusing on these animals since they produce many bioactive compounds, such as the human calcitonin from salmon used for the treatment of postmenopausal osteoporosis [140]. As for amphibians, fishes produce many types of lectin that have been isolated from eggs, skin mucus, plasma, and serum [141–144]. The biologic function of fish lectin is not crystal clear, even if evidence shows some implication on fertilization and morphogenesis and a defense activity versus microorganisms [145], whether for some of them, an antitumor activity has been proven (Table 3).

Aristichthys nobilis, commonly known as bighead carp, belongs to the *Cyprinidae*, from the same family as zebrafish. It produces a lectin, GANL, which causes tumor-type-dependent cell death. GANL is a homo-multimeric glycoprotein that does not require Ca^{2+} ions to perform its functions. The carbohydrate content is approximately 13.4%, while the protein part is enriched in Asp, Glu, Leu, Val, and Lys. These amino acids organize themselves to form α-helices, unordered structures, β-turns, and β-sheets [124,146]. GANL blocked HeLa, SKOV3 (ovarian cancer cells), and HepG2 proliferation in a concentration-dependent manner while no effect was recorded on SMMC-7721 carcinoma cells and BGC803 gastric cancer cells (Table 3). Yao et al. [124] suggested that this behavior could be a consequence of the different glycosylation pattern that characterizes the different cells since it is not known whether the binding site of the lectin is the same for all cell lines and how each cell line expresses it [124].

The α-galoctoside-binding lectins isolated from the eggs of the catfish *Silurus asotus* (SAL) and the chum salmon eggs (CSEL) recognize specifically Gb3 sugar chain. Thanks to Gb3, SAL and CSEL bind cells, are internalized, and exhibit antitumor activity. Accordingly, SAL is not active on the Gb3-devoid K562 cells, while several studies showed its antitumor activity on Burkitt's lymphoma cells [127,128,147] (Table 3). Similarly, CSEL promotes apoptosis on Gb3-positive Caco-2 cells, but it does not exert any effect on Gb3-negative DLD-1 colorectal adenocarcinoma cells [125]. On Raji cells, SAL induces the typic phenotypic changes of apoptosis, such as phosphatidylserine exposition and cell shrinkage [127], but its antitumor activity is more probably linked to its ability to inhibit the cell cycle [128]. A family of kinase proteins regulates rate and progression of the cell cycle. These enzymes consist of a regulating subunit (which takes the name of cyclin) and a catalytic subunit that takes the name of kinase-cyclin-dependent proteins (CDK). CDKs are inactive in the absence of the cyclin subunit and become active only when cyclins bind to the catalytic subunit. CDKs act during particular moments of the cell cycle *via* phosphorylation activity, stimulating or inactivating, the specific proteins that modulate cell-cycle advancement. For their part, synthesis and deactivation of cyclins are strictly regulated within the different phases of the cell cycle in order to allow this process running correctly [148]. After binding Gb3, SAL activated the GTPase Raf and the two kinases MEK and ERK. In turn, they promoted the synthesis of p21 [128], which physiological function is to regulate cell-cycle progression at G1 and S phase [149]. Moreover, SAL increased the expression of CDK4, c-myc, and cyclin D3. The overall effect is a G0-G1 cell-cycle arrest [128].

The Chinook salmon, also known as king salmon for its massive dimensions, is the source of another antitumor lectin, called rhamnose-binding chinook salmon roe lectin (CSRL). CSRL was reported to inhibit MCF-7 and HepG2 cell proliferation. HepG2 resulted in being more sensitive to CSRL than MCF-7 with a double potency (Table 3). As for all lectins mentioned in this review, CSRL as well did not exert any toxic effect on WRL68 non-transformed liver cell line at the same concentration used for MCF-7 and HepG2 cells [126] (Table 3).

Protein arginine methyltransferases (AMTs) are proteins involved in several processes, such as cell proliferation, cell differentiation, and also tumorigenesis. They catalyze the methylation of specific arginines on several nuclear and cytoplasmic substrates. AMT-5′ main targets are specific histones, and the result of their methylation is the silencing of different genes [150]. Recently, the transcription factor E2F-1 has been identified as a specific substrate for symmetric methylation, behaving as AMTs downstream element. E2F-1 modulates both apoptosis and cell-cycle progression, depending on which member of AMT family activates it. In particular, E2F-1 methylation by AMT-5 promotes cell proliferation, while if AMT-1 activates it, apoptosis is triggered [150]. p53 and p73 (the proline 73 polymorphic variant of p53) represent the links between EF2-1 and apoptosis [151,152]. Two different lectins are able to trigger this pathway and promote apoptosis in Hep3b cells. *Dicentrarchus labrax* fucose-binding lectin (DlFBL) and *Anguilla japonica* lectin 1 (AJL1) were harbored in a replication-defective adenovirus, generating respectively DIFBL-FLAG and AJL1-FLAG. DIFBL-FLAG and AJL1-FLAG elicited cytotoxicity in several human liver and lung cancer cell lines [61,123] (Table 3). DIFBL is a Ca^{2+}-independent non-glycosylated lectin. It is a dimeric protein composed by two protein fractions, and, if they are separated, only one of them has lectin activity, while in physiological conditions they are stabilized by disulfide bonds [153]. Additionally, AJL1 has beta-galactoside specific activity in a Ca^{2+}-independent manner. This lectin is composed of 142 amino acid residues having no half-cysteinyl residues, and exists in the form of homodimer without any covalent bonds [144]. On Hep3B cells, both lectins promoted apoptosis through the modulation of the AMT-5 pathway. In particular, the involvement of this enzyme provoked the downregulation of E2F-1 that in turn called Bcl-2 apoptotic protein family into play. Indeed, exogenous DIFBL and AJL1 downregulated Bcl-2 and XIAP proteins [61,123]. AJL1 also downregulated p38 MAPK and ERK protein levels, but it is not clear if this event is linked to its ability to promote apoptosis [123].

Despite the remarkable biological activity of lectins transported by viral vectors, some of them struggle to significantly infect cells, due to the lack of specific receptors such as coxsackie-adenovirus receptor (CAR). To facilitate this process, it is possible to target cell membrane receptors with other oncolytic adenoviruses that conveys a CAR-ligand-expression cassette in its genome. In this way, after infection and replication, the expression of CAR ligand will help the following adenovirus infection, triggering a positive feedback mechanism [57]. For this reason, a CAR-DIFBL was built and used to infect K562/adr doxorubicin-resistant leukemia and U87MG glioblastoma cells. CAR-DIFBL was able to favor cell infection by DIFBL-FLAG and also to synergize its cytotoxic activity [57] (Table 3). However, attention must be paid because another CAR-lectin, CAR-HddSBL, did help DIFBL-FLAG to infect glioblastoma cells, but it counteracted its cytotoxicity probably upregulating E2F-1 transcription levels [57].

6. Conclusions

The role of marine and freshwater lectins as anticancer agents has been discussed in this review and sustained by the reports summarized here. Marine and freshwater lectins exploit their antitumor action triggering apoptosis and other forms of programmed cell death, inhibiting cell cycle and blocking neoangiogenesis (Figure 4).

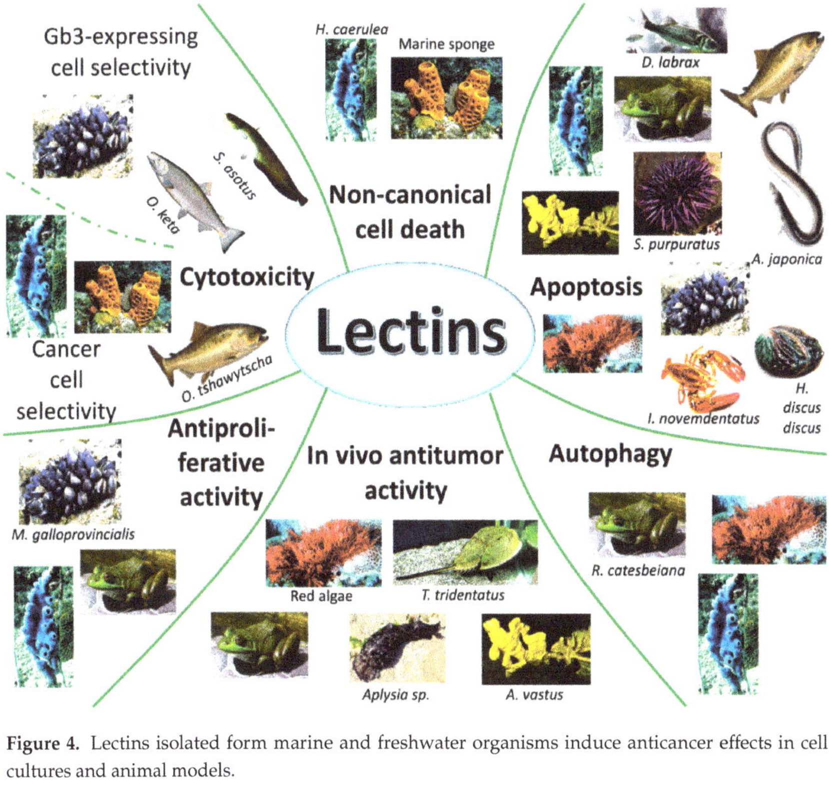

Figure 4. Lectins isolated form marine and freshwater organisms induce anticancer effects in cell cultures and animal models.

Some of them are also able to improve the antineoplastic activity of standard antitumor drugs. Moreover, all the lectins described in this review disclosed distinctive features against different types of tumors and, in the majority of the times, their ability to distinguish between normal and transformed cells. Of note, all lectins tested on animals showed optimal antitumor activity matched by negligible toxicity. Thus, selectivity and cytotoxicity are the two main characteristics that promote lectins to the rank of ideal antitumor agents. Indeed, toxicity of lectins of plant origin has been recorded only after direct ingestion of significant amount of specific lectins, such as phytohemagglutinin, which is particularly enriched in raw kidney beans. Gastric symptoms, such as nausea, vomiting, and diarrhea, are the most common ones after oral acute exposure, while immune system impairment, lung hypertrophy, and other systemic effects are very rare. On the contrary, no significant adverse reactions have been documented if lectins were administered as drug through alternative ways [154].

Another aspect that this review unearthed is that marine lectins are often tumor-type specific. This is not surprising because, going back to tumor heterogenicity, the glycosylation pattern is not the same for all cancer cells. This evidence does not represent a limiting factor for lectins. Indeed, marine biodiversity comes to our aid, and if one tumor is not sensitive to one specific lectin, most likely it would be to another one.

So far, the most concerning issue about exploiting marine compounds for a pharmacological purpose is that lectins, as all marine metabolites, are not produced in vast quantities. To obtain a reasonable amount of them, a large number of organisms has to be harvested. Thus, a significant limiting factor that stands between lectins and their exploitation as therapeutic agents is their availability. In other words, one of the hardest challenges about lectins is moving from laboratory scale to bulk production. However, despite the peculiarity of lectins that does not allow an easy identification of the aminoacidic bone structure and a cheap synthesis de novo, aquaculture or mariculture could bridge the gap. Additionally, the easiest and most efficient way to produce lectins is to exploit heterologous systems, such as bacteria and yeast. Indeed, different synthesis strategies exploit genetic engineering to transfer the genes encoding the lectin of interest to microorganisms, which can be grown in vast quantities, or to virus vectors in order to allow the synthesis of the lectins directly inside tumor cells [26,155,156]. As presented above, the latter approach has been exploited adequately for several lectins and with considerable results.

Despite the outstanding potential of marine lectins demonstrated by preclinical studies, so far, no clinical trials are translating this knowledge to cancer patients. In this context, terrestrial lectins are paving the way in that direction. For instance, mistletoe lectin extract has been subjected to many phase-I clinical trials, with promising, although not conclusive, results [157]. However, bearing in mind that the drug development process can last up to decades, preclinical studies represent the foundation to draw clinical trials.

Taking all this data together, it follows that aquatic organisms represent an important source of lectins and that those proteins represent a bright future for anticancer research. Certainly, further studies are needed to understand the actual antitumor potential of lectins on humans full. To reach this aim, clinical studies are essential to progress cancer research in employing lectins for antineoplastic care, but the premises for outstanding outcomes are all there.

Funding: This research received no external funding.

Conflicts of Interest: The authors declare no conflict of interest.

References

1. Fondation Tara Océan. Available online: https://oceans.taraexpeditions.org/en/m/agenda/tara-museum-national-histoire-naturelle/ (accessed on 18 August 2019).
2. World Register of Marine Species (WORMS). Available online: http://www.marinespecies.org (accessed on 18 August 2019).
3. Vermeulen, N. From Darwin to the census of marine life: marine biology as big science. *PLoS ONE* **2013**, *8*, e54284. [CrossRef] [PubMed]
4. Erwin, P.M.; López-Legentil, S.; Schuhmann, P.W. The pharmaceutical value of marine biodiversity for anti-cancer drug discovery. *Ecol. Econ.* **2010**, *70*, 445–451. [CrossRef]
5. Calcabrini, C.; Catanzaro, E.; Bishayee, A.; Turrini, E.; Fimognari, C. Marine sponge natural products with anticancer potential: An updated review. *Mar. Drugs* **2017**, *15*, 310. [CrossRef] [PubMed]
6. Dyshlovoy, S.A.; Honecker, F. Marine Compounds and Cancer: 2017 Updates. *Mar. Drugs* **2018**, *16*, 41. [CrossRef]

7. Holohan, C.; Van Schaeybroeck, S.; Longley, D.B.; Johnston, P.G. Cancer drug resistance: An evolving paradigm. *Nat. Rev. Cancer* **2013**, *13*, 714–726. [CrossRef]
8. Postovit, L.; Widmann, C.; Huang, P.; Gibson, S.B. Harnessing oxidative stress as an innovative target for cancer therapy. *Oxid. Med. Cell. Longev* **2018**, *2018*, 6135739. [CrossRef]
9. Molecular and Chemical Targets for Tumor-Selective Cancer Treatment. Available online: https://www.frontiersin.org/research-topics/2268/molecular-and-chemical-targets-for-tumor-selective-cancer-treatment (accessed on 20 August 2019).
10. Akhtar, M.J.; Ahamed, M.; Alhadlaq, H.A. Therapeutic targets in the selective killing of cancer cells by nanomaterials. *Clin. Chim. Acta* **2017**, *469*, 53–62. [CrossRef]
11. Hanahan, D.; Weinberg, R.A. Hallmarks of cancer: the next generation. *Cell* **2011**, *144*, 646–674. [CrossRef]
12. Feitelson, M.A.; Arzumanyan, A.; Kulathinal, R.J.; Blain, S.W.; Holcombe, R.F.; Mahajna, J.; Marino, M.; Martinez-Chantar, M.L.; Nawroth, R.; Sanchez-Garcia, I.; et al. Sustained proliferation in cancer: Mechanisms and novel therapeutic targets. *Semin. Cancer Biol.* **2015**, *35*, S25–S54. [CrossRef]
13. Widodo, N.; Priyandoko, D.; Shah, N.; Wadhwa, R.; Kaul, S.C. Selective killing of cancer cells by Ashwagandha leaf extract and its component Withanone involves ROS signaling. *PLoS ONE* **2010**, *5*, e13536. [CrossRef]
14. Kaltner, H.; Toegel, S.; Caballero, G.G.; Manning, J.C.; Ledeen, R.W.; Gabius, H.-J. Galectins: Their network and roles in immunity/tumor growth control. *Histochem. Cell Biol.* **2017**, *147*, 239–256. [CrossRef] [PubMed]
15. Taniguchi, N.; Kizuka, Y. Glycans and cancer: role of *N*-glycans in cancer biomarker, progression and metastasis, and therapeutics. *Adv. Cancer Res.* **2015**, *126*, 11–51. [PubMed]
16. Vajaria, B.N.; Patel, P.S. Glycosylation: A hallmark of cancer? *Glycoconj. J.* **2017**, *34*, 147–156. [CrossRef] [PubMed]
17. Nascimento, K.S.; Cunha, A.I.; Nascimento, K.S.; Cavada, B.S.; Azevedo, A.M.; Aires-Barros, M.R. An overview of lectins purification strategies. *J. Mol. Recognit.* **2012**, *25*, 527–541. [CrossRef] [PubMed]
18. De Araújo, R.M.S.; da Ferreira, R.S.; Napoleão, T.H.; das Carneiro-da-Cunha, M.G.; Coelho, L.C.B.B.; dos Correia, M.T.S.; Oliva, M.L.V.; Paiva, P.M.G. Crataeva tapia bark lectin is an affinity adsorbent and insecticidal agent. *Plant Sci.* **2012**, *183*, 20–26. [CrossRef] [PubMed]
19. Francis, F.; Jaber, K.; Colinet, F.; Portetelle, D.; Haubruge, E. Purification of a new fungal mannose-specific lectin from *Penicillium chrysogenum* and its aphicidal properties. *Fungal Biol.* **2011**, *115*, 1093–1099. [CrossRef]
20. Bhutia, S.K.; Panda, P.K.; Sinha, N.; Praharaj, P.P.; Bhol, C.S.; Panigrahi, D.P.; Mahapatra, K.K.; Saha, S.; Patra, S.; Mishra, S.R.; et al. Plant lectins in cancer therapeutics: Targeting apoptosis and autophagy-dependent cell death. *Pharmacol. Res.* **2019**, *144*, 8–18. [CrossRef]
21. Wang, H.; Ng, T.B. A lectin with some unique characteristics from the samta tomato. *Plant Physiol. Biochem.* **2006**, *44*, 181–185. [CrossRef]
22. Santos, A.F.S.; da Silva, M.D.C.; Napoleão, T.H.; Paiva, P.M.G.; Correia, M.T.S.; Coelho, L.C.B.B. Lectins: Function, structure, biological properties and potential applications. *Curr. Top. Pep. Protein Res.* **2014**, *15*, 41–62.
23. Višnjar, T.; Romih, R.; Zupančič, D. Lectins as possible tools for improved urinary bladder cancer management. *Glycobiology* **2019**, *29*, 355–365. [CrossRef]
24. Shi, Z.; Li, W.; Tang, Y.; Cheng, L. A novel molecular model of plant lectin-induced programmed cell death in cancer. *Biol. Pharm. Bull.* **2017**, *40*, 1625–1629. [CrossRef] [PubMed]
25. Yau, T.; Dan, X.; Ng, C.C.W.; Ng, T.B. Lectins with potential for anti-cancer therapy. *Molecules* **2015**, *20*, 3791–3810. [CrossRef] [PubMed]
26. Martínez-Alarcón, D.; Blanco-Labra, A.; García-Gasca, T. Expression of lectins in heterologous systems. *Int. J. Mol. Sci.* **2018**, *19*, 616. [CrossRef]
27. Drickamer, K. Two distinct classes of carbohydrate-recognition domains in animal lectins. *J. Biol. Chem.* **1988**, *263*, 9557–9560. [PubMed]
28. De Oliveira Figueiroa, E.; Albuquerque da Cunha, C.R.; Albuquerque, P.B.S.; de Paula, R.A.; Aranda-Souza, M.A.; Alves, M.S.; Zagmignan, A.; Carneiro-da-Cunha, M.G.; Nascimento da Silva, L.C.; Dos Santos Correia, M.T. Lectin-carbohydrate interactions: implications for the development of new anticancer agents. *Curr. Med. Chem.* **2017**, *24*, 3667–3680. [CrossRef]
29. Pinho, S.S.; Reis, C.A. Glycosylation in cancer: mechanisms and clinical implications. *Nat. Rev. Cancer* **2015**, *15*, 540–555. [CrossRef]
30. Munkley, J.; Elliott, D.J. Hallmarks of glycosylation in cancer. *Oncotarget* **2016**, *7*, 35478–35489. [CrossRef]

31. Häuselmann, I.; Borsig, L. Altered tumor-cell glycosylation promotes metastasis. *Front. Oncol.* **2014**, *4*, 28. [CrossRef]
32. Schwarz, F.; Aebi, M. Mechanisms and principles of N-linked protein glycosylation. *Curr. Opin. Struct. Biol.* **2011**, *21*, 576–582. [CrossRef]
33. Liu, Y.-C.; Yen, H.-Y.; Chen, C.-Y.; Chen, C.-H.; Cheng, P.-F.; Juan, Y.-H.; Chen, C.-H.; Khoo, K.-H.; Yu, C.-J.; Yang, P.-C.; et al. Sialylation and fucosylation of epidermal growth factor receptor suppress its dimerization and activation in lung cancer cells. *Proc. Natl. Acad. Sci. USA* **2011**, *108*, 11332–11337. [CrossRef]
34. Yang, X.; Huang, B.; Deng, L.; Hu, Z. Progress in gene therapy using oncolytic vaccinia virus as vectors. *J. Cancer Res. Clin. Oncol.* **2018**, *144*, 2433–2440. [CrossRef]
35. Anam, C.; Chasanah, E.; Perdhana, B.P.; Fajarningsih, N.D.; Yusro, N.F.; Sari, A.M.; Nursiwi, A.; Praseptiangga, D.; Yunus, A. Cytotoxicity of Crude Lectins from Red Macroalgae from the Southern Coast of Java Island, Gunung Kidul Regency, Yogyakarta, Indonesia. In Proceedings of the IOP Conference Series: Materials Science and Engineering, Jawa Tengah, Indonesia, 2017; Volume 193, p. 012017.
36. Fukuda, Y.; Sugahara, T.; Ueno, M.; Fukuta, Y.; Ochi, Y.; Akiyama, K.; Miyazaki, T.; Masuda, S.; Kawakubo, A.; Kato, K. The anti-tumor effect of Euchema serra agglutinin on colon cancer cells in vitro and in vivo. *Anticancer Drugs* **2006**, *17*, 943–947. [CrossRef]
37. Hayashi, K.; Walde, P.; Miyazaki, T.; Sakayama, K.; Nakamura, A.; Kameda, K.; Masuda, S.; Umakoshi, H.; Kato, K. Active targeting to osteosarcoma cells and apoptotic cell death induction by the novel lectin *Eucheuma serra* agglutinin isolated from a marine red alga. *J. Drug. Deliv.* **2012**, *2012*, 842785. [CrossRef]
38. Sugahara, T.; Ohama, Y.; Fukuda, A.; Hayashi, M.; Kawakubo, A.; Kato, K. The cytotoxic effect of *Eucheuma serra* agglutinin (ESA) on cancer cells and its application to molecular probe for drug delivery system using lipid vesicles. *Cytotechnology* **2001**, *36*, 93–99. [CrossRef]
39. Omokawa, Y.; Miyazaki, T.; Walde, P.; Akiyama, K.; Sugahara, T.; Masuda, S.; Inada, A.; Ohnishi, Y.; Saeki, T.; Kato, K. In vitro and in vivo anti-tumor effects of novel Span 80 vesicles containing immobilized *Eucheuma serra* agglutinin. *Int. J. Pharm.* **2010**, *389*, 157–167. [CrossRef]
40. Li, Y.; Zhang, X. Recombinant Microcystis viridis lectin as a potential anticancer agent. *Pharmazie* **2010**, *65*, 922–923.
41. Chaves, R.P.; da Silva, S.R.; Nascimento Neto, L.G.; Carneiro, R.F.; da Silva, A.L.C.; Sampaio, A.H.; de Sousa, B.L.; Cabral, M.G.; Videira, P.A.; Teixeira, E.H.; et al. Structural characterization of two isolectins from the marine red alga *Solieria filiformis* (Kützing) P.W. Gabrielson and their anticancer effect on MCF-7 breast cancer cells. *Int. J. Biol. Macromol.* **2018**, *107*, 1320–1329. [CrossRef]
42. Li, G.; Zhao, Z.; Wu, B.; Su, Q.; Wu, L.; Yang, X.; Chen, J. *Ulva pertusa* lectin 1 delivery through adenovirus vector affects multiple signaling pathways in cancer cells. *Glycoconj. J.* **2017**, *34*, 489–498. [CrossRef]
43. Hori, K.; Sato, Y.; Ito, K.; Fujiwara, Y.; Iwamoto, Y.; Makino, H.; Kawakubo, A. Strict specificity for high-mannose type N-glycans and primary structure of a red alga *Eucheuma serra* lectin. *Glycobiology* **2007**, *17*, 479–491. [CrossRef]
44. Hori, K.; Miyazawa, K.; Ito, K. Hemagglutinins in Marine algae. *Nippon Suisan Gakkaishi* **1981**, *47*, 793–798. [CrossRef]
45. UniProt. Available online: https://www.uniprot.org (accessed on 15 August 2019).
46. Hassan, M.; Watari, H.; AbuAlmaaty, A.; Ohba, Y.; Sakuragi, N. Apoptosis and molecular targeting therapy in cancer. *Biomed. Res. Int.* **2014**, *2014*, 150845. [CrossRef] [PubMed]
47. Wang, S.; Zhong, F.-D.; Zhang, Y.-J.; Wu, Z.-J.; Lin, Q.-Y.; Xie, L.-H. Molecular characterization of a new lectin from the marine alga *Ulva pertusa*. *Acta Biochim. Biophys. Sin. (Shanghai)* **2004**, *36*, 111–117. [CrossRef] [PubMed]
48. Atta-ur-Rahman, M.; Choudhary Iqbal, M.; Khan, K. *Frontiers in Natural Product Chemistry*; Bentham Science Publishers: Sharjah, UAE, 2005.
49. Turrini, E.; Calcabrini, C.; Tacchini, M.; Efferth, T.; Sacchetti, G.; Guerrini, A.; Paganetto, G.; Catanzaro, E.; Greco, G.; Fimognari, C. In Vitro study of the cytotoxic, cytostatic, and antigenotoxic profile of Hemidesmus indicus (L.) R.Br. (Apocynaceae) crude drug extract on T lymphoblastic cells. *Toxins* **2018**, *10*, 70. [CrossRef]
50. Voultsiadou, E. Therapeutic properties and uses of marine invertebrates in the ancient Greek world and early Byzantium. *J. Ethnopharmacol.* **2010**, *130*, 237–247. [CrossRef]
51. De Zoysa, M. Medicinal benefits of marine invertebrates: sources for discovering natural drug candidates. *Adv. Food Nutr. Res.* **2012**, *65*, 153–169.

52. Kawsar, S.; Aftabuddin, S.; Yasumitsu, H.; Ozeki, Y. The cytotoxic activity of two D-galactose-binding lectins purified from marine invertebrates. *Arch. biol. sci. (Beogr.)* **2010**, *62*, 1027–1034. [CrossRef]
53. Liao, J.-H.; Chien, C.-T.H.; Wu, H.-Y.; Huang, K.-F.; Wang, I.; Ho, M.-R.; Tu, I.-F.; Lee, I.-M.; Li, W.; Shih, Y.-L.; et al. A multivalent marine lectin from *Crenomytilus grayanus* possesses anti-cancer activity through recognizing globotriose Gb3. *J. Am. Chem. Soc.* **2016**, *138*, 4787–4795. [CrossRef]
54. Chernikov, O.; Kuzmich, A.; Chikalovets, I.; Molchanova, V.; Hua, K.-F. Lectin CGL from the sea mussel *Crenomytilus grayanus* induces Burkitt's lymphoma cells death via interaction with surface glycan. *Int. J. Biol. Macromol.* **2017**, *104*, 508–514. [CrossRef]
55. Li, G.; Cheng, J.; Mei, S.; Wu, T.; Ye, T. *Tachypleus tridentatus* lectin enhances oncolytic vaccinia virus replication to suppress in vivo hepatocellular carcinoma growth. *Mar. Drugs* **2018**, *16*, 200. [CrossRef]
56. Yang, X.; Wu, L.; Duan, X.; Cui, L.; Luo, J.; Li, G. Adenovirus carrying gene encoding Haliotis discus discus sialic acid binding lectin induces cancer cell apoptosis. *Mar. Drugs* **2014**, *12*, 3994–4004. [CrossRef]
57. Wu, B.; Mei, S.; Cui, L.; Zhao, Z.; Chen, J.; Wu, T.; Li, G. Marine lectins DlFBL and HddSBL fused with soluble coxsackie-adenovirus receptor facilitate adenovirus infection in cancer cells but have different effects on cell survival. *Mar. Drugs* **2017**, *15*, 73. [CrossRef]
58. Fujii, Y.; Fujiwara, T.; Koide, Y.; Hasan, I.; Sugawara, S.; Rajia, S.; Kawsar, S.M.A.; Yamamoto, D.; Araki, D.; Kanaly, R.A.; et al. Internalization of a novel, huge lectin from *Ibacus novemdentatus* (slipper lobster) induces apoptosis of mammalian cancer cells. *Glycoconj. J.* **2017**, *34*, 85–94. [CrossRef]
59. Fujii, Y.; Dohmae, N.; Takio, K.; Kawsar, S.M.A.; Matsumoto, R.; Hasan, I.; Koide, Y.; Kanaly, R.A.; Yasumitsu, H.; Ogawa, Y.; et al. A lectin from the mussel Mytilus galloprovincialis has a highly novel primary structure and induces glycan-mediated cytotoxicity of globotriaosylceramide-expressing lymphoma cells. *J. Biol. Chem.* **2012**, *287*, 44772–44783. [CrossRef]
60. Hasan, I.; Sugawara, S.; Fujii, Y.; Koide, Y.; Terada, D.; Iimura, N.; Fujiwara, T.; Takahashi, K.G.; Kojima, N.; Rajia, S.; et al. MytiLec, a mussel R-type lectin, interacts with surface glycan Gb3 on Burkitt's lymphoma cells to trigger apoptosis through multiple pathways. *Mar. Drugs* **2015**, *13*, 7377–7389. [CrossRef]
61. Wu, L.; Yang, X.; Duan, X.; Cui, L.; Li, G. Exogenous expression of marine lectins DlFBL and SpRBL induces cancer cell apoptosis possibly through PRMT5-E2F-1 pathway. *Sci. Rep.* **2014**, *4*, 4505. [CrossRef]
62. Odintsova, N.A.; Belogortseva, N.I.; Khomenko, A.V.; Chikalovets, I.V.; Luk'yanov, P.A. Effect of lectin from the ascidian on the growth and the adhesion of HeLa cells. *Mol. Cell. Biochem.* **2001**, *221*, 133–138. [CrossRef]
63. Wu, T.; Xiang, Y.; Liu, T.; Wang, X.; Ren, X.; Ye, T.; Li, G. Oncolytic Vaccinia Virus Expressing Aphrocallistes vastus Lectin as a Cancer Therapeutic Agent. *Mar. Drugs* **2019**, *17*, 363. [CrossRef]
64. Rabelo, L.; Monteiro, N.; Serquiz, R.; Santos, P.; Oliveira, R.; Oliveira, A.; Rocha, H.; Morais, A.H.; Uchoa, A.; Santos, E. A lactose-binding lectin from the marine sponge *Cinachyrella apion* (Cal) induces cell death in human cervical adenocarcinoma cells. *Mar. Drugs* **2012**, *10*, 727–743. [CrossRef]
65. Queiroz, A.F.S.; Silva, R.A.; Moura, R.M.; Dreyfuss, J.L.; Paredes-Gamero, E.J.; Souza, A.C.S.; Tersariol, I.L.S.; Santos, E.A.; Nader, H.B.; Justo, G.Z.; et al. Growth inhibitory activity of a novel lectin from *Cliona varians* against K562 human erythroleukemia cells. *Cancer Chemother. Pharmacol.* **2009**, *63*, 1023–1033. [CrossRef]
66. Do Nascimento-Neto, L.G.; Cabral, M.G.; Carneiro, R.F.; Silva, Z.; Arruda, F.V.S.; Nagano, C.S.; Fernandes, A.R.; Sampaio, A.H.; Teixeira, E.H.; Videira, P.A. Halilectin-3, a lectin from the marine sponge *Haliclona caerulea*, induces apoptosis and autophagy in human breast cancer MCF7 cells through caspase-9 pathway and LC3-II protein expression. *Anticancer Agents Med. Chem.* **2018**, *18*, 521–528. [CrossRef]
67. Pajic, I.; Kljajic, Z.; Dogovic, N.; Sladic, D.; Juranic, Z.; Gasic, M.J. A novel lectin from the sponge Haliclona cratera: Isolation, characterization and biological activity. *Comp. Biochem. Physiol. C Toxicol. Pharmacol.* **2002**, *132*, 213–221. [CrossRef]
68. Matsumoto, R.; Fujii, Y.; Kawsar, S.M.A.; Kanaly, R.A.; Yasumitsu, H.; Koide, Y.; Hasan, I.; Iwahara, C.; Ogawa, Y.; Im, C.H.; et al. Cytotoxicity and glycan-binding properties of an 18 kDa lectin isolated from the marine sponge *Halichondria okadai*. *Toxins* **2012**, *4*, 323–338. [CrossRef] [PubMed]
69. Hasan, I.; Ozeki, Y. Histochemical localization of *N*-acetylhexosamine-binding lectin HOL-18 in *Halichondria okadai* (Japanese black sponge), and its antimicrobial and cytotoxic anticancer effects. *Int. J. Biol. Macromol.* **2019**, *124*, 819–827. [CrossRef] [PubMed]
70. Ponder, W.; Lindberg, D.R. *Phylogeny and Evolution of the Mollusca*; University of California Press: Berkeley, CA, USA, 2008.

71. Bouchet, P.; Duarte, C.M. The exploration of marine biodiversity scientific and technological challenges. *Fundatiòn BBVA* **2006**, *33*, 1–34.
72. Fredrick, W.S.; Ravichandran, S. Hemolymph proteins in marine crustaceans. *Asian Pac. J. Trop. Biomed* **2012**, *2*, 496–502. [CrossRef]
73. Kovalchuk, S.N.; Chikalovets, I.V.; Chernikov, O.V.; Molchanova, V.I.; Li, W.; Rasskazov, V.A.; Lukyanov, P.A. cDNA cloning and structural characterization of a lectin from the mussel *Crenomytilus grayanus* with a unique amino acid sequence and antibacterial activity. *Fish Shellfish Immunol.* **2013**, *35*, 1320–1324. [CrossRef]
74. Bekri, S.; Lidove, O.; Jaussaud, R.; Knebelmann, B.; Barbey, F. The role of ceramide trihexoside (globotriaosylceramide) in the diagnosis and follow-up of the efficacy of treatment of Fabry disease: A review of the literature. *Cardiovasc. Hematol. Agents Med. Chem.* **2006**, *4*, 289–297. [CrossRef]
75. Behnam-Motlagh, P.; Tyler, A.; Grankvist, K.; Johansson, A. Verotoxin-1 treatment or manipulation of its receptor globotriaosylceramide (gb3) for reversal of multidrug resistance to cancer chemotherapy. *Toxins* **2010**, *2*, 2467–2477. [CrossRef]
76. Qi, M.; Elion, E.A. MAP kinase pathways. *J. Cell. Sci.* **2005**, *118*, 3569–3572. [CrossRef]
77. Keshet, Y.; Seger, R. The MAP kinase signaling cascades: A system of hundreds of components regulates a diverse array of physiological functions. *Methods Mol. Biol.* **2010**, *661*, 3–38.
78. Morrison, D.K. MAP kinase pathways. *Cold Spring Harb. Perspect. Biol.* **2012**, *4*. [CrossRef] [PubMed]
79. Gilboa-Garber, N.; Wu, A.M. Binding properties and applications of *Aplysia gonad* lectin. In *The Molecular Immunology of Complex Carbohydrates—2*; Wu, A.M., Ed.; Springer US: Boston, MA, USA, 2001; pp. 109–126. ISBN 978-1-4615-1267-7.
80. Avichezer, D.; Leibovici, J.; Gilboa-Garber, N.; Michowitz, M.; Lapis, K. New galactophilic lectins reduce tumorigenicity and preserve immunogenicity of Lewis lung carcinoma cells. In Proceedings of the Lectures and Symposia 14th Int. Cancer Cong., II, Budapest, Hungary, 21–27 August 1986; Karger: Basel, Switzerland, 1987; pp. 79–85.
81. Gokudan, S.; Muta, T.; Tsuda, R.; Koori, K.; Kawahara, T.; Seki, N.; Mizunoe, Y.; Wai, S.N.; Iwanaga, S.; Kawabata, S. Horseshoe crab acetyl group-recognizing lectins involved in innate immunity are structurally related to fibrinogen. *Proc. Natl. Acad. Sci. USA* **1999**, *96*, 10086–10091. [CrossRef] [PubMed]
82. Kairies, N.; Beisel, H.G.; Fuentes-Prior, P.; Tsuda, R.; Muta, T.; Iwanaga, S.; Bode, W.; Huber, R.; Kawabata, S. The 2.0-A crystal structure of tachylectin 5A provides evidence for the common origin of the innate immunity and the blood coagulation systems. *Proc. Natl. Acad. Sci. USA* **2001**, *98*, 13519–13524. [CrossRef] [PubMed]
83. Wang, N.; Whang, I.; Lee, J. A novel C-type lectin from abalone, Haliotis discus discus, agglutinates Vibrio alginolyticus. *Dev. Comp. Immunol.* **2008**, *32*, 1034–1040. [CrossRef]
84. Carneiro, R.F.; de Melo, A.A.; de Almeida, A.S.; da Moura, R.M.; Chaves, R.P.; de Sousa, B.L.; do Nascimento, K.S.; Sampaio, S.S.; Lima, J.P.M.S.; Cavada, B.S.; et al. H-3, a new lectin from the marine sponge *Haliclona caerulea*: Purification and mass spectrometric characterization. *Int. J. Biochem. Cell Biol.* **2013**, *45*, 2864–2873. [CrossRef]
85. Xiong, C.; Li, W.; Liu, H.; Zhang, W.; Dou, J.; Bai, X.; Du, Y.; Ma, X. A normal mucin-binding lectin from the sponge *Craniella australiensis*. *Comp. Biochem. Physiol. C Toxicol. Pharmacol.* **2006**, *143*, 9–16. [CrossRef]
86. Miarons, P.B.; Fresno, M. Lectins from tropical sponges. Purification and characterization of lectins from genus *Aplysina*. *J. Biol. Chem.* **2000**, *275*, 29283–29289. [CrossRef]
87. Medeiros, D.S.; Medeiros, T.L.; Ribeiro, J.K.C.; Monteiro, N.K.V.; Migliolo, L.; Uchoa, A.F.; Vasconcelos, I.M.; Oliveira, A.S.; de Sales, M.P.; Santos, E.A. A lactose specific lectin from the sponge *Cinachyrella apion*: purification, characterization, N-terminal sequences alignment and agglutinating activity on Leishmania promastigotes. *Comp. Biochem. Physiol. B Biochem. Mol. Biol.* **2010**, *155*, 211–216. [CrossRef]
88. Frisch, S.M.; Screaton, R.A. Anoikis mechanisms. *Curr. Opin. Cell Biol.* **2001**, *13*, 555–562. [CrossRef]
89. Mihaly, S.R.; Sakamachi, Y.; Ninomiya-Tsuji, J.; Morioka, S. Noncanonical cell death program independent of caspase activation cascade and necroptotic modules is elicited by loss of TGFβ-activated kinase 1. *Sci. Rep.* **2017**, *7*, 2918. [CrossRef]
90. Gundacker, D.; Leys, S.P.; Schröder, H.C.; Müller, I.M.; Müller, W.E. Isolation and cloning of a C-type lectin from the hexactinellid sponge *Aphrocallistes vastus*: A putative aggregation factor. *Glycobiology* **2001**, *11*, 21–29. [CrossRef] [PubMed]

91. Moura, R.M.; Queiroz, A.F.S.; Fook, J.M.S.L.L.; Dias, A.S.F.; Monteiro, N.K.V.; Ribeiro, J.K.C.; Moura, G.E.D.D.; Macedo, L.L.P.; Santos, E.A.; Sales, M.P. CvL, a lectin from the marine sponge *Cliona varians*: Isolation, characterization and its effects on pathogenic bacteria and Leishmania promastigotes. *Comp. Biochem. Physiol. Part A Mol. Integr. Physiol.* **2006**, *145*, 517–523. [CrossRef] [PubMed]
92. Bröker, L.E.; Kruyt, F.A.E.; Giaccone, G. Cell death independent of caspases: A review. *Clin. Cancer Res.* **2005**, *11*, 3155–3162. [CrossRef] [PubMed]
93. Guicciardi, M.E.; Deussing, J.; Miyoshi, H.; Bronk, S.F.; Svingen, P.A.; Peters, C.; Kaufmann, S.H.; Gores, G.J. Cathepsin B contributes to TNF-alpha-mediated hepatocyte apoptosis by promoting mitochondrial release of cytochrome c. *J. Clin. Investig.* **2000**, *106*, 1127–1137. [CrossRef] [PubMed]
94. Foghsgaard, L.; Wissing, D.; Mauch, D.; Lademann, U.; Bastholm, L.; Boes, M.; Elling, F.; Leist, M.; Jäättelä, M. Cathepsin B acts as a dominant execution protease in tumor cell apoptosis induced by tumor necrosis factor. *J. Cell Biol.* **2001**, *153*, 999–1010. [CrossRef] [PubMed]
95. Su, Z.; Yang, Z.; Xie, L.; DeWitt, J.P.; Chen, Y. Cancer therapy in the necroptosis era. *Cell Death Differ.* **2016**, *23*, 748–756. [CrossRef]
96. Das, A.; McDonald, D.G.; Dixon-Mah, Y.N.; Jacqmin, D.J.; Samant, V.N.; Vandergrift, W.A.; Lindhorst, S.M.; Cachia, D.; Varma, A.K.; Vanek, K.N.; et al. RIP1 and RIP3 complex regulates radiation-induced programmed necrosis in glioblastoma. *Tumour Biol.* **2016**, *37*, 7525–7534. [CrossRef]
97. Horton, T.; Kroh, A.; Ahyong, B.; Bailly, N.; Brandão, S.N.; Costello, M.J.; Gofas, S.; Hernandez, F.; Holovachov, O.; Boyko, C.B.; et al. *World Register of Marine Species*; WoRMS Editorial Board: Ostend, Belgium, 2018.
98. Negi, B.; Kumar, D.; Rawat, D.S. Marine Peptides as Anticancer Agents: A Remedy to Mankind by Nature. *Curr. Protein Pept. Sci.* **2017**, *18*, 885–904. [CrossRef]
99. Ankisetty, S.; Khan, S.I.; Avula, B.; Gochfeld, D.; Khan, I.A.; Slattery, M. Chlorinated didemnins from the tunicate *Trididemnum solidum*. *Mar. Drugs* **2013**, *11*, 4478–4486. [CrossRef]
100. Lee, J.; Currano, J.N.; Carroll, P.J.; Joullié, M.M. Didemnins, tamandarins and related natural products. *Nat. Prod. Rep.* **2012**, *29*, 404–424. [CrossRef]
101. Belogortseva, N.; Molchanova, V.; Glazunov, V.; Evtushenko, E.; Luk'yanov, P. *N*-Acetyl-D-glucosamine-specific lectin from the ascidian *Didemnum ternatanum*. *Biochim. Biophys. Acta* **1998**, *1380*, 249–256. [CrossRef]
102. Fang, Y.; Eglen, R.M. Three-dimensional cell cultures in drug discovery and development. *SLAS Discov.* **2017**, *22*, 456–472.
103. Bourgine, P.E.; Klein, T.; Paczulla, A.M.; Shimizu, T.; Kunz, L.; Kokkaliaris, K.D.; Coutu, D.L.; Lengerke, C.; Skoda, R.; Schroeder, T.; et al. In vitro biomimetic engineering of a human hematopoietic niche with functional properties. *Proc. Natl. Acad. Sci. USA* **2018**, *115*, E5688–E5695. [CrossRef] [PubMed]
104. Clarke, B.T. The natural history of amphibian skin secretions, their normal functioning and potential medical applications. *Biol. Rev. Camb. Philos. Soc.* **1997**, *72*, 365–379. [CrossRef] [PubMed]
105. Lu, C.-X.; Nan, K.-J.; Lei, Y. Agents from amphibians with anticancer properties. *Anticancer Drugs* **2008**, *19*, 931–939. [CrossRef] [PubMed]
106. Iwama, M.; Ogawa, Y.; Sasaki, N.; Nitta, K.; Takayanagi, Y.; Ohgi, K.; Tsuji, T.; Irie, M. Effect of modification of the carboxyl groups of the sialic acid binding lectin from bullfrog (*Rana catesbeiana*) oocyte on anti-tumor activity. *Biol. Pharm. Bull.* **2001**, *24*, 978–981. [CrossRef]
107. Nitta, K.; Ozaki, K.; Ishikawa, M.; Furusawa, S.; Hosono, M.; Kawauchi, H.; Sasaki, K.; Takayanagi, Y.; Tsuiki, S.; Hakomori, S. Inhibition of cell proliferation by *Rana catesbeiana* and *Rana japonica* lectins belonging to the ribonuclease superfamily. *Cancer Res.* **1994**, *54*, 920–927.
108. Tatsuta, T.; Hosono, M.; Sugawara, S.; Kariya, Y.; Ogawa, Y.; Hakomori, S.; Nitta, K. Sialic acid-binding lectin (leczyme) induces caspase-dependent apoptosis-mediated mitochondrial perturbation in Jurkat cells. *Int. J. Oncol.* **2013**, *43*, 1402–1412. [CrossRef]
109. Ogawa, Y.; Sugawara, S.; Tatsuta, T.; Hosono, M.; Nitta, K.; Fujii, Y.; Kobayashi, H.; Fujimura, T.; Taka, H.; Koide, Y.; et al. Sialyl-glycoconjugates in cholesterol-rich microdomains of P388 cells are the triggers for apoptosis induced by *Rana catesbeiana* oocyte ribonuclease. *Glycoconj. J.* **2014**, *31*, 171–184. [CrossRef]
110. Tatsuta, T.; Hosono, M.; Takahashi, K.; Omoto, T.; Kariya, Y.; Sugawara, S.; Hakomori, S.; Nitta, K. Sialic acid-binding lectin (leczyme) induces apoptosis to malignant mesothelioma and exerts synergistic antitumor effects with TRAIL. *Int. J. Oncol.* **2014**, *44*, 377–384. [CrossRef]

111. Kariya, Y.; Tatsuta, T.; Sugawara, S.; Kariya, Y.; Nitta, K.; Hosono, M. RNase activity of sialic acid-binding lectin from bullfrog eggs drives antitumor effect via the activation of p38 MAPK to caspase-3/7 signaling pathway in human breast cancer cells. *Int. J. Oncol.* **2016**, *49*, 1334–1342. [CrossRef] [PubMed]
112. Satoh, T.; Tatsuta, T.; Sugawara, S.; Hara, A.; Hosono, M. Synergistic anti-tumor effect of bullfrog sialic acid-binding lectin and pemetrexed in malignant mesothelioma. *Oncotarget* **2017**, *8*, 42466–42477. [CrossRef] [PubMed]
113. Tatsuta, T.; Satoh, T.; Sugawara, S.; Hara, A.; Hosono, M. Sialic acid-binding lectin from bullfrog eggs inhibits human malignant mesothelioma cell growth in vitro and in vivo. *PLoS ONE* **2018**, *13*, e0190653. [CrossRef] [PubMed]
114. Tatsuta, T.; Sato, S.; Sato, T.; Sugawara, S.; Suzuki, T.; Hara, A.; Hosono, M. Sialic acid-binding lectin from bullfrog eggs exhibits an anti-tumor effect against breast cancer cells including triple-negative phenotype cells. *Molecules* **2018**, *23*, 2714. [CrossRef] [PubMed]
115. Yiang, G.-T.; Yu, Y.-L.; Chou, P.-L.; Tsai, H.-F.; Chen, L.-A.; Chen, Y.H.; Su, K.-J.; Wang, J.-J.; Bau, D.-T.; Wei, C.-W. The cytotoxic protein can induce autophagocytosis in addition to apoptosis in MCF-7 human breast cancer cells. *In Vivo* **2012**, *26*, 403–409.
116. Tseng, H.-H.; Yu, Y.-L.; Chen, Y.-L.S.; Chen, J.-H.; Chou, C.-L.; Kuo, T.-Y.; Wang, J.-J.; Lee, M.-C.; Huang, T.-H.; Chen, M.H.-C.; et al. RC-RNase-induced cell death in estrogen receptor positive breast tumors through down-regulation of Bcl-2 and estrogen receptor. *Oncol. Rep.* **2011**, *25*, 849–853.
117. Hu, C.C.; Tang, C.H.; Wang, J.J. Caspase activation in response to cytotoxic *Rana catesbeiana* ribonuclease in MCF-7 cells. *FEBS Lett.* **2001**, *503*, 65–68. [CrossRef]
118. Wei, C.W.; Hu, C.C.A.; Tang, C.H.A.; Lee, M.C.; Wang, J.J. Induction of differentiation rescues HL-60 cells from *Rana catesbeiana* ribonuclease-induced cell death. *FEBS Lett.* **2002**, *531*, 421–426. [CrossRef]
119. Chen, J.-N.; Yiang, G.-T.; Lin, Y.-F.; Chou, P.-L.; Wu, T.-K.; Chang, W.-J.; Chen, C.; Yu, Y.-L. *Rana catesbeiana* ribonuclease induces cell apoptosis via the caspase-9/-3 signaling pathway in human glioblastoma DBTRG, GBM8901 and GBM8401 cell lines. *Oncol. Lett.* **2015**, *9*, 2471–2476. [CrossRef]
120. Tang, C.-H.A.; Hu, C.-C.A.; Wei, C.-W.; Wang, J.-J. Synergism of *Rana catesbeiana* ribonuclease and IFN-gamma triggers distinct death machineries in different human cancer cells. *FEBS Lett.* **2005**, *579*, 265–270. [CrossRef]
121. Hu, C.C.; Lee, Y.H.; Tang, C.H.; Cheng, J.T.; Wang, J.J. Synergistic cytotoxicity of *Rana catesbeiana* ribonuclease and IFN-gamma on hepatoma cells. *Biochem. Biophys. Res. Commun.* **2001**, *280*, 1229–1236. [CrossRef] [PubMed]
122. Cui, L.; Peng, H.; Zhang, R.; Chen, Y.; Zhao, L.; Tang, K. Recombinant hHscFv-RC-RNase protein derived from transgenic tobacco acts as a bifunctional molecular complex against hepatocellular carcinoma. *Biotechnol. Appl. Biochem.* **2012**, *59*, 323–329. [CrossRef] [PubMed]
123. Li, G.; Gao, Y.; Cui, L.; Wu, L.; Yang, X.; Chen, J. Anguilla japonica lectin 1 delivery through adenovirus vector induces apoptotic cancer cell death through interaction with PRMT5. *J. Gene Med* **2016**, *18*, 65–74. [CrossRef] [PubMed]
124. Yao, D.; Pan, S.; Zhou, M. Structural characterization and antitumor and mitogenic activity of a lectin from the gill of bighead carp (*Aristichthys nobilis*). *Fish Physiol. Biochem.* **2012**, *38*, 1815–1824. [CrossRef]
125. Shirai, T.; Watanabe, Y.; Lee, M.; Ogawa, T.; Muramoto, K. Structure of rhamnose-binding lectin CSL3: unique pseudo-tetrameric architecture of a pattern recognition protein. *J. Mol. Biol.* **2009**, *391*, 390–403. [CrossRef]
126. Bah, C.S.F.; Fang, E.F.; Ng, T.B.; Mros, S.; McConnell, M.; Bekhit, A.E.-D.A. Purification and characterization of a rhamnose-binding chinook salmon roe lectin with antiproliferative activity toward tumor cells and nitric oxide-inducing activity toward murine macrophages. *J. Agric. Food Chem.* **2011**, *59*, 5720–5728. [CrossRef]
127. Sugawara, S.; Hosono, M.; Ogawa, Y.; Takayanagi, M.; Nitta, K. Catfish egg lectin causes rapid activation of multidrug resistance 1 P-glycoprotein as a lipid translocase. *Biol. Pharm. Bull.* **2005**, *28*, 434–441. [CrossRef]
128. Sugawara, S.; Im, C.; Kawano, T.; Tatsuta, T.; Koide, Y.; Yamamoto, D.; Ozeki, Y.; Nitta, K.; Hosono, M. *Catfish rhamnose*-binding lectin induces G0/1 cell cycle arrest in Burkitt's lymphoma cells via membrane surface Gb3. *Glycoconj. J.* **2017**, *34*, 127–138. [CrossRef]
129. Titani, K.; Takio, K.; Kuwada, M.; Nitta, K.; Sakakibara, F.; Kawauchi, H.; Takayanagi, G.; Hakomori, S. Amino acid sequence of sialic acid binding lectin from frog (*Rana catesbeiana*) eggs. *Biochemistry* **1987**, *26*, 2189–2194. [CrossRef]

130. Nitta, K.; Takayanagi, G.; Kawauchi, H.; Hakomori, S. Isolation and characterization of *Rana catesbeiana* lectin and demonstration of the lectin-binding glycoprotein of rodent and human tumor cell membranes. *Cancer Res.* **1987**, *47*, 4877–4883.
131. Nitta, K.; Ozaki, K.; Tsukamoto, Y.; Furusawa, S.; Ohkubo, Y.; Takimoto, H.; Murata, R.; Hosono, M.; Hikichi, N.; Sasaki, K. Characterization of a *Rana catesbeiana* lectin-resistant mutant of leukemia P388 cells. *Cancer Res.* **1994**, *54*, 928–934. [PubMed]
132. Irie, M.; Nitta, K.; Nonaka, T. Biochemistry of frog ribonucleases. *Cell. Mol. Life Sci.* **1998**, *54*, 775–784. [CrossRef] [PubMed]
133. Liao, Y.D.; Huang, H.C.; Chan, H.J.; Kuo, S.J. Large-scale preparation of a ribonuclease from *Rana catesbeiana* (bullfrog) oocytes and characterization of its specific cytotoxic activity against tumor cells. *Protein Expr. Purif.* **1996**, *7*, 194–202. [CrossRef] [PubMed]
134. Tatsuta, T.; Hosono, M.; Miura, Y.; Sugawara, S.; Kariya, Y.; Hakomori, S.; Nitta, K. Involvement of ER stress in apoptosis induced by sialic acid-binding lectin (leczyme) from bullfrog eggs. *Int. J. Oncol.* **2013**, *43*, 1799–1808. [CrossRef]
135. Tatsuta, T.; Hosono, M.; Ogawa, Y.; Inage, K.; Sugawara, S.; Nitta, K. Downregulation of Hsp70 inhibits apoptosis induced by sialic acid-binding lectin (leczyme). *Oncol. Rep.* **2014**, *31*, 13–18. [CrossRef]
136. Osborne, C.K.; Wakeling, A.; Nicholson, R.I. Fulvestrant: an oestrogen receptor antagonist with a novel mechanism of action. *Br. J. Cancer* **2004**, *90* (Suppl. 1), S2–S6. [CrossRef]
137. Petrovski, G.; Zahuczky, G.; Katona, K.; Vereb, G.; Martinet, W.; Nemes, Z.; Bursch, W.; Fésüs, L. Clearance of dying autophagic cells of different origin by professional and non-professional phagocytes. *Cell Death Differ.* **2007**, *14*, 1117–1128. [CrossRef]
138. Bayat Mokhtari, R.; Homayouni, T.S.; Baluch, N.; Morgatskaya, E.; Kumar, S.; Das, B.; Yeger, H. Combination therapy in combating cancer. *Oncotarget* **2017**, *8*, 38022–38043. [CrossRef]
139. Lee, Y.-H.; Wei, C.-W.; Wang, J.-J.; Chiou, C.-T. *Rana catesbeiana* ribonuclease inhibits Japanese encephalitis virus (JEV) replication and enhances apoptosis of JEV-infected BHK-21 cells. *Antiviral Res.* **2011**, *89*, 193–198. [CrossRef]
140. McLaughlin, M.B.; Jialal, I. Calcitonin. In *StatPearls*; StatPearls Publishing: Treasure Island, FL, USA, 2019.
141. Jensen, L.E.; Thiel, S.; Petersen, T.E.; Jensenius, J.C. A rainbow trout lectin with multimeric structure. *Comp. Biochem. Physiol. B, Biochem. Mol. Biol.* **1997**, *116*, 385–390. [CrossRef]
142. Ottinger, C.A.; Johnson, S.C.; Ewart, K.V.; Brown, L.L.; Ross, N.W. Enhancement of anti-Aeromonas salmonicida activity in Atlantic salmon (*Salmo salar*) macrophages by a mannose-binding lectin. *Comp. Biochem. Physiol. C, Pharmacol. Toxicol. Endocrinol.* **1999**, *123*, 53–59. [CrossRef]
143. Dong, C.-H.; Yang, S.-T.; Yang, Z.-A.; Zhang, L.; Gui, J.-F. A C-type lectin associated and translocated with cortical granules during oocyte maturation and egg fertilization in fish. *Dev. Biol.* **2004**, *265*, 341–354. [CrossRef] [PubMed]
144. Tasumi, S.; Yang, W.-J.; Usami, T.; Tsutsui, S.; Ohira, T.; Kawazoe, I.; Wilder, M.N.; Aida, K.; Suzuki, Y. Characteristics and primary structure of a galectin in the skin mucus of the Japanese eel, Anguilla japonica. *Dev. Comp. Immunol.* **2004**, *28*, 325–335. [CrossRef] [PubMed]
145. Dutta, S.; Sinha, B.; Bhattacharya, B.; Chatterjee, B.; Mazumder, S. Characterization of a galactose binding serum lectin from the Indian catfish, Clarias batrachus: possible involvement of fish lectins in differential recognition of pathogens. *Comp. Biochem. Physiol. C Toxicol. Pharmacol.* **2005**, *141*, 76–84. [CrossRef]
146. Pan, S.; Tang, J.; Gu, X. Isolation and characterization of a novel fucose-binding lectin from the gill of bighead carp (Aristichthys nobilis). *Vet. Immunol. Immunopathol.* **2010**, *133*, 154–164. [CrossRef]
147. Hosono, M.; Kawauchi, H.; Nitta, K.; Takayanagi, Y.; Shiokawa, H.; Mineki, R.; Murayama, K. Purification and characterization of *Silurus asotus* (catfish) roe lectin. *Biol. Pharm. Bull.* **1993**, *16*, 1–5. [CrossRef]
148. Suryadinata, R.; Sadowski, M.; Sarcevic, B. Control of cell cycle progression by phosphorylation of cyclin-dependent kinase (CDK) substrates. *Biosci. Rep.* **2010**, *30*, 243–255. [CrossRef]
149. Gartel, A.L.; Radhakrishnan, S.K. Lost in transcription: p21 repression, mechanisms, and consequences. *Cancer Res.* **2005**, *65*, 3980–3985. [CrossRef]
150. Shailesh, H.; Zakaria, Z.Z.; Baiocchi, R.; Sif, S. Protein arginine methyltransferase 5 (PRMT5) dysregulation in cancer. *Oncotarget* **2018**, *9*, 36705–36718. [CrossRef]

151. Irwin, M.; Marin, M.C.; Phillips, A.C.; Seelan, R.S.; Smith, D.I.; Liu, W.; Flores, E.R.; Tsai, K.Y.; Jacks, T.; Vousden, K.H.; et al. Role for the p53 homologue p73 in E2F-1-induced apoptosis. *Nature* **2000**, *407*, 645–648. [CrossRef]
152. Wu, X.; Levine, A.J. p53 and E2F-1 cooperate to mediate apoptosis. *Proc. Natl. Acad. Sci. USA* **1994**, *91*, 3602–3606. [CrossRef] [PubMed]
153. Cammarata, M.; Vazzana, M.; Chinnici, C.; Parrinello, N. A serum fucolectin isolated and characterized from sea bass *Dicentrarchus labrax*. *Biochim. Biophys. Acta* **2001**, *1528*, 196–202. [CrossRef]
154. Vasconcelos, I.M.; Oliveira, J.T.A. Antinutritional properties of plant lectins. *Toxicon* **2004**, *44*, 385–403. [CrossRef] [PubMed]
155. Lam, S.K.; Ng, T.B. Lectins: production and practical applications. *Appl. Microbiol. Biotechnol.* **2011**, *89*, 45–55. [CrossRef] [PubMed]
156. Chakraborty, C.; Hsu, C.-H.; Wen, Z.-H.; Lin, C.-S. Anticancer drugs discovery and development from marine organism. *Curr. Top. Med. Chem.* **2009**, *9*, 1536–1545. [CrossRef] [PubMed]
157. Clinicaltrials.gov. Available online: https://clinicaltrials.gov/ct2/results?cond=Breast+Cancer&term=sunitinib&cntry=&state=&city=&dist= (accessed on 16 August 2019).

marine drugs

Article

Oncolytic Vaccinia Virus Expressing *Aphrocallistes vastus* Lectin as a Cancer Therapeutic Agent

Tao Wu, Yulin Xiang, Tingting Liu, Xue Wang, Xiaoyuan Ren, Ting Ye * and Gongchu Li *

Zhejiang Sci-Tech University Hangzhou Gongchu Joint Institute of Biomedicine, College of Life Sciences and Medicine, Zhejiang Sci-Tech University, Hangzhou 310018, China; wutao0920@163.com (T.W.); q522329467@163.com (Y.X.); m13617965853@163.com (T.L.); wx18815610822@163.com (X.W.); imrenxy@163.com (X.R.)

* Correspondence: yeting@zstu.edu.cn (T.Y.); lgc@zstu.edu.cn (G.L.); Tel.: +86-131-7360-7684 (G.L.)

Received: 6 May 2019; Accepted: 17 June 2019; Published: 19 June 2019

Abstract: Lectins display a variety of biological functions including insecticidal, antimicrobial, as well as antitumor activities. In this report, a gene encoding *Aphrocallistes vastus* lectin (AVL), a C-type lectin, was inserted into an oncolytic vaccinia virus vector (oncoVV) to form a recombinant virus oncoVV-AVL, which showed significant in vitro antiproliferative activity in a variety of cancer cell lines. Further investigations revealed that oncoVV-AVL replicated faster than oncoVV significantly in cancer cells. Intracellular signaling elements including NF-κB2, NIK, as well as ERK were determined to be altered by oncoVV-AVL. Virus replication upregulated by AVL was completely dependent on ERK activity. Furthermore, in vivo studies showed that oncoVV-AVL elicited significant antitumor effect in colorectal cancer and liver cancer mouse models. Our study might provide insights into a novel way of the utilization of marine lectin AVL in oncolytic viral therapies.

Keywords: *Aphrocallistes vastus* lectin; oncolytic vaccinia virus; ERK

1. Introduction

Lectins, as a class of specific glycosyl-binding glycoproteins, preferentially recognize and bind carbohydrate complexes [1,2]. Since the first discovery in 1888 [3], hundreds of lectins have been isolated and characterized. Lectins were obtained from microorganisms, animals, and plants. Animal and plant lectins lack primary structural homology [4–6], but they have demonstrated similar carbohydrates-binding activities. Animal lectins are categorized into C-type lectins, Galectins, P-type lectins, L-Type Lectins, R-Type Lectins, and I-type lectins [7]. In past decades, lectins have been developed to form a variety of biological techniques, such as lectin array, lectin blot, as well as lectin-based chromatography, to analyze glycofiles and biomarkers for various cancers, including ovarian cancer [8], pancreatic cancer [9], prostate cancer [10], aggressive breast cancer [11,12], and liver cancer [13]. In addition, lectins such as *Maackia amurensis* seed lectin [14], Concanavalin A [15], *Fenneropenaeus indicus* hemolymph fucose binding lectin [16], *Polygonatum cyrtonema* lectin [17], as well as MytiLec [18–20] were shown to be cytotoxic to cancer cells through inducing apoptosis or autophagy. Furthermore, various lectins delivered through viral vectors elicited anticancer effect in vitro and in vivo [21–26].

Sponges, which are known to be one of the oldest living marine organisms, produce different types of biological molecules, such as okadaic acid and halichondrin B [27]. Halichondrin B from the Japanese black sponge *Halichondria okadai* as an effective anticancer drug [28], can modulate microtubule dynamics. Many types of lectins are also produced in various species of sponges, including galectins, N-acetylamino-carbohydrate-specific lectin, c-type lectin, and so on [29,30]. The *Aphrocallistes vastus* lectin (AVL) is a Ca2+-dependent lectin and inhibited by bird's nest glycoprotein and D-galactose. AVL shows the highest similarity to C-type lectins from higher metazoan phyla [31]. Here, we addressed C-type lectin from *Aphrocallistes vastus* with biological activities in enhancing the replication of oncolytic vaccinia virus and its therapeutic effect for cancers.

Oncolytic viruses have the advantage of high killing effects without drug resistance. Since the first report of a thymidine kinase (TK) deleted herpes simplex virus (HSV) in cancer treatment [32], more than 10 families of oncolytic viruses have entered clinical trials, including adenovirus, coxsackie virus, herpes simplex virus, measles virus, new castle disease virus, parvovirus, poliovirus, reovirus, and vesicular stomatitis virus [33]. Among them, vaccinia virus (VV) has the longest history of use in humans. VV can be used as replicating vectors harboring therapeutic genes to directly lyse tumor cells or as cancer vaccines to stimulate antitumor immunity. JX-594 is an oncolytic vaccinia virus carrying a human granulocyte-macrophage colony-stimulating factor (GM-CSF) gene with the TK gene deletion [34]. Up to now, JX-594 has been advanced to clinical phase III for the treatment of advanced hepatocellular carcinoma (HCC) and clinical phase I trial for renal cell carcinoma [35]. In addition, there are other oncolytic vaccinia viruses that have entered clinical trials, including GL-ONC1 and vvDD-CDSR [36].

Previously, the harboring of a gene encoding *Tachypleus tridentatus* lectin (TTL) was shown to enhance the therapeutic effect of oncolytic vaccinia virus in a hepatocellular carcinoma mouse model [21], suggesting that harboring lectin genes may enhance the therapeutic effect of oncolytic viruses. At present, the therapeutic effect of lectins was determined in animal models and in cancer cell lines. However, there are still no lectins entering clinical trials. In this study, a gene encoding AVL was inserted into an oncolytic vaccinia virus (oncoVV) vector, which is deficient of TK gene for cancer specific replication [37], forming a recombinant virus oncoVV-AVL. The antitumor effect of oncoVV-AVL and the underlying mechanisms were analyzed.

2. Results

2.1. AVL Expression Through a Non-Replicating Adenovirus Showed Cytotoxicity in a Variety of Cancer Cells

In previous studies, adenoviruses (Ad) harboring lectins had shown significant cytotoxicity to a variety of cancer cells [22–25]. To investigate the cytotoxicity of exogenously expressed AVL, Ad-AVL, a non-replicating adenovirus carrying an AVL gene, was assessed by MTT assay in colorectal cancer cell line HCT116, glioma cell line U251, hepatocellular carcinoma cell lines BEL-7404, and MHCC97-H, as well as colon cancer cell line HT-29. Ad-EGFP was used as a control. As shown in the Figure 1, the cell viability after the treatment of Ad-AVL in tumor cells was obviously lower than that of Ad-EGFP. The results indicated that exogenous AVL expression elicited cytotoxicity in various cancer cells.

2.2. AVL Significantly Enhanced the Antiproliferative Effect of Oncolytic Vaccinia Virus

In order to determine the antiproliferative efficacy of oncolytic vaccinia virus oncoVV-AVL, the viability of various cell lines treated with oncoVV-AVL was investigated. The empty vector oncoVV served as the control. After virus infection at different MOIs for 48 h and 72 h, the viability of the cell lines (HCT116, U87, BEL-7404 and 4T1-LUC) was measured by MTT assay. As shown in Figure 2a, the cell viability of the oncoVV-AVL group was obviously lower than that of oncoVV in HCT116 cells, showing a time and dosage dependent manner. As shown in Figure 2b–d, similar results were yielded in U87, 4T1-LUC, and BEL-7404 cells. Taken together, the results indicated that AVL promoted the antiproliferative efficacy of oncolytic vaccinia virus in cancer cells.

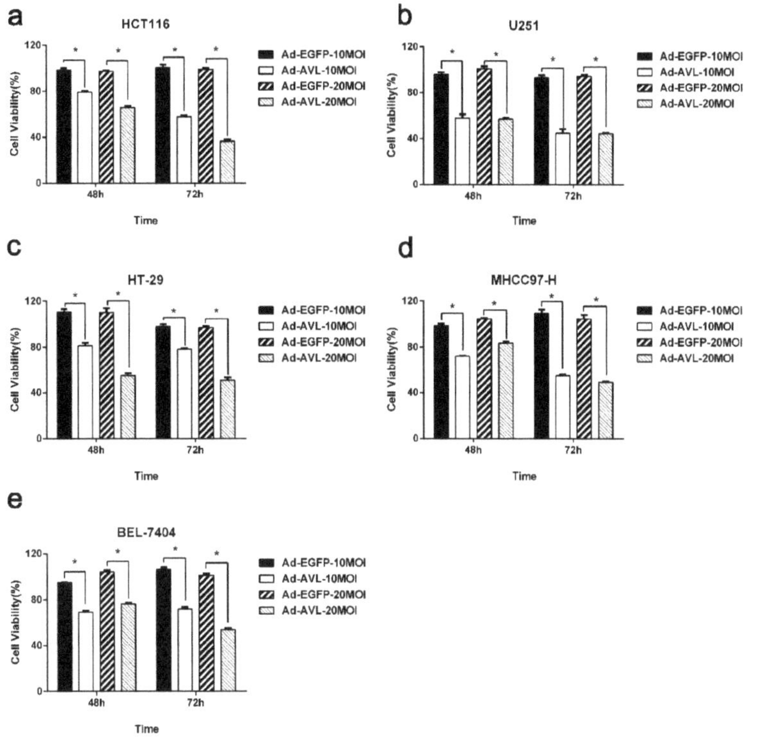

Figure 1. The cytotoxicity of Ad-AVL in various tumor cells. The cytotoxicity of Ad-AVL was measured by MTT assay in HCT116 cells (**a**), U251 cells (**b**), HT-29 cells (**c**), MHCC-97-H cells (**d**), and BEL-7404 cells (**e**). Ad-EGFP served as a control. Data were expressed as the mean ± SEM from at least three separate experiments. (* $p < 0.05$).

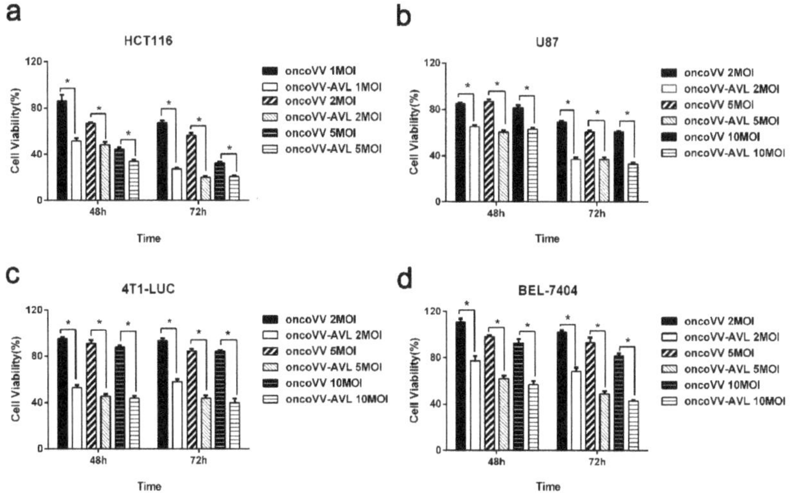

Figure 2. The antiproliferative effect of oncoVV-AVL in cancer cells. The cell viability was measured by MTT assay in HCT116 cells (**a**), U87 cells (**b**), 4T1-LUC (**c**), and BEL-7404 cells (**d**). OncoVV was used as a control virus. Data were expressed as the mean ± SEM from at least three separate experiments. (* $p < 0.05$).

2.3. AVL Improve the Replication Ability of Oncolytic Vaccinia Virus

In order to investigate the underlying mechanism of the enhanced antiproliferative effect of oncoVV-AVL, the replication ability of oncoVV-AVL, oncoVV, and oncoVV-GM-CSF was tested through $TCID_{50}$ method. As shown, oncoVV-AVL replicated significantly faster than control viruses in HCT116 cells, BEL-7404 cells, 4T1-LUC cells, and U87 cells (Figure 3). The results show that oncolytic vaccinia virus harboring AVL gene significantly improved the replication ability in various cancer cells.

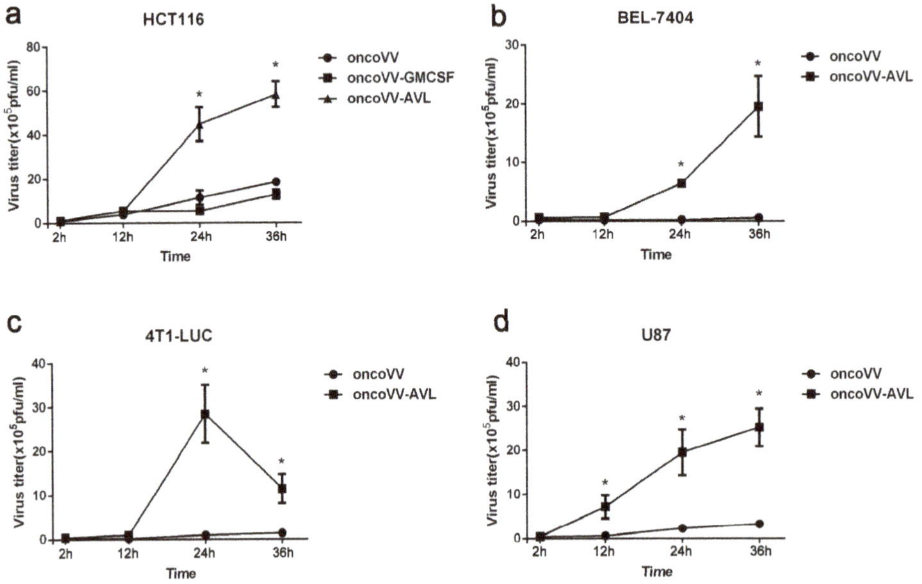

Figure 3. oncoVV-AVL replication in multiple tumor cell lines. The replication of oncoVV-AVL in HCT116 cells (**a**), BEL-7404 cells (**b**), 4T1-LUC cells (**c**), and U87 cells (**d**). Viral replication was determined by $TCID_{50}$ assay. Data were expressed as the mean ± SEM from at least three separate experiments. (* $p < 0.05$).

2.4. OncoVV-AVL Altered ERK and NF-κB Signaling Pathways in Cancer Cells

In order to explore the underlying molecular mechanism of the enhanced cytotoxicity and replication ability of oncoVV-AVL, apoptosis analysis and intracellular signaling elements examination were performed in HCT116 cells. Cells were treated with oncoVV, oncoVV-AVL, or PBS followed by staining with Annexin V-FITC and propidium iodide (PI), a common method for apoptotic cell staining, and analysis under a flow cytometer. As shown in Figure 4a, oncoVV-AVL induced a higher percent of Annexin V^+/PI^- and Annexin V^+/PI^+ cells, as compared to the cells treated with either PBS or oncoVV. Significant differences achieved from three repeats are shown in Figure 4b. Results indicate that oncoVV-AVL induced apoptosis in HCT116 cells. The Western blot results were shown in Figure 4c, and densitometry analysis was shown in Supplementary Figure S1. Melanoma differentiation-associated protein 5 (MDA5), a DDX58 -like receptor, is a dsRNA helicase [38]. Previous studies have shown that MDA5 inhibited the replication of encephalomyocarditis virus (EMCV) and vesicular stomatitis virus (VSV) [39]. Our results showed that there was no significant difference of MDA5 levels among oncoVV-AVL and other treatments, suggesting that oncoVV-AVL did not alter the MDA5 pathway in HCT116 cells.

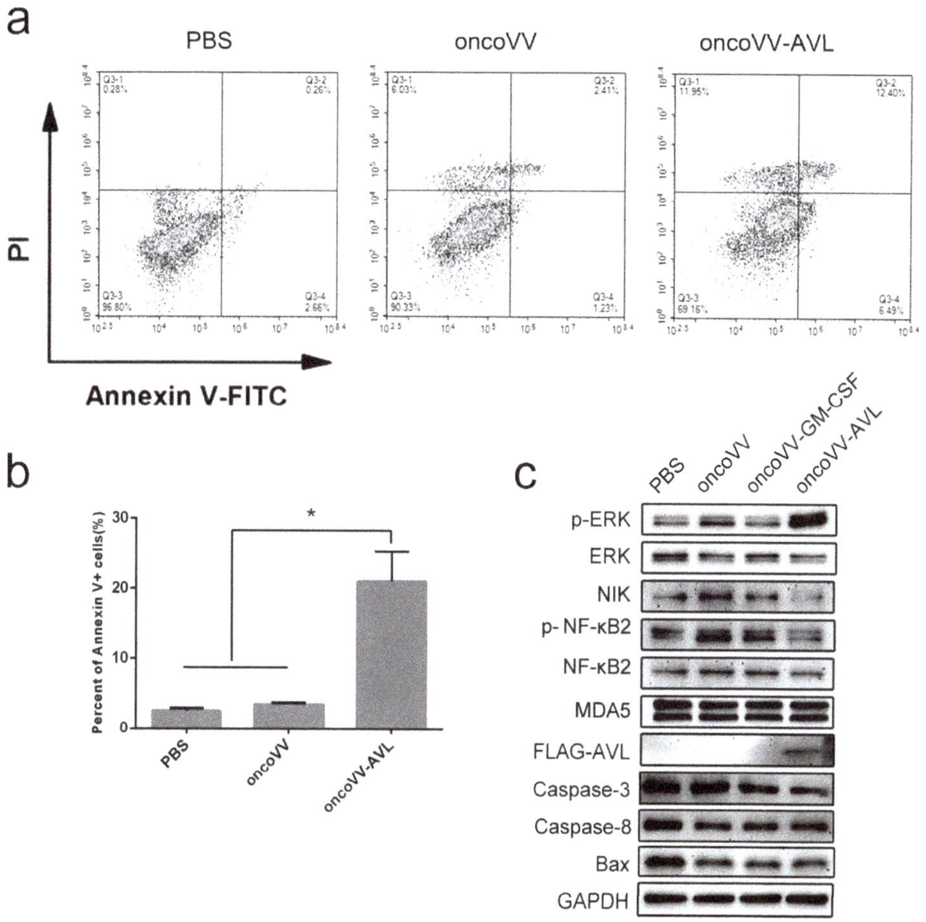

Figure 4. oncoVV-AVL induced apoptosis and altered various intracellular signaling pathways in HCT116 cells. (**a**) HCT116 cells were treated with oncoVV or oncoVV-AVL at 2 MOI as well as PBS control for 24 h. Cells were stained with Annexin V-FITC and PI followed by analysis under a flow cytometer; (**b**) the percent of Annexin V-positive cells from three repeats was shown as mean ± SEM (* $p < 0.05$). (**c**) The levels of p-ERK, ERK, MDA5, NF-κB2, p-NF-κB2, NIK, Caspase-3, Caspase8, Bax and FLAG tagged AVL was detected by Western blot. GAPDH served as a loading control.

Caspase 3, a member of the caspase family, plays a key role in most apoptotic signaling pathways. The appearance of apoptosis leads to the cleavage and activation of Caspase-3 by various proteolytic enzymes [40]. Our results showed that the level of Caspase-3 of oncoVV-AVL treatment was obviously lower than that of oncoVV and PBS, indicating that oncoVV-AVL stimulated caspase cleavage and enhanced apoptotic effect. We further analyzed the effect of oncoVV-AVL treatment on caspase 8 and proapoptotic factor Bax. Our data showed that all three vaccinia viruses slightly downregulated caspase 8 and Bax as compared to PBS control, but no obvious differences were observed among these viruses, indicating that AVL harboring did not affect caspase 8 and Bax.

NIK (NF-κB-inducing kinase), as a kinase of NF-κB, is a member of MAP kinase kinase kinase (MAP3K) [41]. NF-κB2 is a transcription factor in the NF-κB family and plays an important role in NF-κB-mediated inflammation and immunity [42]. In this study, oncoVV upregulated NIK level, which was downregulated by oncoVV-AVL. Furthermore, oncoVV-AVL significantly suppressed the NF-κB2 phosphorylation as compared to control viruses. The results suggested that AVL might enhance viral cytotoxicity in cancer cells by downregulating NF-κB signaling pathway.

It has been reported that extracellular signal-regulated kinase (ERK) was required for virus replication and played a critical role in transmitting signals from surface receptors to the nucleus [43]. In our data, oncoVV-AVL significantly induced ERK phosphorylation as compared to oncoVV and oncoVV-GM-CSF, suggesting that AVL might enhance viral replication by activating ERK.

2.5. The OncoVV Replication Upregulated by AVL Was Completely Dependent on ERK Activation

U0126, a pharmacologic inhibitor, can block MEK1/2-mediated phosphorylation of ERK1/2 [44]. In this study, U0126 was analyzed for its effect on vaccinia virus replication. As shown in Figure 5, U0126 completely suppressed the replication of oncoVV-AVL to the level of oncoVV, which was totally different from an upregulation effect of U0126 on oncoVV, as shown in our data. The results indicated that the oncoVV replication upregulated by AVL was completely dependent on ERK activation.

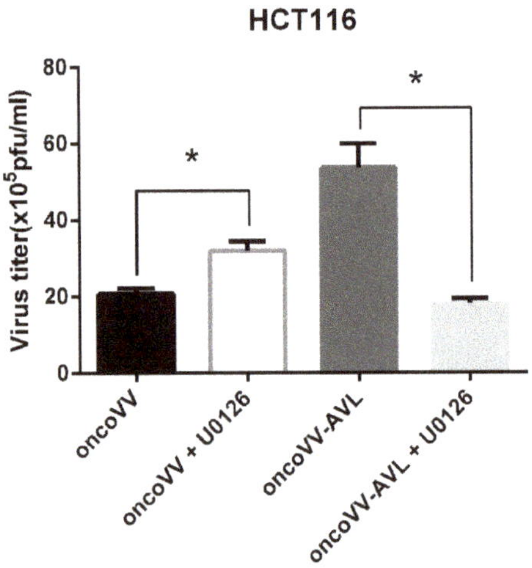

Figure 5. The virus titers of oncoVV and oncoVV-AVL affected by U0126 in HCT116 cells. OncoVV or oncoVV-AVL at 5MOI was used to treat HCT116 cells for 24 h in combination with 10 μM of U0126. Virus titers were measured by $TCID_{50}$ assay. Data were expressed as the mean ± SEM from at least three separate experiment (* $p < 0.05$).

2.6. OncoVV-AVL Has Significant Antitumor Activity in Mice

To assess the efficacy of oncoVV-AVL against tumors in vivo, subcutaneous tumors were established in Balb/c nude mice with BEL-7404 and HCT116 cells. Tumors were then treated with PBS, oncoVV-TTL [21], oncoVV-GM-CSF, as well as oncoVV-AVL. As shown in Figure 6a, oncoVV-TTL and oncoVV-GM-CSF did not significantly suppress HCT116 tumor growth, as compared to PBS control, while oncoVV-AVL significantly inhibited the growth of HCT116 tumors in mice. As shown in Figure 6b, oncoVV-AVL elicited better antitumor effects on BEL-7404 tumors, as compared with oncoVV-TTL and oncoVV-GM-CSF. Therefore, our data demonstrated that oncoVV-AVL achieved significant therapeutic effect on in vivo tumor models.

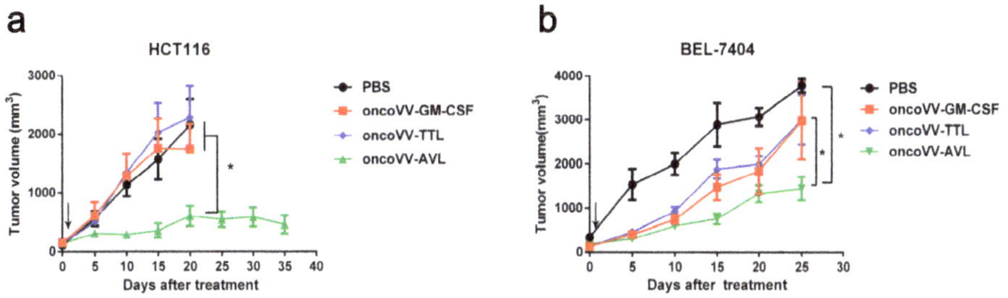

Figure 6. The antitumor effect of oncoVV-AVL on BEL-7404 and HCT116 tumors. (**a**) HCT116 cells or (**b**) BEL7404 cells were injected into the Balb/c nude mice on the back. Tumors were then injected with PBS, oncoVV-GM-CSF, oncoVV-TTL, or oncoVV-AVL. Arrows indicate injections. Data were expressed as the mean ± SEM. (* $p < 0.05$).

3. Discussion

Oncolytic vaccinia virus has the advantages of high expression efficiency, low cost, and good oncolytic effect [45]. Human kind has accumulated abundant clinical experience with vaccinia virus due to its successful use in eliminating smallpox. Since the end of the last century, researchers have explored recombinant VV and other poxviruses as expression vectors to trigger immunization in cancer [46,47]. In this study, oncolytic vaccinia virus harboring the marine lectin AVL gene was evaluated for its anti-tumor effects. Our results showed that AVL significantly enhanced the replication of oncolytic vaccinia virus and improved the antiproliferative efficacy of oncolytic vaccinia virus in various tumor cells lines. Importantly, the in vivo antitumor efficacy of oncoVV-AVL was significantly better than that of oncoVV-GM-CSF and oncoVV-TTL. Our data suggested that oncoVV-AVL may be further developed to be an anticancer agent. However, due to isolated AVL being unavailable in our laboratory at present, we are unable to determine whether isolated AVL affects cancer cells with or without the presence of oncoVV. Furthermore, it is still impossible for us to draw a conclusion that oncoVV-AVL has advantage over isolated AVL in treating cancers.

As reported previously, ERK is required for vaccinia virus replication [43]. In this study, AVL stimulated ERK phosphorylation. Furthermore, U0126 inhibitor analysis showed that AVL upregulated oncoVV replication completely dependent on ERK activation. Interestingly, in our previous study, *Tachypleus tridentatus* plasma lectin TTL was also shown to enhance oncoVV replication in an ERK activation dependent manner [21]. Therefore, we propose here that ERK may serve as a common target for a variety of lectins, pending further investigations.

4. Materials and Methods

4.1. Cell Culture

Human embryonic kidney cells HEK293A, U87 glioma cells, colorectal cancer cells HCT116, human colon cancer cells HT-29, Firefly luciferase-labeled mouse breast cancer cells 4T1-LUC, U251 glioma cells, and hepatocellular carcinoma cells MHCC97-H and BEL-7404 were provided by American Type Culture Collection (Rockville, MD, USA). Cells were cultured in a DMEM medium (Gibco, Thermo Fisher Scientific, Waltham, MA, USA) with 10% fetal bovine serum and 1% Penicillin-Streptomycin Solution.

4.2. Generation of OncoVV-AVL

Aphrocallistes vastus lectin (AVL, GenBank accession No. AJ276450.1) gene was purchased from Shanghai Generay Biotech Co., Ltd., Shanghai, China. Then, the Flag-AVL gene was inserted into the plasmid pCB with Thymidine kinase (TK) gene deletion. After HEK-293A cells were infected with wild type vaccinia virus (Western Reverse) about 2–4 h, pCB-Flag-AVL was transfection into 293A cells. Vaccinia virus oncoVV-AVL was obtained after 48 hrs. Mycophenolic acid, dioxopurine, and hypoxanthine were added to screen effective oncoVV-AVL. Recombinant viruses were gathered from cell culture medium. The virus titers were determined by $TCID_{50}$ (median tissue culture infective dose).

4.3. Cytotoxicity Detection and Flow Cytometry Assay

Cells were seeded in 96-well plates at 5×10^3 per well one day before infection with viruses. Cells were then infected with viruses (Ad-EGFP, Ad-AVL, oncoVV, or oncoVV-AVL), at corresponding multiplicity of infections (MOIs), for the time period as indicated. The cell viability was determined by 3-(4,5-dimethylthiazol-2-yl)-2,5-diphenyltetrazolium bromide (MTT) assay [48]. Meanwhile, cells were infected with oncolytic vaccinia virus at 2MOI for 24 h. Cells were then collected and stained with Annexin V-FITC Apoptosis Detection Kit I (BD Biosciences, San Jose, CA, USA) following the manufacturer's instruction, and analyzed under an ACEA NovoCyte flow cytometry (ACEA Biosciences, San Diego, CA, USA).

4.4. Virus Replication Assay

To determine the viral replication capacity in multiplicity of cells line, cells were plated on 24-well plates at 5×10^4 cells per well one day before treatment with viruses [21]. Cells were infected with oncoVV, oncoVV-GM-CSF or oncoVV-AVL at 5MOI for 2 h, 12 h, 24 h, and 36 h. 10 μM of inhibitor U0126 was used for combination. Cells and culture medium were then collected in a −80 °C refrigerator. The viral titers were determined through tissue culture infectious dose ($TCID_{50}$) assay on 293A cells after three cycles of freeze-thaw in −80 °C and 37 °C.

4.5. Animal Experiments

Female Balb/c nude mice (Shanghai Slack Animal Laboratory, Shanghai, China) of 4–5 weeks ages were used for hepatocellular carcinoma and colorectal cancer tumor bearing mouse models. BEL-7404 and HCT116 cells were injected subcutaneously at 5×10^6 cells/mouse and at 8×10^6 cells/mouse respectively into the mice on the back. Mice were divided into groups as indicated at 6–8 mice per group. When tumor size reached a certain volume, oncolytic vaccinia viruses were injected into mice intratumorally at 1×10^7 plaque-forming units (PFU) each. Then, the volume of tumors was measured every five days. The tumor volume was calculated using the formula: length (mm) × width $(mm)^2 \times 0.5$. Mice were cared for in accordance with the Guide for the Care and Use of Laboratory Animals (Zhejiang Sci-Tech University).

4.6. Western Blotting

The cell extracts were subjected to SDS-PAGE and electroblotted onto nitrocellulose membranes. Then, the membranes were blocked with Tris-buffered saline and Tween -20 containing 5% of bovine serum albumin at room temperature for 2 h [49], following incubation with corresponding antibodies overnight at 4 °C. The membranes were washed and incubated with appropriate dilution of secondary antibodies for 1 h at room temperature. After washing with Tris-buffered saline, the bands were detected under a Tanon 5500 chemiluminescence image system (Tanon Inc., Shanghai, China).

Rabbit anti-MDA5 and mouse anti-Bax antibodies were purchased from Santa Cruz Biotechnology Inc. (Dallas, TX, USA). Rabbit anti-ERK1/2, phospho-ERK1/2, rabbit anti-caspase-3, rabbit anti-caspase-8, rabbit anti-NF-κB2, phosphor-NF-κB2, and rabbit anti-GAPDH antibodies were purchased from Cell Signaling Technology Inc. (Danvers, MA, USA). Rabbit anti-Flag was purchased from Bioss Antibodies (Beijing, China). The HRP conjugated goat anti-rabbit and goat anti-mouse were purchased from MultiSciences (Lianke) Biotech Co., Ltd. (Hangzhou, China).

4.7. Statistical Analysis

Differences among the different treatment groups were determined by student's t-test. $p < 0.05$ and was considered significant.

5. Conclusions

Our studies showed that oncoVV-AVL elicited significant antitumor activity in a hepatocellular carcinoma as well as colorectal cancer mouse models. AVL enhanced viral replication in an ERK activity dependent manner. Our studies might provide insights into the utilization of AVL in oncolytic viral therapies.

Supplementary Materials: The following are available online at http://www.mdpi.com/1660-3397/17/6/363/s1, Figure S1: Densitometry analysis of western blots in HCT116 cells.

Author Contributions: G.L. conceived and designed the experiments; T.W., Y.X., T.L., X.R. and X.W., performed the experiments; G.L. and T.W. analyzed the data; T.Y. and T.W. wrote the paper; G.L. revised the paper.

Funding: This work was funded by Zhejiang Provincial Natural Science Foundation grant LZ16D060002, National Natural Science Foundation of China grant 81572986, Hangzhou Gongchu Biotechnology Co., Ltd., and Zhejiang Provincial Top Key Discipline of Biology.

Conflicts of Interest: The authors declare no conflict of interest.

References

1. Neth, O.; Jack, D.L.; Dodds, A.W.; Holzel, H.; Klein, N.J.; Turner, M.W. Mannose-binding lectin binds to a range of clinically relevant microorganisms and promotes complement deposition. *Infect. Immun.* **2000**, *68*, 688–693. [CrossRef] [PubMed]
2. Weis, W.I.; Drickamer, K. Structural basis of lectin-carbohydrate recognition. *Annu. Rev. Biochem.* **1996**, *65*, 441–473. [CrossRef] [PubMed]
3. Worbs, S.; Köhler, K.; Pauly, D.; Avondet, M.A.; Schaer, M.; Dorner, M.B.; Dorner, B.G. Ricinus communis intoxications in human and veterinary medicine-a summary of real cases. *Toxins* **2011**, *3*, 1332–1372. [CrossRef] [PubMed]
4. Fujita, T. Evolution of the lectin—Complement pathway and its role in innate immunity. *Nat. Rev. Immunol.* **2002**, *2*, 346–353. [CrossRef] [PubMed]
5. Lillie, B.N.; Brooks, A.S.; Keirstead, N.D.; Hayes, M.A. Comparative genetics and innate immune functions of collagenous lectins in animals. *Vet. Immunol. Immunopathol.* **2005**, *108*, 97–110. [CrossRef] [PubMed]
6. Sharon, N. Lectin-carbohydrate complexes of plants and animals: An atomic view. *Essays Biochem.* **1995**, *30*, 221–226. [CrossRef]
7. Varki, A.; Cummings, R.D.; Esko, J.D.; Freeze, H.H.; Stanley, P.; Bertozzi, C.R.; Hart, G.W.; Etzler, M.E. *Essentials of Glycobiology*, 3rd ed.; Cold Spring Harbor Laboratory Press: New York, NY, USA, 2017.
8. Wu, J.; Xie, X.; Liu, Y.; He, J.; Benitez, R.; Buckanovich, R.J.; Lubman, D.M. Identification and confirmation of differentially expressed fucosylated glycoproteins in the serum of ovarian cancer patients using a lectin array and LC-MS/MS. *J. Proteome Res.* **2012**, *11*, 4541–4552. [CrossRef]
9. Li, C.; Simeone, D.M.; Brenner, D.E.; Anderson, M.A.; Shedden, K.A.; Ruffin, M.T.; Lubman, D.M. Pancreatic cancer serum detection using a lectin/glyco-antibody array method. *J. Proteome Res.* **2009**, *8*, 483–492. [CrossRef]
10. Batabyal, S.K.; Majhi, R.; Basu, P.S. Clinical utility of the interaction between lectin and serum prostate specific antigen in prostate cancer. *Neoplasma* **2009**, *56*, 68–71. [CrossRef]
11. Fry, S.A.; Afrough, B.; Lomax-Browne, H.J.; Timms, J.F.; Velentzis, L.S.; Leathem, A.J. Lectin microarray profiling of metastatic breast cancers. *Glycobiology* **2011**, *21*, 1060–1070. [CrossRef]
12. Drake, P.M.; Schilling, B.; Niles, R.K.; Prakobphol, A.; Li, B.; Jung, K.; Cho, W.; Braten, M.; Inerowicz, H.D.; Williams, K.; et al. Lectin chromatography/mass spectrometry discovery workflow identifies putative biomarkers of aggressive breast cancers. *J. Proteome Res.* **2012**, *11*, 2508–2520. [CrossRef] [PubMed]

13. Ahn, Y.H.; Shin, P.M.; Oh, N.R.; Park, G.W.; Kim, H.; Yoo, J.S. A lectin-coupled, targeted proteomic mass spectrometry (MRM MS) platform for identification of multiple liver cancer biomarkers in human plasma. *J. Proteom.* **2012**, *75*, 5507–5515. [CrossRef] [PubMed]
14. Ochoa-Alvarez, J.A.; Krishnan, H.; Shen, Y.; Acharya, N.K.; Han, M.; McNulty, D.E.; Hasegawa, H.; Hyodo, T.; Senga, T.; Geng, J.G.; et al. Plant lectin can target receptors containing sialic acid, exemplified by podoplanin, to inhibit transformed cell growth and migration. *PLoS ONE* **2012**, *7*, e41845. [CrossRef] [PubMed]
15. Chang, C.P.; Yang, M.C.; Liu, H.S.; Lin, Y.S.; Lei, H.Y. Concanavalin a induces autophagy in hepatoma cells and has a therapeutic effect in a murine in situ hepatoma model. *Hepatology* **2007**, *45*, 286–296. [CrossRef] [PubMed]
16. Chatterjee, B.; Ghosh, K.; Yadav, N.; Kanade, S.R. A novel l-fucose-binding lectin from *Fenneropenaeus indicus* induced cytotoxicity in breast cancer cells. *J. Biochem.* **2016**, *161*, 87–97. [CrossRef] [PubMed]
17. Liu, B.; Cheng, Y.; Bian, H.J.; Bao, J.K. Molecular mechanisms of *Polygonatum cyrtonema* lectin-induced apoptosis and autophagy in cancer cells. *Autophagy* **2009**, *5*, 253–255. [CrossRef] [PubMed]
18. Terada, D.; Kawai, F.; Noguchi, H.; Unzai, S.; Hasan, I.; Fujii, Y.; Park, S.Y.; Ozeki, Y.; Tame, J.R. Crystal structure of mytilec, a galactose-binding lectin from the mussel mytilus galloprovincialis with cytotoxicity against certain cancer cell types. *Sci. Rep.* **2016**, *6*, 28344. [CrossRef]
19. Hasan, I.; Sugawara, S.; Fujii, Y.; Koide, Y.; Terada, D.; Iimura, N.; Fujiwara, T.; Takahashi, K.G.; Kojima, N.; Rajia, S.; et al. Mytilec, a mussel r-type lectin, interacts with surface glycan gb3 on burkitt's lymphoma cells to trigger apoptosis through multiple pathways. *Mar. Drugs* **2015**, *13*, 7377–7389. [CrossRef]
20. Fujii, Y.; Dohmae, N.; Takio, K.; Kawsar, S.M.; Matsumoto, R.; Hasan, I.; Koide, Y.; Kanaly, R.A.; Yasumitsu, H.; Ogawa, Y.; et al. A lectin from the mussel mytilus galloprovincialis has a highly novel primary structure and induces glycan-mediated cytotoxicity of globotriaosylceramide-expressing lymphoma cells. *J. Biol. Chem.* **2012**, *287*, 44772–44783. [CrossRef]
21. Li, G.; Cheng, J.; Mei, S.; Wu, T.; Ye, T. Tachypleus tridentatus lectin enhances oncolytic vaccinia virus replication to suppress in vivo hepatocellular carcinoma growth. *Mar. Drugs* **2018**, *16*, 200. [CrossRef]
22. Wu, L.; Yang, X.; Duan, X.; Cui, L.; Li, G. Exogenous expression of marine lectins DlFBL and SpRBL induces cancer cell apoptosis possibly through PRMT5-E2F-1 pathway. *Sci. Rep.* **2014**, *4*, 4505. [CrossRef] [PubMed]
23. Lu, Q.; Li, N.; Luo, J.; Yu, M.; Huang, Y.; Wu, X.; Wu, H.; Liu, X.Y.; Li, G. Pinellia pedatisecta agglutinin interacts with the methylosome and induces cancer cell death. *Oncogenesis* **2012**, *1*, e29. [CrossRef] [PubMed]
24. Yang, X.; Wu, L.; Duan, X.; Cui, L.; Luo, J.; Li, G. Adenovirus carrying gene encoding Haliotis discus discus sialic acid binding lectin induces cancer cell apoptosis. *Mar. Drugs* **2014**, *12*, 3994–4004. [CrossRef] [PubMed]
25. Li, G.; Gao, Y.; Cui, L.; Wu, L.; Yang, X.; Chen, J. Anguilla japonica lectin 1 delivery through adenovirus vector induces apoptotic cancer cell death through interaction with PRMT5. *J. Gene Med.* **2016**, *18*, 65–74. [CrossRef] [PubMed]
26. Wu, B.; Mei, S.; Cui, L.; Zhao, Z.; Chen, J.; Wu, T.; Li, G. Marine Lectins DlFBL and HddSBL Fused with Soluble Coxsackie-Adenovirus Receptor Facilitate Adenovirus Infection in Cancer Cells BUT Have Different Effects on Cell Survival. *Mar. Drugs* **2017**, *15*, 73. [CrossRef]
27. Bai, R.; Nguyen, T.L.; Burnett, J.C.; Atasoylu, O.; Munro, M.H.; Pettit, G.R.; Smith, A.B.; Gussio, R.; Hamel, E. Interactions of halichondrin B and eribulin with tubulin. *J. Chem. Inf. Model.* **2011**, *51*, 1393–1404. [CrossRef]
28. Konoki, K.; Okada, K.; Kohama, M.; Matsuura, H.; Saito, K.; Cho, Y.; Nishitani, G.; Miyamoto, T.; Fukuzawa, S.; Tachibana, K. Identification of okadaic acid binding protein 2 in reconstituted sponge cell clusters from Halichondria okadai and its contribution to the detoxification of okadaic acid. *Toxicon* **2015**, *108*, 38–45. [CrossRef]
29. Schröder, H.C.; Boreiko, A.; Korzhev, M.; Tahir, M.N.; Tremel, W.; Eckert, C.; Ushijima, H.; Müller, I.M.; Müller, W.E. Co-expression and functional interaction of silicatein with galectin: Matrix-guided formation of siliceous spicules in the marine demosponge Suberites domuncula. *J. Biol. Chem.* **2006**, *281*, 12001–12009. [CrossRef]
30. Dresch, R.R.; Zanetti, G.D.; Kanan, J.H.; Mothes, B.; Lerner, C.B.; Trindade, V.M.; Henriques, A.T.; Vozárihampe, M.M. Immunohistochemical localization of an N-acetyl amino-carbohydrate specific lectin (ACL-I) of the marine sponge Axinella corrugata. *Acta Histochem.* **2011**, *113*, 671–674. [CrossRef]
31. Gundacker, D.; Leys, S.P.; Schroder, H.C.; Muller, I.M.; Muller, W.E. Isolation and cloning of a C-type lectin from the hexactinellid sponge Aphrocallistes vastus: A putative aggregation factor. *Glycobiology* **2001**, *11*, 21–29. [CrossRef]

32. Lundstrom, K. New frontiers in oncolytic viruses: Optimizing and selecting for virus strains with improved efficacy. *Biol. Targets Ther.* **2018**, *12*, 43–60. [CrossRef] [PubMed]
33. Veyer, D.L.; Carrara, G.; Maluquer de Motes, C.; Smith, G.L. Vaccinia virus evasion of regulated cell death. *Immunol. Lett.* **2017**, *186*, 68–80. [CrossRef] [PubMed]
34. Park, B.H.; Hwang, T.; Liu, T.C.; Sze, D.Y.; Kim, J.S.; Kwon, H.C.; Oh, S.Y.; Han, S.Y.; Yoon, J.H.; Hong, S.H. Use of a targeted oncolytic poxvirus, JX-594, in patients with refractory primary or metastatic liver cancer: A phase I trial. *Lancet Oncol.* **2008**, *9*, 533–542. [CrossRef]
35. Yamada, T.; Hamano, Y.; Hasegawa, N.; Seo, E.; Fukuda, K.; Yokoyama, K.K.; Hyodo, I.; Abei, M. Oncolytic virotherapy and gene therapy strategies for hepatobiliary cancers. *Curr. Cancer Drug Targets* **2018**, *18*, 188–201. [CrossRef] [PubMed]
36. Mell, L.K.; Yu, Y.A.; Brumund, K.T.; Advani, S.J.; Onyeama, S.; Daniels, G.A.; Weisman, R.A.; Martin, P.; Szalay, A.A. Phase 1 Trial of attenuated vaccinia virus (GL-ONC1) delivered intravenously with concurrent cisplatin and radiation therapy in patients with locoregionally advanced head-and-neck carcinoma: Definitive management of head-and-neck squamous cell carcinoma. *Int. J. Radiat. Oncol. Biol. Phys.* **2014**, *88*, 477–478. [CrossRef]
37. Mackett, M.; Smith, G.L.; Moss, B. Vaccinia virus: A selectable eukaryotic cloning and expression vector. *Proc. Natl. Acad. Sci. USA* **1982**, *79*, 7415–7419. [CrossRef]
38. Takeuchi, O.; Akira, S. MDA5/RIG-I and virus recognition. *Curr. Opin. Immunol.* **2008**, *20*, 17–22. [CrossRef]
39. Barral, P.M.; Sarkar, D.; Su, Z.Z.; Barber, G.N.; DeSalle, R.; Racaniello, V.R.; Fisher, P.B. Functions of the cytoplasmic RNA sensors RIG-I and MDA-5: Key regulators of innate immunity. *Pharmacol. Ther.* **2009**, *124*, 219–234. [CrossRef]
40. Kuribayashi, K.; Mayes, P.A.; El-Deiry, W.S. What are caspases 3 and 7 doing upstream of the mitochondria? *Cancer Biol. Ther.* **2006**, *5*, 763–765. [CrossRef]
41. Woronicz, J.D.; Gao, X.; Cao, Z.; Rothe, M.; Goeddel, D.V. IκB Kinase-β: NF-κB activation and complex formation with IκB Kinase-α and NIK. *Science* **1997**, *278*, 866–869. [CrossRef]
42. Xiao, G.; Harhaj, E.W.; Sun, S.C. NF-κB-inducing kinase regulates the processing of NF-κB2 p100. *Mol. Cell* **2001**, *7*, 401–409. [CrossRef]
43. Kim, Y.; Lee, C. Extracellular signal-regulated kinase (ERK) activation is required for porcine epidemic diarrhea virus replication. *Virology* **2015**, *484*, 181–193. [CrossRef] [PubMed]
44. Namura, S.; Iihara, K.; Takami, S.; Nagata, I.; Kikuchi, H.; Matsushita, K.; Moskowitz, M.A.; Bonventre, J.V.; Alessandrini, A. Intravenous administration of MEK inhibitor U0126 affords brain protection against forebrain ischemia and focal cerebral ischemia. *Proc. Natl. Acad. Sci. USA* **2001**, *98*, 11569–11574. [CrossRef] [PubMed]
45. Guse, K.; Cerullo, V.; Hemminki, A. Oncolytic vaccinia virus for the treatment of cancer. *Expert Opin. Biol. Ther.* **2011**, *11*, 595–608. [CrossRef] [PubMed]
46. Carroll, M.W.; Kovacs, G.R. Virus-based vectors for gene expression in mammalian cells: Vaccinia virus. *New Compr. Biochem.* **2003**, *38*, 125–136.
47. Broder, C.C.; Earl, P.L. Recombinant vaccinia viruses. *Mol. Biotechnol.* **1999**, *13*, 223–245. [CrossRef]
48. Fotakis, G.; Timbrell, J.A. In vitro cytotoxicity assays: Comparison of LDH, neutral red, MTT and protein assay in hepatoma cell lines following exposure to cadmium chloride. *Toxicol. Lett.* **2006**, *160*, 171–177. [CrossRef] [PubMed]
49. Ren, R.; Sun, D.J.; Yan, H.; Wu, Y.P.; Zhang, Y. Oral exposure to the herbicide simazine induces mouse spleen immunotoxicity and immune cell apoptosis. *Toxicol. Pathol.* **2013**, *41*, 63–72. [CrossRef]

marine drugs

MDPI

Article

Functional Characterization of OXYL, A SghC1qDC LacNAc-specific Lectin from The Crinoid Feather Star *Anneissia Japonica*

Imtiaj Hasan [1,2], Marco Gerdol [3], Yuki Fujii [4] and Yasuhiro Ozeki [1,*]

1. Graduate School of NanoBio Sciences, Yokohama City University, 22-2 Seto, Kanazawa-ku, Yokohama 236-0027, Japan; hasanimtiaj@yahoo.co.uk
2. Department of Biochemistry and Molecular Biology, Faculty of Science, University of Rajshahi, Rajshahi 6205, Bangladesh
3. Department of Life Sciences, University of Trieste, Via Licio Giorgieri 5, 34127 Trieste, Italy; mgerdol@units.it
4. Graduate School of Pharmaceutical Sciences, Nagasaki International University, 2825-7 Huis Ten Bosch, Sasebo, Nagasaki 859-3298, Japan; yfujii@niu.ac.jp

* Correspondence: ozeki@yokohama-cu.ac.jp; Tel.: +81-45-787-2221; Fax: +81-45-787-2413

Received: 29 January 2019; Accepted: 18 February 2019; Published: 25 February 2019

Abstract: We identified a lectin (carbohydrate-binding protein) belonging to the complement 1q(C1q) family in the feather star *Anneissia japonica* (a crinoid pertaining to the phylum Echinodermata). The combination of Edman degradation and bioinformatics sequence analysis characterized the primary structure of this novel lectin, named OXYL, as a secreted 158 amino acid-long globular head (sgh)C1q domain containing (C1qDC) protein. Comparative genomics analyses revealed that OXYL pertains to a family of intronless genes found with several paralogous copies in different crinoid species. Immunohistochemistry assays identified the tissues surrounding coelomic cavities and the arms as the main sites of production of OXYL. Glycan array confirmed that this lectin could quantitatively bind to type-2 N-acetyllactosamine (LacNAc: Galβ1-4GlcNAc), but not to type-1 LacNAc (Galβ1-3GlcNAc). Although OXYL displayed agglutinating activity towards *Pseudomonas aeruginosa*, it had no effect on bacterial growth. On the other hand, it showed a significant anti-biofilm activity. We provide evidence that OXYL can adhere to the surface of human cancer cell lines BT-474, MCF-7, and T47D, with no cytotoxic effect. In BT-474 cells, OXYL led to a moderate activation of the p38 kinase in the MAPK signaling pathway, without affecting the activity of caspase-3. Bacterial agglutination, anti-biofilm activity, cell adhesion, and p38 activation were all suppressed by co-presence of LacNAc. This is the first report on a type-2 LacNAc-specific lectin characterized by a C1q structural fold.

Keywords: N-Acetyllactosamine (LacNAc); *Anneissia japonica*; anti-biofilm activity; cell adhesion; crinoid; Echinoderm; feather star; lectin; signal transduction; sghC1qDC

---

1. Introduction

Molecular recognition is one of the most essential mechanisms used to regulate cell systems. In primitive organisms, in specific cell types and in early developmental stages, glycans (monosaccharide chains) located on the cell surface act as a fundamental sugar code for recognition. Indeed, specific glycan structures are initially recognized by glycans (via carbohydrate–carbohydrate interaction) [1] and, subsequently, by lectins (via carbohydrate–protein interaction). Virtually all living organisms are endowed with lectins (glycan-binding proteins) that enable such molecular interactions in different biological contexts.

Echinodermata are the second largest phylum in the Deuterostomia lineage and, due to their position in the metazoan tree of life, they can be considered as relatives of chordates and vertebrates.

To date, a large number of lectins have been isolated from members of the subphylum Eleutherozoa, containing mobile echinoderms belonging to the classes Echinoidea (sea urchins), Asteroidea (sea stars), Holothuroidea (sea cucumbers), and Ophiuroidea (brittle stars). These structurally different molecules pertain to different protein families, such as galectin [2], C-type [3,4], SUEL/RBL-type [5,6], R-type [7], and HSP110-type [8] lectins. Numerous studies have reported that invertebrate lectins are able to inhibit or to promote mammalian cell growth [9–12]. Moreover, some lectins are capable of killing carcinoma cells and microorganisms through the binding to N-acetylhexosamines, such as N-acetyl D-galactosamine (GalNAc), N-acetyl D-glucosamine (GlcNAc), N-acetyl D-mannosamine (ManNAc) and N-acetylneuramic acid (NeuAc) which in turn activates signal transduction, leading to cell death. Lectins obtained from echinoderms can regulate cell growth in different ways. For example, echinoidin, a C-type lectin found in the coelomic fluid of sea urchins, exerts its cell adhesion activity via the tripeptide motif RGD, known as a cell adhesive signal [13]; on the other hand, the SUEL/RBL-type lectin found in sea urchin venom promotes mitogenesis [14] and an R-type lectin isolated in sea cucumber has hemolytic properties [7].

Compared with other echinoderms, the class Crinoidea (subphylum Pelmatozoa) has been almost completely neglected, as far as lectin research is concerned. The morphology of crinoids resembles that of flowering plants due to the presence of hundred feather-like pinnules attached to their crown of arms. Fossil record, as well as molecular phylogeny, indicate that crinoids are the most primitive type of existing echinoderms. The oldest crinoid representative had already emerged in the mid-early Ordovician period of the Paleozoic era (542–251 million years ago). After becoming nearly extinct at the end of the Permian period mass extinction, crinoids recovered in the early Triassic, in the Mesozoic era (251–66 million years ago), and have subsequently undergone diversification to the present level [15,16]. The two extant forms or crinoids, which include feather stars (free-swimming organisms) and sea lilies (sessile animals that possess a stalk to attach to rocks), are both regarded as living fossils. Despite their ancient evolutionary origins, crinoids have a complex nervous system and possess an excellent ability of tissue-regeneration, which is particularly evident in their arms [17]. Since these properties are potentially useful for biological and medicinal studies, large efforts have been endeavored to develop protocols that could ensure a large-scale annual supply of individuals as an experimental animal model [18]. The improved study of lectins in this class of echinoderms could enable to clarify some aspects of deuterostome evolution through an improved view on lectin-glycan interactions.

In a previous study, we reported the discovery of OXYL, a Ca^{2+}-independent lectin isolated from the arms of the feather star *Oxycomanthus japonicus* (family Comatulidae), a species whose scientific name has been recently updated to *Anneissia japonica*. This family includes four subfamilies, 21 genera and approximately 95 species. Comatulidae is the most commonly encountered and species-rich crinoid family in coastal regions and tropical coral reefs, particularly in the Indo-Western Pacific region [19]. The isolated lectin was soluble and consisted of a 14 kDa polypeptide. Frontal affinity chromatography showed that this molecule recognized type-2 N-acetyllactosamine (LacNAc: Galβ1-4GlcNAc), a common structure of complex-type glycans found in vertebrate Asn(N)-type glycoproteins and glycosphingolipids [20]. However, the primary structure of the lectin, its tissue localization and effects in vertebrate cell proliferation were unknown at the time of its initial discovery.

We took advantage of the availability of genome and transcriptome sequence data from *A. japonica*, resulting from a study carried out by the Brown University in 2014 [21], to investigate in detail the primary structure of OXYL, which was revealed as a protein belonging to the sgh(secreted globular head) C1qDC (complement (C)1q-domain-containing) protein family. We also found that OXYL was mainly localized in the tissues surrounding the coelom (the main body cavity) and in the arms, which suggests a possible function as an innate immune molecule.

In this study, we further show that OXYL pertains to a lineage-specific subfamily of C1qDC proteins that specifically evolved in crinoids, but not in other Eleutherozoa. Despite its ability to recognize target glycans on the cell surface, OXYL did not to show antibacterial or cytotoxic effects. This suggests that the activation of metabolic pathways by the interaction between the lectin and its ligands did not lead to the activation of apoptotic pathways in the target cells. We discuss the evolutionary implications of the glycan-binding properties of OXYL in this basal deuterostome, keeping in mind the fundamental role C1q proteins carry out at the crossroads between acquired (antibody-based) and innate immunity in vertebrates.

2. Results

2.1. Structural Characterization of OXYL as a sghC1qDC Protein

The N-terminal region of OXYL, including the first 40 amino acids of the mature protein, was determined by Edman degradation, with a repetitive yield of 87.59% (Figure S1). This partial amino acid sequence found a perfect hit with a virtually-translated 778-nucleotide long genomic contig, which contained a single 480 nucleotides-long open reading frame (Figure 1), with no introns. Unfortunately, the high fragmentation of the *A. japonica* genome assembly, which might be due to a combination of the high heterozygosity of echinoderms and the low sequencing coverage, prevented a reconstruction of the complete locus, covering the entire 5′ and 3′ untranslated regions (UTRs). At the present stage, the OXYL contig includes a likely incomplete 58 bp 5′UTR region and 240 bp downstream to the translation termination signal, which includes a potential polyadenylation signal in position 617–622, followed shortly thereafter by a CA sequence (position 631–632) and a T-rich region, which represent the canonical consensus for recognition by polyadenylation factors [22]. Altogether, these observations suggest that the 3′UTR of the OXYL mRNA is 93 bp long. The predicted sequence of the OXYL mRNA has been deposited in GenBank under the accession ID MK434202. No sequence corresponding to OXYL could be identified in the ovary transcriptome of *A. japonica*, indicating the lack of expression of this gene in this tissue.

```
1     T  TCT CTT AAT ATT TTT TGC TAT AGG CCT CTA ATT TAA TTG TTA TAT TTC CGT CTA GAA   58
                                                                                 *
59   ATG GCT TTG TTT ACT GGT GCT CTG CTT GCG GCA GTT ATA ATA GTA GTA GCA GGT GAT TGT  118
1     M   A   L   F   T   G   A   L   L   A   A   V   I   I   V   V   A   G   D   C    20
119  AAA CCA CGG TCA GCA TTT ACG GCA AAG ATG ACA AGC AGC ATG GAG GGA GAG CCG GAG AAA  178
21    K   P   R   S   A   F   T   A   K   M   T   S   S   M   E   G   E   P   E   K    40
179  GAT TTA CCT ATC GTC TTC GAT TCA GCC TTG ACA AAC GTC GGC AAT GAC TAC AAT CCT ACA  238
41    D   L   P   I   V   F   D   S   A   L   T   N   V   G   N   D   Y   N   P   T    60
239  GAT GGC AAA TTT ACT ACC AGA ATC GCA GGA ACG TAC GTT TTC TCC TTA ACT CTA ATG AGT  298
61    D   G   K   F   T   T   R   I   A   G   T   Y   V   F   S   L   T   L   M   S    80
299  AAG CTC TCA TCT GAC AGC ACT GCG TAT GGT TGT TTG CTT TTG AAT GGC GAA GAA CAG GTC  358
81    K   L   S   S   D   S   T   A   Y   G   C   L   L   L   N   G   E   E   Q   V   100
359  TGC GTC TGG TCC AAT GGA ATC AAC GGC CAC GAG ATG TCA AGC AAT ACA ATC ACA GTT AAA  418
101   C   V   W   S   N   G   I   N   G   H   E   M   S   S   N   T   I   T   V   K   120
419  TTG AAA AGT GGT GAT GTC GTT GAC AAT ATC CTA AAG GCT GGC ACT CGC TTT GCA CTT CAC  478
121   L   K   S   G   D   V   V   D   N   I   L   K   A   G   T   R   F   A   L   H   140
479  AAC GCC GGT CAA GCT GGA TAC TGT TCA TTT ACT GGC TTC TTG CTT TAT CCA GAA GTA TAA  538
141   N   A   G   Q   A   G   Y   C   S   F   T   G   F   L   L   Y   P   E   V  stop 160
539  GTA GGC ATC TAG GCA TAC GGT CAC TCA GTA GCC TTA CTT GGT GGT TGT GAA TTG ACA GAC  598

599  AGG CAC ACG ACA TGA TAT AAT AAA ATT CAT TGC                                      631
```

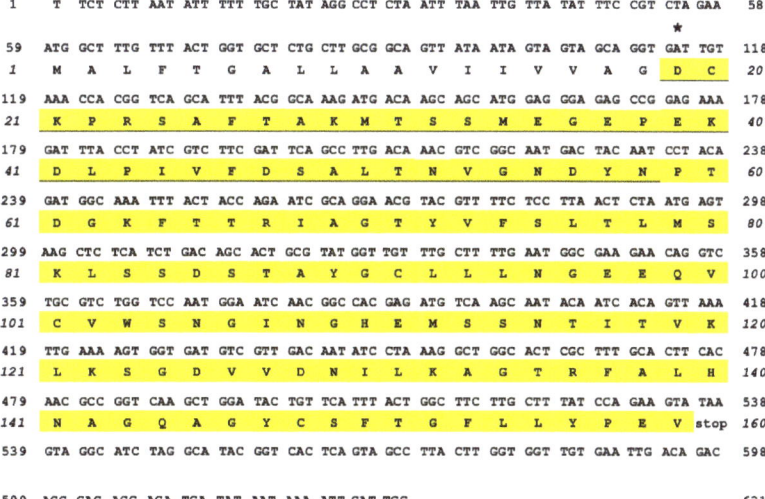

Figure 1. cDNA sequence and deduced amino acid sequence of OXYL. The asterisk (^{19}Asp) indicates the first N-terminal amino acid of the mature lectin (yellow). The amino acid sequence identified by Edman degradation is underlined.

The encoded protein consists of 159 amino acids and contains a N-terminal signal sequence (^{1}Met-^{18}Gly), which was inferred to undergo proteolytic cleavage through the secretory pathway, as the N-terminal residue of the mature polypeptide was determined to be ^{19}Asp (Figure 1, asterisk). This is consistent with SignalP prediction, which identified a putative cleavage site with high confidence in the same position. The polypeptide displayed a candidate site for N-glycosylation (Asn-X-Thr), including the residues ^{58}Asn-^{60}Thr. However, the yield of ^{58}Asn (126 pmol) quantitatively obtained by Edman degradation (Figure S1) indicates that this Asn residue did not undergo translational modification. The secreted OXYL polypeptide was therefore predicted to include, following single peptide cleavage, 141 amino acids, with a molecular mass of 15193.2.

A C1q domain (Pfam family: *C1q* (PF00386)) could be recognized with high confidence by Hmmer (e-value = 5.9E^{-22}), starting immediately after the signal peptide, and comprising all the remaining part of the polypeptide. The 8 residues (^{26}Phe, ^{46}Phe, ^{52}Asn, ^{64}Phe, ^{70}Gly, ^{72}Tyr, ^{150}Phe, and ^{152}Gly) that are typically conserved in all C1q proteins [23] were also present in OXYL (Figure 2A, asterisks). The precursor protein lacks both collagen-like regions (found in many vertebrate C1qDC proteins) and coiled-coiled regions (found in many invertebrate C1qDC proteins). Therefore, based on a previously suggested classification scheme for C1qDc proteins, OXYL should be categorized as a sgh(secreted globular head)C1q protein [24,25].

2.2. OXYL is Part of a Multigenic Family of sghC1qDC Proteins Restricted to Comatulida

Our investigation permitted to identify several additional genes encoding proteins sharing high similarity with OXYL in the *A. japonica* genome, even though the high fragmentation of this resource did not allow an exhaustive discrimination between paralogous gene copies and allelic variants, also preventing the full-length reconstruction of two gene sequences. In detail, the *A. japonica* genome contains four other complete genes closely related to OXYL, all predicted to be intronless. These genes encode precursors of similar size (159–180 amino acids), all classifiable as sghC1qDC proteins, sharing 58–87% sequence homology in pairwise comparisons (Figure 2A). Such a divergence, together with the origins of the genome data from a single individual, is consistent with the presence of multiple paralogous gene copies.

Previous studies have revealed that marine invertebrates can contain an extremely variable number of C1qDC genes, ranging from just a very few to several hundred [25], as this family underwent multiple lineage-specific expansion events along its evolution. The screening of echinoderm genomes indicated that these animals possess a moderate number of C1qDC genes, compared with other metazoans. Namely, we could detect 6 genes in the sea cucumber *Apostichopus japonicus*, 14 genes in the sea urchin *Strongylocentrotus purpuratus*, and a much larger number (59) in the sea star *Acanthaster planci*. Although the number of C1qDC sequences found in a transcriptome most certainly depends on multiple factors, e.g., tissue of origin, sequencing depth, and others, the data collected from crinoid transcriptomes can provide a rough estimate of the number of sghC1qDC genes present in this class of echinoderms. This number varied considerably from species to species, ranging from 2 (in the *A. japonica* ovary transcriptome) to 50 (in the *Notocrinus virilis* arm transcriptome). While the domain architecture of the encoded proteins varied to some extent, we could detect sequences orthologous to OXYL in seven other species pertaining to the order Comatulida, namely *Antedon mediterranea*, *Aporometra wilsoni*, *Cenolia trichoptera*, *Isometra vivipara*, *Notocrinus virilis*, *Oligometra serripinna*, and *Phrixometra nutrix* (Figure 2A). With the exception of *A. wilsoni*, where three paralogous sequences were found, the transcriptomes of all the other species only possessed a single OXYL-like sequence. All OXYL-like sequences share, at the amino acid level, pairwise sequence identity higher than 40% and possess very similar length and domain organization. On the other hand, no expressed sequence closely related to OXYL was found in members of the orders Cyrtocrinida, Hyocrinida and Isocrinida.

Sequence clustering approaches placed all Comatulida OXYL-like sequences in a well-supported clade (bootstrap support = 94, indicated in light blue in the radial tree in Figure 2C). This clade did not include any C1qDC sequence neither from other crinoids, nor from Echinozoa and Asterozoa. Bayesian inference analysis (Figure 2B): (i) confirmed the lack of closely related sequences in other non-crinoid echinoderms (for simplicity's sake, only sea urchin sequences have been included in this tree); (ii) identified a small subgroup of sequences (Isovip, Phrnut and Antmed in Figure 2B) with peculiar features, all pertaining to species in the Antedonoidea superfamily; (iii) strongly supported the monophyly of the five full-length OXYL-like genes identified in *A. Japonica*.

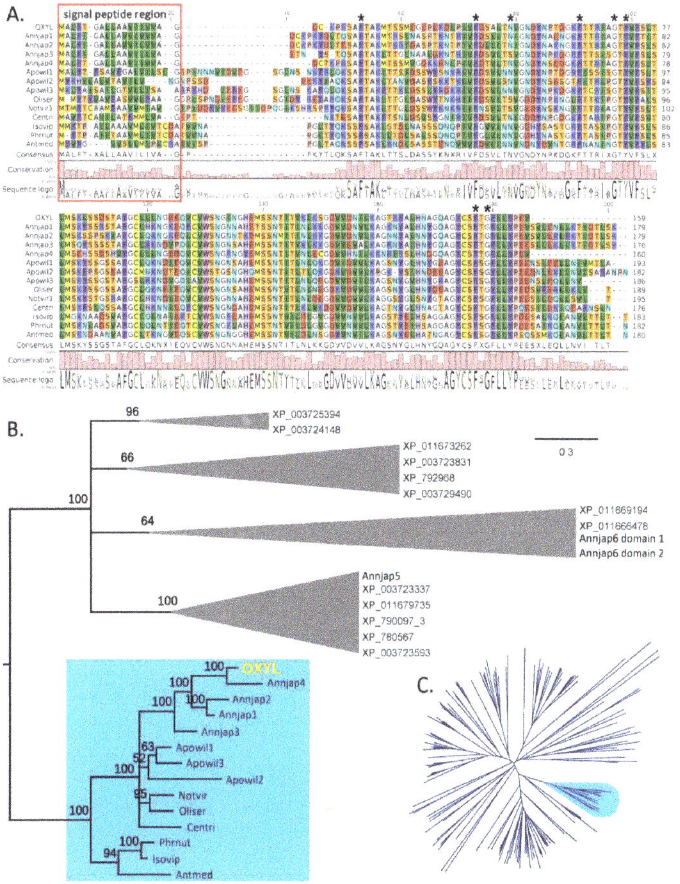

Figure 2. Panel **A**: Multiple sequence alignment of the full-length OXYL precursor and homologous sequences from other crinoid species. Histogram bars represent amino acid conservation in each position of the alignment. The signal peptide region is indicated in a red box. The 8 residues typically conserved in C1qDC proteins are marked with an asterisk. Panel **B**: Bayesian phylogenetic tree of OXYL-like sequences. Multiple sequences obtained from the same species are indicated with progressive numbers. The tree was rooted by using the C1qDC protein sequences from the genome of the sea urchin *Strongylocentrotus purpuratus* (indicated with GenBank accession codes, starting with "XP_") and the three C1q domains from the two C1qDC proteins identified in the *A. japonica* transcriptome, but unrelated to OXYL. For simplicity's sake, outgroup sequences have been collapsed in cartoons. Numbers close to each node represent posterior probabilities. The OXYL clade is indicated with a blue background. Panel **C**: neighbor joining tree of all C1qDC sequences from echinoderms, based on the multiple sequence alignment of the C1q domain (see materials and methods for details). The OXYL clade, supported by bootstrap value = 94, is indicated with a blue background. Each sequence is designated with a six letter code, indicating the first three letters of the genus and species name (see materials and methods for a complete list of sequences).

2.3. Tetrameric Structure

The analytical ultracentrifugation showed that OXYL was mainly found in a tetrameric form (Figure 3), although sghC1qDC family proteins are generally known to associate as trimers. Among the peaks of c(s), the large peak observed near 0 S was most likely originated from the gradient of buffer solution and/or the high concentration salt (Figure 3a). The main peak of c(s) was evident at 4.88 S (Figure 3c, 65.8 kDa: 70.3%). In addition, minor peaks of c(s) were also present at 3.28 S (Figure 3b, 37 kDa: 13.5%), 7.25 S (Figure 3d, 121 kDa: 27%), and 9.16 S (Figure 3e, 178 kDa: 13.5%), respectively (Figure S2). Besides the main tetrameric form, these results highlighted that the tetrameric OXYL was further associated to form octamer and dodecamer, in addition to a minor fraction of lectin associated as a dimer. The finding that OXYL is predominantly found as a tetramer is further supported by observations previously reported based on gel permeation chromatography [20]. Furthermore, this analysis suggested that OXYL is in dynamic equilibrium between tetramer and octamer. The frictional ratio (f/f_0) of OXYL is equal to 1.23, indicating that the protein is globular supporting the organization of OXYL as a sghC1qDC protein.

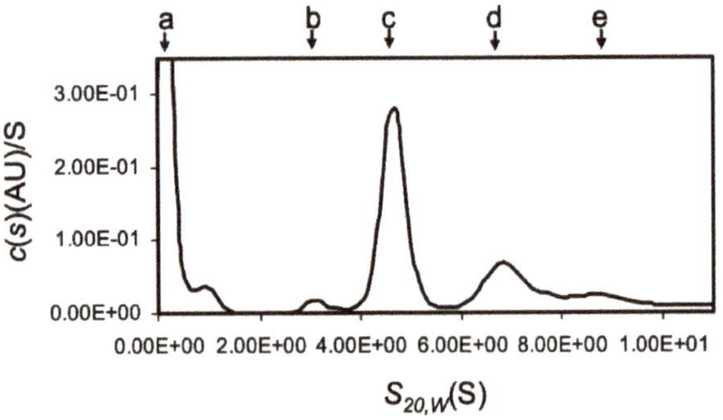

Figure 3. Multimeric structure of OXYL by using analytical ultracetrifugation. Distribution of sedimentation coefficient c(s20,w) by sedimentation velocity AUC. Calculated c(s) was plotted versus s20,w the sedimentation coefficients corrected to 20°C in water. **a**: very likely come from salt and/or buffer, **b**: 3.28 S (37 kDa, dimer), **c**: 4.88 S (65.8 kDa, tetramer), **d**: 7.25 S (121 kDa, octemer), and **e**: 9.19 S (178 kDa, dodecamer). The experiment was performed with 0.4 mg/mL protein.

2.4. Generation of Antiserum Against OXYL

Anti-OXLY antiserum obtained from immunized rabbits was used to identify OXYL by western blotting. Proteins obtained from the crude extract from the arms of *A. japonica* were separated by SDS-PAGE and blotted on a PVDF membrane. A band with a molecular mass of 14,000, detected by HRP-conjugated goat anti-rabbit IgG (Figure 4, crude extract), was evident at the same molecular weight of purified OXYL, stained with Coomassie brilliant blue (Figure 4, OXYL). The obtained antiserum allowed to identify the arms as the main site of localization of OXYL in the feather star.

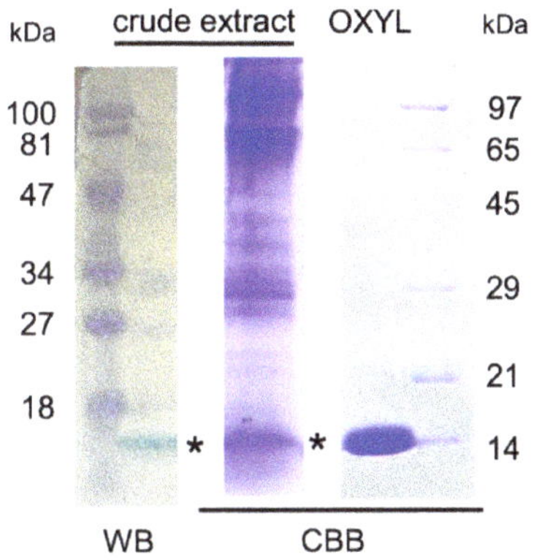

Figure 4. Detection of OXYL by antiserum. *A. japonica* arm extract (crude extract) and purified lectin (OXYL) were separated by SDS-PAGE and transferred to a membrane. OXYL (asterisk) was detected by peroxidase staining of HRP-conjugated goat anti-rabbit IgG raised against OXYL and by Coomassie Brilliant Blue R-250. Numbers at left (pre-stained) and right: molecular mass (×1000) standard.

2.5. Tissue Localization of OXYL

In this study, the arm tissue was decalcified during the preparation of tissue sections from adult feather stars, since the raw tissue displayed a remarkable hardness. Even though such procedure could have produced some artifacts, the expression of OXYL appeared to display a remarkable pattern of distribution. The detection of OXYL signal by the antiserum indicated its presence in the regions surrounding the coelom (Figure 5A,E) and in spicules (Figure 5G,M). These signals were overlapping with the DAPI signal (Figure 5D,H,K,N), indicating that the lectin was produced by proliferating cells. Since OXYL is a secretory protein (Figure 1) that shows high solubility (Figure 4), its presence in such tissues can be explained even in absence of its glycan ligands. The SUEL/RBL-type D-Gal-binding lectin from the sea urchin *Heliocidaris crassispina*, which is expressed in unfertilized eggs and secreted in the extracellular matrix, is known to play a key regulatory role in early embryonal development, from the fertilization to the gastrulation stage [26]. A different D-Gal-binding lectin (echinonectin), isolated in another sea urchin species (*Lytechinus variegatus*) and displaying Del-1/lactadherin-like (discoidin-like) primary structure, was also found to be expressed in the early embryo and secreted into the extracellular matrix. Its ability to adhere to cells suggested a possible function in the control of cell movement [27]. In contrast to these lectins, which are exclusively expressed in early developmental phases, OXYL was also expressed in adult tissues. This observation supports the involvement of lectins pertaining to different structural families in specific roles, which may be covered in different life stages and different tissues in the phylum Echinodermata.

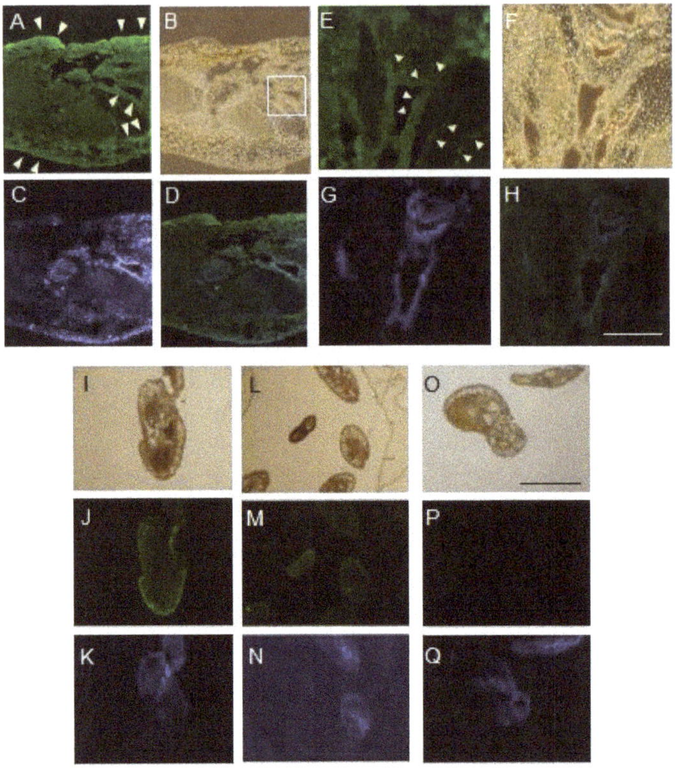

Figure 5. Localization of OXYL in *A. japonica*. Paraffin-embedded serial sections were stained by hematoxylin-eosin and anti-OXYL rabbit antiserum (**A–N**) and pre-immune rabbit serum (**O–Q**) followed by FITC-conjugated secondary anti-rabbit IgG goat antibody, and observed by fluorescence microscopy. Detections: FITC (**A,D,E,H,J,M,P**), DAPI (**C,D,G,H,K,N,Q**) and hematoxylin-eosin staining (**B,F,I,L,O**). The square in panel B was zoomed-in E-H. Scale bars: 100 μm (white) and 300 μm (black), respectively.

2.6. *OXYL Quantitatively Binds to Type-2 N-LacNAc*

Glycan array analysis, carried out to investigate the binding specificity of OXYL, led to the same results previously obtained by frontal affinity chromatography [20]. The glycans analyzed in the microarrays in this study are listed in Table 1 (see also Figure S3). OXYL was shown to bind to N-neo tetraose (LNnT: Galβ1-4GlcNAcβ1-3Galβ1-4Glc), bi- (NA2), tri- (NA3) and tetra-antennary (NA4) N-glycans (Figure 6). All these sugars possess type-2 LacNAc (Galβ1-4GlcNAc) (Figure 6). On the other hand, OXYL did not bind neither to lacto N-tetraose (LNT: Galβ1-3GlcNAcβ1-3Galβ1-4Glc), with type-1 LacNAc (Galβ1-3GlcNAc), nor to lactose (Galβ1-4Glc), without an N-acetyl group (Figure 6, LNT). OXYL could bind to 3-sialylated LacNAc more strongly than to 6-sialylated LacNAc (Figure 6, S3LN versus S6LN). However, since this interaction was much weaker than that observed with type-2 LacNAc (Figure 6, S3LN versus LNnT), it can be hypothesized that the free C-6 hydroxyl group of Gal in type-2 LacNAc is essential for OXYL binding. The specific binding of OXLY to type-2 LacNAc was further confirmed by hemagglutination inhibition assays (Table 2, Galβ1-4GlcNAc versus Galβ1-4Glc). Hemagglutination was not inhibited by monosaccharides (Gal and NeuAc), β-galactoside disaccharides without an N-acetyl group (lactose). In addition, neither the addition of glycans, such as chondroitin sulphate and heparine, nor the presence of *P. aeruginosa* lipopolysaccharide or *M. luteus* peptidoglycan interfered with hemagglutionation (Table 2).

Table 1. Structure of glycan for array analysis.

No.	Name	Structures
1.	Lac	Galβ1-4Glc
2.	S3LN	NeuAcα2-3Galβ1-4GlcNAc
3.	S6LN	NeuAcα2-6Galβ1-4GlcNAc
4.	LNnT	Galβ1-3GlcNAcβ1-3Galβ1-4Glc
5.	LNT	Galβ1-3GlcNAcβ1-3Galβ1-4Glc
6.	NA2	Galβ1-4GlcNAcβ1-2Manα1 6 3 Manβ1-4GlcNAcβ1-4GlcNAc Galβ1-4GlcNAcβ1-2Manα1
7.	NA3	Galβ1-4GlcNAcβ1-2Manα1 6 3 Manβ1-4GlcNAcβ1-4GlcNAc Galβ1-4GlcNAcβ1 4 Manα1 Galβ1-4GlcNAcβ1 2
8.	NA4	Galβ1-4GlcNAcβ1 6 Galβ1-4GlcNAcβ1 2 Manα1 6 3 Manβ1-4GlcNAcβ1-4GlcNAc Galβ1-4GlcNAcβ1 4 Manα1 Galβ1-4GlcNAcβ1 2

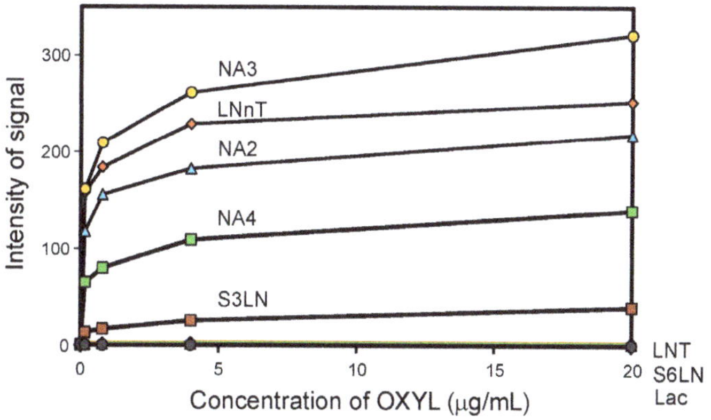

Figure 6. Glycan-binding properties of OXYL. Fluorescence-labeled OXYL was subjected to glycan array analysis using glycochip, where 8 glycan structures were immobilized. Fluorescence signals for the 8 glycans (listed in Table 1) are represented as signal intensities.

Table 2. Carbohydrate-binding specificity of OXYL [1].

Saccharides	Minimum inhibitory conc. (mM)
N-acetyllactosamine (Galβ1-4GlcNAc)	3.13
lactose (Galβ1-4Glc)	>100 [2,3]
D-galactose (Gal)	>100 [2,3]
N-acetylneuramic acid (NeuAc)	>100 [2]
Glycosamino glycans (GAG)	Minimum inhibitory conc. (mg/mL)
chondroitin sulphate	>50 [2]
heparin sodium	>50 [2]
Lipopolysaccharide (LPS) and Peptidoglycan (PG)	Minimum inhibitory conc. (mg/mL)
P. aeruginosa lipopolysaccharide	>50 [2]
M. luteus peptidoglycan	>50 [2]

[1] The titer of OXYL was previously diluted to 16 hemagglutination unit. [2] Inhibition was not observed at 100 mM (saccharides) or 50 mg/mL (GAG, LPS, and PG). [3] From previous results [20].

2.7. OXYL Displays Bacteria Agglutination Properties

An assay carried out with HiLyte-555 fluoro-labeled OXYL showed that this lectin led to the strong agglutination of *P. aeruginosa* bacterial cells (Figure 7). This agglutination was specifically inhibited by the co-presence of LacNAc, which, as previously demonstrated, is one of the saccharides recognized by OXYL (Figure 7B,b). This specificity was confirmed by the lack of any inhibitory effect on agglutination in the co-presence of lactose (Figure 7C,c). At the same time, bacterial lipopolysaccharide did not inhibit the agglutination of *P. aeruginosa* (Figure 7D,d).

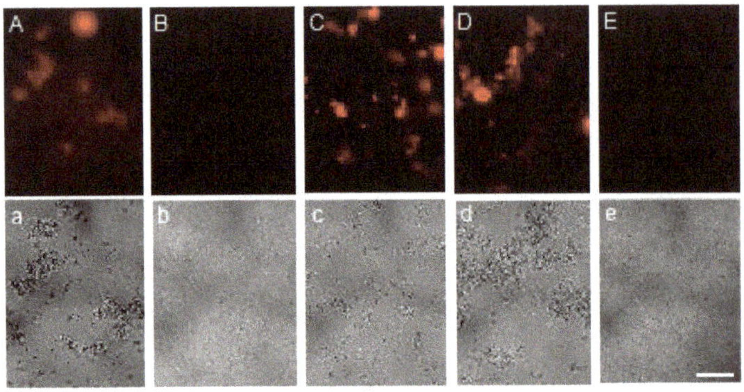

Figure 7. Agglutination of *Pseudomonas aeruginosa* by OXYL. Two micrograms of OXYL were administrated to the bacteria (**A–D**,**a–d**), which were observed by fluorescence microscopy (**A–E**; $\lambda_{ex/em}$ = 550/566 nm) and phase-contrast microscopy (**a–e**). Ten mM LacNAc (**B**,**b**), lactose (**C**,**c**) or 0.5 mg/mL lipopolysaccharide (**D**,**d**) were added in co-presence with OXYL. E and e are negative controls without OXYL. The bar indicates 100 μm.

2.8. Antibiofilm Activity and Influence on Bacterial Growth of OXYL

The presence of OXYL inhibited biofilm formation in *P. aeruginosa* (Figure 8A, solid line), even though it did not affect bacterial growth. As in the case of bacterial agglutination, anti-biofilm activity was inhibited by the co-presence of LacNAc (Figure 8B, black bar). We have previously reported a similar reduction of *P. aeruginosa* biofilm formation by another N-acetylhexosamine-binding lectin extracted from a sponge (*Halichondria okadai*) [12].

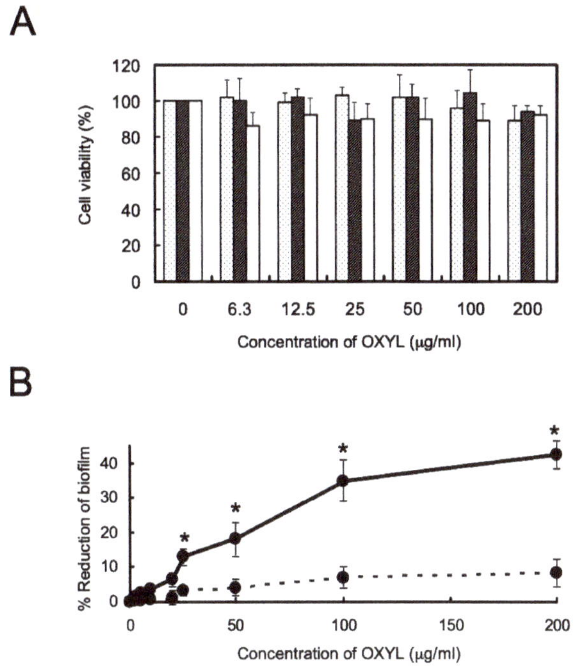

Figure 8. Influence of OXYL on bacterial growth, and its anti-biofilm activity. **A:** Bacteriostatic activity. Dotted, hatched, and white columns indicate the growth of *L. monocytogenes*, *S. boydii* and *P. aeruginosa*, respectively. Error bars: SE from three independent experiments (replicates). The data reported here are the mean ± SE ($n = 3$). **B:** Anti-biofilm activity. *P. aeruginosa* was cultured for 24 h with various OXYL concentrations (0–200 µg/mL) in 96-well plates in the absence (solid line) or presence (dotted line) of LacNAc (10 mM). * $p < 0.05$.

2.9. *OXYL Adhered to Cells by Binding to LacNAc on the Cell Surface*

HiLyte Fluoro 555-labeled OXYL could bind to BT-474 (breast), MCF-7 (breast), and T47D (breast) and HeLa (cervical), cancer cells (Figure 9Aa–d) cell lines. The fluorescent signal clearly outlined the contour shape of each cell, indicating that the lectin was bound to the cell surface via glycans containing LacNAc structures, which are commonly present on the membranes of mammalian cells. In agreement with the results reported in the previous section, OXYL binding was inhibited by the addition of LacNAc (Figure 9Ae–h). The amount and type of glycans on cell surface greatly differ depending on the cell line. MCF-7 and HeLa, the two cell lines where OXYL binding could not be completely inhibited by the addition of LacNAc, may be rich in branched N-type sugar chains and/or lacto-neo-tetraose glyco sphingolipid. Alternatively, these cells may be rich in NeuAcα2-3-linked glycans. Moreover, the fluorescence signal remained on the cell surface even when incubation lasted for a relatively long period of time (2 to 12 h). This behavior was clearly different from that observed for other N-acetylhexosamine-binding lectins, such as iNoL [10] and HOL-18 [12], which migrated inside the cells over time, triggering apoptosis. This observation is consistent with the complete lack of cytotoxicity of OXYL on bound cells (Figure 9B). These results provide useful evidence about the cellular function exerted by this lectin in crinoids, pointing out that cell binding is not necessarily associated with cytotoxicity.

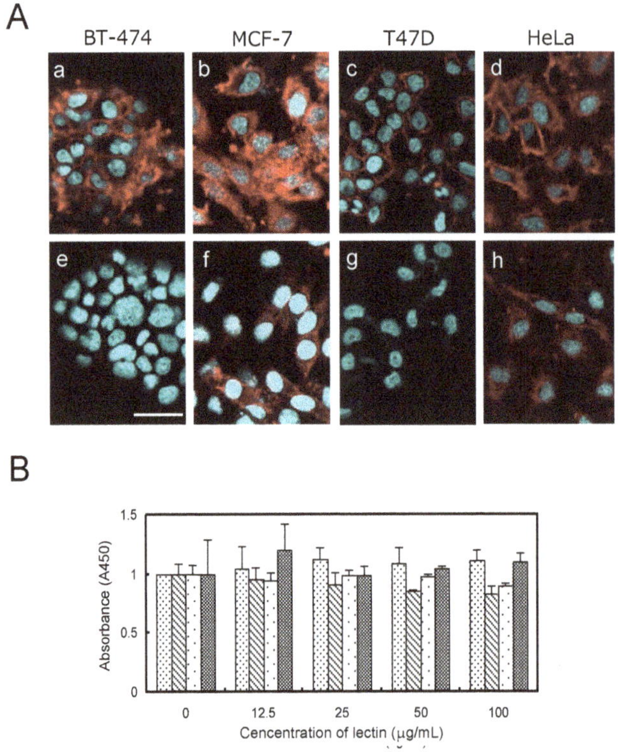

Figure 9. Binding and growth effect of OXYL to cancer cells. **A:** HiLyte Fluoro 555-conjugated OXYL (10 μg per cells) was administrated to BT-474 (**a**,**e**), MCF-7 (**b**,**f**), T47D (**c**,**g**), and HeLa (**d**,**h**) without (**a**–**d**) or with (**e**–**h**) LacNAc (10 mM). **B:** Cell viability of OXYL against four types of cancer cells. Cells were treated with OXYL at various concentrations (0–100 μg/mL) for 48 h, and viability was determined by WST-8 assay. Values for BT-474, MCF-7, T47D, and HeLa are shown by fine dotted, hatched, rough dotted and mesh bars, respectively.

2.10. *OXYL LacNAc-Dependently Activated the p38 of BT-474 Cancer Cells*

The activation of p38, one of the most important kinases in the MAPK pathway, was moderately induced by OXYL in BT-474 cells in a dose-dependent manner (Figure 10, P-p38 versus p38). This behavior may be explained by the involvement of p38 in a secondary signaling pathway, only indirectly activated by OXYL binding, which would be consistent with the relatively weak degree of phosphorylation observed in this study. This effect was cancelled by the co-presence of LacNAc, which has been demonstrated as the ligand of OXYL by western blotting (Figure 10, LacNAc (−) versus LacNAc (+)). Nevertheless, OXYL did not induce the activation of caspase-3 (Figure 10, Pro-caspase-3 versus Cleaved caspase-3). Moderate levels of phosphorylated p38 may not have been sufficient to trigger activation of caspase-3 and apoptosis. A number of animal lectins are able to induce signal transduction, with the activation of both MAPK pathway and caspases [10,12,28]. On the other hand, ability of some lectins to activate the MAPK pathway without affecting caspases has been also reported in literature. For example, Gb3-binding lectins do not necessarily produce cytotoxic effects against Gb3-expressing cells. Indeed, SAL (*Silurus asotus* lectin), a lectin pertaining to the SUEL/RBL family, isolated from catfish eggs, activated MAP kinases in Burkitt's lymphoma cells through Gb3 binding, but did not trigger apoptosis [29]. OXYL proved to be another example of a lectin that binds to specific glycans exposed on cell surface, activating the metabolic system without induction of cell death.

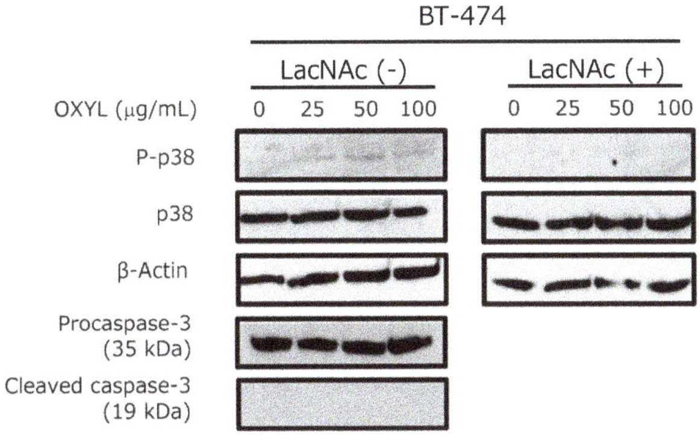

Figure 10. Effects of OXYL on phosphorylation and expression levels of p38 and caspase-3 in the BT-474 cell line. Cells (4 × 10^5 in each experiment) were treated with various concentrations of OXYL (0–100 μg/mL), and activation levels were evaluated by western blotting of lysates in absence (−) and presence (+) of LacNAc (10 mM). Experiments were performed in triplicates.

3. Discussion

We identified secreted globular head C1q (SghC1q) protein as a major lectin family in the crinoid class (Figure 1). This result is in contrast with the previous observation that C-type lectins, such as "echinoidin", represent the main glycan-recognition molecules in the coelomic fluid of species from the subphylum Eleutherozoa [3,4]. This suggests that a certain degree of flexibility exists in echinoderms about the structural architecture of secretory lectins that may cover important functions, such as immune recognition. The primary structure obtained from Edman degradation and bioinformatics analysis of genome data highlighted several interesting points of contact and divergence with other C1q domain-containing (C1qDC) proteins previously identified in other invertebrates. Indeed, immunostaining indicated a certain degree of tissue-specificity and the fact that OXYL mRNA was not expressed in the ovary tissue confirms that the lectin is produced only in specific tissues. Furthermore, unlike most bivalve C1qDC genes [25], the crinoid gene encoding OXYL was intronless, as no intron was found in the region separating the signal sequence from the globular C1q domain. Whether this is a general feature of all crinoid C1qDC genes, or it is a specific characteristic of OXYL will be clarified by further bioinformatics analyses, once a complete crinoid genome will become available.

Our study indicated the presence of at least four other paralogous OXYL gene copies in this species, as well as several orthologous sequences in other species of the order Comatulida, suggesting that this molecule plays an important role in crinoids (Figure 2). Future studies should be directed at clarifying the functional meaning of this gene family expansion event, aiming to investigate whether any of these genes is transcriptionally responsive to environmental changes, such as infection and change of water quality. C1qDC genes are well known to be responsive to immune challenges and stress conditions in bivalves, where they are part of largely expanded gene families [30–32], but additional evidence has linked them to the complex inter-specific interactions in other invertebrates, such as the response to predator kairomones in marine snails [33]. The echinoderm immune defense system is one of the best characterized in invertebrates, as multiple key components of immune recognition, signal transduction and antimicrobial effectors have been identified over the years [34,35]. This wealth of information may enable to chase down the possible function of OXYL and its interaction with other immune receptors and effectors.

The C1q domain is found in a large number of metazoan proteins, such as the main chains of the C1q complex, collagens, adiponectins, hibernation-specific protein, regulator of synapse organizer and also several lectins [36,37]. This domain is characterized by a jelly-roll fold of β-sandwich structure consisting of 10 β-strands. This topology is also found in proteins pertaining to the TNF superfamily, even though the primary structures of TNF and C1q do not share any significant similarity [38], in addition to galectins and legume lectins [39], and TNF-like bacterial lectins [40]. It seems likely that this fold may be used as a common structural scaffold with high potential for intermolecular interaction. Although the origins of the C1q domain are still unclear, this fold is found in both protostome and deuterostome animals, and it is also present in some bacteria. Since C1qDC lectins have been isolated in different animals, from the comparative glycobiology viewpoint metazoans have used C1q as a recurrent tool for glycan-binding. In contrast with most C1qDC lectins described in literature, which require divalent cations for their activity, OXYL did not require them at all for its carbohydrate-binding activity.

The sghC1qDC architecture, which characterizes proteins with an N-terminal signal sequence for secretion and without a collagen tail, have been reported in a number of vertebrate and invertebrate animals, and in some cases they work as lectins [24]. In mollusks, such proteins have been isolated as lectins binding to NeuAc, D-Man and lipopolysaccharide [41–43]. In chordates, D-Fuc [44] and D-GlcNAc [45] binding lectins have been purified from Petromyzontiformes and Osteichthyes, respectively. However, several studies have recently started to also unveil the glycan-binding properties of human C1qDC proteins. For example, the galactosylation of complex-type N-linked glycans in the Fc region of human IgG has been demonstrated to improve the binding of this immunoglobulin with C1q [46]. Moreover, C1qDC proteins are known to bind to monosaccharides [47] and glycosaminoglycan [48], and the human adiponectin can bind to bacterial lipopolysaccharides [49]. Considering the weight of such evidence, the ancestor of C1q was likely to be a carbohydrate-binding protein, which may have acquired other characteristic features along with evolution in different lineages.

Glycan array assays and the inhibition test demonstrated that OXYL specifically recognized type-2 LacNAc (Figure 6, Tables 1 and 2). On the other hand, the lectin did not bind to monosaccharide or lactose, proving further evidence in support of its specific recognition of the N-acetyl group in GlcNAc. These recognition properties are clearly different from those of RCA 120, ECA and galectin, which are all known as LacNAc-binding lectin being also able to bind to lactose. This result resembles the observations made for echinoidin, which can bind to the core structure of the mucin-type glycan, without binding to the monosaccharide [3]. The diverse carbohydrate-binding properties of lectins derived from marine invertebrates, might be exploited in the future with the aim to discover new drugs or to develop new diagnostic tools for cellular biology.

Another peculiar feature of OXYL was the finding that it adopts tetrameric structure, as revealed by ultracentrifugation, unlike most C1qDC proteins, which are mostly organized as trimers [34]. This uncommon result found strong experimental confirmation in the results obtained by the gel permeation chromatography results reported in a previous study [20]. In addition to the tetrameric form (Figure 3c), OXYL seemed to be able to reversibly adopt other associative forms, from dimers to dodecamers (Figure 3b,d,e). The ability to create homo- or heterotrimeric assemblies is a well-known property of many C1qDC proteins, including the three main chains of the human C1q complex, cerebellin, and adiponectin [50–52]. Such proteins can also create higher-order hexametric complexes, whose formation is usually mediated by intermolecular disulfide bonds. The associative state of the C1q subunits has a strong impact on the biological properties of such complexes. For example, adiponectin is able to stimulate the activity of AMPK only in its trimeric form, whereas the hexameric form is the only one capable of triggering NF-kB signaling [53,54]. The assembly of higher-order "bouquet of tulips" super-structures, mediated by N-terminal collagen regions, is also of primary importance to enable the activation of the classical pathway of the complement system by the C1q complex [55]. In light of these observations, the presence of multiple associative forms of OXYL is of great interest, as the predominant tetrameric form may exert different biological activities compared to the less abundant dimeric, octameric, or dodecameric OXYL. Moreover, it is presently unknown whether heteromeric complexes between OXYL and its paralogous proteins (Figure 2A) can be created, or such assemblies are strictly homomeric. Altogether, the complex and still poorly characterized associative properties of OXYL may translate into a fine regulation of immune response upon pathogen recognition in this species.

Although several sghC1qDC proteins have been so far identified in invertebrates, very few studies have investigated the status of association of their subunits. There is little doubt that the accumulation of physicochemical knowledge on the structural properties of C1qDC proteins with carbohydrate-binding activity, as well as the elucidation of the structural features of their multimeric assemblies will improve our understanding of the biological role of these molecules.

Immunohistochemistry showed that the expression of OXYL was mainly detectable around the coelomic cavities and in the peripheral areas of skin in the arms. This suggests that this lectin exerts its biological functions in the outer and humoral environments, implying a possible involvement in body defense (Figure 5). Although only limited information is presently available concerning the localization of the different lectins isolated so far in the phylum Echinodermata [26,27], different types (C1q, SUEL/RBL, discoidin-like and C-type) of lectins display a variegated tissue distribution and pattern of expression during development. The accumulation of such information in the years to come will most likely help to better understand the evolutionary processes that have led to the functional diversification of invertebrate lectins compared with those found in vertebrates. The localization of OXYL in the close proximity of the coeloms and stromal tissues in the arms of the feather star is reminiscent of that of the mammalian C1q, which is expressed in the stroma and vascular endothelium of several human malignant tumors [56].

Although the C1q domain is mostly known as the characterizing domain of the C1q chains, primary mediators of immune response, which connect antigen-complexed IgGs with other complement factors in human, some studies have reported complement-independent antibacterial [57,58] or cancer growth promoting [59] activities by C1qDC proteins purified from fish and human, respectively. Although OXYL showed no direct antibacterial activity (Figure 8A), it strongly agglutinated *P. aeruginosa* cells (Figure 7A). Lipopolysaccharides are typical cell wall components and well-known virulence factors in Gram-negative bacteria [60], which may be recognized by some of C1qDC proteins acting as pattern recognition receptors [41–43]. However, the results of the assays carried out in this study demonstrate that the agglutination of *P. aeruginosa* by OXYL was LPS-independent, as this activity could be only inhibited by LacNAc (Figure 7B,D). The identification of the ligands specifically recognized on the bacterial cell surface by OXYL might help to understand the functional importance of the interaction between LacNAc-binding lectins and bacterial glycans in metazoan immune system.

OXYL, on the other hand, showed a remarkable anti-biofilm activity (Figure 8B). Similar properties have been previously observed in galactose-binding lectins isolated from snake venom and sea hare eggs [61,62], as both displayed anti-biofilm, but not antibacterial activity. Even though OXYL is not able to kill bacterial cells by itself, we hypothesize that its role in the context of crinoid immunity might be to facilitate the recognition of agglutinated bacteria by opsonins, enabling the elimination of the invading cells and allowing the intervention of circulating phagocytes thanks to the disruption of biofilm structure.

It is known that lectins found in plants, invertebrates and fishes, generally associated in dimeric or higher-order multimeric complexes, can influence the proliferation of mammalian cells. For example MytiLec-1, a dimeric lectin isolated from the mantle of mussels, can bind to human lymphoma cells, induced apoptosis. However, when the same lectin was dissociated in its monomeric state, it lost its agglutination and apoptosis-inducing properties [63]. Although OXYL displays a similar multimeric structure (Figure 3) and also shows a significant cell adhesion activity (Table 2, Figure 9A), it has no effect on cell growth (Figure 9B). SAL, a trimeric lectin isolated from catfish eggs, is another example of a lectin which can modulate cellular pathways upon the interaction with target cells. This lectin can trigger the activation of the intracellular MAP kinase system in Burkitt's lymphoma cells, leading to a decreased expression of members of the ATP-binding cassette family, even though it cannot prevent cell proliferation [64]. The combined observation that (i) OXYL possess can activate MAP kinases, without affecting the activity of caspases (Figure 9), and that (ii) its cell-recognition properties depend on the binding of glycans having LacNAc structure, with no effect on cell proliferation (Figure 8), can be interpreted as follows. We hypothesize that the interaction between OXYL and the LacNAc glycans located on the surface of its target cells may trigger the activation of intracellular metabolism. This interaction probably activates presently unknown pathways unrelated to apoptosis though it is still linked with a moderate activation of the p38 kinase.

A comparative overview on the properties of OXYL and members of the galectin family might be important for understanding the role of this lectin. While some galectins can have a cytotoxic effect on the cells they recognize [65], OXYL did not suppress the growth of cancer cells. On the other hand, OXYL and galectins share a jelly roll topology, in spite of a complete lack of primary structure similarity. Galectins basically display binding properties to type-2 LacNAc. They selectively also recognize NeuAcα2-3LacNAc but not NeuAcα2-6LacNAc [66]. These properties reveal a glycan-binding specificity similar to that of OXYL (Figure 6). Functional studies have revealed that galectins interact with the hydroxyl group of the galactose found in C-6 position in lactose and did not interact with the hydroxyl group at the C-3 position [67]. In a similar fashion to galectins, OXYL may also interact with the hydroxyl group at the C-6 position, as explained by its stronger binding to LNnT, compared to S3LN (Figure 6). The detailed bonding structure will be clarified by the elucidation of the 3D structure of this molecule. After all, these two types of lectins share similar LacNAc-binding properties and the same stereoscopic topology.

In detail, C1q/TNF-family has been shown to bind to various glycans [46–48], to modify activation of MAP kinases, such as ERK and p38 [68,69], and to regulate the proliferation of cells. Although OXYL could not regulate cell growth by itself, it led to a moderate activation of p38 (Figure 10), maybe as a part of a secondary signaling pathway. The study of the immune system of lamprey, a key organism in the evolution of vertebrates, has brought new impulse to the study of C1q in animal immunity. This primitive jawless fish was found to produce a characteristic C1q protein (LC1q), which was discovered as a LacNAc-binding lectin during the search for the ancestral C1q molecule of chordates. LC1q was elucidated to specifically activate the complement system, activating cell lytic pathways in a different way compared with mammals [70]. In light of these observations, it can be suggested that the peculiar lamprey complement system evolved independently from that of jawed vertebrates, using a proto-typical C1q protein that acquired both immune and glycan-binding activities [45]. Although several sialic acid (NeuAc)-binding lectins described in invertebrate animals share a C1q-like fold [41], recent studies have revealed that the same binding function can be also achieved by lectins (SigLec) with an immunoglobulin-like fold [71]. The glycan-binding properties of lectins have long been considered to be strictly dependent on their unique primary structures. However, these results, together with the elucidation of the structural properties of C1q-like and immunoglobulin-like lectins, provide a new framework of conceiving the lectin structure/function relationship. Indeed, the widespread nature of these recurrent structural folds suggests an ancient origin in biological systems as an example of molecular bricolage [72]. Thanks to their remarkable binding properties, these structures may have been recruited as lectins, components of the complement system or recognition molecules of the adaptive immune system along evolution, based on the increasing need to recognize specific ligands with the colonization of new environments.

Although OXYL has been first isolated by basic biochemical methods in a previous study [20], we here provide the characterization of its structure as a member of the C1qDC family as a type-2 LacNAc-binding lectin. The comparative view among OXYL, LC1q, and other C1qDC proteins isolated in various invertebrates may lead to new relevant findings concerning the evolution of innate immunity in Deuterostomia. Moreover, in light of the emerging role of human C1q as cancer proliferation regulating factor, OXYL derivatives may also be also studied for the development of novel drugs in the field of oncology.

4. Experimental Design

4.1. Materials

A. japonica was supplied from Mrs Hisanori Kohtsuka and Mamoru Sekifuji in the Misaki Marine Biological Station, The University of Tokyo, Miura city, Sagami Bay, Kanagawa Prefecture, Japan, and stored at −80 °C until the beginning of this study. Strains of three bacteria (*Pseudomonas aeruginosa, Shigella boydii* and *Listeria monocytogenes*) were obtained from the Dept. of Biochemistry and Molecular Biology, University of Rajshahi, Bangladesh. Human cell lines BT-474 (breast cancer), MCF-7 (breast cancer), T47D (breast cancer), and HeLa (cervical cancer) were obtained from the American Type Culture Collection (Manassas, VA, USA). Lipopolysaccharide (from *P. aeruginosa*), peptidoglycan (from *Micrococcus luteus*), bovine serum albumin (BSA), Pathoprep-568, cell lysis buffer M, Mayer's hematoxylin solution, eosin alcohol solution, 4% paraformaldehyde phosphate buffer, 20% glutaraldehyde solution, cacodylate buffer, Canada balsam, crystal violet solution, Penicillin-Streptomycin solution, and horseradish peroxidase (HRP)-conjugated β-actin mAb were all acquired from FUJIFILM Wako Pure Chemical Corp. (Osaka, Japan). Standard protein markers for SDS-PAGE were purchased from Takara Bio Inc. (Kyoto, Japan). HRP-conjugated goat anti-rabbit IgG was obtained from Tokyo Chemical Industry Co. (Tokyo, Japan), Type-2 LacNAc was obtained from Dextra Laboratories Ltd. (Reading, England). Prestained protein marker was from Bio-Rad Laboratories (Hercules, CA, USA). FITC-labeled goat anti-rabbit IgG from Abcam (Cambridge, UK), anti-P38 mAbs, anti-phosphorylated P38 (^{180}Thr/^{182}Tyr) mAbs and anti-caspase-3 mAb from Cell Signaling Technology (Danvers, MA, USA). Can Get Signal Immunoreaction Enhancer solutions A and B were purchased from Toyobo Co. (Osaka, Japan). 4′,6-diamidino-2-phenylindole (DAPI), RPMI 1640 medium, and fetal bovine serum (FBS) were acquired from Gibco/ Thermo Fisher (Waltham, MA, USA). Poly-L-lysine-coated slides used in this study were from MilliporeSigma (Darmstadt, Germany). Cell Counting Kit-8 (including WST-8[2-(2-methoxy-4-nitrophenyl)-3-(4-nitrophenyl)-5-(2,4-disulfophenyl)-2H-tetrazolium monosodium salt]) and HiLyte555 Labeling kit-NH_2 were provided by Dojindo Laboratories (Kumamoto, Japan). PVDF membranes for electroblotting and peroxidase substrate EzWestBlue were obtained from ATTO Corp. (Tokyo, Japan).

4.2. Purification of the OXYL Lectin

OXYL was purified from stored *A. japonica* specimens as previously described [20], with minor modifications. Arms were homogenized with 10 volumes (*w*/*v*) Tris-buffer (10 mM tris(hydroxymethyl)aminomethane-HCl, pH 7.4). Homogenates were filtered through gauze and centrifuged (model Suprema 21, TOMY Co.; Tokyo Japan) at 27,500× *g* for 1 h at 4 °C. Crude supernatant was loaded in a Q-Sepharose-packed column (10 mL), which was then washed with Tris-buffer. The OXYL-containing fraction was gradually eluted with 50–300 mM NaCl in Tris-buffer, and dialyzed with Tris-buffered saline (TBS: 10 mM Tris-HCl and 150 mM NaCl, pH 7.4). The crude fraction was loaded in a fetuin-agarose column (5 mL), and OXYL was eluted with TBS containing 6 M urea. The purity of the lectin was evaluated by SDS-PAGE [73] using 15% (*w*/*v*) acrylamide gel under reducing conditions. Protein was quantified using a bicinchoninic acid protein assay kit with BSA as standard protein. Absorbance was measured at 562 nm using a 96-well microplate photoreader (SmartSpec 3000, Bio-Rad Laboratories; Hercules, CA, USA) [74,75].

4.3. Identification of the OXYL Coding Sequence

The N-terminal amino acid sequence of OXYL was determined with automated Edman degradation by using a protein/peptide sequencer (Shimadzu Co. Ltd., Kyoto, Japan) [5]. The partial amino acid sequence of OXYL was used as a query in a tBLASTN search [76] against two publicly available sequence resources for *A. japonica*: (i) the de novo assembled ovary transcriptome, obtained from the NCBI TSA database (mater record: GAZO00000000.1). (ii) a 30X coverage partial genome, which expected size is 650 Mbp (BioProject ID: PRJNA236227). In this case, raw sequencing data was imported in the CLC Genomics Workbench v.11 (Qiagen, Hilden, Germany), trimmed to remove sequencing adapters, low quality bases and failed reads, and de novo assembled using automatically estimated word size and bubble size parameters, and allowing a minimal contig length of 200 nucleotides. Nearly perfect matches were initially searched, by selecting hits with BLAST matches scoring e-value lower than 1×10^{-10}. Since no significant hit could be found in the ovary transcriptome, we extended the initial BLAST hit identified in the genome to the overlapping open reading frame. The presence of possible donor and acceptors splicing sites was predicted with Genie [77], and the possible location of the poly-adenylation signal was identified by the presence of the typical eukaryotic sequence consensus AATAAA [78]. Gene structure was further confirmed by comparative means, through the multiple sequence alignment of the *A. japonica* genomic DNA with orthologous full-length cDNA sequences from other crinoids (see Section 4.4).

4.4. Phylogeny of OXYL

To track the evolutionary history of OXYL, all the sequences encoding a protein containing the characterizing functional domain of this lectin, the C1q domain were isolated from the available transcriptomes of crinoids and, for comparative purpose, from the fully sequenced genomes of *Strongylocentrotus purpuratus* (Echinozoa, Echinoidea), *Apostichopus japonicus* (Echinozoa. Holothuroidea) and *Acanthaster planci* (Asterozoa, Asteroidea). For Crinoidea, the following species were selected: *Antedon mediterranea*, *Cenolia trichoptera*, *Democrinus brevis*, *Dumetocrinus antarcticus*, *Florometra serratissima*, *Isometra vivipara*, *Metacrinus rotundus*, *Notocrinus virilis*, *Oligometra serripinna*, *Phrixometra nutrix*, *Promachocrinus kerguelensis*, *Psathyrometra fragilis*, and *Ptilometra australis*. All transcriptomes were de novo assembled following the protocol described in Section 4.3 and protein predictions were carried out with TransDecoder v.5.3. C1q domain containing (C1qDC) proteins were identified with hmmer v.3.1b [79], based on the detection of the C1q Pfam domain (PF00386) with a significant e-value (lower than 1×10^{-5}). OXYL paralogous gene copies were also identified in the *A. japonica* genome assembly with a BLASTx approach (the e-value threshold was set at 1×10^{-5}). Partial proteins with an incomplete C1q domain were discarded prior to further analysis.

All the C1qDC protein sequences identified as detailed above were used to generate a multiple sequence alignment with MUSCLE [80], which was further refined to only retain the evolutionarily informative region corresponding to the C1q domain. This dataset was preliminarily used to assess, with a Neighbor Joining clustering approach, the placement of OXYL within the phylogenetic tree of all echinoderm C1qDc proteins. A selection sequence including all the sequences putatively pertaining to a cluster of OXYL orthologous and paralogous genes was subjected to a more rigorous Bayesian inference phylogenetic analysis with MrBayes v.3.2 [81], including in the MSA the C1qDC sequences *from S. purpuratus* and the two additional C1qDC proteins detected in the *A. japonica* transcriptome as outgroups. This analysis was carried out by running two independent Monte Carlo Markov Chains in parallel for 160,000 generations, under a WAG+G+I model of molecular evolution, estimated by ModelTest-NG as the best fitting one for this dataset [82]. Run convergence was assessed by the reaching of an effective sample size higher than 200 for all the parameters of the model with Tracer (https://github.com/beast-dev/tracer/).

4.5. Molecular Mass Determination of OXYL

Sample concentration was estimated to be 1.0 mg/mL based on A280 measurement. Sedimentation velocity experiments were performed using an Optima XL-I analytical ultracentrifuge (Beckman Coulter; Brea, CA, USA) with An-50 Ti rotor. Standard Epon two-channel centerpiece with quartz windows were loaded with 400 μL sample and 420 μL reference solution (50 mM potassium phosphate, pH 7.4, 0.1 M NaCl). Prior to each run, the rotor was kept in a stationary state at 293 K in vacuum chamber for 1 h for temperature equilibration. A280 scans were performed without time intervals during sedimentation at 50,000 rpm, and analyzed using SEDFIT with the continuous distribution (c(s)) analysis module [83,84]. Frictional ratio (f/f_0) was allowed to float during fitting. c(s) distribution was converted to molar mass distribution c(M).

4.6. Generation of Antiserum and Evaluation of Anti-OXYL Antibody Specificity

The antiserum against OXYL was raised in rabbit serum by Sigma-Aldrich. Antigen (500 μg synthesized peptide consisted of 18 amino acids of ^{6}Ser-^{23}Lys in OXYL) was injected 2x during 25 days, and antiserum was collected using saturated NH_3SO_4. Crude feather star extract and purified OXYL separated by SDS-PAGE were electroblotted on a PVDF membrane [85]. The blotted membrane was masked with TBS containing 1% (*w*/*v*) BSA, soaked with 0.2% Triton X-100 at room temperature (RT), treated with anti-OXYL rabbit serum (1:1000 dilution) (primary antibody) and HRP-conjugated goat anti-rabbit IgG (secondary antibody) for 1 h each, and colored with EzWestBlue as per the manufacturer's instructions.

4.7. Immunohistochemistry of OXYL in Feather Star Tissues

Paraffin embedded sections were prepared according to the Shibata's protocol [86]. Arms were fixed in phosphate buffered saline (PBS) containing 10% formalin overnight at RT. After washing with PBS, the specimens were decalcified in a solution containing 10% EDTA, 5% HCl, 1% formic acid and 10% sodium citrate overnight, dehydrated through a graded ethanol series, embedded in paraffin, sectioned (8 mm) and then mounted on slides. After deparaffination in xylene and hydration through the ethanol series, the sections were stained with hematoxylin-eosin. Sections were blocked with 1% (*w*/*v*) BSA containing TBS overnight at RT, treated with anti-OXYL antiserum (diluted 1:500 with PBS) and FITC-labeled anti-rabbit goat IgG for 1 h, counterstained with hematoxylin-eosin, mounted in Canada balsam, and observed by fluorescence ($\lambda_{ex/em}$ = 494/520 nm for FITC) and optical microscopy. Nuclei were stained by DAPI (364/454 nm for DAPI).

4.8. Sugar and Glycoconjugates-Binding Specificity of OXYL

The binding specificity of OXLY to sugars and glycoconjugate derivatives was estimated using 96-well V-bottom plates. A volume of twenty μL of each solution was serially diluted in TBS, mixed with 20 μL lectin solution (previously adjusted to titer 16 [87]), trypsinized, and applied to glutaraldehyde-fixed rabbit erythrocytes in TBS containing 0.2% Triton X-100. Each plate was kept at RT for 1 h to form a dot (no agglutination; inhibited; effective) or sheet (agglutination; not inhibited; ineffective) at the bottom. For each saccharide, the minimum concentration at which each dot turned into a sheet was defined as the binding affinity to OXYL.

4.9. Glycan Array Analysis

Glycan array analysis was performed by Sumitomo Bakelite Co. (Tokyo, Japan). OXYL was fluorescence-labeled ($\lambda_{ex/em}$ = 555/570 nm) using HiLyte Fluor 555 labeling kit-NH_2 (Dojindo) as per the manufacturer's instructions. A total of 8 glycans were immobilized on wells of a glycan chip (1 mM of each). Fluorescence-labeled OXYL, at concentrations ranging from 0 to 20 μg/mL, was incubated overnight at 4 °C with shielding from light. OXYL-binding glycans were detected by a Bio-REX Scan 300 evanescent fluorescence scanner (Rexxam Co. Ltd.; Osaka, Japan).

4.10. Bacteriostatic Assay

Bacteriostatic assays were performed as in our previous studies [12]. Gram-positive (*L. monocytogenes*) and gram-negative (*S. boydii* and *P. aeruginosa*,) bacteria were grown overnight in LB medium, harvested, and washed with phosphate-buffered saline (PBS). Fifty μL bacterial suspension (turbidity adjusted to OD_{600} = 0.6) in PBS was mixed with serial dilutions of OXYL for a quantitative assay. Bacterial suspensions were washed, and OD_{600} adjusted to 1.0. Bacteria were mixed with OXYL to final concentrations equal to 6.25, 12.5, 25, 50, 100, and 200 μg/mL in 96-well flat-type microtiter plates, and cultured at 37 °C, with OD_{600} measured every 4 h. The growth suppressive activity (%) of OXYL was calculated as (1 – ($OD_{600\ experiment}/OD_{600\ control}$)) × 100%.

4.11. Anti-Biofilm Activity of OXYL

Anti-biofilm activity was evaluated as in our previous study [12]. *P. aeruginosa* cells were grown in nutrient broth for 24 h at 30 °C. Colonies were transferred into test tubes and centrifuged at 3500× *g* for 3 min. Turbidity of bacterial cell suspensions was adjusted to OD_{640} = 1.0. Fifty µL bacterial suspension was mixed with the serial dilution of purified lectin (final volume 100 µL) in a 96-well microtiter plate, and incubated for 24 h at 37 °C. In each well, the biofilm was stained by exposure to 20 µL of 0.1% (*w*/*v*) crystal violet solution (filtered through pore size 0.45 µm filter paper) for 10 min at RT. Each well was washed 3x with TBS to remove free dye, then treated with 150 µL of 95% ethanol for 10 min at RT to release crystal violet. Extracted dye was transferred to another 96-well plate, and OD_{640} values were recorded by automated microtiter plate reader at 640 nm. The percentage of reduction of biofilm formation resulting from lectin treatment, relative to control, was calculated as follows:

$$\%\ \text{reduction of biofilm formation} = (1 - (OD_{640\ \text{experiment}}/OD_{640\ \text{control}})) \times 100\%$$

4.12. Binding of the Surface of Bacteria and Cultured Cancer Cells by OXYL

OXYL was labeled by HiLyte Fluor 555-labeling kit as per the manufacturer's instructions. Cancer cell lines BT-474, MCF-7, T47D and HeLa were cultured and maintained in RPMI 1640 supplemented with heat-inactivated FBS (10%, *v*/*v*), penicillin (100 IU/mL), and streptomycin (100 µg/mL) at 37 °C in 95% air/5% CO_2 atmosphere. Cells were washed 3× with PBS and incubated 2 h with 25 µg/mL HiLyte 555 Fluoro-labeled OXYL. Nuclei were stained by DAPI, and cells were fixed with 4% paraformaldehyde and observed by fluorescence microscopy ($\lambda_{ex/em}$ = 555/570 nm for HiLyte Fluor 555; 364/454 nm for DAPI). The cytotoxic activity of OXYL (in concentrations ranging from 0–100 µg/mL) against the different cancer cell lines was determined using a Cell Counting Kit-8 containing WST-8 [16].

4.13. Detection of Activated Signal Transduction Molecules and their Phosphorylated Forms in BT-474 Cells in the Presence of OXYL

The human breast carcinoma cell line BT-474 (3×10^5 cells) was cultured with OXYL (0–100 µg/mL) for 24 h, and cells were lysed with 200 µL cell lysis buffer M. The cell lysate was separated by SDS-PAGE and electroblotted on a PVDF membrane. The primary antibodies used in this study were directed at P38 (mouse mAb; dilution 1:3000), phospho-P38 (mouse mAb; dilution 1:3000) and caspase-3 (rabbit mAb; dilution 1:5000). Membrane was masked with TBS containing 1% BSA, soaked with 2% Triton X-100 at RT, incubated with HRP-conjugated goat anti-mouse IgG (for mouse mAb) or anti-rabbit IgG (for rabbit mAb) for 1 h [12], and colored with EzWestBlue. Experiments were performed in triplicates.

4.14. Statistical Analysis

For each of the studied parameters, experimental results were presented as mean ± standard error (SE) for three replicates. Data were subjected to one-way analysis of variance (ANOVA) followed by Dunnett's test, using the SPSS Statistics software package (Chicago, IL, USA), v. 10 (www.ibm.com/products/spss-statistics). Differences with $p < 0.05$ were considered as statistically significant.

5. Conclusions

In this study, we characterized a type-2 LacNAc-binding 14 kDa lectin, named OXYL, from the feather star *A. japonica* (phylum Echinodermata, class Crinoidea), which proved to belong to the C1qDC family. We identified several paralogous genes in the same species, in addition to a variety of orthologs in other species of the class Crinoidea. Its tissue localization and primary structure suggest that

OXYL covers a role related with immunity in this animal. Although C1qDC proteins generally create trimeric structures though a collagen tail, OXYL quaternary structure was estimated to be a multimer with tetrameric and octameric organization by analytical ultracentrifugation. OXYL caused a strong aggregation of bacterial cells, but did not affect their growth. On the other hand, the lectin displayed a LacNAc recognition-dependent anti-biofilm activity. We demonstrated that LacNAc glycans were also fundamental to enable adhesion to the surface of mammalian cultured cells. This interaction activated the MAP kinase p38, but it did not affect cell growth or affected caspase activity. The novel primary structure and the unique activities of this lectin, discovered in an organism regarded as a living fossil, highlights once again the structural and functional diversification of metazoan lectins, opening new questions about the biological role of LacNAc-binding lectin with jelly roll topology in marine invertebrates.

Supplementary Materials: The following are available online at http://www.mdpi.com/1660-3397/17/2/136/s1, Figure S1: Primary structure determination of the N-terminal region of OXYL, Figure S2: Analysis of OXYL by SEDFIT program, Figure S3: Glycan-array analysis.

Author Contributions: I.H. did the majority of the study including protein sequencing and wrote the MS. M.G. supported genomic analysis. Y.F. supported purification of OXYL. Y.O. conducted the study including technical and writing assistance.

Funding: This study was supported in part by a grant for academic research from Yokohama City University, Japan. The study was partly done by research and developmental support for life sciences from the City of Yokohama (life innovation platform, LIP.Yokohama).

Acknowledgments: The authors are grateful to Koji Akasaka, Hisanori Kohtsuka, Mariko Kondo, Akihito Omori and Mamoru Sekifuji from the Misaki Marine Biological Station, The University of Tokyo for supplying experimental animals and discussion; Tatsuo Oji from Nagoya University Museum for useful discussion on the evolution of crinoids; Tomoko F. Shibata from the Yokohama City University for help in tissue section preparation; Fumio Arisaka from the Tokyo Institute of Technology for assistance and discussion on the analytical ultracentrifugation; Nicolò Fogal from the University of Trieste for assistance in bioinformatics analysis.

Conflicts of Interest: The authors declare no conflict of interest.

References

1. Vilanova, E.; Santos, G.R.; Aquino, R.S.; Valle-Delgado, J.J.; Anselmetti, D.; Fernàndez-Busquets, X.; Mourão, P.A. Carbohydrate-carbohydrate interactions mediated by sulfate esters and calcium provide the cell adhesion required for the emergence of early metazoans. *J. Biol. Chem.* **2016**, *291*, 9425–9437. [CrossRef] [PubMed]
2. Karakostis, K.; Costa, C.; Zito, F.; Matranga, V. Heterologous expression of newly identified galectin-8 from sea urchin embryos produces recombinant protein with lactose binding specificity and anti-adhesive activity. *Sci. Rep.* **2015**, *5*, 17665. [CrossRef] [PubMed]
3. Giga, Y.; Ikai, A.; Takahashi, K. The complete amino acid sequence of echinoidin, a lectin from the coelomic fluid of the sea urchin *Anthocidaris crassispina*. Homologies with mammalian and insect lectins. *J. Biol. Chem.* **1987**, *262*, 6197–6203. [PubMed]
4. Flores, R.L.; Livingston, B.T. The skeletal proteome of the sea star *Patiria miniata* and evolution of biomineralization in echinoderms. *BMC Evol. Biol.* **2017**, *17*, 125. [CrossRef] [PubMed]
5. Ozeki, Y.; Matsui, T.; Suzuki, M.; Titani, K. Amino acid sequence and molecular characterization of a D-galactoside-specific lectin purified from sea urchin (*Anthocidaris crassispina*) eggs. *Biochemistry* **1991**, *30*, 2391–2394. [CrossRef] [PubMed]
6. Carneiro, R.F.; Teixeira, C.S.; de Melo, A.A.; de Almeida, A.S.; Cavada, B.S.; de Sousa, O.V.; da Rocha, B.A.; Nagano, C.S.; Sampaio, A.H. L-Rhamnose-binding lectin from eggs of the *Echinometra lucunter*: Amino acid sequence and molecular modeling. *Int. J. Biol. Macromol.* **2015**, *78*, 180–188. [CrossRef] [PubMed]
7. Uchida, T.; Yamasaki, T.; Eto, S.; Sugawara, H.; Kurisu, G.; Nakagawa, A.; Kusunoki, M.; Hatakeyama, T. Crystal structure of the hemolytic lectin CEL-III isolated from the marine invertebrate *Cucumaria echinata*: implications of domain structure for its membrane pore-formation mechanism. *J. Biol. Chem.* **2004**, *279*, 37133–37141. [CrossRef] [PubMed]

8. Maehashi, E.; Sato, C.; Ohta, K.; Harada, Y.; Matsuda, T.; Hirohashi, N.; Lennarz, W.J.; Kitajima, K. Identification of the sea urchin 350-kDa sperm-binding protein as a new sialic acid-binding lectin that belongs to the heat shock protein 110 family: implication of its binding to gangliosides in sperm lipid rafts in fertilization. *J. Biol. Chem.* **2003**, *278*, 42050–42057. [CrossRef] [PubMed]
9. Pohleven, J.; Renko, M.; Magister, Š.; Smith, D.F.; Künzler, M.; Štrukelj, B.; Turk, D.; Kos, J.; Sabotiè, J. Bivalent carbohydrate binding is required for biological activity of *Clitocybe nebularis* lectin (CNL), the *N,N′*-diacetyllactosediamine (GalNAcβ1-4GlcNAc, LacdiNAc)-specific lectin from basidiomycete *C. nebularis*. *J. Biol. Chem.* **2012**, *287*, 10602–10612. [CrossRef] [PubMed]
10. Fujii, Y.; Fujiwara, T.; Koide, Y.; Hasan, I.; Sugawara, S.; Rajia, S.; Kawsar, S.M.; Yamamoto, D.; Araki, D.; Kanaly, R.A.; et al. Internalization of a novel, huge lectin from *Ibacus novemdentatus* (slipper lobster) induces apoptosis of mammalian cancer cells. *Glycoconj. J.* **2017**, *34*, 85–94. [CrossRef] [PubMed]
11. Carvalho, E.V.M.M.; Oliveira, W.F.; Coelho, L.C.B.B.; Correia, M.T.S. Lectins as mitosis stimulating factors: Briefly reviewed. *Life Sci.* **2018**, *207*, 152–157. [CrossRef] [PubMed]
12. Hasan, I.; Ozeki, Y. Histochemical localization of N-acetylhexosamine-binding lectin HOL-18 in *Halichondria okadai* (Japanese black sponge), and its antimicrobial and cytotoxic anticancer effects. *Int. J. Biol. Macromol.* **2019**, *124*, 819–827. [CrossRef] [PubMed]
13. Ozeki, Y.; Matsui, T.; Titani, K. Cell adhesive activity of two animal lectins through different recognition mechanisms. *FEBS Lett.* **1991**, *289*, 145–147. [CrossRef]
14. Hatakeyama, T.; Ichise, A.; Unno, H.; Goda, S.; Oda, T.; Tateno, H.; Hirabayashi, J.; Sakai, H.; Nakagawa, H. Carbohydrate recognition by the rhamnose-binding lectin SUL-I with a novel three-domain structure isolated from the venom of globiferous pedicellariae of the flower sea urchin *Toxopneustes pileolus*. *Protein Sci.* **2017**, *26*, 1574–1583. [CrossRef] [PubMed]
15. Zamora, S.; Rahman, I.A.; Ausich, W.I. Palaeogeographic implications of a new iocrinid crinoid (Disparida) from the Ordovician (Darriwillian) of Morocco. *PeerJ.* **2015**, *3*, e1450. [CrossRef] [PubMed]
16. Oji, T.; Twitchett, R.J. The oldest post-Palaeozoic Crinoid and Permian-Triassic origins of the Articulata (Echinodermata). *Zoolog. Sci.* **2015**, *32*, 211–215. [CrossRef] [PubMed]
17. Kondo, M.; Akasaka, K. Regeneration in crinoids. *Dev. Growth Differ.* **2010**, *52*, 57–68. [CrossRef] [PubMed]
18. Shibata, T.F.; Sato, A.; Oji, T.; Akasaka, K. Development and growth of the feather star *Oxycomanthus japonicus* to sexual maturity. *Zoolog. Sci.* **2008**, *25*, 1075–1083. [CrossRef] [PubMed]
19. Summers, M.M.; Messing, C.G.; Rouse, G.W. Phylogeny of Comatulidae (Echinodermata: Crinoidea: Comatulida): a new classification and an assessment of morphological characters for crinoid taxonomy. *Mol. Phylogenet. Evol.* **2014**, *80*, 319–339. [CrossRef] [PubMed]
20. Matsumoto, R.; Shibata, T.F.; Kohtsuka, H.; Sekifuji, M.; Sugii, N.; Nakajima, H.; Kojima, N.; Fujii, Y.; Kawsar, S.M.; Yasumitsu, H.; et al. Glycomics of a novel type-2 N-acetyllactosamine-specific lectin purified from the feather star, *Oxycomanthus japonicus* (Pelmatozoa: Crinoidea). *Comp. Biochem. Physiol. B Biochem. Mol. Biol.* **2011**, *158*, 266–273. [CrossRef] [PubMed]
21. Sequencing The Genome of An Early Branching Echinoderm; The Crinoid Oxycomanthus Japonicus. Available online: https://www.ncbi.nlm.nih.gov/bioproject/236227 (accessed on 24 February 2019).
22. Takagaki, Y.; Manley, J.L. R.N.A. recognition by the human polyadenylation factor CstF. *Mol. Cell. Biol.* **1997**, *17*, 3907–3914. [CrossRef] [PubMed]
23. Ghai, R.; Waters, P.; Roumenina, L.; Gadjeva, M.; Kojouharova, M.S.; Reid, K.B.; Sim, R.B.; Kishore, U. C1q and its growing family. *Immunobiology* **2007**, *212*, 253–266. [CrossRef] [PubMed]
24. Carland, T.M.; Gerwick, L. The C1q domain containing proteins: Where do they come from and what do they do? *Dev. Comp. Immunol.* **2010**, *34*, 785–790. [CrossRef] [PubMed]
25. Gerdol, M.; Venier, P.; Pallavicini, A. The genome of the Pacific oyster *Crassostrea gigas* brings new insights on the massive expansion of the C1q gene family in Bivalvia. *Dev. Comp. Immunol.* **2015**, *49*, 59–71. [CrossRef] [PubMed]
26. Ozeki, Y.; Yokota, Y.; Kato, K.H.; Titani, K.; Matsui, T. Developmental expression of D-galactoside-binding lectin in sea urchin (*Anthocidaris crassispina*) eggs. *Exp. Cell Res.* **1995**, *216*, 318–324. [CrossRef] [PubMed]
27. Alliegro, M.C.; Alliegro, M.A. Echinonectin is a Del-1-like molecule with regulated expression in sea urchin embryos. *Gene Expr. Patterns* **2007**, *7*, 651–656. [CrossRef] [PubMed]

28. Hasan, I.; Sugawara, S.; Fujii, Y.; Koide, Y.; Terada, D.; Iimura, N.; Fujiwara, T.; Takahashi, K.G.; Koima, N.; Rajia, S.; et al. MytiLec, a mussel R-type lectin, interacts with surface glycan Gb3 on Burkitt's lymphoma cells to trigger apoptosis through multiple pathways. *Mar. Drugs.* **2015**, *13*, 7377–7389. [CrossRef] [PubMed]
29. Sugawara, S.; Im, C.; Kawano, T.; Tatsuta, T.; Koide, Y.; Yamamoto, D.; Ozeki, Y.; Nitta, K.; Hosono, M. Catfish rhamnose-binding lectin induces G(0/1) cell cycle arrest in Burkitt's lymphoma cells via membrane surface Gb3. *Glycoconj. J.* **2017**, *34*, 127–138. [CrossRef] [PubMed]
30. Gerdol, M.; Manfrin, C.; De Moro, G.; Figueras, A.; Novoa, B.; Venier, P.; Pallavicini, A. The C1q domain containing proteins of the Mediterranean mussel *Mytilus galloprovincialis*: a widespread and diverse family of immune-related molecules. *Dev. Comp. Immunol.* **2011**, *35*, 635–643. [CrossRef] [PubMed]
31. Allam, B.; Pales Espinosa, E.; Tanguy, A.; Jeffroy, F.; Le Bris, C.; Paillard, C. Transcriptional changes in Manila clam (*Ruditapes philippinarum*) in response to Brown Ring Disease. *Fish Shellfish Immunol.* **2014**, *41*, 2–11. [CrossRef] [PubMed]
32. Liu, H.H.; Xiang, L.X.; Shao, J.Z. A novel C1q-domain-containing (C1qDC) protein from *Mytilus coruscus* with the transcriptional analysis against marine pathogens and heavy metals. *Dev. Comp. Immunol.* **2014**, *44*, 70–75. [CrossRef] [PubMed]
33. Tills, O.; Truebano, M.; Feldmeyer, B.; Pfenninger, M.; Morgenroth, H.; Schell, T.; Rundle, S.D. Transcriptomic responses to predator kairomones in embryos of the aquatic snail *Radix balthica*. *Ecol. Evol.* **2018**, *8*, 11071–11082. [CrossRef] [PubMed]
34. Gross, P.S.; Al-Sharif, W.Z.; Clow, L.A.; Smith, L.C. Echinoderm immunity and the evolution of the complement system. *Dev. Comp. Immunol.* **1999**, *23*, 429–442. [CrossRef]
35. Buckley, K.M.; Smith, L.C. Extraordinary diversity among members of the large gene family, 185/333, from the purple sea urchin, *Strongylocentrotus purpuratus*. *BMC Mol. Biol.* **2007**, *8*, 68. [CrossRef] [PubMed]
36. Ressl, S.; Vu, B.K.; Vivona, S.; Martinelli, D.C.; Südhof, T.C.; Brunger, A.T. Structures of C1q-like proteins reveal unique features among the C1q/TNF superfamily. *Structure.* **2015**, *23*, 688–699. [CrossRef] [PubMed]
37. Kishore, U.; Gaboriaud, C.; Waters, P.; Shrive, A.K.; Greenhough, T.J.; Reid, K.B.; Sim, R.B.; Arlaud, G.J. C1q and tumor necrosis factor superfamily: modularity and versatility. *Trends Immunol.* **2004**, *25*, 551–561. [CrossRef] [PubMed]
38. Shapiro, L.; Scherer, P.E. The crystal structure of a complement-1q family protein suggests an evolutionary link to tumor necrosis factor. *Curr. Biol.* **1998**, *8*, 335–338. [CrossRef]
39. Surolia, A.; Swaminathan, C.P.; Ramkumar, R.; Podder, S.K. Unusual structural stability and ligand induced alterations in oligomerization of a galectin. *FEBS Lett.* **1997**, *409*, 417–420. [CrossRef]
40. Sulák, O.; Cioci, G.; Delia, M.; Lahmann, M.; Varrot, A.; Imberty, A.; Wimmerová, M.A. TNF-like trimeric lectin domain from *Burkholderia cenocepacia* with specificity for fucosylated human histo-blood group antigens. *Structure* **2010**, *18*, 59–72. [CrossRef] [PubMed]
41. Li, C.; Yu, S.; Zhao, J.; Su, X.; Li, T. Cloning and characterization of a sialic acid binding lectins (SABL) from Manila clam *Venerupis philippinarum*. *Fish Shellfish Immunol.* **2011**, *30*, 1202–1206. [CrossRef] [PubMed]
42. Kong, P.; Zhang, H.; Wang, L.; Zhou, Z.; Yang, J.; Zhang, Y.; Qiu, L.; Wang, L.; Song, L. AiC1qDC-1, a novel gC1q-domain-containing protein from bay scallop *Argopecten irradians* with fungi agglutinating activity. *Dev. Comp. Immunol.* **2010**, *34*, 837–846. [CrossRef] [PubMed]
43. Lv, Z.; Qiu, L.; Wang, M.; Jia, Z.; Wang, W.; Xin, L.; Liu, Z.; Wang, L.; Song, L. Comparative study of three C1q domain containing proteins from pacific oyster *Crassostrea gigas*. *Dev. Comp. Immunol.* **2018**, *78*, 42–51. [CrossRef] [PubMed]
44. Nakamura, O.; Wada, Y.; Namai, F.; Saito, E.; Araki, K.; Yamamoto, A.; Tsutsui, S. A novel C1q family member with fucose-binding activity from surfperch, Neoditrema ransonnetii (*Perciformes, Embiotocidae*). *Fish Shellfish Immunol.* **2009**, *27*, 714–720. [CrossRef] [PubMed]
45. Matsushita, M.; Matsushita, A.; Endo, Y.; Nakata, M.; Kojima, N.; Mizuochi, T.; Fujita, T. Origin of the classical complement pathway: Lamprey orthologue of mammalian C1q acts as a lectin. *Proc. Natl. Acad. Sci. USA* **2004**, *101*, 10127–10131. [CrossRef] [PubMed]
46. Peschke, B.; Keller, C.W.; Weber, P.; Quast, I.; Lünemann, J.D. Fc-galactosylation of human immunoglobulin gamma isotypes improves C1q binding and enhances. Complement-dependent cytotoxicity. *Front. Immunol* **2017**, *8*, 646. [CrossRef] [PubMed]

47. Païdassi, H.; Tacnet-Delorme, P.; Lunardi, T.; Arlaud, G.J.; Thielens, N.M.; Frachet, P. The lectin-like activity of human C1q and its implication in DNA and apoptotic cell recognition. *FEBS Lett.* **2008**, *582*, 3111–3116. [CrossRef] [PubMed]
48. Garlatti, V.; Chouquet, A.; Lunardi, T.; Vivès, R.; Païdassi, H.; Lortat-Jacob, H.; Thielens, N.M.; Arlaud, G.J.; Gaboriaud, C. Cutting edge: C1q binds deoxyribose and heparan sulfate through neighboring sites of its recognition domain. *J. Immunol.* **2010**, *185*, 808–812. [CrossRef] [PubMed]
49. Peake, P.W.; Shen, Y.; Campbell, L.V.; Charlesworth, J.A. Human adiponectin binds to bacterial lipopolysaccharide. *Biochem. Biophys. Res. Commun.* **2006**, *341*, 108–115. [CrossRef] [PubMed]
50. Gaboriaud, C.; Juanhuix, J.; Gruez, A.; Lacroix, M.; Darnault, C.; Pignol, D.; Verger, D.; Fontecilla-Camps, J.C.; Arlaud, G.J. The Crystal structure of the globular head of complement protein C1q provides a basis for its versatile recognition properties. *J. Biol. Chem.* **2003**, *278*, 46974–46982. [CrossRef] [PubMed]
51. Pajvani, U.B.; Du, X.; Combs, T.P.; Berg, A.H.; Rajala, M.W.; Schulthess, T.; Engel, J.; Brownlee, M.; Scherer, P.E. Structure-function studies of the adipocyte-secreted hormone Acrp30/adiponectin. Implications fpr metabolic regulation and bioactivity. *J. Biol. Chem.* **2003**, *278*, 9073–9085. [CrossRef] [PubMed]
52. Bao, D.; Pang, Z.; Morgan, J.I. The structure and proteolytic processing of Cbln1 complexes. *J. Neurochem.* **2005**, *95*, 618–629. [CrossRef] [PubMed]
53. Tsao, T.-S.; Murrey, H.E.; Hug, C.; Lee, D.H.; Lodish, H.F. Oligomerization state-dependent activation of NF-kappa B signaling pathway by adipocyte complement-related protein of 30 kDa (Acrp30). *J. Biol. Chem.* **2003**, *277*, 29359–29362. [CrossRef] [PubMed]
54. Tsao, T.-S.; Tomas, E.; Murrey, H.E.; Hug, C.; Lee, D.H.; Ruderman, N.B.; Heuser, J.E.; Lodish, H.F. Role of disulfide bonds in Acrp30/adiponectin structure and signaling specificity. Different oligomers activate different signal transduction pathways. *J. Biol. Chem.* **2003**, *278*, 50810–50817. [CrossRef] [PubMed]
55. Venkatraman Girija, U.; Gingras, A.R.; Marshall, J.E.; Panchal, R.; Sheikh, M.A.; Harper, J.A.J.; Gál, P.; Schwaeble, W.J.; Mitchell, D.A.; Moody, P.C.E.; et al. Structural basis of the C1q/C1s interaction and its central role in assembly of the C1 complex of complement activation. *Proc. Natl. Acad. Sci. USA* **2013**, *110*, 13916–13920. [CrossRef] [PubMed]
56. Agostinis, C.; Vidergar, R.; Belmonte, B.; Mangogna, A.; Amadio, L.; Geri, P.; Borelli, V.; Zanconati, F.; Tedesco, F.; Confalonieri, M.; et al. Complement protein C1q binds to hyaluronic acid in the malignant pleural mesothelioma microenvironment and promotes tumor growth. *Front. Immunol.* **2017**, *8*, 1559. [CrossRef] [PubMed]
57. Zeng, Y.; Xiang, J.; Lu, Y.; Chen, Y.; Wang, T.; Gong, G.; Wang, L.; Li, X.; Chen, S.; Sha, Z. sghC1q, a novel C1q family member from half-smooth tongue sole (*Cynoglossus semilaevis*): identification, expression and analysis of antibacterial and antiviral activities. *Dev. Comp. Immunol.* **2015**, *48*, 151–163. [CrossRef] [PubMed]
58. Wang, L.; Fan, C.; Xu, W.; Zhang, Y.; Dong, Z.; Xiang, J.; Chen, S. Characterization and functional analysis of a novel C1q-domain-containing protein in Japanese flounder (*Paralichthys olivaceus*). *Dev Comp. Immunol.* **2017**, *67*, 322–332. [CrossRef] [PubMed]
59. Bulla, R.; Tripodo, C.; Rami, D.; Ling, G.S.; Agostinis, C.; Guarnotta, C.; Zorzet, S.; Durigutto, P.; Botto, M.; Tedesco, F. C1q acts in the tumour microenvironment as a cancer-promoting factor independently of complement activation. *Nat. Commun.* **2016**, *7*, 10346. [CrossRef] [PubMed]
60. Pier, G.B. *Pseudomonas aeruginosa* lipopolysaccharide: a major virulence factor, initiator of inflammation and target for effective immunity. *Int. J. Med. Microbiol.* **2007**, *297*, 277–295. [CrossRef] [PubMed]
61. Carneiro, R.F.; Torres, R.C.; Chaves, R.P.; de Vasconcelos, M.A.; de Sousa, B.L.; Goveia, A.C.; Arruda, F.V.; Matos, M.N.; Matthews-Cascon, H.; Freire, V.N.; et al. Purification, biochemical characterization, and amino acid sequence of a novel type of lectin from *Aplysia dactylomela* eggs with antibacterial/antibiofilm potential. *Mar. Biotechnol. (NY)* **2017**, *19*, 49–64. [CrossRef] [PubMed]
62. Klein, R.C.; Fabres-Klein, M.H.; de Oliveira, L.L.; Feio, R.N.; Malouin, F.; Ribon Ade, O.A. C-type lectin from *Bothrops jararacussu* venom disrupts Staphylococcal biofilms. *PLoS One* **2015**, *10*, e0120514. [CrossRef] [PubMed]
63. Terada, D.; Voet, A.R.D.; Noguchi, H.; Kamata, K.; Ohki, M.; Addy, C.; Fujii, Y.; Yamamoto, D.; Ozeki, Y.; Tame, J.R.H.; et al. Computational design of a symmetrical β-trefoil lectin with cancer cell binding activity. *Sci. Rep.* **2017**, *7*, 5943. [CrossRef] [PubMed]

64. Fujii, Y.; Sugawara, S.; Araki, D.; Kawano, T.; Tatsuta, T.; Takahashi, K.; Kawsar, S.M.; Matsumoto, R.; Kanaly, R.A.; Yasumitsu, H.; et al. MRP1 expressed on Burkitt's lymphoma cells was depleted by catfish egg lectin through Gb3-glycosphingolipid and enhanced cytotoxic effect of drugs. *Protein J.* **2012**, *31*, 15–26. [CrossRef] [PubMed]
65. Wada, J.; Ota, K.; Kumar, A.; Wallner, E.I.; Kanwar, Y.S. Developmental regulation, expression, and apoptotic potential of galectin-9, a beta-galactoside binding lectin. *J. Clin. Invest.* **1997**, *99*, 2452–2461. [CrossRef] [PubMed]
66. Hirabayashi, J.; Hashidate, T.; Arata, Y.; Nishi, N.; Nakamura, T.; Hirashima, M.; Urashima, T.; Oka, T.; Futai, M.; Muller, W.E.; et al. Oligosaccharide specificity of galectins: a search by frontal affinity chromatography. *Biochim. Biophys. Acta* **2002**, *1572*, 232–254. [CrossRef]
67. Iwaki, J.; Hirabayashi, J. Carbohydrate-binding specificity of human galecins: An overview by frontal affinity chromatography. *Trends Glycosci. Glycotechnol* **2018**, *30*, 137–153. [CrossRef]
68. Canesi, L.; Ciacci, C.; Fabbri, R.; Balbi, T.; Salis, A.; Damonte, G.; Cortese, K.; Caratto, V.; Monopoli, M.P.; Dawson, K.; et al. Interactions of cationic polystyrene nanoparticles with marine bivalve hemocytes in a physiological environment: Role of soluble hemolymph proteins. *Environ. Res.* **2016**, *150*, 73–81. [CrossRef] [PubMed]
69. Schmid, A.; Kopp, A.; Hanses, F.; Karrasch, T.; Schäffler, A. C1q/TNF-related protein-3 (CTRP-3) attenuates lipopolysaccharide (LPS)-induced systemic inflammation and adipose tissue Erk-1/-2 phosphorylation in mice *in vivo*. *Biochem. Biophys. Res. Commun.* **2014**, *452*, 8–13. [CrossRef] [PubMed]
70. Matsushita, M. The complement system of agnathans. *Front. Immunol.* **2018**, *18*, 1405. [CrossRef] [PubMed]
71. Liu, C.; Jiang, S.; Wang, M.; Wang, L.; Chen, H.; Xu, J.; Lv, Z.; Song, L. A novel siglec (CgSiglec-1) from the Pacific oyster (*Crassostrea gigas*) with broad recognition spectrum and inhibitory activity to apoptosis, phagocytosis and cytokine release. *Dev. Comp. Immunol.* **2016**, *61*, 136–144. [CrossRef] [PubMed]
72. Pallen, M.J.; Gophna, U. Bacterial flagella and Type III secretion: case studies in the evolution of complexity. *Genome Dyn.* **2007**, *3*, 30–47. [PubMed]
73. Laemmli, U.K. Cleavage of structural proteins during the assembly of the head of bacteriophage T4. *Nature* **1970**, *227*, 680–685. [CrossRef] [PubMed]
74. Smith, P.K.; Krohn, R.I.; Hermanson, G.T.; Mallia, A.K.; Krohn, R.I.; Gartner, F.T.; Provenzano, M.D.; Fujimoto, E.K.; Goeke, N.M.; Olson, B.J.; Klenk, D.C. Measurement of protein using bicinchoninic acid. *Anal. Biochem.* **1985**, *150*, 76–85. [CrossRef]
75. Wiechelman, K.J.; Braun, R.D.; Fitzpatrick, J.D. Investigation of the bicinchoninic acid protein assay: Identification of the groups responsible for color formation. *Anal. Biochem.* **1988**, *75*, 231–237. [CrossRef]
76. Altschul, S.F.; Gish, W.; Miller, W.; Myers, E.W.; Lipman, D.J. Basic local alignment search tool. *J. Mol. Biol.* **1990**, *215*, 403–410. [CrossRef]
77. Reese, M.G.; Eeckman, F.H.; Kulp, D.; Haussler, D. Improved splice site detection in Genie. *J. Comput. Biol.* **1997**, *4*, 311–323. [CrossRef] [PubMed]
78. Proudfoot, N.J. Ending the message: poly(A) signals then and now. *Genes Dev.* **2011**, *25*, 1770–1782. [CrossRef] [PubMed]
79. Finn, R.D.; Clements, J.; Eddy, S.R. HMMER web server: interactive sequence similarity searching. *Nucleic Acids Res.* **2011**, *39*, W29–W37. [CrossRef] [PubMed]
80. Edgar, R.C. MUSCLE: multiple sequence alignment with high accuracy and high throughput. *Nucleic Acids Res.* **2004**, *32*, 1792–1797. [CrossRef] [PubMed]
81. Huelsenbeck, J.P.; Ronquist, F. MRBAYES: Bayesian inference of phylogenetic trees. *Bioinforma. Oxf. Engl.* **2001**, *17*, 754–755. [CrossRef]
82. Abascal, F.; Zardoya, R.; Posada, D. ProtTest: selection of best-fit models of protein evolution. *Bioinforma. Oxf. Engl.* **2005**, *21*, 2104–2105. [CrossRef] [PubMed]
83. Schuck, P. Sedimentation analysis of noninteracting and self-associating solutes using numerical solutions to the Lamm equation. *Biophys. J.* **1998**, *75*, 1503–1512. [CrossRef]
84. Schuck, P. Size-distribution analysis of macromolecules by sedimentation velocity ultracentrifugation and lamm equation modeling. *Biophys. J.* **2000**, *78*, 1606–1619. [CrossRef]
85. Goldman, A.; Ursitti, J.A.; Mozdzanowski, J.; Speicher, D.W. Electroblotting from polyacrylamide gels. *Curr. Protoc. Protein. Sci.* **2015**, *82*, 1–16.

86. Shibata, T.F.; Oji, T.; Akasaka, K.; Agata, K. Staging of regeneration process of an arm of the feather star *Oxycomanthus japonicus* focusing on the oral-aboral boundary. *Dev. Dyn.* **2010**, *239*, 2947–2961. [CrossRef] [PubMed]
87. Gourdine, J.P.; Cioci, G.; Miguet, L.; Unverzagt, C.; Silva, D.V.; Varrot, A.; Gautier, C.; Smith-Ravin, E.J.; Imberty, A. High affinity interaction between a bivalve C-type lectin and a biantennary complex-type *N*-glycan revealed by crystallography and microcalorimetry. *J. Biol. Chem.* **2008**, *283*, 30112–30120. [CrossRef] [PubMed]

marine drugs

MDPI

Article

In Vitro Anticancer and Proapoptotic Activities of Steroidal Glycosides from the Starfish *Anthenea aspera*

Timofey V. Malyarenko [1,2,*], Olesya S. Malyarenko [1], Alla A. Kicha [1], Natalia V. Ivanchina [1], Anatoly I. Kalinovsky [1], Pavel S. Dmitrenok [1], Svetlana P. Ermakova [1] and Valentin A. Stonik [1,2]

1 G.B. Elyakov Pacific Institute of Bioorganic Chemistry, Far Eastern Branch of the Russian Academy of Sciences, Pr. 100-let Vladivostoku 159, 690022 Vladivostok, Russia; malyarenko.os@gmail.com (O.S.M.); kicha@piboc.dvo.ru (A.A.K.); ivanchina@piboc.dvo.ru (N.V.I.); kaaniv@piboc.dvo.ru (A.I.K.); paveldmt@piboc.dvo.ru (P.S.D.); swetlana_e@mail.ru (S.P.E.); stonik@piboc.dvo.ru (V.A.S.)

2 Far Eastern Federal University, Sukhanova Str. 8, 690000 Vladivostok, Russia

* Correspondence: malyarenko-tv@mail.ru; Tel.: +7-423-2312-360; Fax: +7-423-2314-050

Received: 8 October 2018; Accepted: 25 October 2018; Published: 1 November 2018

Abstract: New marine glycoconjugates—the steroidal glycosides designated as anthenosides V–X (**1–3**)—and the seven previously known anthenosides E (**4**), G (**5**), J (**6**), K (**7**), S1 (**8**), S4 (**9**), and S6 (**10**) were isolated from the extract of the tropical starfish *Anthenea aspera*. The structures of **1–3** were elucidated by extensive NMR and ESIMS techniques. Glycoside **1** contains a rare 5α-cholest-8(14)-ene-3α,7β,16α-hydroxysteroidal nucleus. Compounds **2** and **3** were isolated as inseparable mixtures of epimers. All investigated compounds (**1–10**) at nontoxic concentrations inhibited colony formation of human melanoma RPMI-7951, breast cancer T-47D, and colorectal carcinoma HT-29 cells to a variable degree. The mixture of **6** and **7** possessed significant anticancer activity and induced apoptosis of HT-29 cells. The molecular mechanism of the proapoptotic action of this mixture was shown to be associated with the regulation of anti- and proapoptotic protein expression followed by the activation of initiator and effector caspases.

Keywords: starfish; *Anthenea aspera*; steroidal glycosides; colony formation; apoptosis; Bcl-2 proteins

1. Introduction

Over the last decades, oncological diseases have become the second leading cause of death in the world, after cardiovascular diseases [1]. Nowadays, there are many therapeutic strategies applied to cancer treatment, including chemotherapy, surgery, radio- and immunotherapy, gene therapy, and target therapy. The last one is a type of cancer treatment that targets the changes in cancer cells to inhibit their growth, proliferation, and invasion [2]. Cancer cells are proved to be able to proliferate uncontrollably and avoid programmed cell death, apoptosis.

Apoptosis is a general biological process responsible for maintaining the constancy of the number of cells in cellular populations, as well as regulating the formation and elimination of defective cells. The main mediators of the intrinsic (mitochondrial) pathway of apoptosis are proteins of the Bcl-2 family, which are represented by antiapoptotic (Bcl-xL, Bcl-W, Mcl-1) and proapoptotic (Bax, Bak, Bad, Bim) proteins. They are considered to promote the alterations in mitochondrial membrane permeability required for the release of cytochrome c and other apoptogenic proteins that lead to the activation of caspases and the induction of apoptosis [3,4]. The search for and development of effective marine natural compounds that induce the apoptosis of cancer cells by targeting proteins of Bcl-2 family is a prospective strategy to control and stop cancer growth.

Some marine natural compounds have already been shown to possess significant anticancer activity in vitro and in vivo with relatively low toxicity, so marine organisms remain a highly productive source of promising cancer preventive or anticancer therapeutic agents [5]. It would be of particular interest to find natural compounds with relatively low cytotoxicity, but good cancer preventive properties: for example, the effective inhibition of microcolony formation of tumor cells at noncytotoxic doses.

Starfishes are known to contain diverse types of secondary metabolites possessing a wide spectrum of biological activities, including cytotoxic, antibacterial, neuritogenic, antifungal, cancer preventive, anti-inflammatory, and other effects [6–12]. At the same time, steroidal glycosides (glycosides of polyhydroxysteroids, cyclic glycosides, and asterosaponins) are common secondary metabolites of starfish. In most cases, polyhydroxysteroidal glycosides from starfish contain polyoxygenated steroidal aglycons (usually having from four to nine hydroxyl groups) and pentose (β-D-xylopyranosyl or α-L-arabinofuranosyl) or hexose (β-D-glucopyranosyl or β-D-galactofuranosyl) monosaccharide residues.

A series of glycosides of polyhydroxysteroids, named anthenosides, were previously isolated from starfishes of the genus *Anthenea* [13–17]. These compounds have a range of unusual structural features: $\Delta^{8(14)}$-3β,4β,6β,7β,16α-pentahydroxy, $\Delta^{8(14)}$-3β(α),6β,7β,16α-tetrahydroxy, or $\Delta^{8(14)}$-3α,7β,16α- trihydroxy steroidal nuclei; unoxidized side chains; and β-D-galactofuranosyl, 6-*O*-methyl-β-D- galactofuranosyl, 3-*O*-methyl-β-D-galactofuranosyl, 3-*O*-methyl-β-D-glucopyranosyl, 4-*O*-methyl-β-D-glucopyranosyl, or 2-acetamido-2-deoxy-4-*O*-methyl-β-D-glucopyranosyl residues.

It was previously reported that polyhydroxysteroidal glycosides from the starfish *Anthenea chinensis* exhibited significant activity against the promotion of tubulin polymerization in vitro, inhibiting the proliferation of human leukemia K-562, hepatoma BEL-7402, and spongioblastoma U87MG cell lines [14]. Recently, we demonstrated that some anthenosides from *Anthenea sibogae* slightly inhibited the proliferation and decreased the colony size of human breast cancer T-47D cells [16], whereas the anthenosides A_1 and A_2 from *A. aspera* slightly inhibited the cell viability of human cancer T-47D cells and did not show cytotoxic effects against human melanoma RPMI-7951 cells [17].

A few studies devoted to the proapoptotic activity of steroidal glycosides from starfishes have been reported. These compounds were proved to be able to induce mitochondrial apoptosis in glioblastoma U87MG cells [18], cause growth inhibition of human lung cancer A549 cells through the regulation of endoplasmic reticulum-induced apoptosis (ER-apoptosis) [19], and induce p53-dependent apoptosis by inhibition of AP-1, NF-κB, and ERK activities in human leukemia HL-60 and THP-1 cells [20]. These data confirm that polar steroids from starfishes are prospective anticancer compounds with proapoptotic activity. The aim of this work is to describe the isolation and structures of three new minor anthenosides V–X (**1–3**) and investigate the effects of these glycosides, as well as of a series of earlier known anthenosides isolated from *A. aspera*, on cell viability, colony formation, and induction of apoptosis in human melanoma RPMI-7951 as well as breast adenocarcinoma T-47D and colorectal carcinoma HT-29 cells.

2. Results and Discussion

The concentrated ethanol extract of *A. aspera* was subjected to sequential separation by chromatography on columns with Polychrom-1 and silica gel, followed by HPLC on semipreparative Diasorb-130-C16T and analytical Discovery C_{18} and Diasorb-130 Si gel columns to yield three new steroidal biglycosides (**1**−**3**), named anthenosides V–X (Figure 1), and seven previously known steroidal glycosides. The known compounds were identified by comparison of their ^{1}H- and ^{13}C-NMR and MS spectra with those reported for anthenoside E (**4**), anthenoside G (**5**), and the epimer mixture of the anthenosides J and K (**6** and **7**, ratio of 3:1, respectively) from *A. chinensis* [14], and anthenoside S1 (**8**), anthenoside S4 (**9**), and anthenoside S6 (**10**) from *A. sibogae* [16].

The molecular formula of compound **1** was determined to be of $C_{41}H_{68}O_{13}$ from the $[M + Na]^+$ sodiated adduct ion peak at m/z 791.4550 in the (+)HRESIMS spectrum (Figure S1). The ^{1}H- and ^{13}C-NMR spectral data belonging to the tetracyclic moiety of the aglycon of **1** showed the resonances of protons and carbons of two angular methyl groups, CH_3-18 and CH_3-19 (δ_H 0.90 s, 0.67 s; δ_C 20.2, 11.6), an 8(14) double bond (δ_C 128.9, 144.9), one oxygenated methine HC-3 (δ_H 3.98 t (J = 2.6); δ_C 67.1), and two oxygenated methine groups bearing the monosaccharide residues HC-7 (δ_H 4.40 t (J = 2.8); δ_C 74.0) and HC-16 (δ_H 4.43 td (J = 8.7, 4.6); δ_C 78.2) (Table 1, Figures S2 and S3). The ^{1}H-^{1}H COSY and HSQC correlations attributable to a steroidal nucleus revealed the corresponding sequences of protons from C-1 to C-7, C-9 to C-12 through C-11, and C-15 to C-17 (Figure 2A, Figures S4 and S5). Key HMBC cross-peaks, such as H-7/C-9; H-15/C-8, C-13, C-14; H_3-18/C-12, C-13, C-14, C-17; and H_3-19/C-1, C-5, C-9, C-10, confirmed the overall structure of the steroidal moiety of **1** (Figure 2A, Figure S6). The ROESY cross-peaks showed the common 5α/9α/10β/13β stereochemistry of the steroidal nucleus and 7β,16α-configurations of oxygenated substituents in **1** (Figure 2B, Figure S7).

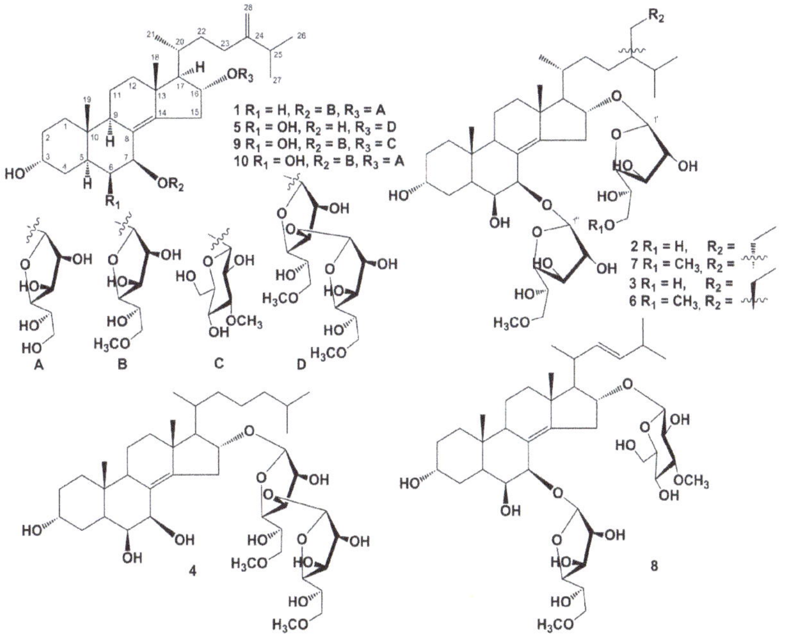

Figure 1. The structures of new compounds **1–10** isolated from *Anthenea aspera*.

The NMR data attributable to the side chain of **1** indicated the existence of three secondary methyls, CH_3-21 (δ_H 1.05 d (J = 6.6); δ_C 21.4), CH_3-26 (δ_H 1.04 d (J = 6.7); δ_C 22.5), and CH_3-27 (δ_H 1.04 d (J = 6.7); δ_C 22.3), and a 24(28) double bond (δ_H 4.76 brs, 4.72 brd (J = 1.3); δ_C 157.7, 107.1) (Table 1, Figures S2 and S3). The proton sequences from H-17 to H_3-21 and H-23 and from H_3-26 to H_3-27 through H-25, correlating with the corresponding carbon atoms of the side chain of **1**, were assigned using the ^{1}H-^{1}H COSY and HSQC experiments (Figure 2A, Figures S4 and S5). The HMBC correlations H_3-21/C-17, C-20; H-23/C-24; H_3-26/C-24, C-25, C-27; H_3-27/C-24, C-25, C-26; and H_2-28/C-23, C-25 and the ROESY correlations H_3-21/H-22; H-23/H-25; H_2-28/H_3-26, H_3-27 supported the total structure of the $\Delta^{24(28)}$-24-methyl-cholestane side chain (Figure 2A,B, Figures S6 and S7) previously found in anthenosides F, G, and S4 [13,14]. A 20*R*-configuration was assumed on the basis of ROESY correlations of H_3-18/H-20, H_3-21; H_β-16/H-22, and H_3-21/H_β-12. On the basis of all the above-mentioned data, the structure of the steroidal aglycon of **1** was determined to be (20*R*)-24-methyl-5α-cholesta-8(14),24(28)-diene-3α,7β,16α-triol.

The ^{1}H-NMR spectrum of **1** showed two resonances at δ_H 4.94 and 4.99 in the deshielded region belonging to anomeric protons of the monosaccharide units that correlated in the HSQC spectrum with carbon atom signals at δ_C 107.6 and 107.3, respectively, as well as one resonance due to the

O-methyl protons of the monosaccharide unit at δ_H 3.38, which correlated in the HSQC experiment with a carbon signal at δ_C 59.3 (Table 1, Figures S2 and S3). The fragment ion peaks at *m/z* 597 [(M + Na) − $C_7H_{14}O_6$]$^+$, 217 [$C_7H_{14}O_6$ + Na]$^+$, and 203 [$C_6H_{12}O_6$ + Na]$^+$ in the (+)ESIMS/MS spectrum from the precursor ion at *m/z* 791 [M + Na]$^+$ showed the presence of *O*-methylhexose and hexose units in **1**. The chemical shifts and coupling constants of H-1−H-6 of the *O*-methylhexose and hexose units were determined by the irradiation of the anomeric proton atoms in the 1D TOCSY experiment. The application of ^{1}H-^{1}H COSY, HSQC, HMBC, and ROESY experiments led to the assignment of all carbon and proton resonances of the carbohydrate moieties in the NMR spectra of **1** (Table 1, Figure 2A,B). The carbon and proton signals and the corresponding coupling constants of the monosaccharide units agreed well with those of the terminal 6-*O*-methyl-β-galactofuranosyl residue and a β-galactofuranosyl residue in the NMR spectra of the mixture of anthenosides T and U [15]. The D-series of both monosaccharide units were expected by analogy with co-occurring anthenosides M and Q [15]. The attachments of the 6-*O*-methyl-β-D-galactofuranosyl and β-D-galactofuranosyl units to the steroidal aglycon were determined by the HMBC and ROESY spectra, in which the HMBC cross-peaks between H-1′ of Gal$_f$ and C-16 of the aglycon, H-1″ of 6-OMe-Gal$_f$ and C-7 of the aglycon were observed, and the ROESY cross-peaks between H-1′ of Gal$_f$ and H-16 of the aglycon and H-1″ of 6-OMe-Gal$_f$ and H-7 of the aglycon were observed. All these data allowed for the establishment of the structure of anthenoside V (**1**) as (20*R*)-7-*O*-(6-*O*-methyl-β-D-galactofuranosyl)-16-*O*-(β-D-galactofuranosyl)-24-methyl-5α-cholesta-8(14),24(28)-diene-3α,7β,16α-triol.

Table 1. ^{1}H- (700.13 MHz) and ^{13}C-NMR (176.04 MHz) chemical shifts of **1** and the mixture of **2** and **3** in CD_3OD at 30 °C; δ in ppm, *J* values in Hz.

Position	1			Mixture of 2 and 3		
	DEPT	δ_H	δ_C	DEPT	δ_H	δ_C
1β	CH_2	1.52 dd (12.3, 5.1)	32.8	CH_2	1.53 m	34.5
1α		1.39 m			1.30 m	
2	CH_2	1.62 m	29.6	CH_2	1.62 m	29.9
3	CH	3.98 t (2.6)	67.1	CH	4.08 m	67.5
4β	CH_2	1.44 m	36.3	CH_2	1.96 td (13.6, 2.7)	33.3
4α		1.37 m			1.37 m	
5	CH	2.17 m	33.1	CH	2.12 dt (13.6, 2.7)	38.0
6	CH_2	1.55 dt (14.2, 2.8)	34.3	CH	3.62 t (2.7)	75.2
		1.23 m				
7	CH	4.40 t (2.8)	74.0	CH	4.22 d (2.7)	78.3
8	C		128.9	C		127.0
9	CH	2.25 m	46.3	CH	2.26 m	45.9
10	C		38.6	C		38.8
11β	CH_2	1.65 m	19.8	CH_2	1.65 m	19.5
11α		1.45 m			1.54 m	
12β	CH_2	1.81 m	37.2	CH_2	1.80 dt (12.5, 3.6)	37.1
12α		1.26 m			1.24 m	
13	C		44.8	C		45.0
14	C		144.9	C		147.3
15β	CH_2	2.77 ddd (16.8, 8.7, 3.0)	33.7	CH_2	2.88 ddd (17.1, 9.1, 3.2)	33.7
15α		2.56 ddd (16.8, 4.6, 1.8)			2.59 ddd (17.1, 5.5, 2.0)	
16	CH	4.43 td (8.7, 4.6)	78.2	CH	4.49 td (9.1, 5.5)	77.3
17	CH	1.46 t (4.6)	62.9	CH	1.47 dd (9.1, 3.9)	63.0
18	CH_3	0.90 s	20.2	CH_3	0.93 s	20.2
19	CH_3	0.67 s	11.6	CH_3	0.85 s	15.4
20	CH	1.67 m	33.3	CH	1.60 m	33.5 33.4
21	CH_3	1.05 d (6.6)	21.4	CH_3	1.05 d (6.9)	21.7 21.6
22	CH_2	1.81 m	33.9	CH_2	1.65 m	33.2
		1.43 m			1.26 m	

Table 1. *Cont.*

Position	1			Mixture of 2 and 3		
23	CH_2	2.23 m 1.95 m	33.7	CH_2	1.47 m 1.09 m	29.7 29.4
24	C		157.7	CH	1.05 m	47.2 47.0
25	CH	2.26 m	35.0	CH	1.75 m	30.2 30.5
26	CH_3	1.04 d (6.7)	22.5	CH_3	0.86 d (6.8)	20.1
27	CH_3	1.04 d (6.7)	22.3	CH_3	0.85 d (6.8)	19.3 19.4
28	CH_2	4.76 brs 4.72 brd (1.3)	107.1	CH_2	1.37 m 1.20 m	24.0 24.2
29				CH_3	0.89 t (7.4)	12.7 12.4
			β-D-Gal*f*			
1′	CH	4.94 brd (2.4)	107.6		4.96 brd (2.2)	107.7
2′	CH	3.95 m	83.6		3.96 dd (4.6, 2.2)	83.7
3′	CH	4.03 dd (7.0, 4.8)	78.3		4.04 dd (7.0, 4.6)	78.4
4′	CH	3.88 dd (7.0, 2.9)	84.4		3.89 dd (7.0, 3.0)	84.6
5′	CH	3.72 m	72.4		3.73 ddd (7.6, 4.5, 3.0)	72.5
6′	CH_2	3.61 dd (11.3, 7.5) 3.58 dd (11.3, 4.8)	65.5		3.64 dd (11.2, 7.6) 3.61 dd (11.2, 4.5)	65.4
			6-OMe-β-D-Gal*f*			
1′	CH	4.99 brd (2.0)	107.3		4.99 brd (1.9)	108.4
2′	CH	3.93 dd (3.9, 2.0)	83.5		3.91 dd (3.9, 1.9)	83.4
3′	CH	3.95 m	78.8		3.95 dd (6.1, 3.9)	78.7
4′	CH	3.90 dd (6.1, 3.7)	84.8		3.87 dd (6.1, 3.6)	85.0
5′	CH	3.83 m	70.8		3.84 m	70.7
6′	CH_2	3.53 dd (10.1, 4.6) 3.50 dd (10.1, 7.2)	75.4		3.53 d (6.0)	75.5
OCH_3	CH_3	3.38 s	59.3		3.39 s	59.3

Compounds **2** and **3** were not separated by repeated reversed-phase HPLC. They have the same molecular formula, $C_{42}H_{72}O_{14}$, determined from the $[M + Na]^+$ sodiated adduct ion peak at *m/z* 823.4811 in the (+)HRESIMS spectrum (Figure S8). The ^1H- and ^{13}C-NMR spectroscopic data referring to the steroidal nucleus of the mixture of **2** and **3** revealed the chemical shifts of proton and carbon atoms of two angular methyl groups CH_3-18 (δ_H 0.93 s, δ_C 20.2) and CH_3-19 (δ_H 0.85 s, δ_C 15.4), an 8(14) double bond (δ_C 127.0, 147.3), two oxygenated methines HC-3 (δ_H 4.08 m; δ_C 67.5) and HC-6 (δ_H 3.62 t (J = 2.7); δ_C 75.2), as well as two oxygenated methines HC-7 (δ_H 4.22 d (J = 2.7); δ_C 78.3) and HC-16 (δ_H 4.49 td (J = 9.1, 5.5); δ_C 77.3) bearing *O*-monosaccharide residues (Table 1, Figures S9 and S10). The proton and carbon atom resonances of CH_3-18, CH_3-19, HC-3, HC-6, HC-7, and HC-16 were similar to the corresponding signals in the NMR spectra of most of the anthenosides from *A. chinensis* [14], *A. aspera* [15], and *A. sibogae* [16] and testified to a $\Delta^{8(14)}$-3α,6β,7β,16α-tetrahydroxysteroidal nucleus glycosylated at the C-7 and C-16 positions in **2** and **3** (Figures S11–S14). The NMR data of the side chains indicated the existence of three secondary methyls, CH_3-21 (δ_H 1.05 d (J = 6.9); δ_C 21.7, 21.6), CH_3-26 (δ_H 0.86 d (J = 6.8); δ_C 20.1), and CH_3-27 (δ_H 0.85 d (J = 6.8); δ_C 19.3, 19.4), and one primary CH_3-29 (δ_H 0.89 t (J = 7.4); δ_C 12.7, 12.4) (Table 1, Figures S9 and S10). The complexity of the methyl region of the ^{13}C NMR data clearly showed the presence of two epimeric glycosides with 24*R* and 24*S* configurations. The proton and carbon resonances of the side chains, established by 2D NMR experiments, were similar to those of the side chains in the previously known anthenosides T and U [15] (Figures S11–S14). The chemical shift of C-29 in an ethyl group of the 24*S* epimer was more deshielded than that in the 24*R* epimer, while the chemical shifts of C-25, C-27, and C-28 of the 24*S* epimer were more shielded than those in the 24*R* epimer (Table 1, Figures S9 and S10). Therefore, the structures of the aglycon moieties of **2** and **3** were defined as (20*R*,24*S*)- and (20*R*,24*R*)-24-ethyl-5α-cholest-8(14)-ene-3α,6β,7β,16α-tetraols, respectively. Evaluation of the intensities of the C-29 resonances revealed a ratio of 3:1 for the **2**:**3** in the mixture, showing the predominance of the 24*S* epimer (**2**).

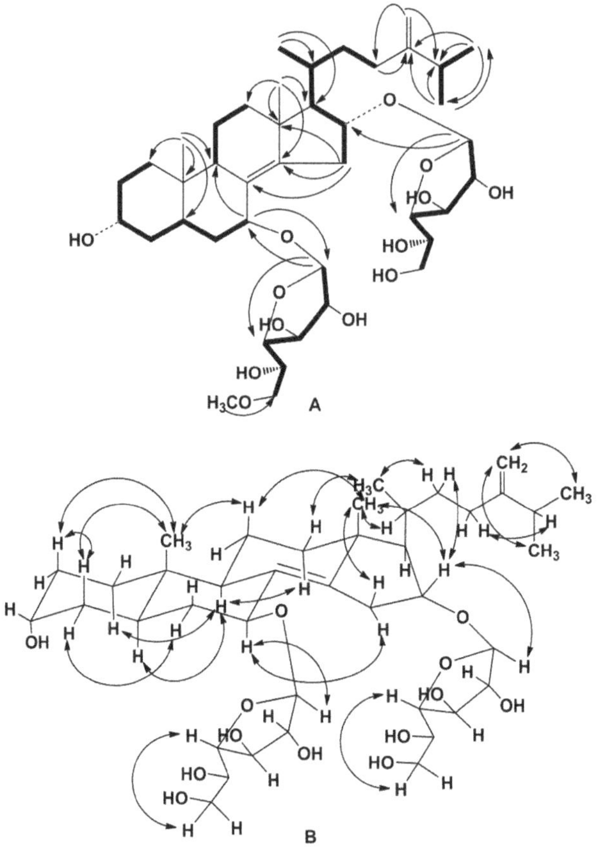

Figure 2. (**A**) ^{1}H-^{1}H COSY and key HMBC correlations for compound **1**. (**B**) Key ROESY correlations for compound **1**.

The ^{1}H-NMR spectrum of the mixture of **2** and **3** showed two resonances at δ_H 4.96 and 4.99 in the deshielded region belonging to anomeric protons of the monosaccharide units that correlated in the HSQC spectrum with carbon atom signals at δ_C 107.7 and 108.4, respectively, as well as one resonance due to *O*-methyl protons of the monosaccharide unit at δ_H 3.39, correlated in the HSQC experiment with a carbon signal at δ_C 59.3 (Table 1, Figures S9 and S10). A detailed comparison of the ^{1}H- and ^{13}C-NMR data of monosaccharide units with those of **1** clearly indicated that compounds **2** and **3** contained the same 6-*O*-methyl-β-galactofuranosyl and β-galactofuranosyl residues (Table 1). The attachments of the 6-*O*-methyl-β-D-galactofuranosyl and β-D-galactofuranosyl units to the steroidal aglycon were determined by the HMBC and ROESY spectra, where the HMBC cross-peaks between H-1′ of Gal$_f$ and C-16 of the aglycon, H-1″ of 6-OMe-Gal$_f$ and C-7 of the aglycon, as well as the ROESY cross-peaks between H-1′ of Gal$_f$ and H-16 of the aglycon and H-1″ of 6-OMe-Gal$_f$ and H-7 of the aglycon, were observed. Therefore, the structures of anthenosides W (**2**) and X (**3**) were determined to be (20*R*,24*S*)- and (20*R*,24*R*)-7-*O*-(6-*O*-methyl-β-D-galactofuranosyl)-16-*O*-(β-D-galactofuranosyl)-24-ethyl-5α-cholest-8(14)-ene-3α,6β,7β,16α-tetraols, respectively.

The cytotoxic activity of compounds 1–10 from *A. aspera*. The study on the biological activities of the tested compounds involved the determination of their cytotoxicity as the first step. The MTS method was used to assess the cytotoxic activity of glycosides **1–10** against the human melanoma

RPMI-7951, breast cancer T-47D, and colorectal carcinoma HT-29 cell lines. Cells were treated with compounds **1–10** in concentrations ranging from 20 to 160 μM and incubated for 24 h, as described in the section titled "Experimental".

All investigated compounds, except the mixture of **6** and **7**, did not cause inhibition of the cell viability of RPMI-7951, T-47D, and HT-29 cells at concentrations up to 160 μM. The mixture of **6** and **7** possessed a comparable cytotoxic effect against RPMI-7951, T-47D, and HT-29 cell lines with IC_{50} values of 89, 91, and 85 μM, respectively.

The inhibitory activity of glycosides 1–10 against colony formation of human cancer cells. In the last decade, the influence of polar steroids from starfishes on the formation and growth of colonies of various types of cancer cells has been intensively studied [21–23].

In the present work, the effects of compounds **1–10** on the formation and growth of colonies of human melanoma, breast cancer, and colorectal carcinoma were determined using the soft agar method, which is considered to be the most accurate type of in vitro assay for detecting the malignant transformation of cells [24]. Compounds **1–10** at a noncytotoxic concentration of 40 μM were shown to inhibit spontaneous colony formation and growth of RPMI-7951, T-47D, and HT-29 cells to a variable degree (Figure 3A–C). All compounds, except the mixture of **6** + **7**, inhibited colony formation of the investigated cells by less than 30%, compared to non-treated cells (control). Anthenosides J and K possessed significant inhibitory activity against all tested cell lines; these glycosides decreased the number of colonies of RPMI-7951, T-47D, and HT-29 cancer cells by 64%, 55%, and 83%, respectively, compared to non-treated cells (control). The chemotherapeutic drug cisplatin, used as a positive control in this study, at a noncytotoxic dose (1 μM) inhibited colony formation of RPMI-7951, T-47D, and HT-29 cells by 24%, 48%, and 59%, respectively, compared to controls (Figure 3A–C).

Compounds **4**, **5**, and **10** (40 μM) effectively inhibited colony formation of HT-29 and T-47D cells to a comparable degree by 34% and 33%; 41% and 43%; and 30% and 41%, respectively. Compounds **1**, **2** + **3**, **8**, and **9** at the same concentration had a moderate inhibitory activity and suppressed colony formation of melanoma, breast cancer, and colorectal adenocarcinoma cells by less than 30%. The colorectal carcinoma cells HT-29 were found to be the most sensitive to the effect of the anthenosides, while the melanoma cells RPMI-7951 were the most chemoresistant. Recently, similar effects of sulfated polar steroids from the Far Eastern starfish *Leptasterias ochotensis*, as well as monoglycosides, anthenosides A1 and A2, and anthenoside A from the starfish *Anthenea aspera*, were reported. These compounds were able to slightly inhibit colony formation of melanoma cell lines and effectively decrease the number and size of colonies of breast and colorectal cancer cells [17,25]. We suggest that the further search for cancer preventive agents in the form of natural products from starfish may lead to more active compounds.

It is likely that the effective inhibitory activity of the investigated compounds against colorectal carcinoma cells can be explained by the regulation of specific molecules involved in a signaling cascade which is activated in these cancer cells. Since the mixture of anthenosides J (**6**) and K (**7**) exerted the greatest anticancer activity against the colorectal carcinoma cells HT-29, they were chosen for further investigation of their proapoptotic activity in this cell line.

The proapoptotic activity of the mixture of anthenosides J (6) and K (7) in human colon carcinoma cells. The induction of apoptosis of cancer cells is known to be one of the accepted strategies of modern chemotherapy. In the present study, we supposed that the significant inhibitory activity of colony formation observed for colon carcinoma cells by the mixture of anthenosides J (**6**) and K (**7**) was due to the induction of apoptosis.

Apoptosis is a multistage process and can be realized by several signaling pathways. The mechanism of cell death mediated through the activation of cell death receptors on the cell surface is the extrinsic pathway of apoptosis, while the signaling cascade initiated through mitochondria is named the intrinsic pathway of apoptosis [26].

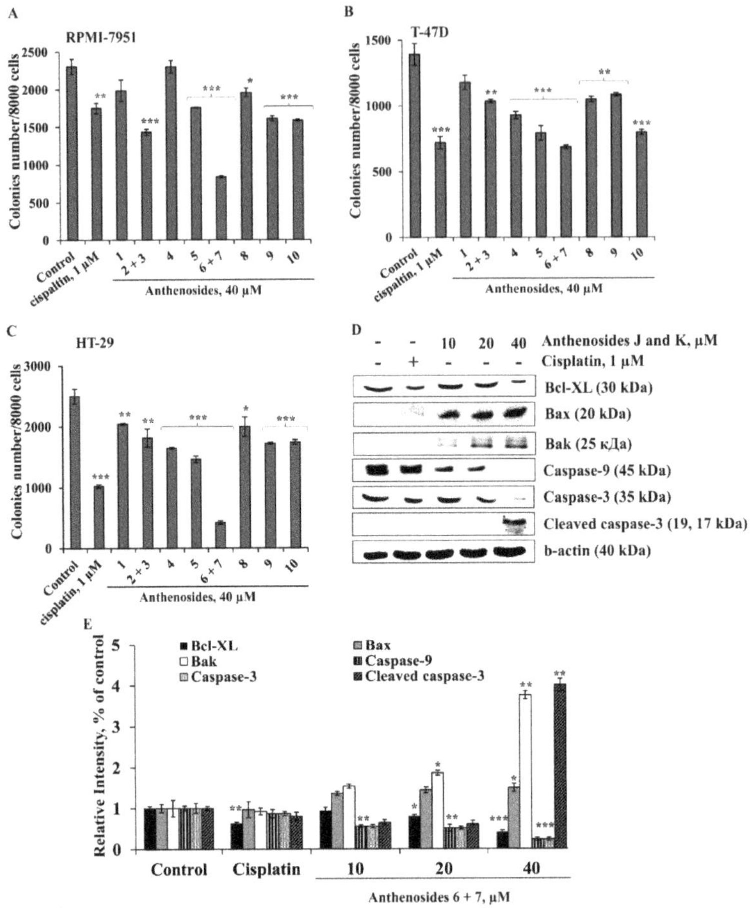

Figure 3. The effects of anthenosides **1–10** on colony formation of human melanoma, breast cancer, and colorectal carcinoma cells and on induction of apoptosis of HT-29 cells. RPMI-7951 (**A**), T-47D (**B**), and HT-29 (**C**) cells (2.4×10^4/mL) treated with/without the investigated compounds (40 µM) or cisplatin (1 µM) (positive control) were exposed to 1 mL of 0.3% Basal Medium Eagle (BME) agar containing 10% FBS and overlaid with 3.5 mL of 0.5% BME agar containing 10% FBS. The culture was maintained at 37 °C in a 5% CO_2 atmosphere for 2 weeks. The colonies were counted under a microscope with the aid of the ImageJ software program. The significant differences were evaluated using Student's *t* test. The asterisks indicate a significant decrease in colony formation of cancer cells treated with the tested compounds or cisplatin compared to the non-treated cells (control), * $p < 0.05$, ** $p < 0.01$, *** $p < 0.001$. (**D**) Regulation of anti- and proapoptotic protein expression, as well as initiator and effector caspase activity, by the mixture of anthenosides J (**6**) and K (**7**) in HT-29 cells. HT-29 cells were either treated by cisplatin (1 µM) or the mixture of **6** + **7** (10, 20, and 40 µM) for 24 h. After drug treatment, total protein lysates were prepared. The protein samples (30 µg) were subjected to SDS-PAGE, followed by detection with immunoblotting using antibodies against Bcl-XL (30 kDa), Bax (20 kDa), Bak (25 kDa), caspase-9 (45 kDa) and -3 (35 kDa), cleaved caspase-3 (17, 19 kDa), and b-actin (40 kDa) proteins. (**E**) Relative band intensity was measured using Image Lab™ Software ("Bio Rad", Hercules, CA, USA). The quantitative results are presented as the mean value from three independent experiments. The significant differences were evaluated using Student's *t* test. The asterisks indicate a significant alteration of the proteins' expression in cells treated by cisplatin or the mixture of **6** + **7** compared with the non-treated cells (control), * $p < 0.05$, ** $p < 0.01$, *** $p < 0.001$.

Among the major regulators of the intrinsic pathway are the caspases and Bcl-2 protein family members. Caspases are cysteine aspartate-specific protein kinases. Based on their function, caspases are divided into two groups: the initiator and the effector caspases. The initiator caspases (caspase-2, -8, -9, and -10) activate the effector caspases, while the effector caspases (caspase-6, -7, and -3) cause the degradation of specific substrates, disrupting the integration of cellular subsystems [4].

It is known that the Bcl-2 protein family controls the membrane permeability of mitochondria and is represented by the proapoptotic BH3-only proteins (Bid, Bim, Puma, Noxa, Bad, Bmf, Hrk, and Bik); the prosurvival Bcl-2-like proteins (Bcl-2, Bcl-XL, Bcl-XS, Bcl-w, BAG); and the pore-forming Bax and Bak proteins [27]. The BH3-only proteins may directly bind and activate Bax and Bak, and also bind to the prosurvival Bcl-2-like proteins to indirectly activate Bax and Bak. Once activated, Bax and Bak oligomerize to form pores in the mitochondrial outer membrane that release cytochrome c. Cytosolic cytochrome c leads to caspase activation and subsequent cell death. These proteins have special significance since they can determine whether the cell commits to apoptosis or aborts the process [28].

In the present study, the capability of the mixture of anthenosides J (**6**) and K (**7**) to regulate the expression of proteins in the Bcl-2 family, as well as activate the initiator or effector caspases, was first determined.

Anthenosides J (**6**) and K (**7**) were shown to downregulate the expression of the antiapoptotic Bcl-XL protein and upregulate the expression of the proapoptotic proteins Bax and Bak that lead to the activation of the initiator caspase-9. Activated caspase-9, in turn, induced the upregulation of the effector caspase-3 expression in a dose-dependent manner. As a result, the mixture of the investigated compounds **6** + **7** at 40 μM caused proteolytic cleavage of caspase-3 and led to apoptosis of colorectal carcinoma cells (Figure 3D). Recently, we demonstrated that luzonicosides from the starfish *Echinaster luzonicus* induced the apoptosis of human melanoma cells by the regulation of p21, cyclin D1, caspase-3, Bcl-2, and Survivin protein expression levels [29]. It was also shown that novaeguinoside II from the starfish *Culcita novaeguineae* induced apoptosis in human glioblastoma U87MG cells by increasing cytochrome-c release from mitochondria, depolarizing of ΔΨm, and by activating caspase 3, followed by DNA degradation [18]. Asterosaponin 1 was found to cause the inhibition of proliferation of human lung cancer A549 cells through the regulation of endoplasmic reticulum-induced apoptosis (ER-apoptosis) [19]. Leviusculoside G from the starfish *Henricia leviuscula* induced p53-dependent apoptosis by inhibition of AP-1, NF-κB, and ERK activities in human leukemia HL-60 and THP-1 cells [20].

It is interesting to note that plancitoxin I, the major lethal factor from the crown-of-thorns starfish *Acanthaster planci* venom, also possesses strong proapoptotic activity that is realized by a caspase 3-independent apoptotic pathway in rat liver TRL 1215 cells [30]. Moreover, this toxin was found to increase the reactive oxygen species (ROS) formation, which induces mitochondrial depolarization, and the elevation of p38 expression; it then can induce the fragmentation of nuclear DNA, triggering apoptosis [31].

In conclusion, our data provide evidence that the mixture of **6** + **7** significantly inhibited colony formation of human melanoma, breast cancer, and colorectal carcinoma cells and induced apoptosis in colorectal carcinoma HT-29 cells through the regulation of anti- and proapoptotic protein expression and the activation of initiator and effector caspases.

3. Materials and Methods

3.1. General Procedures

Optical rotations were determined on a PerkinElmer 343 polarimeter (Waltham, MA, USA). The ^{1}H- and ^{13}C-NMR spectra were recorded on a Bruker Avance III 700 spectrometer (Bruker, Germany) at 700.13 and 176.04 MHz, respectively, and chemical shifts were referenced to the corresponding residual solvent signal (δ_H 3.30/δ_C 49.0 for CD_3OD). The HRESIMS spectra were recorded on a Bruker Impact II Q-TOF mass spectrometer (Bruker, Germany); the samples were dissolved in MeOH (*c* 0.001 mg/mL). HPLC separations were carried out on an Agilent 1100 Series chromatograph (Agilent Technologies, Santa Clara, CA, USA) equipped with a differential refractometer; Diasorb-130-C16T (11 μm, 250 × 16 mm, Biochemmack, Moscow, Russia), Discovery C_{18} (5 μm, 250 × 4 mm, Supelco, North Harrison, PA, USA), and Diasorb-130 Si gel columns (6 μm, 250 × 4.6 mm, Biochemmack, Moscow, Russia) were used. Low-pressure liquid column chromatography was carried out with Polychrom-1 (powdered Teflon, 0.25–0.50 mm; Biolar, Olaine, Latvia) and Si gel KSK (50–160 μm, Sorbpolimer, Krasnodar, Russia). Sorbfil Si gel plates (4.5 × 6.0 cm, 5–17 μm, Sorbpolimer, Krasnodar, Russia) were used for thin-layer chromatography.

3.2. Animal Material

Specimens of *Anthenea aspera* Döderlein, 1915 (order Valvatida, family Oreasteridae) were collected at a depth of 3–20 m by hand via scuba at Tu Long Bay near Khuan Lan Island in the South China (East) Sea during the research vessel Akademik Oparin's 34th scientific cruise in May 2007. Species identification was carried out by Dr. T.I. Antokhina (Severtsov Institute of Ecology and Evolution, RAS, Moscow). A voucher specimen (no. 034-142) is on deposit at the marine specimen collection of the G.B. Elyakov Pacific Institute of Bioorganic Chemistry of the FEB RAS, Vladivostok, Russia.

3.3. Extraction and Isolation

The fresh specimens of *A. aspera* (1.3 kg, crude weight) were chopped into small pieces and extracted thrice with EtOH. The H_2O/EtOH layer was evaporated, and the residue was dissolved in H_2O (1.0 L). The H_2O-soluble material was passed through a Polychrom-1 column (7 × 26.5 cm), eluted with distilled H_2O (4.0 L) until a negative chloride ion reaction was obtained, and then eluted with EtOH (3.5 L). The combined EtOH eluate was evaporated to give a reddish residue (7.0 g). This material was chromatographed over a Si gel column (6 × 18.5 cm) using $CHCl_3$/EtOH (stepwise gradient, 6:1–EtOH, *v*/*v*) to yield eight fractions, 1–8, which were then analyzed by TLC in the eluent system BuOH/EtOH/H_2O (4:1:2, *v*/*v*/*v*). Fractions 1–6 mainly contained the polyhydroxysteroids and related glycosides and admixtures of pigments and concomitant lipids. HPLC separation of fractions 2 (113 mg) and 5 (212 mg) on a Diasorb-130-C16T column (2.5 mL/min) with EtOH/H_2O (75:25, *v*/*v*) as an eluent system yielded pure **4** (7.5 mg, R_t 22.1 min) and **5** (3.5 mg, R_t 26.7 min) and subfractions 2.4 (4.5 mg) and 5.7 (13.0 mg), which were additionally submitted for purification on a Discovery C_{18} analytical column (1.0 mL/min) with EtOH/H_2O (70:30, *v/v*) as an eluent system to give subfractions 2.42 (2.5 mg) and 5.72 (8.5 mg), respectively. HPLC separation of subfractions 2.42 and 5.72 on a Diasorb-130 Si gel analytical column (1.0 mL/min) with EtOAc/EtOH (30:1, *v*/*v*) as an eluent system yielded pure **1** (0.5 mg, R_t 24.3 min), a mixture of **2** and **3** (1.0 mg, R_t 54.7 min), and pure **8** (1.5 mg, R_t 28.5 min), **9** (2.8 mg, R_t 31.7 min), and **10** (3.5 mg, R_t 36.6 min). HPLC separation of fraction 4 (183 mg) on a Diasorb-130-C16T column (2.5 mL/min) with EtOH/H_2O (70:30, *v*/*v*) as an eluent system yielded a mixture of **6** and **7** (9.3 mg, R_t 50.2 min).

3.4. Compound Characterization Data

Anthenoside V [(20*R*)-7-*O*-(6-*O*-methyl-β-D-galactofuranosyl)-16-*O*-(β-D-galactofuranosyl)-24-methyl-5α-cholest-8(14),24(28)-diene-3α,7β,16α-triol] (**1**): Amorphous powder; $[\alpha]^{D}_{25}$: −22.8 (*c* 0.05, MeOH); IR ($CDCl_3$) ν_{max} 3418, 3021, 2856, 1603, 1260, 1216, 1034 cm^{-1}; HRESIMS *m*/*z* 791.4550 [M + Na]$^+$ (calcd. for $C_{41}H_{68}O_{13}$, 791.4549); ESIMS/MS of the ion at *m*/*z* 791: 597 [(M + Na)–$C_7H_{14}O_6$]$^+$, 217 [$C_7H_{14}O_6$ + Na]$^+$, 203 [$C_6H_{12}O_6$ + Na]$^+$; ^{1}H- and ^{13}C-NMR data, see Table 1.

Mixture of the anthenosides W and X (20*R*,24*S*)- and (20*R*,24*R*)-7-*O*-(6-*O*-methyl-β-D-galactofuranosyl)-16-*O*-(β-D-galactofuranosyl)-24-ethyl-5α-cholest-8(14)-ene-3α,6β,7β,16α-tetraols (**2** + **3**, ratio of 3:1): Amorphous powder; $[\alpha]^{D}_{25}$: −33.1 (*c* 0.1, MeOH); IR ($CDCl_3$) ν_{max} 3420, 3020, 2857, 1603, 1261, 1216, 1034 cm^{-1}; HRESIMS *m*/*z* 823.4811 [M + Na]$^+$ (calcd. for $C_{42}H_{72}O_{14}$, 823.4814); ESIMS/MS of the ion at *m*/*z* 823: 629 [(M + Na)–$C_7H_{14}O_6$]$^+$, 217 [$C_7H_{14}O_6$ + Na]$^+$, 203 [$C_6H_{12}O_6$ + Na]$^+$; ESIMS/MS of the ion at m/z 799: 605 $[(M-H)–C_7H_{14}O_6]^-$, 443 $[(M-H)–C_7H_{14}O_6–C_6H_{12}O_6]^-$; ^{1}H- and ^{13}C-NMR data, see Table 1.

3.5. *Bioactivity Assay*

3.5.1. Reagents

Phosphate-buffered saline (PBS), L-glutamine, penicillin–streptomycin solution (10,000 U/mL, 10 µg/mL) were from the "Sigma-Aldrich" company (St. Louis, MO, USA).

MTS reagent—3-[4,5-dimethylthiazol-2-yl]-2,5-diphenyltetrazolium bromide—was purchased from "Promega" (Madison, WI, USA).

Basal Medium Eagle (BME), Dulbecco's Modified Eagle's Medium (DMEM), Minimum Essential Medium Eagle (MEM), McCoy's 5A Modified Medium (McCoy's 5A), trypsin, fetal bovine serum (FBS), agar, and the protein marker "PageRuler Plus Prestained Protein Ladder" were purchased from "ThermoFisher Scientific" (Waltham, MA, USA).

Primary antibodies against caspase-9, caspase-3, cleaved caspase-3, Bcl-XL, Bax, Bak, and b-actin, and horseradish peroxidase (HRP)-conjugated secondary antibody from rabbit and mouse were obtained from "Cell Signaling Technology" (Danvers, MA, USA).

3.5.2. Cell Lines and Culture Conditions

Human malignant melanoma RPMI-7951 cells (ATCC® no. HTB-66™), breast cancer T-47D cells (ATCC® no. HTB-133™), and colorectal carcinoma HT-29 cells (ATCC® no. HTB-38™) were obtained from the American Type Culture Collection (Manassas, WV, USA).

RPMI-7951, T-47D, and HT-29 cells were cultured in 200 µL of complete DMEM/10% FBS and RPMI-1640/10% FBS, and McCoy's 5A medium, respectively, and 1% penicillin–streptomycin solution. The cell cultures were maintained at 37 °C in a humidified atmosphere containing 5% CO_2. Every 3–4 days cells were rinsed with PBS, detached from the tissue culture flask by 0.25% trypsin/0.05 M EDTA, and 10–20% of the harvested cells were transferred to a new flask containing fresh culture media.

3.5.3. Cell Viability Assay

RPMI-7951 (1.0×10^4), T-47D (1.2×10^4), and HT-29 (1.0×10^4) cells were cultured in 200 µL of complete DMEM/10% FBS and RPMI-1640/10% FBS, and McCoy's 5A medium, respectively, for 24 h at 37 °C in 5% CO_2 incubator. The cell monolayer was treated with PBS (control), cisplatin (1 µM), and various concentrations of compounds **1–10** (20, 40, 80, and 160 µM) for 24 h. Subsequently, the cells were incubated with 15 µL MTS reagent for 3 h, and the absorbance of each well was measured at 490/630 nm using a Power Wave XS microplate reader ("BioTek", Wynusky, VT, USA).

3.5.4. The Soft Agar Colony Formation Assay

Cells (2.4×10^4/mL) were seeded in a 6-well plate and treated with compounds (20 and 40 µM) in 1 mL of 0.3% Basal Medium Eagle (BME) agar containing 10% FBS, 2 mM L-glutamine, and 25 µg/mL gentamicin. The cultures were maintained at 37 °C in a 5% CO_2 incubator for 14 days, and the cell colonies were scored using a microscope (Motic AE 20, XiangAn, Xiamen, China) and the ImageJ software.

3.5.5. Western Blotting

HT-29 cells (6×10^5) were seeded in a 10 cm dish overnight and cultured for 24 h. Then, they were treated with cisplatin (1 µM) or the mixture of compounds **6** and **7** (10, 20, and 40 µM) for 48 h. Then, the cells were harvested and centrifuged at 5000 rpm for 5 min. The harvested cells were lysed with lysis buffer (50 mM Tris–HCl (pH 7.4), 150 mM NaCl, 1 mM EDTA, 1 mM EGTA, 10 mg/mL aprotinin, 10 mg/mL leupeptin, 5 mM phenylmethanesulfonyluoride (PMSF), 1 mM dithiolthreitol (DTT) containing 1% Triton X-100) and centrifuged at 12,000 rpm for 15 min to removed insoluble debris. The protein content was determined using Bradford reagent ("Bio-Rad", Hercules, CA, USA). Lysate protein (20–40 µg) was subjected to 12% SDS-PAGE and electrophoretically transferred to polyvinylidene difluoride membranes (PVDF) ("Millipore", Burlington, MA, USA). The membranes were blocked with 5% non-fat milk for 1 h and then incubated with the respective specific primary antibody at 4 °C overnight. Protein bands were visualized using an enhanced chemiluminescence reagent (ECL) ("Bio Rad", Hercules, CA, USA) after hybridization with an HRP-conjugated secondary antibody. Band density was quantified using the ImageJ software.

3.5.6. Statistical Analysis

All assays were performed in at least three independent experiments. Results are expressed as the mean ± standard deviation (SD). Student's t test was used to evaluate the data with the following significance levels: * $p < 0.05$, ** $p < 0.01$, *** $p < 0.001$.

4. Conclusions

Three new minor steroidal glycosides—anthenosides V−X (**1**–**3**)—and the seven previously known anthenosides E (**4**), G (**5**), J (**6**), K (**7**), S1 (**8**), S4 (**9**), and S6 (**10**) were isolated from the extract of the tropical starfish *Anthenea aspera*. Anthenoside V (**1**) has the rare steroidal aglycon without the OH-group at C-6. Previously, only five steroidal glycosides with a non-oxygenated C-6 atom—anthenosides T and U from *A. aspera* and kurilensosides E, F, and G from *Hippasteria kurilensis* [32]—had been found in starfishes. Anthenosides W and X (**2** and **3**) were isolated as unseparated mixtures of epimers. Previously, glycoside pairs with the same side chains—anthenosides H and I, anthenosides J and K, and anthenosides T and U—were found in the starfishes *A. chinensis* and *A. aspera* [14,15] also as unseparated mixtures of epimers. Anthenoside **1**–**10** at noncytotoxic concentrations inhibited the formation and growth of colonies of human melanoma RPMI-7951, breast cancer T-47D, and colorectal carcinoma HT-29 cells to a variable degree. The mixture of the anthenosides **6** + **7** possessed the most significant inhibitory activity against all tested cancer cell lines among the investigated compounds. The molecular mechanism of anticancer activity of these compounds was associated with the induction of apoptosis of colorectal carcinoma HT-29 cells through the regulation of anti- and proapoptotic protein expression followed by the activation of initiator and effector caspases. Further investigations on the detailed molecular mechanism of the anticancer effect of anthenosides from starfishes are needed to provide new approaches to the development of effective cancer therapy regimens.

Supplementary Materials: The following are available online at http://www.mdpi.com/1660-3397/16/11/420/s1. Copies of HRESIMS (Figures S1 and S8), ^{1}H-NMR (Figures S2 and S9), ^{13}C-NMR (Figures S3 and S10), ^{1}H-^{1}H-COSY (Figures S4 and S11), HSQC (Figures S5 and S12), HMBC (Figures S6 and S13), and ROESY (Figures S7 and S14) spectra of compounds **1** and the mixture of **2** and **3**, respectively. This material is available free of charge online.

Author Contributions: T.V.M. isolated the metabolites, elucidated their structure, and prepared the manuscript; O.S.M. investigated proapoptotic activity of compounds and prepared the manuscript. S.P.E. determined the effect of the investigated metabolites on cell viability and the colony formation and growth of human melanoma, breast cancer, and colorectal carcinoma cells; A.A.K. and N.V.I. analyzed the compounds and edited the manuscript; A.I.K. performed the acquisition and interpretation of NMR spectra; P.S.D. did the acquisition and interpretation of mass spectra; V.A.S. edited the manuscript.

Funding: The isolation, the establishment of the chemical structures, and the determination of cytotoxic activity was partially supported by Grant No. 18-53-54002 Viet-a from the RFBR. The study of the anticancer and proapoptotic activities of the anthenosides was supported by Grant No. 18-74-10028 from the RSF (Russian Science Foundation).

Acknowledgments: We are grateful to T.I. Antokhina (Severtsov Institute of Ecology and Evolution, RAS, Moscow, Russia) for species identification of the starfish.

Conflicts of Interest: The authors declare no conflict of interest.

References

1. Fitzmaurice, C. Global, regional and national cancer incidence, mortality, years of life lost, years lived with disability, and disability adjusted life-years for 32 cancer groups, 1990 to 2015: A systematic analysis for the global burden of disease study. *JAMA Oncol.* **2017**, *3*, 524–548. [PubMed]
2. Ke, X.; Shen, L. Molecular targeted therapy of cancer: The progress and future prospect. *Front. Lab. Med.* **2017**, *1*, 69–75. [CrossRef]
3. Hassan, M.; Watari, H.; AbuAlmaaty, A.; Ohba, Y.; Sakuragi, N. Apoptosis and molecular targeting therapy in cancer. *BioMed Res. Int.* **2014**, *2014*, 150845. [CrossRef] [PubMed]
4. Pfeffer, C.M.; Singh, A.T.K. Apoptosis: A target for anticancer therapy. *Int. J. Mol. Sci.* **2018**, *19*, 448. [CrossRef] [PubMed]
5. Schumacher, M.; Kelkel, M.; Dicato, M.; Diederich, M. Gold from the sea: Marine compounds as inhibitors of the hallmarks of cancer. *Biotechnol. Adv.* **2011**, *29*, 531–547. [CrossRef] [PubMed]
6. Minale, L.; Riccio, R.; Zollo, F. Steroidal oligoglycosides and polyhydroxysteroids from Echinoderms. *Fortschr. Chem. Org. Naturst.* **1993**, *62*, 75–308. [PubMed]
7. Stonik, V.A. Marine polar steroids. *Russ. Chem. Rev.* **2001**, *70*, 673–715. [CrossRef]
8. Iorizzi, M.; De Marino, S.; Zollo, F. Steroidal oligoglycosides from the Asteroidea. *Curr. Org. Chem.* **2001**, *5*, 951–973. [CrossRef]
9. Stonik, V.A.; Ivanchina, N.V.; Kicha, A.A. New polar steroids from starfish. *Nat. Prod. Commun.* **2008**, *3*, 1587–1610.
10. Dong, G.; Xu, T.H.; Yang, B.; Lin, X.P.; Zhou, X.F.; Yang, X.W.; Liu, Y.H. Chemical constituents and bioactivities of starfish. *Chem. Biodivers.* **2011**, *8*, 740–791. [CrossRef] [PubMed]
11. Ivanchina, N.V.; Kicha, A.A.; Stonik, V.A. Steroid glycosides from marine organisms. *Steroids* **2011**, *76*, 425–454. [CrossRef] [PubMed]
12. Ivanchina, N.V.; Kicha, A.A.; Malyarenko, T.V.; Stonik, V.A. *Advances in Natural Products Discovery*; Gomes, A.R., Rocha-Santos, T., Duarte, A., Eds.; Nova Science Publishers: Hauppauge, NY, USA, 2017; pp. 191–224.
13. Ma, N.; Tang, H.F.; Qiu, F.; Lin, H.W.; Tian, X.R.; Zhang, W. A new polyhydroxysteroidal glycoside from the starfish *Anthenea chinensis*. *Chin. Chem. Lett.* **2009**, *20*, 1231–1234. [CrossRef]
14. Ma, N.; Tang, H.F.; Qiu, F.; Lin, H.W.; Tian, X.R.; Yao, M.N. Polyhydroxysteroidal glycosides from the starfish *Anthenea chinensis*. *J. Nat. Prod.* **2010**, *73*, 590–597. [CrossRef] [PubMed]
15. Malyarenko, T.V.; Kharchenko, S.D.; Kicha, A.A.; Ivanchina, N.V.; Dmitrenok, P.S.; Chingizova, E.A.; Pislyagin, E.A.; Evtushenko, E.V.; Antokhina, T.I.; Minh, C.V.; et al. Anthenosides L-U, steroidal glycosides with unusual structural features from the starfish *Anthenea aspera*. *J. Nat. Prod.* **2016**, *79*, 3047–3056. [CrossRef] [PubMed]
16. Kicha, A.A.; Ha, D.T.; Ivanchina, N.V.; Malyarenko, T.V.; Kalinovsky, A.I.; Dmitrenok, P.S.; Ermakova, S.P.; Malyarenko, O.S.; Hung, N.A.; Thuy, T.T.T.; et al. Six new polyhydroxysteroidal glycosides, anthenosides S1–S6, from the starfish *Anthenea sibogae*. *Chem. Biodivers.* **2018**, *15*, 1700553. [CrossRef] [PubMed]
17. Malyarenko, T.V.; Ivanchina, N.V.; Malyarenko, O.S.; Kalinovsky, A.I.; Dmitrenok, P.S.; Evtushenko, E.V.; Minh, C.V.; Kicha, A.A. Two new steroidal monoglycosides, anthenosides A_1 and A_2, and revision of the structure of known anthenoside A with unusual monosaccharide residue from the starfish *Anthenea aspera*. *Molecules* **2018**, *23*, 1077. [CrossRef] [PubMed]
18. Zhou, J.; Cheng, G.; Tang, H.F.; Zhang, X. Novaeguinoside II inhibits cell proliferation and induces apoptosis of human brain glioblastoma U87MG cells through the mitochondrial pathway. *Brain Res.* **2011**, *1372*, 22–28. [CrossRef] [PubMed]
19. Zhao, Y.; Zhu, C.; Li, X.; Zhang, Z.; Yuan, Y.; Ni, Y.; Liu, T.; Deng, S.; Zhao, J.; Wang, Y. Asterosaponin 1 induces endoplasmic reticulum stress-associated apoptosis in A549 human lung cancer cells. *Oncol. Rep.* **2011**, *26*, 919–924. [CrossRef] [PubMed]
20. Fedorov, S.N.; Shubina, L.K.; Kicha, A.A.; Ivanchina, N.V.; Kwak, J.Y.; Jin, J.O.; Bode, A.M.; Dong, Z.; Stonik, V.A. Proapoptotic and anticarcinogenic activities of leviusculoside G from the starfish Henricia leviuscula and probable molecular mechanism. *Nat. Prod. Commun.* **2008**, *3*, 1575–1580.

21. Kicha, A.A.; Kalinovsky, A.I.; Ivanchina, N.V.; Malyarenko, T.V.; Dmitrenok, P.S.; Ermakova, S.P.; Stonik, V.A. Four new asterosaponins, hippasteriosides A–D, from the Far Eastern starfish *Hippasteria kurilensis*. *Chem. Biodivers.* **2011**, *8*, 166–175. [CrossRef] [PubMed]
22. Malyarenko, T.V.; Kicha, A.A.; Ivanchina, N.V.; Kalinovsky, A.I.; Popov, R.S.; Vishchuk, O.S.; Stonik, V.A. Asterosaponins from the Far Eastern starfish *Leptasterias ochotensis* and their anticancer activity. *Steroids* **2014**, *87*, 119–127. [CrossRef] [PubMed]
23. Malyarenko, T.V.; Kicha, A.A.; Ivanchina, N.V.; Kalinovsky, A.I.; Dmitrenok, P.S.; Ermakova, S.P.; Stonik, V.A. Cariniferosides A-F and other steroidal biglycosides from the starfish *Asteropsis carinifera*. *Steroids* **2011**, *76*, 1280–1287. [CrossRef] [PubMed]
24. Borowicz, S.; Van Scoyk, M.; Avasarala, S.; Rathinam, M.K.K.; Tauler, J.; Bikkavilli, R.K.; Winn, R.A. The soft agar colony formation assay. *J. Vis. Exp.* **2014**, *92*, 51998. [CrossRef] [PubMed]
25. Malyarenko, T.V.; Malyarenko (Vishchuk), O.S.; Ivanchina, N.V.; Kalinovsky, A.I.; Popov, R.S.; Kicha, A.A. Four new sulfated polar steroids from the Far Eastern starfish *Leptasterias ochotensis*: Structures and activities. *Mar. Drugs* **2015**, *13*, 4418–4435. [CrossRef] [PubMed]
26. Jin, Z.; El-Deiry, W.S. Overview of cell death signaling pathways. *Cancer Biol. Ther.* **2005**, *4*, 139–163. [CrossRef]
27. Westphal, D.; Dewson, G.; Czabotar, P.E.; Kluck, R.M. Molecular biology of Bax and Bak activation and action. *Biochim. Biophys. Acta* **2011**, *1813*, 521–531. [CrossRef] [PubMed]
28. Vervloessem, H.A.T.; Kiviluoto, S.; Bittremieux, M.; Parys, J.B.; De Smedt, H.; Bultyn, G. A dual role for the anti-apoptotic Bcl-2 protein in cancer: Mitochondria versus endoplasmic reticulum. *Biochim. Biophys. Acta* **2014**, *1843*, 2240–2252.
29. Malyarenko, O.S.; Dyshlovoy, S.A.; Kicha, A.A.; Ivanchina, N.V.; Malyarenko, T.V.; Carsten, B.; von Gunhild, A.; Stonik, V.A.; Ermakova, S.P. The Inhibitory Activity of luzonicosides from the starfish *Echinaster luzonicus* against human melanoma cells. *Mar. Drugs* **2017**, *15*, 227. [CrossRef] [PubMed]
30. Ota, E.; Nagashima, Y.; Shiomi, K.; Sakurai, T.; Kojima, C.; Waalkes, M.P.; Himeno, S. Caspase-independent apoptosis induced in rat liver cells by plancitoxin I, the major lethal factor from the crown-of-thorns starfish Acanthaster planci venom. *Toxicon* **2006**, *48*, 1002–1010. [CrossRef] [PubMed]
31. Lee, C.C.; Hsieh, H.J.; Hwang, D.F. Cytotoxic and apoptotic activities of the plancitoxin I from the venom of crown-of-thorns starfish (*Acanthaster planci*) on A375.S2 cells. *J. Appl. Toxicol.* **2015**, *35*, 407–417. [CrossRef] [PubMed]
32. Kicha, A.A.; Ivanchina, N.V.; Kalinovsky, A.I.; Dmitrenok, P.S.; Agafonova, I.G.; Stonik, V.A. Steroidal triglycosides, kurilensosides A, D, and C, and other polar steroids from the Far Eastern starfish *Hippasteria kurilensis*. *J. Nat. Prod.* **2008**, *71*, 793–798. [CrossRef]

marine drugs

MDPI

Article

The Distribution of Asterosaponins, Polyhydroxysteroids and Related Glycosides in Different Body Components of the Far Eastern Starfish *Lethasterias fusca*

Roman S. Popov, Natalia V. Ivanchina, Alla A. Kicha, Timofey V. Malyarenko, Boris B. Grebnev, Valentin A. Stonik and Pavel S. Dmitrenok *

G.B. Elyakov Pacific Institute of Bioorganic Chemistry, Far Eastern Branch of Russian Academy of Sciences, 159 Prospect 100-letiya Vladivostoku, Vladivostok 690022, Russia

* Correspondence: paveldmt@piboc.dvo.ru; Tel.: +7-423-231-1132

Received: 15 August 2019; Accepted: 30 August 2019; Published: 6 September 2019

Abstract: Glycoconjugated and other polar steroids of starfish have unique chemical structures and show a broad spectrum of biological activities. However, their biological functions remain not well established. Possible biological roles of these metabolites might be indicated by the studies on their distribution in the organism–producer. In order to investigate the localization of polar steroids in body components of the Far Eastern starfish *Lethasterias fusca*, chemical constituents of body walls, gonads, stomach, pyloric caeca, and coelomic fluid were studied by nanoflow liquid chromatography/mass spectrometry with captive spray ionization (nLC/CSI–QTOF–MS). It has been shown that the levels of polar steroids in the studied body components are qualitatively and quantitatively different. Generally, the obtained data confirmed earlier made assumptions about the digestive function of polyhydroxysteroids and protective role of asterosaponins. The highest level of polar steroids was found in the stomach. Asterosaponins were found in all body components, the main portion of free polyhydroxysteroids and related glycosides were located in the pyloric caeca. In addition, a great inter-individual variability was found in the content of most polar steroids, which may be associated with the peculiarities in their individual physiologic status.

Keywords: starfish; *Lethasterias fusca*; asterosaponins; polyhydroxysteroids; glycosides; body components; distribution

1. Introduction

Starfish are characterized by a high content of polar, frequently glycoconjugated steroids. Polar steroid compounds from starfish have put together an important class of biologically active marine metabolites, including steroids bearing four to nine hydroxyl groups, related polyhydroxysteroid mono-, bi- and triosides, and oligoglycosides known as asterosaponins [1–6]. Structural diversity of these metabolites is significant. Several hundred new individual natural compounds were discovered from different species of Asteroidea up to date.

Glycoconjugated polyhydroxysteroids usually have one or two monosaccharide units in steroid nucleus, side chain or both steroid nucleus and side chain. Aglycones of asterosaponins are proved to be $\Delta^{9(11)}$-3β,6α-dihydroxysteroids with a sulfate group at C-3, while their oligosaccharide chains, affixed to C-6 of aglycone, contain four to six sugar units with one or two branchings. The side chains of asterosaponin aglycones and their oligosaccharide fragments show significant structural diversity [1–6].

Polar steroids from starfish demonstrate cytotoxic, antibacterial, antiviral, antifungal, anticancer, anti-inflammatory, analgesic, and neuritogenic effects [1–6]. At the same time, the knowledge about their biological functions is limited. To understand biological roles of these substances in the organism–producer, accurate information of the polar steroid distribution is helpful because it is assumed that the biological roles of the steroid metabolites may be related to organ and tissue distribution. In addition, the data concerning the localization of these metabolites in animals might be useful in the targeted isolation of new natural compounds.

Previously, the studies on the asterosaponin distribution were based on the determination of sugar composition and the hemolytic activities of the corresponding extracts due to high hemolytic properties of asterosaponins, while polyhydroxysteroids and related glycosides show only weak or moderate hemolytic activity. Asterosaponins were found in both inner organs and body walls of starfish *Asterias amurensis*, *Asterias rubens*, and *Marhtasterias glacialis* [7–9]. Analysis of the asterosaponin fractions isolated from aboral and oral body walls, stomach, gonads, and pyloric caeca of the starfish *Leptasterias polaris* by liquid chromatography demonstrated the presence of characteristic mixtures of asterosaponins in the different body parts [10]. Recently, a combination of matrix-assisted laser desorption/ionization mass spectrometry (MALDI–MS), MALDI–MS imaging and liquid chromatography coupled to mass spectrometry (LC/MS) was applied to study the diversity, body distribution and localization of asterosaponins in *A. rubens* [11,12]. Asterosaponins from aboral and oral body walls, pyloric caeca, gonads, and stomach from four animals were extracted and analyzed separately, allowing comparisons to be made between body parts and between individuals [11]. Asterosaponins were found in all body parts of all individual starfish, but each organ was characterized by a distinctive set of asterosaponins and their concentrations varied significantly among organs as well as among individuals. Pyloric caeca were shown to be organs presenting the lowest asterosaponin concentrations. Although extracts of body walls and gonads showed high asterosaponin concentrations, significant differences between individuals were noted. MALDI–MS imaging was used to clarify inter- and intra-organ spatial distributions of asterosaponins that had been early found in *A. rubens* extracts [12]. This approach, performed at different spatial resolutions, revealed the complicated character of distributions of asterosaponins and showed that some asterosaponins are located not only inside the body walls of the starfish but also within the outer mucus layer, where they probably protect the animals against predators. Thus, the body distribution of asterosaponins indicates that the participation in the defense of starfish against predatory fish can be the main biological function of these glycosides. In addition, it is suggested that asterosaponins could be involved in reproduction processes and interspecific chemical signaling [1,6].

The distribution of free sterols, polyhydroxysteroids and steroid glycosides in aboral and oral body walls, gonads, stomach, and pyloric caeca of the starfish *Patiria* (=*Asterina*) *pectinifera* has been studied [13]. It was shown that all these body components contained asterosaponins. However, polyhydroxysteroids and glycosides of polyhydroxysteroids were located mainly in the stomach and pyloric caeca [13]. It was suggested that these metabolites are involved in the digestion of food based on the presence of these compounds in digestive organs of the starfish and their structural resemblance to certain bile alcohols. It has also been established that the concentration of polyhydroxylated steroids and polyhydroxysteroid glycosides in stomach and pyloric caeca of *P. pectinifera* depends on the season and is related to the periods of active feeding [14,15].

The Far Eastern starfish *Lethasterias fusca* has been recently studied by our group and 14 polar steroid compounds, including three polyhydroxysteroids, six glycosides of polyhydroxysteroids, and five asterosaponins, were isolated [16,17]. The cancer-preventive activity of the asterosaponins was investigated. It has been shown that lethasterioside A demonstrates a considerable inhibition of the human breast T-47D (97%), melanoma RPMI-7951 (90%), and colorectal carcinoma HCT-116 (90%) cell colony formations in a soft agar clonogenic assay [16]. Recently, we have applied a nanoflow liquid chromatography/tandem mass spectrometry with captive spray ionization (nLC/CSI–QTOF–MS/MS) for the profiling and characterization of the polar steroid metabolites of *L. fusca*. In total, 207 steroid compounds, including 106 asterosaponins, six native aglycones of asterosaponins, 81 polyhydroxysteroid glycosides, and 14 non-glycoconjugated polyhydroxysteroids, were found. According to the obtained data, exact structures of twenty steroids were determined and tentative structures for previously undescribed steroid compounds were proposed [18].

Herein, we describe the application of nLC/CSI–QTOF–MS for the profiling of purified fractions of polar steroids from body walls, coelomic fluid, gonads, stomach, and pyloric caeca to establish the content of these metabolites in different body components of the starfish *L. fusca*.

2. Results and Discussion

Previously, we have studied polar steroid metabolites of the starfish *L. fusca* by nanoflow liquid chromatography coupled with a captive spray ionization time-of-flight tandem mass spectrometer (nLC/CSI–QTOF–MS/MS). As a result, 207 polar steroids, including 81 glycosides of polyhydroxysteroids, 14 polyhydroxylated steroids, 106 asterosaponins, and six native aglycones of asterosaponins were detected. Although exact structures cannot be deduced from mass spectrometry data only, a detailed fragmentation analysis with accurate mass measurement allowed for characterizing and to proposing tentative structures for the detected compounds. Among the identified asterosaponins, new compounds with unusual monosaccharide units and new aglycone types were found. Polyhydroxysteroids and related glycosides were found in both sulfated and non-sulfated forms and demonstrated great structural diversity [18].

2.1. Distribution of Polar Steroids in Body Components of the Starfish L. fusca

The present study was undertaken to establish the distribution of previously detected polar steroid compounds in different body components of the starfish *L. fusca*. For this purpose, we separately extracted polar steroids from body walls (BW), gonads (G), stomach (S), and pyloric caeca (PC) from five animals. Starfish coelomic fluids (CF), which are in contact with all internal organs, were also collected and analyzed. To obtain a purified fraction of polar steroid compounds from complicated ethanolic extracts, the two-stage liquid–liquid extraction followed by desalting by solid-phase extraction was used. The quality of extraction of polar steroid metabolites was controlled by ESI MS and LC/ESI MS at every stage (data not shown). Finally, polar steroid metabolites from different body components of the starfish *L. fusca* were analyzed qualitatively and semi-quantitatively by nLC/CSI–QTOF–MS. Based on the previous results, the mass spectra were recorded in negative ion mode. Some typically obtained base-peak chromatograms are shown in Figure S1. Amounts of every compound were semi-quantified using lethasterioside A as a reference standard for asterosaponins, fuscaside A as a reference standard for sulfated polyhydroxysteroid glycosides and sulfated polyhydroxysteroids and 5α-cholestan-3β,4β,6α,7α,8,15β,16β,26-octaol as a reference standard for non-sulfated polyhydroxysteroid compounds and non-sulfated polyhydroxysteroid glycosides and are shown as ng/g of the wet weight of the organs (Table S4).

The profiling of polar steroids from various body components showed that the distribution of these compounds is qualitatively and quantitatively different (Table 1, Table S4, Figure 1). The maximal content of the sum of all polar steroids was observed in the stomach when compared with other body components of *L. fusca* (577.5 μg/g of wet weight, Table 1). The main part of polar steroids in the stomach was proved to be asterosaponins and native aglycones of asterosaponins (97% of all polar steroids content in the stomach). It can be assumed that asterosaponins in the stomach have a protective function against predators, the same as in body walls. It is known that starfish have the so-called "external" nutrition, they evert the stomach from the organism's body and engulf the food. Probably, at such moments, the protection of this organ is especially important. At the same time, it is known that asterosaponins exhibit antimicrobial properties [1–6] and can protect the organism of a starfish from pathogenic microorganisms coming from food. Toxic properties of asterosaponins also can help starfish immobilize or kill living creatures in food.

Table 1. Concentrations of asterosaponins, polyhydroxysteroids, and related glycosides in different organs and coelomic fluid of the starfish *L. fusca* (μg/g wet weight of the organs for body walls, gonads, pyloric caeca and stomach and μg/mL for coelomic fluid).

	Asterosaponins and Native Aglycones	Polyhydroxysteroids	Glycosides of Polyhydroxysteroids	Total
Body walls (μg/g)	59.7	0.5	6.3	66.5
Gonads (μg/g)	68.4	0.7	9.8	78.9
Pyloric caeca (μg/g)	23.8	23.1	267.5	314.4
Stomach (μg/g)	561.1	1.9	14.5	577.5
Coelomic fluid (μg/mL)	0.14	-	0.02	0.16

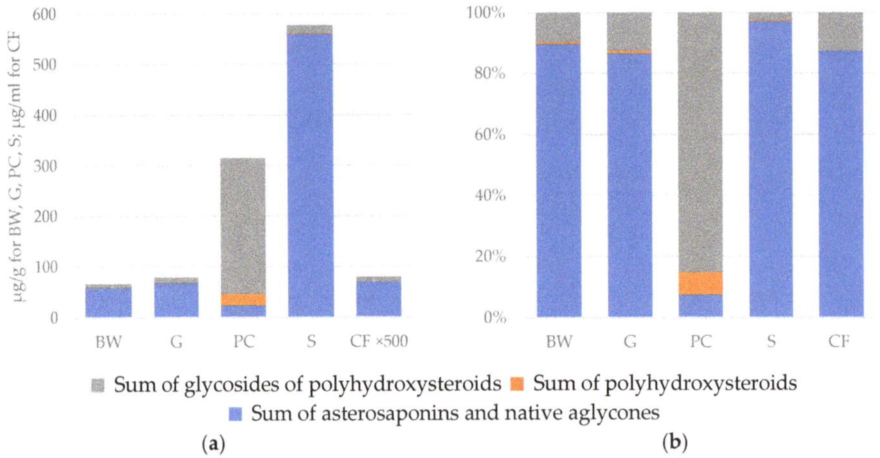

Figure 1. (**a**) concentrations of asterosaponins, polyhydroxysteroids, and related glycosides in different organs and coelomic fluid of the starfish *L. fusca* (μg/g wet weight of the organs for body walls (BW), gonads (G), pyloric caeca (PC) and stomach (S) and μg/mL for coelomic fluid (CF); CF values multiplied 500-fold); (**b**) ratios of concentrations of asterosaponins, polyhydroxysteroids, and related glycosides in different organs and coelomic fluid of the starfish *L. fusca*.

Polar steroids content in the pyloric caeca was 314.4 μg/g. The main compounds in PC were glycosides of polyhydroxysteroids (85%); asterosaponins and polyhydroxysteroids presented a less significant part (both about 7%) (Table 1, Figure 1). These data are in good agreement with the previously hypothesized digestive role of polyhydroxysteroids and their related glycosides [13]. The body walls and gonads contained smaller concentrations of polar steroid compounds (66.5 and 78.9 μg/g, respectively). Major steroid constituents of both organs were identified as asterosaponins (90% for BW and 87% for G), a less substantial part was presented by polyhydroxysteroid glycosides (9% for BW and 12% for G) (Table 1, Figure 1).

Profiling of the coelomic fluid obtained from all five individuals showed that the content of polar steroids in all CF samples was minimal among analyzed samples. The average content of polar steroids in CF was 0.16 μg/mL of coelomic fluid, which is below 0.02% of the sum of all polar steroid content in the starfish (Table 1, Figure 1). Most of the compounds were either detected in trace amounts or not detected. Only 29 compounds showed a concentration higher than 1 ng/mL of coelomic fluid. Among them were 23 asterosaponins, four sulfated glycosides, and the maximum concentrations were shown by two native aglycones of asterosaponins (3-*O*-sulfothornasterol A (**205**) (20.9 ng/mL) and 3-*O*-sulfo-24,25-dihydromarthasterone (**207**) (10.9 ng/mL)) (Table S4). Due to the low content of the studied compounds, the CF samples were excluded for further analysis of the distribution of polar steroids.

2.2. Distribution of Asterosaponins

Previously, 112 asterosaponins, including 66 pentaosides, 28 hexaosides, 7 triosides, five "shortened" asterosaponins with one-monosaccharide units at C-6 and six native aglycones of asterosaponins were found and structurally characterized in *L. fusca* [18]. According to the proposed structures of aglycones, all the detected asterosaponins of *L. fusca* were divided into twenty-three groups (I—XXIII) according to the types of aglycones (AG I—AG XXIII) (structure of aglycones given in Figure 2).

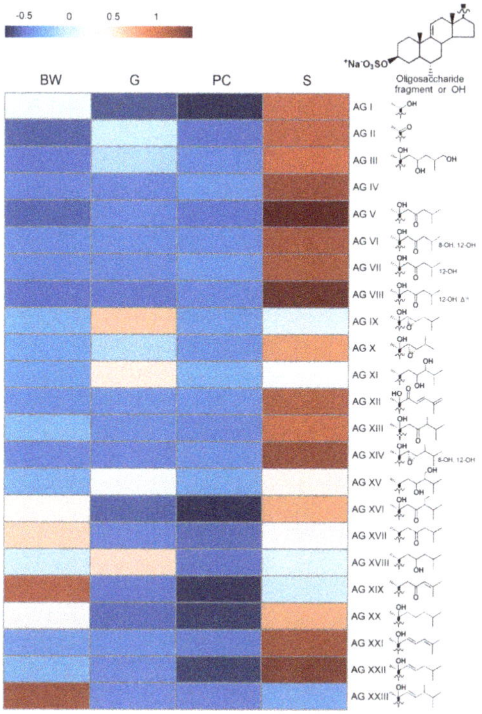

Figure 2. Heatmap demonstrating relative content of asterosaponins in different organs of the starfish *L. fusca*. Each row represents the sum content of asterosaponins with the same aglycone type (AG I–AG XXIII). Color bars represent auto-scaled relative abundances of asterosaponin groups, going from less (blue) to more (red).

Asterosaponins were detected in all studied body parts. In the body walls, gonads, and stomach asterosaponins were the main class of detected polar steroids, but the stomach had a maximal concentration of these compounds (561.1, 68.4, and 59.7 μg/g in the stomach, gonads, and body walls, respectively). In addition, each organ was characterized by its specific mixture of asterosaponins. The majority of asterosaponins were present at a maximum concentration in the stomach; other asterosaponins were more characteristic of the body walls or gonads. The content of forty-two asterosaponins was found to be statistically different between the various organs by ANOVA and Tukey's post hoc test (Table 2, Table S4); the content of other seventy asterosaponins showed a large inter-individual variability.

Table 2. Asterosaponins that display significant differences (q-value < 0.05) between contents in the various body parts of the starfish *L. fusca* (content is given as a mean ± SD ng/g wet weight of the organs).

N	Content in Different Organs (Mean ± SD, ng/g)				Proposed Structures
	BW	G	PC	S	
3	27.5 ± 23.5	0.7 ± 0.9	0.8 ± 1	6.2 ± 1.5	Hex–dHex–Hex–Xyl(-Qui)–Qui–AG I
16	0.1 ± 0.2	0 ± 0	3.8 ± 3.5	87.7 ± 74.5	dHex–dHex–Qui(-Qui)–Hex–AG IV
17	50.1 ± 53.2	39.7 ± 24	79.3 ± 61.6	658.1 ± 232.1	Hex–dHex–Hex–Xyl(-Qui)–Qui–AG II
32	0.8 ± 0.8	3.5 ± 2.5	7.4 ± 3.4	818 ± 552.4	dHex–dHex–Qui(-Qui)–DXU–AG VI
33	45.4 ± 65.8	295.6 ± 526.5	20.3 ± 17.6	2749.1 ± 1759	dHex–dHex–Qui(-Qui)–DXU–AG II
34	93.2 ± 58.6	10.4 ± 6.5	13.8 ± 6.7	48.5 ± 32.5	AG I
40	23.9 ± 25.3	12.9 ± 18.7	18.3 ± 16.1	615.5 ± 394.5	Hex–dHex–Hex–Xyl(-Qui)–Qui–AG VII
41	0.9 ± 0.6	5.7 ± 9.6	5.8 ± 6.1	325.3 ± 204.6	Hex–dHex–Hex–Xyl(-Qui)–Qui–AG VIII
44	64.4 ± 70.9	8.9 ± 8.4	5.6 ± 4	350.2 ± 240.5	Hex–dHex–Hex–Xyl(-Qui)–Hex–AG V
51	86.1 ± 168.1	26.4 ± 22.3	45.4 ± 15.4	1059.8 ± 733.7	Qui–Xyl–Qui–AG I
60	80.7 ± 102.3	52 ± 39.6	67.3 ± 59.3	1232 ± 639.6	dHex–Hex–Xyl(-Qui)–Hex–AG V
65	0.5 ± 0.3	1.2 ± 2.1	3.9 ± 2.8	375.3 ± 370	dHex–dHex–Qui(-Qui)–Hex–AG XII
70	34.8 ± 15.4	39.4 ± 18.5	24 ± 11.3	2805.1 ± 929.5	dHex–dHex–Glc(-Qui)–Hex–AG V
74	4.8 ± 3.3	12 ± 6.4	11.7 ± 5.2	1468.7 ± 949.4	dHex–dHex–Qui(-Qui)–DXU–AG VIII
77	13 ± 4.9	30.6 ± 25.6	23.3 ± 8.2	2888.6 ± 1744.3	dHex–dHex–Qui(-Qui)–DXU–AG VII
78	2468.3 ± 2319.4	1881.1 ± 1119	953.2 ± 857.2	23,431.5 ± 7708.7	Hex–dHex–Hex–Xyl(-Qui)–Qui–AG V
88	77.5 ± 66.8	110.5 ± 52.2	33.4 ± 25.3	4674.9 ± 1961	Hex–dHex–dHex–Glc(-Qui)–Qui–AG V
90	0.2 ± 0.2	0 ± 0.1	2.6 ± 2.5	83.8 ± 67.6	dHex–dHex–Glc(-Qui)–Qui–AG XIV
93	104 ± 125.9	121.9 ± 83.6	71.4 ± 61	4978 ± 4835.3	dHex–dHex–Qui(-Qui)–Hex–AG V
105	8.8 ± 10.7	11.6 ± 12	10.1 ± 5.5	789.8 ± 563.5	dHex–dHex–Glc(-Qui)–Hex–AG XIII
106	29.7 ± 24.6	21.4 ± 12.8	39.8 ± 31.9	566 ± 187.1	dHex–dHex–Hex–Xyl(-Qui)–Qui–AG V
111	304.4 ± 169.6	153.5 ± 38	111.4 ± 65.9	44,57.6 ± 3242.3	dHex–Hex–Glc(-Qui)–Qui–AG V
112	53.5 ± 51.3	16.2 ± 15.1	7.5 ± 5.7	655.2 ± 516.7	Hex–dHex–Hex–Xyl(-Qui)–Hex–AG XVII
118	831.6 ± 848.4	2816.5 ± 1624.1	2521.9 ± 2733.1	10,972.3 ± 4393.7	Fuc–Gal–Xyl(-Qui)–Qui–AG V (thornasteroside A)
122	2553.5 ± 1926.5	6991.5 ± 3200.9	3934.5 ± 2011.3	173,398.3 ± 148,397.4	Fuc–Qui–Glc(-Qui)–Qui–AG V (luidiaquinoside)
123	2846.4 ± 2424.5	162.9 ± 111.1	42.8 ± 46.9	333.7 ± 264.9	Hex–dHex–Hex–Xyl(-Qui)–Qui–AG XX
129	37.4 ± 27.4	68.1 ± 40.1	25.7 ± 12.6	2677.6 ± 1160.6	Hex–dHex–dHex–Qui(-Qui)–Qui–AG V
135	437.3 ± 207.6	2343.4 ± 2062	273.3 ± 196.3	64,570.3 ± 23,737.3	dHex–dHex–Qui(-Qui)–DXU–AG V
150	1249.7 ± 1349	1098.9 ± 860.2	224 ± 215.9	28,498.5 ± 24,555.7	dHex–dHex–Qui(-Qui)–Hex–AG XXI
152	399.9 ± 462.7	931 ± 1346	114 ± 78.6	6581 ± 1828.8	dHex–dHex–Glc(-Qui)–Hex–AG XXII
159	1114.2 ± 860	305.4 ± 187.7	66.7 ± 57.1	364.7 ± 155.4	Fuc–Gal–Xyl(-Qui)–Qui–AG XVIII (lethasterioside B)
174	1.7 ± 2	13.6 ± 10.1	91.8 ± 61.2	818.9 ± 580.7	AG VIII
175	141.9 ± 90.5	355 ± 174.4	225.1 ± 120.2	7623.4 ± 3936.2	dHex–dHex–Qui(-Qui)–Qui–AG V
178	4.7 ± 5.2	25.3 ± 21.7	219.6 ± 166.5	1023.6 ± 609.4	AG VII
180	51 ± 40.9	725.5 ± 1061.2	95 ± 52	4869.7 ± 3668.5	dHex–dHex–Qui(-Qui)–($C_6H_6O_3$)–AG V
182	3440.1 ± 3370.9	1962.9 ± 1574	614.6 ± 567.6	36,581.4 ± 30,632.4	dHex–dHex–Qui(-Qui)–Hex–AG XXII
186	7.8 ± 5.5	633.3 ± 1139.3	50 ± 35.4	3777.4 ± 3427.8	dHex–dHex–Qui(-Qui)–($C_7H_9NO_4$)–AG V
194	295 ± 239.4	9.3 ± 5.8	4.7 ± 4.2	30.3 ± 38.5	Hex–dHex–Hex–Xyl(-Qui)–Qui–AG XXIII
195	1935.9 ± 1321.3	228.1 ± 163.7	14.6 ± 5.5	746.4 ± 802.9	dHex–dHex–Qui(-Qui)–DXU–AG XIX
199	4.8 ± 6.4	61.6 ± 76	68.6 ± 88.9	7221.9 ± 7112.6	dHex–dHex–Qui(-Qui)–($C_7H_9NO_4$)–AG V
205	1096.5 ± 566.1	10,113.8 ± 13,871.4	7737.1 ± 4085.7	4,5330.5 ± 21,263.7	3-*O*-sulfothornasterol A (AG V)
207	477.9 ± 255.2	154.4 ± 155.8	320.3 ± 99	664.4 ± 284.8	3-*O*-sulfo-24,25-dihydromarthasterone (AG XVII)

Often, the localization of asterosaponins was associated with the type of aglycone (Figure 2). At the same time, the structure of the oligosaccharide chains did not affect the distribution of asterosaponins. All asterosaponins with aglycones having additional hydroxy groups in steroid nucleus (AG VI, AG VII, AG VIII, and AG XIV), asterosaponins **16** (AG IV) and **65** (AG XII), most of the asterosaponins with aglycones AG V, AG XXI, and AG XXII as well as their native aglycones (**205** (3-*O*-sulfothornasterol A (AG V)), **174** (AG VIII), and **178** (AG VII) were found mainly in the stomach. On the contrary, it was established that concentrations of asterosaponins with aglycone AG XIX and asterosaponin **194** (AG XXIII) were significantly higher in the body walls. Asterosaponins with AG I were shown to be characteristic of body walls and stomach, and asterosaponins with AG II (3-*O*-sulfoasterone) were characteristic of gonads and stomach. Levels of other compounds, belonging to this class, did not show significant differences and had a large inter-individual variability. Asterosaponins with AG XIII, AG XVI, AG XVII, and AG XX were distributed between body walls and stomach, and asterosaponins with AG III, AG IX, AGX, AG XI, AG XV, and AG XVIII were distributed between body walls and gonads.

Previous investigations of the asterosaponins distribution have also shown that these compounds presented in all starfish organs and each organ had a characteristic mixture of asterosaponins. For example, in *P. pectinifera*, asterosaponins were found in maximal concentrations in the aboral and oral body walls and gonads [13]. In addition, in *A. rubens*, the highest concentrations of asterosaponins were measured in the body walls and gonads [11,12]. Otherwise, it was found that asterosaponin content in the stomach of *L. polaris* was several times greater than in the body walls and pyloric caeca [19]. Our results also indicate that the asterosaponins content prevails in the stomach rather than in the body walls or gonads. The fact that some asterosaponins are found only in one organ and not in others may indicate quite specific biological function of individual asterosaponins in starfish.

Thus, our data along with previously obtained results demonstrate the presence of asterosaponins in all starfish body components. The fact that toxic triterpene glycosides, which are considered to be part of chemical defense system in sea cucumbers, were also found in all body components of sea cucumbers [20–22] may support the notion of the protective function of asterosaponins.

2.3. Distribution of Polyhydroxysteroids and Glycosides of Polyhydroxysteroids

The highest concentrations of both non-sulfated and sulfated polyhydroxysteroid compounds were observed in extracts of the pyloric caeca (23.1 μg/g) (Table 1, Figure 1). The content of non-sulfated polyhydroxysteroids 5α-cholestane-3β,4β,6α,7α,8,15α,16β,26-octaol (**4**), 5α-cholestane-3β,6α,7α,8,15α,16β,26-heptaol (**11**), and 5α-cholestane-3β,4β,6α,7α,8,15β,16β,26-octaol (**15**) and sulfated polyhydroxysteroids 5α-cholestane-3β,4,6,7,8,15α,16β,26-octaol 6-*O*-sulfate (**63**), 5α-cholestane-3β,6,8,15α,16β,26-hexaol 6-*O*-sulfate (**117**), 5α-cholestane-3β,6,8,15β,16β,26-hexaol 6-*O*-sulfate (**133**), 5α-cholestane-3β,6,8,15,24-pentaol 24-*O*-sulfate (**136**), 5α-ergost-22-ene-3β,6,7,8,15α,16β,26-heptaol 6-*O*-sulfate (**146**), and 5α-ergost-22-ene-3β,6,8,15,16β,26-hexaol 6-*O*-sulfate (**153**) were found to be statistically different between the pyloric caeca and other samples by ANOVA with Tukey's post hoc test (Table 3). Polyhydroxysteroids were about 10-fold less concentrated in the stomach (about 2 μg/g) than in the pyloric caeca. However, it should be noted that samples S#2, S#3, and S#4 contained large concentrations of sulfated polyhydroxysteroid **30** (the structure was not assigned), while all other polyhydroxysteroids showed maximum concentrations in the pyloric caeca samples. In addition,, stomach extracts showed a significant amount of 5α-cholestane-3β,6,8,15α,16β,26-hexaol 6-*O*-sulfate (**117**), 5α-cholestane-3β,6,8,15β,16β,26-hexaol 6-*O*-sulfate (**133**), and 5α-ergost-22-ene-3β,6,8,15,16β,26-hexaol 6-*O*-sulfate (**153**). The analysis of the body walls and gonads showed significantly lower concentrations of polyhydroxysteroid compounds (0.5 and 0.7 μg/g, respectively).

Table 3. Polyhydroxysteroids and related glycosides that display significant differences (q-value < 0.05) between contents in the various body parts of the starfish *L. fusca* (content is given as mean ± SD ng/g wet weight of the organs for BW, G, PC, and S).

	Content in Different Organs (Mean ± SD, ng/g)				Proposed Structures
	BW	G	PC	S	
1	1.1 ± 0.7	4.3 ± 7.3	369.8 ± 244.8	9.4 ± 7.6	24-*O*-pentosyl-5α-cholestane-3β,6,8,15α,16β,24-hexaol
2	1.5 ± 1.5	3.5 ± 5.2	421.5 ± 320.8	11.2 ± 7.2	24-*O*-hexosyl-5α-cholestane-3β,6,7,8,15α,16β,24-heptaol
4	0.1 ± 0.1	0.2 ± 0.5	123.8 ± 53.7	1.1 ± 0.4	5α-cholestane-3β,4β,6α,7α,8,15α,16β,26-octaol
11	0.4 ± 0.2	1 ± 1.7	325.9 ± 257.2	8.2 ± 6.4	5α-cholestane-3β,6α,7α,8,15α,16β,26-heptaol
15	2.2 ± 1.4	7.8 ± 11.3	1016.8 ± 582	15.3 ± 9.8	5α-cholestane-3β,4β,6α,7α,8,15β,16β,26-octaol
18	22.9 ± 18	70.6 ± 108.7	3276.3 ± 2782.6	167.5 ± 142.1	pycnopodioside A
21	17.8 ± 8.5	41.1 ± 51.9	2153.6 ± 1149.2	80.3 ± 53.5	desulfated minutoside A
23	26.3 ± 26.9	82.4 ± 148	1579.6 ± 1499.7	60.4 ± 54.8	3-*O*-pentosyl-5α-cholestane-3β,6,7,8,15,16β,26-heptaol 26-*O*-sulfate
53	10.6 ± 19.3	8.3 ± 8.3	164.8 ± 109	27.4 ± 42.9	24-*O*-sulfohexosyl-5α-cholestane-3β,6,8,15,16β,24-hexaol
58	64.8 ± 36.7	53.9 ± 62.7	2845.4 ± 1755.4	125.4 ± 81.2	-
59	49.8 ± 52.5	38.8 ± 40	2129.6 ± 1784.9	101.8 ± 62.5	3-*O*-pentosyl-5α-cholestane-3β,6,8,15,26-pentaol 26-*O*-sulfate
63	12.2 ± 12.5	16.9 ± 28.8	455.5 ± 388.1	22.8 ± 14	5α-cholestane-3β,4,6,7,8,15α,16β,26-octaol 6-*O*-sulfate
71	10.7 ± 6.9	4.5 ± 1.7	509.3 ± 373	19.8 ± 15.5	26-*O*-sulfohexosyl-27-nor-5α-ergost-22-ene-3β,6,8,15,16β,26-hexaol
81	177.8 ± 126.7	337.9 ± 578.6	7196.8 ± 3778.9	309.9 ± 243.4	24-*O*-pentosyl-5α-cholestane-3β,6,8,15,24-pentaol 3-*O*-sulfate
83	18.2 ± 18.4	8.9 ± 2.9	837.4 ± 721	35.5 ± 28	28-*O*-sulfohexosyl-5α-ergostane-3β,6,8,15,16β,28-hexaol
91	11.1 ± 10	5.4 ± 2.8	654.6 ± 581	20.4 ± 11.8	26-*O*-sulfohexosyl-27-nor-5α-ergost-22-ene-3β,6,8,15,16β,26-hexaol
92	35.2 ± 23.6	24.6 ± 32.1	924.1 ± 715.2	26.3 ± 9.1	3-*O*-pentosyl-24-*O*-sulfopentosyl-5α-cholestane-3β,6,8,15,24-pentaol
94	16.3 ± 14.1	19.1 ± 22.9	551.8 ± 333.7	23.8 ± 12.6	3-*O*-pentosyl-26-*O*-sulfohexosyl-5α-ergost-22-ene-3β,6,8,15,26-pentaol
99	10.4 ± 6.9	16.7 ± 21.8	582.9 ± 330	26.8 ± 14.5	28-*O*-sulfohexosyl-5α-ergost-20(22)-ene-3β,6,8,15,16β,28-hexaol
108	31.1 ± 18.9	53.7 ± 81.4	1191.2 ± 499.9	58.5 ± 26.7	3-*O*-pentosyl-24-*O*-methylsulfopentosyl-5α-cholestane-3β,6,8,15,24-pentaol
109	255.3 ± 194.8	213.7 ± 276.4	9046.4 ± 7355.2	476.4 ± 512.3	24-*O*-pentosyl-5α-cholest-22-ene-3β,6,8,15,24-pentaol 3-*O*-sulfate
113	68.3 ± 82.6	26.2 ± 17.1	2005.6 ± 1910.1	173.9 ± 243	26-*O*-sulfohexosyl-5α-ergost-22-ene-3β,6,8,15,16β,26-hexaol
115	33.9 ± 14.4	48.1 ± 68.6	1687.4 ± 1086.5	94.9 ± 64.4	24-*O*-sulfohexosyl-5α-cholest-20(22)-ene-3β,6,8,15,24-pentaol
117	21 ± 12.8	12.8 ± 7.8	994.9 ± 636	109 ± 89.2	5α-cholestane-3β,6,8,15α,16β,26-hexaol 6-*O*-sulfate
120	66.3 ± 74.4	24.5 ± 6.8	2870.3 ± 2423.8	122.3 ± 97.6	28-*O*-sulfohexosyl-5α-ergostane-3β,6,8,15,16β,28-hexaol
125	29.5 ± 13	30.4 ± 45.5	1280.1 ± 177.9	269.9 ± 315.4	-
126	63.6 ± 61.2	142 ± 260.4	2519.8 ± 1747.5	105.1 ± 107.1	28-*O*-sulfohexosyl-5α-ergostane-3β,6,8,15,28-pentaol
131	108.1 ± 73.9	124.3 ± 152.8	4377.2 ± 4373.3	194.2 ± 39.6	26-*O*-sulfohexosyl-27-nor-5α-ergost-20(22)-ene-3β,6,8,15,26-pentaol
133	21.1 ± 18.2	4.2 ± 1.4	620.1 ± 555.7	62 ± 59.2	5α-cholestane-3β,6,8,15β,16β,26-hexaol 6-*O*-sulfate
136	274.8 ± 165.5	549.3 ± 866.2	12,619.1 ± 6096.6	785.4 ± 489.5	5α-cholestane-3β,6,8,15,24-pentaol 24-*O*-sulfate
138	82 ± 41.9	185.6 ± 252.1	3811 ± 1281.6	122.3 ± 84.6	26-*O*-sulfohexosyl-27-nor-5α-ergost-20(22)-ene-3β,6,8,15,26-pentaol
146	10.8 ± 10.2	5.6 ± 4.1	584.6 ± 501.6	26.7 ± 25.3	5α-ergost-22-ene-3β,6,7,8,15α,16β,26-heptaol 6-*O*-sulfate
147	25.1 ± 22.7	14.4 ± 8.8	1053.4 ± 813.6	60 ± 48.5	29-*O*-sulfohexosyl-5α-stigmast-20(22)-ene-3β,6,8,15,16β,29-hexaol
148	14.9 ± 8.3	9.1 ± 6.5	777.4 ± 500.7	27.2 ± 19.5	28-*O*-sulfohexosyl-5α-ergost-22-ene-3β,6,8,15,16β,28-hexaol
151	76.2 ± 95.2	41.2 ± 14	3204.2 ± 2986.5	206.6 ± 204.7	28-*O*-sulfopentosyl-5α-ergost-20(22)-ene-3β,6,8,15,16β,28-hexaol
153	82.6 ± 58.9	40.5 ± 18.6	4829.9 ± 3939.8	342.7 ± 214.2	5α-ergost-22-ene-3β,6,8,15,16β,26-hexaol 6-*O*-sulfate
154	248.2 ± 182.6	488.2 ± 798.4	9881.2 ± 6654.5	629.1 ± 380.1	28-*O*-sulfohexosyl-5α-ergost-22-ene-3β,6,8,15,28-pentaol
163	70.5 ± 33.4	105.7 ± 164.8	3006.8 ± 503.1	172.5 ± 49.5	29-*O*-sulfohexosyl-5α-stigmastane-3β,6,8,15,16β,29-hexaol
164	210.1 ± 127.7	256.7 ± 404.1	6606.4 ± 2715.5	565.2 ± 205	26-*O*-sulfohexosyl-5α-ergost-22-ene-3β,6,8,15,26-pentaol
165	130.1 ± 93.8	86.5 ± 38.4	5780.8 ± 4339.3	315.8 ± 242.6	28-*O*-sulfohexosyl-5α-ergost-22-ene-3β,6,8,15,16β,28-hexaol
171	216.3 ± 130.9	419.9 ± 643.2	7547.9 ± 3816.2	478.3 ± 231.9	28-*O*-sulfohexosyl-5α-ergost-22-ene-3β,6,8,15,28-pentaol
172	18.1 ± 10	38.3 ± 54.7	1052 ± 570.4	123.6 ± 116.4	24-*O*-methylsulfopentosyl-5α-cholestane-3β,6,8,15,24-pentaol
179	276.4 ± 232.1	536.3 ± 873.4	10,891.4 ± 6548.5	681.7 ± 327.8	26-*O*-sulfohexosyl-5α-ergost-22-ene-3β,6,8,15,26-pentaol
181	53.5 ± 47.7	21.9 ± 4.9	2603.9 ± 2287.1	130.8 ± 110	29-*O*-sulfohexosyl-5α-stigmastane-3β,6,8,15,16β,29-hexaol
183	25 ± 27.3	10.4 ± 4.5	1081.5 ± 1033.7	67.5 ± 59.6	29-*O*-sulfohexosyl-5α-stigmast-22-ene-3β,6,8,15,16β,29-hexaol
189	24.3 ± 20.3	17.7 ± 11.1	1287.3 ± 919.2	65.8 ± 38.2	29-*O*-sulfohexosyl-5α-stigmast-22-ene-3β,6,8,15,16β,29-hexaol
196	335.5 ± 360.2	827.8 ± 1416.6	14,640.2 ± 14,245.6	1044.1 ± 925.3	24-*O*-sulfopentosyl-5α-cholest-22-ene-3β,6,8,15,24-pentaol

Analysis of the starfish extracts from different body components showed that the pyloric caeca are characterized by maximal level of non-sulfated and sulfated glycosides of polyhydroxysteroids among all samples. Concentrations of four non-sulfated and thirty-four sulfated glycosides were found to be significantly higher in the pyloric caeca than in other organs from the ANOVA analysis (Table 3). Concentrations of other glycosides of this structural group were also higher in the pyloric caeca samples; however, there was a large inter-individual variability. For example, the extract of the pyloric caeca of animal #2 (PC#2) was characterized by a high concentrations of non-sulfated 3-O-Xyl-24-O-Xyl-glycosylated 5α-cholest-pentaol derivatives—fuscaside B (**5**), distolasteroside D_1 (**10**), and distolasteroside D_2 (**14**)—and by low concentrations of 28-O-pentosyl-5α-ergostane-3β,6,8,15,16β,28-hexaol (**7**) and 24-O-hexosyl-5α-cholestane-3β,6,8,15β,24-pentaol (**12**). On the contrary, sample PC#5 had high concentrations of **7** and **12** and a low concentration of **5**, **10**, and **14**. Sample PC#4 had a maximal concentration of 24-O-pentosyl-5α-cholestane-3β,6,8,15α,16β,24-hexaol (**1**), 24-O-hexosyl-5α-cholestane-3β,6,7,8,15α,16β,24-heptaol (**2**), pycnopodioside A (**18**), and desulfated minutoside A (**21**); all of these compounds are 24-O-glycosylated polyhydroxysteroids.

For sulfated glycosides, a similar distribution was observed. While all the glycosides were more concentrated in the pyloric caeca samples, each animal was characterized by its own ratio of these compounds. The PC#2 is characterized by a high content of most of glycosides having pentose unit at C-3 and sulfated monosaccharide unit at the side chain: 3-O-pentosyl-24-O-sulfohexosyl-5α-cholestane-3β,6,8,15,24-pentaol (**46**), 3-O-pentosyl-28-O-sulfohexosyl-5α-ergostane-3β,6,8,15,28-pentaol (**57**), 3-O-pentosyl-26-O-sulfohexosyl-27-nor-5α-ergost-22-ene-3β,6,8,15,26-pentaol (**69**), 3-O-pentosyl-28-O-sulfohexosyl-5α-ergost-22-ene-3β,6,8,15,28-pentaol (**85**), 3-O-pentosyl-24-O-sulfopentosyl-5α-cholestane-3β,6,8,15,24-pentaol (**92**), 3-O-pentosyl-26-O-sulfohexosyl-5α-ergost-22-ene-3β,6,8,15,26-pentaol (**94**), 3-O-pentosyl-28-O-sulfopentosyl-5α-ergost-20(22)-ene-3β,6,8,15,28-pentaol (**114**), and 3-O-pentosyl-24-O-methylsulfopentosyl-5α-cholestane-3β,6,8,15,24-pentaol (**145**). PC#5 had a high concentration of certain pentaol derivatives with monosaccharide units at the side chains: 24-O-hexosyl-5α-cholestane-3β,6,8,15,24-pentaol 3-O-sulfate (**48**), 28-O-sulfohexosyl-5α-ergost-20(22)-ene-3β,6,8,15,28-pentaol (**103**), 28-O-[sulfohexosyl-hexosyl]-5α-ergost-22-ene-3β,6,8,15,28-pentaol (**119**), 28-O-sulfohexosyl-5α-ergostane-3β,6,8,15,28-pentaol (**126**), 24-O-sulfopentosyl-5α-cholestane-3β,6,8,15,24-pentaol (**155**), 29-O-sulfohexosyl-5α-stigmastane- 3β,6,8,15,29-pentaol (**185**), and 28-O-sulfopentosyl-5α-ergost-20(22)-ene-3β,6,8,15,28-pentaol (**188**). PC#5 also had a minimal concentration of glycosylated hexaols among all pyloric caeca samples while samples PC#1 and PC#2 had a higher content of glycosylated hexaols. In addition, sample PC#5 contained minimal concentrations of glycosides of polyhydroxysteroids with stigmastane side chains (except glycoside **185**).

In the stomach, body walls and gonads, non-sulfated glycosides presented in trace amounts. Sulfated glycosides were found in the other organs but only in small quantities. Localization of polyhydroxysteroids and glycosides of polyhydroxysteroids in the *L. fusca* pyloric caeca confirmed data previously obtained by Kicha et al. on the distribution of polar steroids of *P. pectinifera*, demonstrated that polyhydroxysteroids and polyhydroxysteroid glycosides were localized mainly in the stomach and pyloric caeca [13–15]. It was suggested that the location of these compounds in starfish digestive organs along with their definite structural resemblance to bile alcohols of hagfishes and amphibians and their ability to solubilize a lipid suspension connected with roles of these compounds in digestion processes. The observed high inter-individual variability may be associated with the biogenesis of these compounds. It is known that only a part of cholestane steroids is synthesized de novo in the starfish [23,24]. Other cholestane steroids, as well as all steroids with stigmastane and ergostane side chains, are biosynthesized from dietary phytosterols and dietary cholesterol [25]. Therefore, different ratios of polyhydroxysteroids and related glycosides in individual animals may be related to their diet.

3. Materials and Methods

3.1. Chemicals

Water (LC/MS grade) and acetonitrile (UHPLC grade) were purchased from Panreac (Barcelona, Spain), methanol (HPLC grade) was purchased from J.T. Baker (Deventer, the Netherlands). All other chemicals were of analytical grade or equivalent. Lethasterioside A, fuscaside A, and 5α-cholestan-3β,4β,6α,7α,8,15β,16β,26-octaol isolated early from the starfish *L. fusca* were used as standards of polar steroid glycosides. Structures of these compounds were established using different methods including high-resolution NMR [16,17].

3.2. Animal Material

Individuals of *L. fusca* (order Forcipulatida, family Asteriidae) were collected at the coastal area of the Posyet Gulf, the Sea of Japan, in August 2017, from a depth of 1–3 m. Identification of the species was carried out by B.B. Grebnev (G.B. Elyakov Pacific Institute of Bioorganic Chemistry of the Far Eastern Branch of Russian Academy of Sciences (PIBOC FEB RAS), Vladivostok, Russia). All animals were sexually mature and ranged in diameter from 10 to 18 cm; the identification of sex of the animals was not performed. The voucher specimen No. PIBOC-2017-08-LF is preserved in the collection of PIBOC FEB RAS.

3.3. Sample Preparation and Solid-Phase Extraction (SPE)

Five freshly caught animals were rapidly dissected and separated into body walls (BW), gonads (G), stomach (S) and pyloric caeca (PC). The wet weights of the starfish, as well as weights of collected body parts, are listed in Table S1. In addition, the coelomic fluid (CF) of the individuals was obtained by puncturing at the arm tip and collected by gravity into separate cold tubes. BW, G, S, and PC were undergoing the triple extraction with ten folds volumes of ethanol during 10 h. The extracts from different body components as well as CF were filtered and evaporated in vacuo. For removing lipid contaminations, dried samples were dissolved with a solvent combination of chloroform: methanol: water ($CHCl_3$/MeOH/H_2O 8:4:3, *v*/*v*/*v*) to a final dilution 30-fold in relation to the weight of the dried sample. After centrifugation (2000 *g* for 10 min) 100 µL of the upper water-methanolic layer were transferred to another vial and subjected to the second liquid–liquid extraction with a solvent combination of $CHCl_3$/MeOH/H_2O (1000:450:250 µL). The 400 µL of the upper layer were transferred to another vial, evaporated in vacuo, dissolved in 50% methanol in water (*v*/*v*, 150 µL) and were subjected to the solid-phase extraction for desalting. Sorbent of SPE cartridges (Bond Elut C18 Cartridges, 100 mg/1 mL, Agilent Technologies, Santa Clara, CA, USA) was moistened by 3 mL of acetonitrile (ACN) and equilibrated with 3 mL of 0.1% formic acid (FA) in water. The 100 µL sample was loaded into the SPE cartridge by drops. After washing the cartridge with 0.5 mL of 0.1% FA, polar steroid metabolites were eluted with 1.5 mL of 100% ACN, evaporated and dissolved in 200 µL 50% ACN. Samples were centrifuged (15,000 *g* for 10 min) and supernatant was placed in 200 µL glass micro insert (Agilent Technologies, Santa Clara, CA, USA) in an autosampler vial (2 mL) (Agilent Technologies, Santa Clara, CA, USA) for LC/MS.

3.4. LC/MS Analysis

All samples were subjected to nLC/CSI–QTOF–MS analysis using an UltiMate 3000 RSLCnano System (Dionex, Sunnyvale, CA, USA) connected to a Bruker Impact II Q-TOF mass spectrometer (Bruker Daltonics, Bremen, Germany) equipped with a CaptiveSpray ionization source (Bruker Daltonics, Bremen, Germany). Analysis conditions were similar to what was described previously [18]. Acclaim PepMap RSLC column (75 µm × 150 mm, nanoViper, C18, 2 µm, 100 A; Thermo Scientific, City, US State abbrev. if applicable, Country) with a cartridge-based trap column µ-Precolumn (300 µm × 5 mm, C18, 5 µm, 100 A; Thermo Scientific) was applied for separation of polar steroids at 40 °C, and the injection volume was 0.2 µL. Chromatographic separation was done with water containing 0.1% FA as solvent A and ACN containing 0.1% FA as solvent B using the following gradient profile: 0–5 min 34% B; 5–20 min ramping to 58% B; 20–70 min ramping to 80% B; 70–71 min ramping to 99% B; 71–80 min hold at 99% B; 80–81 min return to 34% B; and equilibration at 34% B during 15 min; flow rate of 400 nl/min. The mass spectrometer was operated in a negative ion mode using the following parameters: mass range 100–2000 *m*/*z*; capillary voltage, 1300 V; the drying gas temperature, 150 °C; the drying gas flow rate, 3 l/min. ESI-L Low Concentration Tuning Mix (Agilent Technologies, Santa Clara, CA, USA) and hexakis(1H,1H,3H-tetrafluoropropoxy)phosphazine (966.0007 *m*/*z* in negative mode; Agilent Technologies, Santa Clara, CA, USA) were used for calibration and lock-mass calibration, respectively. The otofControl (ver. 4.0, Bruker Daltonics, Bremen, Germany) and DataAnalysis Software (ver. 4.3, Bruker Daltonics, Bremen, Germany) were using for obtaining and analyzing of MS data.

A series of four consecutive injections of a pooled sample of *L. fusca* were run prior to the sample analysis to conditions of the chromatographic column. The analysis of replicate injections showed little retention time shift following this conditioning procedure. Samples were analyzed in random order. Quality control samples (pooled samples of starfish) and blank samples (50% ACN) were analyzed after every three samples throughout the batch analysis in order to check the performance of data acquisition and to ensure reproducibility. Method blanks demonstrated no carry-over between sample runs.

The raw spectra were converted to mzML files by the open-source msConvert tool of the ProteoWizard library [26]. Data preprocessing was performed using MZmine 2.40 [27]. Applied parameters are given as supplementary material (Table S2). The peak detection batch step was carried out via a targeted peak detection module using the peak list file containing retention times and *m*/*z* values of previously detected steroid metabolites of *L. fusca* [18] (Table S3). Sulfated steroid metabolites were detected as $[M - Na]^-$ ions, non-sulfated polyhydroxysteroids and related glycosides were detected as $[M - H]^-$ and $[M + FA]^-$ ions. Resulted data (*m*/*z* values, retention times and peak area) were exported into a csv file. Compound identification was performed by comparing their retention time, elemental composition and MS/MS spectrum with those data obtained previously [18]. A mass accuracy tolerance between the measured mass and the theoretical mass calculated from the molecular formula did not exceed 3 ppm. Numbers of detected compounds were assigned according to compound numbers from our previous work (Table S3).

3.5. Semi-Quantitative Analysis of Detected Polar Steroid Compounds

For semi-quantitative analysis, we used lethasterioside A as a reference standard for asterosaponins (R^2 = 0.9922), fuscaside A as a reference standard for sulfated polyhydroxysteroid glycosides and sulfated polyhydroxysteroids (R^2 = 0.9939) and 5α-cholestan-3β,4β,6α,7α,8,15β,16β,26-octaol as a reference standard for non-sulfated polyhydroxysteroid compounds and non-sulfated polyhydroxysteroid glycosides (R^2 = 0.9888). Standard solutions at concentrations of 0.1, 1.0, 2.5, and 5.0 μg/mL were used for building calibration curves (Figures S2–S4). All experiments were carried out at least three times, LC/MS conditions are identical to those described above. As a result, the amounts of detected compounds were calculated through calibration curves. Results are shown in Table S4 as concentration in ng/g wet weight of animal organs.

Statistical analysis (ANOVA followed by Tukey's HSD (honestly significant difference) test of multiple comparisons ($\alpha = 0.05$)) was performed using Metaboanalyst 4.0 (www.metaboanalyst.ca, free software updated and maintained by Xia Lab at McGill University [28]). Data were auto-scaled prior to ANOVA tests. To avoid false-positive results, multiple comparisons were compensated for using false discovery rate (FDR) calculations [29], and FDRs were estimated using the *q*-value method [30]. A q-value < 0.05 was considered statistically significant.

4. Conclusions

The distribution of polar steroid compounds in different body components of the Far Eastern starfish *L. fusca* was investigated using a modern nLC/CSI–QTOF–MS technique. Comparison of the sum of polar steroid content and individual asterosaponins, polyhydroxysteroids and related glycosides from the body walls, coelomic fluid, gonads, stomach, and pyloric caeca was performed. It was shown that the distribution of individual asterosaponins, polyhydroxysteroids, and polyhydroxysteroid glycosides is qualitatively and quantitatively different. The toxic asterosaponins were found in all organs of the starfish. The maximal concentration of the sum of all polar steroids was observed in the stomach and most of them were asterosaponins and native aglycones of asterosaponins. The comparison of the content of individual asterosaponins in different organs of the starfish probably suggests different biological roles of these metabolites in the starfishes. This may be due to the toxic, protective or antimicrobial properties of these compounds. The main part of glycosides of polyhydroxysteroids was located in the pyloric caeca and this confirmed the digestive function of these steroids in starfishes. At the same time, the levels of these steroids can vary greatly depending on the individual. This observed high inter-individual variability may be associated with different physiological statuses of the animals and partly with the biogenesis of some these compounds from dietary steroids.

Supplementary Materials: The following are available online at http://www.mdpi.com/1660-3397/17/9/523/s1, Table S1: Description of studied individuals of *L. fusca*, Figure S1: Typical nLC/CSI–QTOF–MS base-peak chromatograms of purified fractions of polar steroid compounds of different body components of the starfish *L. fusca* in negative ion mode: (a) body walls, sample BW#5; (b) coelomic fluid, sample CF#5; (c) gonads, sample G#5; (d) stomach, sample S#1; (e) pyloric caeca, sample PC#1, Table S2: Batch steps and parameters used for data preprocessing in MZmine, Table S3: Polar steroids of the starfish *L. fusca* detected by nLC/CSI–QTOF–MS, Figure S2. Calibration curve for fuscaside A (ion $[M - Na]^-$ at m/z 795.38), Figure S3: Calibration curve for lethasterioside A (ion $[M - Na]^-$ at m/z 1227.54), Figure S4: Calibration curve for 5α-cholestan-3β,4β,6α,7α,8,15β,16β,26-octaol (for the sum of ion intensities $[M - H]^-$ at m/z 499.33 and $[M + FA]^-$ at m/z 545.3), Table S4: Content of detected compounds in different organs and coelomic fluid of the starfish *L. fusca* (ng/g wet weight of the organs for BW, G, PC, and S and ng/mL for CF) and the result of statistical analysis (ANOVA followed by Tukey HSD test of multiple comparisons was performed for BW, G, PC, and S groups; q-value < 0.05 was considered statistically significant).

Author Contributions: Investigation, R.S.P., N.V.I., A.A.K., T.V.M., and B.B.G.; writing—original draft preparation, R.S.P. and N.V.I.; writing—review and editing V.A.S. and P.S.D.

Funding: This research was funded by Russian Foundation for Basic Research (RFBR), Grant No. 17-04-00034.

Acknowledgments: The study was carried out with the equipment of the Collective Facilities Center "The Far Eastern Center for Structural Molecular Research (NMR/MS) PIBOC FEB RAS".

Conflicts of Interest: The authors declare no conflict of interest.

References

1. Minale, L.; Riccio, R.; Zollo, F. Steroidal Oligoglycosides and Polyhydroxysteroids from Echinoderms. *Fortschr. Chem. Org. Naturst.* **1993**, *62*, 75–308.
2. Stonik, V.A. Marine polar steroids. *Russ. Chem. Rev.* **2001**, *70*, 673–715. [CrossRef]
3. Stonik, V.A.; Ivanchina, N.V.; Kicha, A.A. New polar steroids from starfish. *Nat. Prod. Commun.* **2008**, *3*, 1587–1610. [CrossRef]
4. Ivanchina, N.V.; Kicha, A.A.; Stonik, V.A. Steroid glycosides from marine organisms. *Steroids* **2011**, *76*, 425–454. [CrossRef]
5. Dong, G.; Xu, T.H.; Yang, B.; Lin, X.P.; Zhou, X.F.; Yang, X.W.; Liu, Y.H. Chemical Constituents and Bioactivities of Starfish. *Chem. Biodivers.* **2011**, *8*, 740–791. [CrossRef]
6. Ivanchina, N.V.; Kicha, A.A.; Malyarenko, T.V.; Stonik, V.A. Recent studies of polar steroids from starfish: Structures, biological activities and biosynthesis. In *Advances in Natural Products Discovery*; Gomes, R., Rocha-Santos, T., Duarte, A., Eds.; Nova Sci.: New York, NY, USA, 2017; pp. 191–224.
7. Yasumoto, T.; Tanaka, M.; Hashimoto, Y. Distribution of Saponin in Echinoderms. *Bull. Jpn. Soc. Sci. Fish* **1966**, *32*, 673–676. [CrossRef]
8. Mackie, A.M.; Singh, H.T.; Owen, J.M. Studies on the distribution, biosynthesis and function of steroidal saponins in echinoderms. *Comp. Biochem. Physiol. Part B Comp. Biochem.* **1977**, *56*, 9–14. [CrossRef]
9. Voogt, P.A.; van Rheenen, J.W.A. Carbohydrate content and composition of asterosaponins from different organs of the sea star *Asterias rubens*: Relation to their haemolytic activity and implications for their biosynthesis. *Comp. Biochem. Physiol. Part B Comp. Biochem.* **1982**, *72*, 683–688. [CrossRef]

10. Garneau, F.X.; Harvey, C.; Simard, J.L.; Apsimon, J.W.; Burnell, D.J.; Himmelman, J.H. The distribution of asterosaponins in various body components of the starfish *Leptasterias polaris*. *Comp. Biochem. Physiol. Part B Comp. Biochem.* **1989**, *92*, 411–416. [CrossRef]
11. Demeyer, M.; De Winter, J.; Caulier, G.; Eeckhaut, I.; Flammang, P.; Gerbaux, P. Molecular diversity and body distribution of saponins in the sea star *Asterias rubens* by mass spectrometry. *Comp. Biochem. Physiol. Part B Biochem. Mol. Biol.* **2014**, *168*, 1–11. [CrossRef]
12. Demeyer, M.; Wisztorski, M.; Decroo, C.; De Winter, J.; Caulier, G.; Hennebert, E.; Eeckhaut, I.; Fournier, I.; Flammang, P.; Gerbaux, P. Inter- and intra-organ spatial distributions of sea star saponins by MALDI imaging. *Anal. Bioanal. Chem.* **2015**, *407*, 8813–8824. [CrossRef]
13. Kicha, A.A.; Ivanchina, N.V.; Gorshkova, I.A.; Ponomarenko, L.P.; Likhatskaya, G.N.; Stonik, V.A. The distribution of free sterols, polyhydroxysteroids and steroid glycosides in various body components of the starfish *Patiria (=Asterina) pectinifera*. *Comp. Biochem. Physiol. Part B Biochem. Mol. Biol.* **2001**, *128*, 43–52. [CrossRef]
14. Kicha, A.A.; Ivanchina, N.V.; Stonik, V.A. Seasonal variations in the levels of polyhydroxysteroids and related glycosides in the digestive tissues of the starfish *Patiria (Asterina) pectinifera*. *Comp. Biochem. Physiol. Part B Biochem. Mol. Biol.* **2003**, *136*, 897–903. [CrossRef]
15. Kicha, A.A.; Ivanchina, N.V.; Stonik, V.A. Seasonal variations in polyhydroxysteroids and related glycosides from digestive tissues of the starfish *Patiria (=Asterina) pectinifera*. *Comp. Biochem. Physiol. Part B Biochem. Mol. Biol.* **2004**, *139*, 581–585. [CrossRef]
16. Ivanchina, N.V.; Malyarenko, T.V.; Kicha, A.A.; Kalinovskii, A.I.; Dmitrenok, P.S. Polar steroidal compounds from the Far-Eastern starfish *Lethasterias fusca*. *Russ. Chem. Bull.* **2008**, *57*, 204–208. [CrossRef]
17. Ivanchina, N.V.; Kalinovsky, A.I.; Kicha, A.A.; Malyarenko, T.V.; Dmitrenok, P.S.; Ermakova, S.P.; Stonik, V.A. Two new asterosaponins from the Far Eastern starfish *Lethasterias fusca*. *Nat. Prod. Commun.* **2012**, *7*, 853–858. [CrossRef]
18. Popov, R.S.; Ivanchina, N.V.; Kicha, A.A.; Malyarenko, T.V.; Dmitrenok, P.S. Structural characterization of polar steroid compounds of the Far Eastern starfish *Lethasterias fusca* by nanoflow liquid chromatography coupled to quadrupole time-of-flight tandem mass spectrometry. *J. Am. Soc. Mass Spectrom.* **2019**, *30*, 743–764. [CrossRef]
19. Harvey, C.; Garneau, F.-X.; Himmelman, J. Chemodetection of the predatory sea-star *Leptasterias polaris* by the whelk *Buccinum undatum*. *Mar. Ecol. Prog. Ser.* **1987**, *40*, 79–86. [CrossRef]
20. Van Dyck, S.; Flammang, P.; Meriaux, C.; Bonnel, D.; Salzet, M.; Fournier, I.; Wisztorski, M. Localization of secondary metabolites in marine invertebrates: Contribution of MALDI MSI for the study of saponins in cuvierian tubules of *H. forskali*. *PLoS ONE* **2010**, *5*, e13923. [CrossRef]
21. Popov, R.S.; Ivanchina, N.V.; Silchenko, A.S.; Avilov, S.A.; Kalinin, V.I.; Dolmatov, I.Y.; Stonik, V.A.; Dmitrenok, P.S. Metabolite profiling of triterpene glycosides of the Far Eastern sea cucumber *Eupentacta fraudatrix* and their distribution in various body components using LC-ESI QTOF-MS. *Mar. Drugs* **2017**, *15*, 302. [CrossRef]
22. Van Dyck, S.; Gerbaux, P.; Flammang, P. Qualitative and quantitative saponin contents in five sea cucumbers from the Indian Ocean. *Mar. Drugs* **2010**, *8*, 173–189. [CrossRef]
23. Goad, L.J. Sterol biosynthesis and metabolism in marine invertebrates. *Pure Appl. Chem.* **1981**, *53*, 837–852. [CrossRef]
24. Goad, L.J. The sterols of marine invertebrates: Composition, biosynthesis, and metabolites. In *Marine Natural Products. Chemical and Biological Perspectives*; Scheuer, P.J., Ed.; Academic Press: New York, NY, USA, 1978; pp. 76–173.
25. Ivanchina, N.V.; Kicha, A.A.; Malyarenko, T.V.; Kalinovsky, A.I.; Dmitrenok, P.S.; Stonik, V.A. Biosynthesis of polar steroids from the Far Eastern starfish *Patiria (=Asterina) pectinifera*. Cholesterol and cholesterol sulfate are converted into polyhydroxylated sterols and monoglycoside asterosaponin P1 in feeding experiments. *Steroids* **2013**, *78*, 1183–1191. [CrossRef]
26. Chambers, M.C.; MacLean, B.; Burke, R.; Amodei, D.; Ruderman, D.L.; Neumann, S.; Gatto, L.; Fischer, B.; Pratt, B.; Egertson, J.; et al. A cross-platform toolkit for mass spectrometry and proteomics. *Nat. Biotechnol.* **2012**, *30*, 918–920. [CrossRef]

27. Pluskal, T.; Castillo, S.; Villar-Briones, A.; Orešič, M. MZmine 2: Modular framework for processing, visualizing, and analyzing mass spectrometry-based molecular profile data. *BMC Bioinform.* **2010**, *11*, 395. [CrossRef]
28. Chong, J.; Soufan, O.; Li, C.; Caraus, I.; Li, S.; Bourque, G.; Wishart, D.S.; Xia, J. MetaboAnalyst 4.0: Towards more transparent and integrative metabolomics analysis. *Nucleic Acids Res.* **2018**, *46*, W486–W494. [CrossRef]
29. Benjamini, Y.; Hochberg, Y. Controlling the false discovery rate: A practical and powerful approach to multiple testing. *J. R. Stat. Soc. Ser. B* **1995**, *57*, 289–300. [CrossRef]
30. Storey, J.D.; Tibshirani, R. Statistical significance for genomewide studies. *Proc. Natl. Acad. Sci. USA* **2003**, *100*, 9440–9445. [CrossRef]

marine drugs

Article

Structures and Bioactivities of Six New Triterpene Glycosides, Psolusosides E, F, G, H, H_1, and I and the Corrected Structure of Psolusoside B from the Sea Cucumber *Psolus fabricii*

Alexandra S. Silchenko, Anatoly I. Kalinovsky, Sergey A. Avilov, Vladimir I. Kalinin *, Pelageya V. Andrijaschenko, Pavel S. Dmitrenok, Roman S. Popov, Ekaterina A. Chingizova, Svetlana P. Ermakova and Olesya S. Malyarenko

G.B. Elyakov Pacific Institute of Bioorganic Chemistry, Far Eastern Branch of the Russian Academy of Sciences, Pr. 100-letya Vladivostoka 159, Vladivostok 690022, Russia; sialexandra@mail.ru (A.S.S.); kaaniv@pidoc.dvo.ru (A.I.K.); avilov-1957@mail.ru (S.A.A.); pandryashchenko@mail.ru (P.V.A.); paveldmt@piboc.dvo.ru (P.S.D.); rs.popov@outlook.com (R.S.P.); martyyas@mail.ru (E.A.C.); svetlana_ermakova@hotmail.com (S.P.E.); malyarenko.os@gmail.com (O.S.M.)

* Correspondence: kalininv@piboc.dvo.ru; Tel./Fax: +7(423)2-31-40-50

Received: 27 May 2019; Accepted: 11 June 2019; Published: 14 June 2019

Abstract: Seven sulfated triterpene glycosides, psolusosides B (**1**), E (**2**), F (**3**), G (**4**), H (**5**), H_1 (**6**), and I (**7**), along with earlier known psolusoside A and colochiroside D have been isolated from the sea cucumber *Psolus fabricii* collected in the Sea of Okhotsk. Herein, the structure of psolusoside B (**1**), elucidated by us in 1989 as a monosulfated tetraoside, has been revised with application of modern NMR and particularly MS data and proved to be a disulfated tetraoside. The structures of other glycosides were elucidated by 2D NMR spectroscopy and HR-ESI mass-spectrometry. Psolusosides E (**2**), F (**3**), and G (**4**) contain holostane aglycones identical to each other and differ in their sugar compositions and the quantity and position of sulfate groups in linear tetrasaccharide carbohydrate moieties. Psolusosides H (**5**) and H_1 (**6**) are characterized by an unusual sulfated trisaccharide carbohydrate moiety with the glucose as the second sugar unit. Psolusoside I (**7**) has an unprecedented branched tetrasaccharide disulfated carbohydrate moiety with the xylose unit in the second position of the chain. The cytotoxic activities of the compounds **2**–**7** against several mouse cell lines—ascite form of Ehrlich carcinoma, neuroblastoma Neuro 2A, normal epithelial JB-6 cells, and erythrocytes—were quite different, at that hemolytic effects of the tested compounds were higher than their cytotoxicity against other cells, especially against the ascites of Ehrlich carcinoma. Interestingly, psolusoside G (**4**) was not cytotoxic against normal JB-6 cells but demonstrated high activity against Neuro 2A cells. The cytotoxic activity against human colorectal adenocarcinoma HT-29 cells and the influence on the colony formation and growth of HT-29 cells of compounds **1**–**3**, **5**–**7** and psolusoside A was checked. The highest inhibitory activities were demonstrated by psolusosides E (**2**) and F (**3**).

Keywords: *Psolus fabricii*; triterpene glycosides; psolusosides; sea cucumber; cytotoxic activity

1. Introduction

The sea cucumbers triterpene glycosides are long-time investigated natural compounds characterized by significant structural diversity, exhibiting a broad spectrum of biological activity [1–9]. Some of them are under study as marine drugs.

The investigation of a complicated glycoside composition of the sea cucumber *Psolus fabricii* (Psolidae, Dendrochirotida) was started in the 1980s of XX century. Only two main compounds, psolusosides A [10,11] and B [12,13], had been isolated in that time. Recently, we have recommenced the studies on the glycosides of *P. fabricii* that resulted in the isolation of eight new hexaosides, psolusosides C_1–C_3 and D_1–D_5, as well as five previously known compounds [14,15]. Herein, we report the isolation and structural elucidation of six new glycosides, psolusosides E (**2**), F (**3**), G (**4**), H (**5**), H_1 (**6**), and I (**7**), as well as an earlier known psolusoside B (**1**), whose structure has been revised based on the modern NMR and HR MS techniques. Earlier known glycosides, psolusoside A and colochiroside D, were also isolated and identified. The structures of the glycosides were established based on ^{1}H, ^{13}C NMR, and 1D TOCSY spectra and 2D NMR (^{1}H,^{1}H-COSY, HMBC, HSQC, ROESY) and confirmed by HR-ESI mass spectrometry. The hemolytic activities against mouse erythrocytes and cytotoxic activities against mouse Ehrlich carcinoma cells (ascite form), neuroblastoma Neuro 2A cells and normal epithelial JB-6 cells of **2**–**7** have been studied. Psolusoside I (**7**) demonstrated moderate hemolytic activity when compounds **2**–**6** were highly hemolytic, but none of them, with the exception of known psolusoside A, which was used as control, were not cytotoxic against mouse Ehrlich carcinoma cells. Psolusoside G (**4**) was not cytotoxic against normal JB-6 cells but demonstrated high activity against Neuro 2A cells. Psolusosides E (**2**) and F (**3**), with the holostane aglycones and linear tetrasaccharide monosulfated sugar chains, demonstrated the highest in the series of tested compounds inhibitory activity on the colony formation and growth of H-29 cells.

2. Results and Discussion

2.1. Structural Elucidation of the Glycosides

The concentrated ethanolic extract of *P. fabricii* was re-extracted with $CHCl_3$/MeOH, concentrated, and delipidized with EtOAc/H_2O. The water layer was chromatographed on a Polychrom-1 (powdered Teflon, Biolar, Latvia) in 50% EtOH and on Si gel columns using $CHCl_3$/EtOH/H_2O (100:75:10), (100:100:17) and (100:125:25) as mobile phases to give fractions I–VIII. The obtained fractions III–VIII were subjected to HPLC on reversed-phase or silica-based columns to give psolusosides: B (**1**) (67 mg), E (**2**) (10 mg), F (**3**) (1.4 mg), G (**4**) (46.5 mg), H (**5**) (1.4 mg), H_1 (**6**) (1.4 mg), and I (**7**) (1.1 mg) (Figure 1) as well as two known earlier compounds, psolusoside A (36.5 mg) found earlier in this species of sea cucumbers [10,11] and colochiroside D (2.5 mg) isolated first from *Colochirus robustus* [16]. The known compounds were identified by comparison of their ^{1}H and ^{13}C NMR spectra with those reported for psolusoside A (3β-*O*-[6-*O*-sodium sulfate-3-*O*-methyl-β-D-glucopyranosyl-(1→3)-6-*O*-sodium-sulfate-β-D-glucopyranosyl-(1→4)-β-D-quinovopyranosyl-(1→2)-β-D-xylopyranosyl]-16-ketoholosta-9(11),25-diene) and colochiroside D (3β-*O*-[3-*O*-methyl-β-D-glucopyranosyl-(1→3)-6-*O*-sodium-sulfate-β-D-glucopyranosyl-(1→4)-β-D-glucopyranosyl-(1→2)-β-D-xylopyranosyl]-16-ketoholosta-9(11),25-diene).

The structure of psolusoside B assigned earlier [12,13] was shown to be monosulfated branched tetraoside with non-holostane aglycone, namely 3β-*O*-{β-D-glucopyranosyl-(1→4)-β-D-glucopyranosyl-(1→2)-[6-*O*-sodium-sulfate-β-D-glucopyranosyl-(1→4)]-β-D-xylopyranosyl}-9βH,20(*S*)-acetoxylanosta-7,25-diene-18(16)-lactone.

However, the reinvestigation has shown that this glycoside has two sulfate groups instead of the one reported earlier. In fact, the more accurate molecular formula of psolusoside B (**1**) was determined to be $C_{55}H_{84}O_{30}S_2Na_2$ from the $[M_{2Na} + Na]^+$ ion peak at *m/z* 1357.4169 (calc. 1357.4176) and $[M_{2Na} + 2Na]^{2+}$ at *m/z* 690.2039 (calc. 690.2034) in the (+)HR-ESI-MS and indicated the presence of two sulfate groups in **1**. The comparison of 1D and 2D NMR spectra of the aglycone part of psolusoside B (**1**) (Table 1, Figures S1–S8) with those of the aglycone part of colochiroside E, isolated from the sea cucumber *Colochirus robustus* [17], has confirmed their identity with the aglycone of psolusoside B elucidated earlier [12]. Thus, psolusoside B (**1**), isolated by us, actually contains non-holostane aglycone with 18(16)-lactone and *O*-acetic group at C-20, which was described earlier as onekotanogenin.

Table 1. ^{13}C and ^{1}H NMR chemical shifts and HMBC and ROESY correlations of aglycone moiety of psolusoside B (**1**). [a] Recorded at 176.04 MHz in C_5D_5N/D_2O (4/1). [b] Recorded at 700.13 MHz in C_5D_5N/D_2O (4/1).

Position	δ_C mult. [a]	δ_H mult. [b] (J in Hz)	HMBC	ROESY
1	35.6 CH_2	1.41 m		
		1.36 m		
2	26.7 CH_2	1.96 m		
		1.78 m		H-19, H-30
3	89.3 CH	3.14 (dd, 3.9; 11.8)	C: 4, 30, 31, C-1 Xyl1	H-1, H-5, H-31, H-1 Xyl1
4	39.2 C			
5	47.6 CH	0.84 (dd, 3.8; 11.8)	C: 4, 10, 19, 30, 31	H-3, H-31
6	23.1 CH_2	1.87 m	C: 5, 10	H-31
		1.75 m		
7	122.8 CH	5.56 (brd, 6.8)	C: 6, 9	H-15, H-32
8	147.0 C			
9	45.9 CH	2.97 (brd, 13.9)		H-19
10	35.4 C			
11	21.9 CH_2	1.99 m		
		1.47 m		
12	20.0 CH_2	2.33 (d, 12.9)	C: 13, 14, 18	
		2.02 m		
13	54.9 C			
14	45.6 C			
15	44.2 CH_2	2.10 m	C: 8, 16, 17	H-7
		2.07 m	C: 14, 32	
16	79.7 CH	4.93 brs	C: 13, 14, 18	H-21, H-22, H-23
17	60.5 CH	3.05 s	C: 13, 14, 18, 20, 21, 22	H-15, H-21, H-22, H-23
18	182.3 C			
19	23.8 CH_3	0.88 s	C: 1, 5, 9, 10	H-2, H-6, H-9, H-30
20	84.1 C			
21	23.6 CH_3	1.62 s	C: 17, 20, 22	H-16, H-17
22	37.7 CH_2	2.23 (dt, 4.5; 13.2)		
		1.82 m	C: 17, 21, 23	H-16
23	21.7 CH_2	1.47 m	C: 22, 24, 25	
24	37.7 CH_2	1.97 (dd, 6.9; 13.1)	C: 22, 23, 25, 26	H-26
25	145.4 C			
26	110.7 CH_2	4.73 brs	C: 24, 25, 27	
27	22.1 CH_3	1.65 s	C: 24, 25, 26	H-26
30	17.1 CH_3	0.98 s	C: 3, 4, 5, 31	H-2, H-6, H-19
31	28.5 CH_3	1.12 s	C: 3, 4, 5, 30	H-3, H-5, H-6, H-1 Xyl1
32	34.2 CH_3	1.39 s	C: 8, 13, 14, 15	H-7, H-15, H-17
OAc	170.9 C			
	21.6 CH_3	2.06 s	OAc	

In the ^{1}H and ^{13}C NMR spectra (Table 2, Figures S1 and S2) of the carbohydrate part of **1**, four characteristic doublets at δ (H) 4.56–5.11 (J = 7.3–7.9 Hz) and, corresponding to them, four signals of anomeric carbons at δ(C) 100.9–104.8 (Figure S4) were indicative of a tetrasaccharide chain and *β*-configurations of glycosidic bonds. The ^{1}H,^{1}H-COSY and 1D TOCSY spectra of **1** showed the signals of the isolated spin systems assigned to one xylose and three glucose residues (Figures S3, S7 and S8). The positions of interglycosidic linkages were established by the ROESY and HMBC spectra of **1** (Table 2, Figures S5 and S6) where the cross-peaks between H(1) of the xylose and H(3) (C(3)) of an aglycone, H(1) of the second residue (glucose) and H(2) (C(2)) of the xylose, H(1) of the third residue (glucose) and H(4) (C(4)) of the second residue (glucose), and H(1) of the fourth residue (glucose) and H-4 (C(4)) of the first residue (xylose), were observed. These data indicated the same architecture (tetrasaccharide branched chain) and monosaccharide composition of sugar chain of **1** as it has been reported earlier [13]. Thorough analysis of the NMR spectra of **1** showed the glucose residue (the third sugar unit, in which signals were deduced by ^{1}H,^{1}H-COSY, and confirmed by 1D TOCSY) attached to C(4) of the second sugar unit (glucose) was sulfated by C(6) due to α- and β-shifting effects observed in the ^{13}C NMR spectrum. Really, the signal of C(6) was observed at δ(C) 67.5 and the signal of C(5) at δ(C) 75.5. Hence, these signals were shifted in comparison with corresponding signals ($\delta C_{(6)}$ at 62.1, $\delta C_{(5)}$ at 77.8) in non-sulfated glucose residue in the same position of carbohydrate chains of psolusosides, belonging to the group D [15]. The analogous shifting effects were observed for the signals C(2) and C(1) of the glucose occupying the fourth position of the carbohydrate chain of **1** and attached to C(4) of the xylose unit. The signal of C(2) of this residue was observed at δ 80.6 due to the attachment of a sulfate group to this position while the signal of C(1) was shifted upfield to δ(C) 100.9 due to β-effect of a sulfate group. Moreover, the signals in the ^{13}C NMR spectrum of **1** assigning to this glucose residue were coincident with the corresponding signals of glucose residue sulfated by C-2 and attached to C-4 of the first xylose unit in the spectrum of colochiroside E [17] corroborating the unusual position of one of sulfate groups in psolusoside B (**1**). So, the both spectroscopic methods—HR-ESI-MS and NMR—confirmed the presence of two sulfate groups in the carbohydrate chain of psolusoside B (**1**).

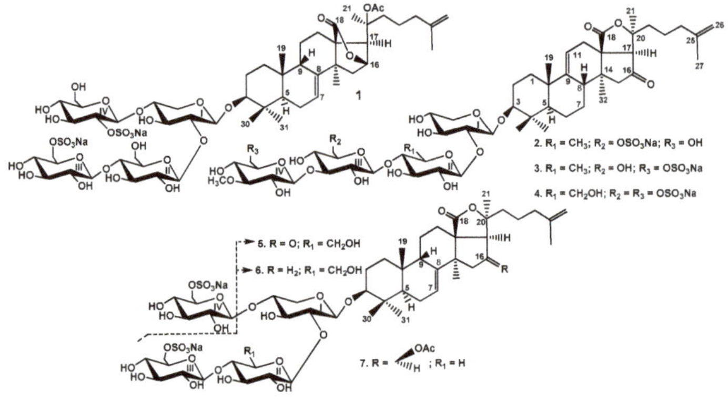

Figure 1. Chemical structure of the glycosides isolated from *Psolus fabricii:* **1**—psolusoside B; **2**—psolusoside E; **3**—psolusoside F; **4**—psolusoside G; **5**—psolusoside H; **6**—psolusoside H_1; **7**—psolusoside I.

The structure of **1** was also confirmed by the (+)ESI-MS/MS of the $[M_{2Na} + Na]^+$ ion at *m/z* 1357.4, in which the peaks of fragment ions were observed at *m/z* 1297.4 $[M_{2Na} + Na - CH_3COOH]^+$, 1237.4 $[M_{2Na} + Na - NaHSO_4]^+$, 1177.4 $[M_{2Na} + Na - CH_3COOH - NaHSO_4]^+$, 1117.4 $[M_{2Na} + Na - 2NaHSO_4]^+$, 1075.5 $[M_{2Na} + Na - C_6H_{10}O_9SNa\ (GlcSO_3Na)]^+$, 913.4 $[M_{2Na} + Na - GlcSO_3Na - Glc]^+$, 863.1 $[M_{2Na} + Na - C_{32}H_{47}O_4\ (Agl) + H]^+$, 743.1 $[M_{2Na} + Na - C_{32}H_{47}O_4\ (Agl) - NaHSO_4]^+$, 581.1 $[M_{2Na} + Na - C_{32}H_{47}O_4\ (Agl) - C_6H_{10}O_9SNa\ (GlcSO_3Na)]^+$, 449.0 $[M_{2Na} + Na - C_{32}H_{47}O_4\ (Agl) - C_6H_{10}O_9SNa\ (GlcSO_3Na) - Xyl]^+$, 287.0 $[M_{2Na} + Na - C_{32}H_{47}O_4\ (Agl) - C_6H_{10}O_9SNa\ (GlcSO_3Na) - Xyl - Glc]^+$.

Based on these results, the structure of psolusoside B (**1**) was determined as 3β-*O*-{6-*O*-sodium sulfate-β-D-glucopyranosyl-(1→4)-β-D-glucopyranosyl-(1→2)-[2-*O*-sodium sulfate-β-D-glucopyranos yl-(1→4)]-β-D-xylopyranosyl}-9βH,20(*S*)-acetoxylanosta-7,25-diene-18(16)-lactone.

Colochiroside E [17], having trisaccharide sugar chain with terminal (glucose) residue sulfated by C(2), differed from psolusoside B (**1**) only by the lack of a terminal glucose residue attached to C(4) of the glucose (the second unit in the chain). This fact indicates the biogenetic interconnection of these compounds: colochiroside E seems to be a biosynthetic precursor of psolusoside B (**1**) that, additionally, corroborates the new structure of **1**. The incorrect structure elucidation of psolusoside B in 1989 [13] could be explained by an ambiguity of interpretation of the ^{13}C NMR signals. The use of FAB-MS for the molecular formula calculation obviously resulted in the desulfation of the glycoside during the spectrum registration.

The ^{1}H and ^{13}C NMR spectra of aglycone parts of psolusosides E (**2**), F (**3**), and G (**4**) were coincident to each other showing the identity of the aglycones in these glycosides. In the aglycone part of the ^{13}C NMR spectra of **2–4,** the signals characteristic of 18(20)-lactone (δ(C) 175.8 C(18) and 82.9 (C(20)), 9(11)- (δ(C) 151.2 C(9) and 110.9 C(11)), and 25(26)-double bonds (δ(C) 145.4 C(25) and 110.3 C(26)), as well as the signal of C-16 keto-group (δ(C) 212.9) were observed (Table 3, Figures S9, S17 and S25). Based on the analysis of the NMR spectra, the aglycone of compounds **2–4** was identified as earlier known 16-ketoholosta-9(11),25-dien-3β-ol, found first in holotoxins A_1 and B_1 from the sea cucumbers belonging to the family Stichopodidae [18], and frequently occurred in the glycosides of different sea cucumber taxa.

Table 2. ^{13}C and ^{1}H NMR chemical shifts and HMBC and ROESY correlations of carbohydrate moiety of psolusoside B (**1**). [a] Recorded at 176.04 MHz in C_5D_5N/D_2O (4/1). [b] Bold is interglycosidic positions. [c] Italic is sulphate position. [d] Recorded at 700.13 MHz in C_5D_5N/D_2O (4/1). Multiplicity by 1D TOCSY.

Atom	δ_C mult. [a,b,c]	δ_H mult. [d] (J in Hz)	HMBC	ROESY
Xyl1 (1→C-3)				
1	104.8 CH	4.56 (d, 7.3)	C: 3; C: 5 Xyl1	H-3; H-3, 5 Xyl1
2	**81.0** CH	4.01 (t, 8.0)	C: 1 Glc2; C: 1, 3 Xyl1	H-1 Glc2
3	75.2 CH	4.20 (t, 8.8)	C: 2, 4 Xyl1	H-1 Xyl1
4	**78.6** CH	4.08 (dt, 5.6; 9.6)	C: 1 Clc4; C: 5 Xyl1	H-1 Glc4
5	63.6 CH_2	4.43 (dd, 5.2; 12.1)	C: 1, 3, 4 Xyl1	
		3.73 (brt, 11.3)	C: 1 Xyl1	H-1 Xyl1
Glc2 (1→2Xyl1)				
1	104.1 CH	5.11 (d, 7.8)	C: 2 Xyl1; C: 5 Glc2	H-2 Xyl1; H-3, 5 Glc2
2	75.1 CH	3.82 (t, 7.8)	C: 1, 3 Glc2	
3	75.2 CH	3.96 (t, 8.7)	C: 2, 4 Glc2	H-1 Glc2
4	**82.2** CH	3.87 (t, 8.7)	C: 1 Glc3; C: 5, 6 Glc2	H-1 Glc3
5	75.9 CH	3.70 (dt, 2.9; 9.7)		H-1, 3 Glc2
6	61.4 CH_2	4.30 (dd, 2.9; 12.3)		
		4.25 (dd, 4.6; 12.2)	C: 4, 5 Glc2	
Glc3 (1→4Glc2)				
1	104.5 CH	4.81 (d, 7.9)	C: 4 Glc2	H-4 Glc2; H-5 Glc3
2	74.1 CH	3.79 (t, 9.2)	C: 1, 3, 4 Glc3	H-4 Glc3
3	**76.8** CH	4.07 (t, 9.2)	C: 2, 4 Glc3	H-1 Glc3
4	70.7 CH	3.90 (t, 9.2)	C: 3, 5, 6 Glc3	
5	75.5 CH	4.03 (dd, 4.6; 10.1)		H-1 Glc3
6	*67.5* CH_2	5.01 (d, 10.1)	C: 4 Glc3	
		4.64 (dd, 6.7; 11.1)	C: 5 Glc3	
Glc4 (1→4Xyl1)				
1	100.9 CH	4.92 (d, 7.8)	C: 4 Xyl1	H-4 Xyl1; H-3, 5 Glc4
2	*80.6* CH	4.74 (t, 8.9)	C: 1, 3 Glc4	
3	76.8 CH	4.28 (t, 8.9)	C: 2, 4 Glc4	H-1, 5 Glc4
4	70.7 CH	3.90 (t, 8.9)	C: 3, 5, 6 Glc4	
5	77.4 CH	3.84 (dd, 4.6; 10.2)	C: 4 Glc4	H-1 Glc4
6	61.8 CH_2	4.32 (dd, 2.5; 12.1)	C: 4 Glc4	
		4.01 (dd, 6.4; 12.1)	C: 4, 5 Glc4	

The molecular formula of psolusoside E (**2**) was determined to be $C_{54}H_{83}O_{25}SNa$ from the $[M_{Na} - Na]^-$ ion peak at *m/z* 1163.4945 (calc. 1163.4950) in the (−)HR-ESI-MS. In the ^{1}H and ^{13}C NMR spectra of the carbohydrate part of psolusoside E (**2**), four characteristic doublets at δ(H) 4.77–5.22 (J = 7.3–7.7 Hz) and, corresponding to them, signals of anomeric carbons at δ(C) 104.6–105.4 were indicative of a tetrasaccharide chain and β-configurations of glycosidic bonds (Table 4, Figures S9, S10 and S12). The ^{1}H,^{1}H-COSY, and 1D TOCSY spectra of **2** showed the signals of four isolated spin systems assigned to the xylose, quinovose, glucose, and 3-*O*-methylglucose residues (Figures S11, S15 and S16). The positions of interglycosidic linkages were elucidated by the ROESY and HMBC spectra of **2** (Table 4, Figures S13 and S14) by same manner as for **1,** indicating the presence of linear tetrasaccharide chain in psolusoside E (**2**). The signals of C(6) and C(5) of the glucose residue (the third sugar), observed at δ(C) 67.6 and 75.0, correspondingly, were characteristic of the sulfated by C(6) glucopyranose residue. Thus, psolusoside E (**2**) is a monosulfated tetraoside, with the glucose residue, sulfated by C(6), as the third monosaccharide unit. Such carbohydrate chain was not found earlier in the glycosides from sea cucumbers.

Table 3. ^{13}C and ^{1}H NMR chemical shifts and HMBC and ROESY correlations of aglycone moiety of psolusosides E (**2**), F (**3**), G (**4**). [a] Recorded at 176.04 MHz in C_5D_5N/D_2O (4/1). [b] Recorded at 700.13 MHz in C_5D_5N/D_2O (4/1).

Position	δ_C mult. [a]	δ_H mult. [b] (*J* in Hz)	HMBC	ROESY
1	36.2 CH_2	1.89 m		H-11, H-19
		1.52 m		H-3, H-5, H-11
2	27.0 CH_2	2.30 m		
		2.02 m		H-19, H-30
3	88.7 CH	3.31 (dd, 4.8; 11.6)	C: 4, 30, 31, C-1 Xyl1	H-1, H-5, H-31, H-1 Xyl1
4	39.6 C			
5	52.8 CH	0.99 (brd, 12.0)	C: 4, 10, 19, 30	H-1, H-3, H-7, H-31
6	21.0 CH_2	1.75 m		
		1.57 m		H-19, H-30
7	28.4 CH_2	1.62 m		H-15
		1.27 m		H-5, H-32
8	38.6 CH	3.29 m	C: 9	H-15, H-19
9	151.2 C			
10	39.8 C			
11	110.9 CH	5.35 m	C: 8, 13	H-1
12	32.0 CH_2	2.48 m	C: 14	H-21
		2.52 m	C: 9, 11, 13, 14, 18	H-17, H-32
13	55.6 C			
14	41.9 C			
15	51.8 CH_2	2.39 d (15.6)	C: 13, 16, 17, 32	H-7, H-32
		2.23 d (15.6)	C: 14, 16, 32	H-8
16	212.9 C			
17	61.2 CH	2.80 s	C: 12, 13, 16, 18, 20, 21	H-12, H-21, H-22, H-32
18	175.8 C			
19	21.9 CH_3	1.43 s	C: 1, 5, 9, 10	H-1, H-2, H-8, H-30
20	82.9 C			
21	26.6 CH_3	1.40 s	C: 17, 20, 22	H-12, H-17, H-22
22	38.3 CH_2	1.81 m		H-12, H-17, H-21
		1.66 m		
23	22.1 CH_2	1.81 m		
		1.53 m		
24	37.8 CH_2	1.99 m	C: 25, 26, 27	H-27
25	145.4 C			
26	110.3 CH_2	4.78 brs	C: 24, 25, 27	H-27
27	22.2 CH_3	1.70 s	C: 24, 25, 26	
30	16.5 CH_3	1.11 s	C: 3, 4, 5, 31	H-2, H-6, H-19
31	27.9 CH_3	1.31 s	C: 3, 4, 5, 30	H-3, H-5, H-6, H-1 Xyl1
32	20.5 CH_3	0.92 s	C: 8, 13, 14, 15	H-7, H-12, H-15, H-17

The (−)ESI-MS/MS of **2** demonstrated the fragmentation of $[M_{Na} - Na]^-$ ion at *m/z* 1163.5. The peaks of fragment ions were observed at *m/z*: 987.4 $[M_{Na} - Na - MeGlc + H]^-$, 695.2 $[M_{Na} - Na - C_{30}H_{43}O_4 (Agl) - H]^-$, 563.1 $[M_{Na} - Na - C_{30}H_{43}O_4 (Agl) - Xyl]^-$, 417.1 $[M_{Na} - Na - C_{30}H_{43}O_4 (Agl) - Xyl - Qui]^-$, 241.0 $[M_{Na} - Na - C_{30}H_{43}O_4 (Agl) - Xyl - Qui - MeGlc]^-$ (corresponds to desodiated sulfated glucose residue) corroborating the structure of psolusoside E (**2**).

All these data indicate that psolusoside E (**2**) is 3β-*O*-[3-*O*-methyl-β-D-glucopyranosyl-(1→3)-6-*O*-sodium-sulfate-β-D-glucopyranosyl-(1→4)-β-D-quinovopyranosyl-(1→2)-β-D-xylopyranosyl]-16-ketoholosta-9(11),25-diene.

Table 4. ^{13}C and ^{1}H NMR chemical shifts and HMBC and ROESY correlations of carbohydrate moiety of psolusoside E (**2**). [a] Recorded at 176.04 MHz in C_5D_5N/D_2O (4/1). [b] Bold is interglycosidic positions. [c] Italic is sulphate position. [d] Recorded at 700.13 MHz in C_5D_5N. Multiplicity by 1D TOCSY.

Atom	δ_C mult. [a, b c]	δ_H mult. [d] (J in Hz)	HMBC	ROESY
Xyl1 (1→C-3)				
1	105.4 CH	4.77 d (7.4)	C-3	H-3; H-3, 5 Xyl1
2	**83.1** CH	4.00 t (8.8)	C: 1, 3 Xyl1; C: 1 Qui2	H-1 Qui2
3	77.6 CH	4.18 t (8.8)	C: 4 Xyl1	H-1 Xyl1
4	70.8 CH	4.12 m		
5	66.5 CH_2	4.26 dd (5.4; 11.5)		
		3.61 t (10.9)		H-1, 3 Xyl1
Qui2 (1→2Xyl1)				
1	104.6 CH	5.12 d (7.7)	C: 2 Xyl1	H-2 Xyl1; H-3, 5 Qui2
2	75.7 CH	3.96 t (8.9)	C: 1, 3 Qui2	H-4 Qui2
3	75.1 CH	4.08 t (8.9)	C: 2, 4 Qui2	H-1, 5 Qui2
4	**87.7** CH	3.51 t (8.9)	C: 5 Qui2, 1 Glc3	H-1 Glc3; H-2 Qui2
5	71.3 CH	3.71 dd (5.9; 8.9)		H-1, 3 Qui2
6	17.8 CH_3	1.67 d (5.9)	C: 4, 5 Qui2	H-4, 5 Qui2
Glc3 (1→4Qui2)				
1	104.9 CH	4.81 d (7.3)	C: 4 Qui2	H-4 Qui2; H-3,5 Glc3
2	73.2 CH	3.99 t (8.1)	C: 3 Glc3	H-4 Glc3
3	**87.3** CH	4.16 t (8.8)	C: 2, 4 Glc3; 1 MeGlc4	H-1 MeGlc4; H-1 Glc3
4	69.9 CH	3.88 t (8.8)	C: 3, 5, 6 Glc3	H-6 Glc3
5	75.0 CH	4.22 t (8.8)		H-1 Glc3
6	*67.6* CH_2	5.16 brd (10.3)		
		4.77 t (8.8)		H-4 Glc3
MeGlc4(1→3Glc3)				
1	105.4 CH	5.22 d (7.3)	C: 3 Glc3	H-3 Glc3; H-3, 5 MeGlc4
2	74.8 CH	3.94 t (8.8)	C: 1, 3 MeGlc4	
3	87.8 CH	3.68 t (8.8)	C: 2, 4 MeGlc4, OMe	H-1, 5 MeGlc4; OMe
4	70.4 CH	4.11 t (8.8)	C: 3, 5, 6 MeGlc4	
5	78.2 CH	3.91 m	C: 3 MeGlc4	H-1, 3 MeGlc4
6	61.9 CH_2	4.43 dd (2.6; 11.7)		
		4.24 dd (5.1; 11.7)		
OMe	60.5 CH_3	3.85 s	C: 3 MeGlc4	

The molecular formula of psolusoside F (**3**) was determined to be $C_{54}H_{83}O_{25}SNa$ from the $[M_{Na} - Na]^-$ ion peak at *m/z* 1163.4952 (calc. 1163.4950) in the (−)HR-ESI-MS and was coincident with the formula of psolusoside E (**2**). In the ^{1}H and ^{13}C NMR spectra of the carbohydrate part of psolusoside F (**3**), four characteristic doublets at δ(H) 4.71–5.12 (J = 7.3–8.2 Hz) and corresponding signals of anomeric carbons at δ(C) 104.0–105.0 were indicative of a tetrasaccharide chain and β-configurations of glycosidic bonds (Figures S17, S18 and S20). The positions of interglycosidic linkages were elucidated by the ROESY and HMBC spectra of **3** (Table 5, Figures S21 and S22) as described above, indicating the presence of linear tetrasaccharide carbohydrate chain. The monosaccharide composition of **3**, deduced from the ^{1}H,^{1}H-COSY, and 1D TOCSY spectra (Figures S19, S23 and S24), was the same as in **2**. The comparison of the ^{13}C NMR spectra of these compounds showed the coincidence of their signals corresponding to xylose and quinovose residues. The signals of C(6) and C(5) of the glucose residue (the third unit in the chain) in the ^{13}C NMR spectrum of **3** were observed at δ(C) 61.7 (shielded as compared with corresponding signal in the spectrum of **2**) and 77.1 (de-shielded as compared with C(5) of the glucose in the spectrum of **2**), correspondingly, indicating the absence of a sulfate group in this residue. The signal of C(6) of 3-*O*-methyl-glucose residue was observed at δ(C) 67.0 and the signal C(5) of the same residue—at δ(C) 75.6 in the ^{13}C NMR spectrum of **3** indicating the attachment of a sulfate group to C(6) of terminal 3-*O*-methyl-glucose unit in the carbohydrate chain of psolusoside F (**3**). So, psolusosides E (**2**) and F (**3**) differed from each other only in the position of a sulfate group. The carbohydrate chain of **3** is a new one.

Table 5. ^{13}C and ^{1}H NMR chemical shifts and HMBC and ROESY correlations of carbohydrate moiety of psolusoside F (3). [a] Recorded at 176.04 MHz in C_5D_5N/D_2O (4/1). [b] Bold is interglycosidic positions. [c] Italic is sulphate position. [d] Recorded at 700.13 MHz in C_5D_5N. Multiplicity by 1D TOCSY.

Atom	δ_C mult. [a,b,c]	δ_H mult. [d] (*J* in Hz)	HMBC	ROESY
Xyl1 (1→C-3)				
1	105.0 CH	4.71 d (7.3)	C: 3	H-3; H-3, 5 Xyl1
2	**83.1** CH	3.95 t (8.6)	C: 1, 3 Xyl1; C: 1 Qui2	H-1 Qui2
3	77.2 CH	4.15 t (8.6)	C: 2, 4 Xyl1	H-1, 5 Xyl1
4	70.2 CH	4.10 m		
5	66.0 CH_2	4.25 dd (4.9; 11.6)	C: 3 Xyl1	
		3.63 t (11.0)		H-1 Xyl1
Qui2 (1→2Xyl1)				
1	104.8 CH	5.02 d (7.3)	C: 2 Xyl1	H-2 Xyl1; H-3, 5 Qui2
2	75.7 CH	3.94 t (8.7)	C: 1, 3 Qui2	
3	75.3 CH	3.99 t (8.7)		H-1 Qui2
4	**86.4** CH	3.58 t (8.7)	C: 3, 5 Qui2, 1 Glc3	H-1 Glc3
5	71.5 CH	3.67 dd (5.8; 10.2)		H-1 Qui2
6	17.9 CH_3	1.65 d (5.8)		H-4, 5 Qui2
Glc3 (1→4Qui2)				
1	104.0 CH	4.88 d (8.2)	C: 4 Qui2	H-4 Qui2; H-3, 5 Glc3
2	73.5 CH	3.92 t (8.2)	C: 1, 3 Glc3	
3	**87.6** CH	4.13 t (8.8)	C: 4 Glc3; 1 MeGlc4	H-1 MeGlc4; H-1 Glc3
4	69.4 CH	3.84 t (8.8)	C: 3, 5 Glc3	
5	77.1 CH	3.93 m		
6	61.7 CH_2	4.36 dd (2.9; 12.3)		
		4.03 dd (7.0; 12.3)		
MeGlc4 (1→3Glc3)				
1	104.8 CH	5.12 d (7.8)	C: 3 Glc3	H-3 Glc3; H-3, 5 MeGlc4
2	74.3 CH	3.79 t (9.4)	C: 1 MeGlc4	H-4 MeGlc4
3	86.4 CH	3.65 t (9.4)	C: 2, 4 MeGlc4, OMe	H-1 MeGlc4, OMe
4	69.9 CH	3.98 t (9.4)	C: 3, 5, 6 MeGlc4	H-6 MeGlc4
5	75.6 CH	4.04 dd (7.8; 10.9)		H-1 MeGlc4
6	*67.0* CH_2	4.98 d (9.4)		
		4.74 dd (5.5; 10.9)		
OMe	60.5 CH_3	3.76 s	C: 3 MeGlc4	

The (−)ESI-MS/MS of **3** demonstrated the fragmentation of $[M_{Na} - Na]^-$ ion at *m/z* 1163.5. The peaks of fragment ions were observed at *m/z*: 695.2 $[M_{Na} - Na - C_{30}H_{43}O_4$ (Agl) $- H]^-$, 563.1 $[M_{Na} - Na - C_{30}H_{43}O_4$ (Agl) $-$ Xyl$]^-$, 417.1 $[M_{Na} - Na - C_{30}H_{43}O_4$ (Agl) $-$ Xyl $-$ Qui$]^-$, 255.0 $[M_{Na} - Na - C_{30}H_{43}O_4$ (Agl) $-$ Xyl $-$ Qui $-$ Glc$]^-$, confirming the sequence of monosaccharides in the sugar chain.

All these data indicate that psolusoside F (**3**) is 3β-*O*-[6-*O*-sodium-sulfate-3-*O*-methyl-β-D-glucopyranosyl-(1→3)-β-D-glucopyranosyl-(1→4)-β-D-quinovopyranosyl-(1→2)-β-D-xylopyranosyl]-16-ketoholosta-9(11),25-diene.

The molecular formula of psolusoside G (**4**) was determined to be $C_{54}H_{82}O_{29}S_2Na_2$ from the $[M_{2Na} - Na]^-$ ion peak at *m/z* 1281.4313 (calc. 1281.4286) in the (−)HR-ESI-MS indicating the presence of two sulfate groups in this glycoside. In the ^{1}H and ^{13}C NMR spectra of the carbohydrate part of psolusoside G (**4**), four characteristic doublets at δ(H) 4.72–5.16 (*J* = 7.2–8.4 Hz) and, corresponding to them, signals of anomeric carbons at δ(C) 103.8–105.0 were indicative of a tetrasaccharide chain and β-configurations of glycosidic bonds (Table 6, Figures S25, S26 and S28).

Analysis of the ^{1}H,^{1}H-COSY and 1D TOCSY spectra of psolusoside G (**4**) showed the availability of one xylose, two glucose, and one 3-O-methyl-glucose residues (Figures S27, S31 and S32). So, the quinovose unit was absent in the chain of **4** that was corroborated by the lack of characteristic doublet of methyl group of quinovose residue at δ(H) ≈1.70 in the ^{1}H NMR spectrum and the corresponding signal at δ(C) ≈18.0 in the ^{13}C NMR spectrum of **4**. It was supposed that the second position of carbohydrate moiety was occupied by the glucose residue and confirmed by the appearance of the additional signal at δ(C) 61.2 corresponding to C(6) of a glucopyranose residue. Two signals at δ(C) 67.4 and 67.1 corresponding to sulfated hydroxy-methylene carbons of glucopyranose residues were observed in the

^{13}C NMR spectrum of **4** indicating the presence of two sulfate groups. The positions of interglycosidic linkages and the consequence of monosaccharides in the chain of **4** were established by analysis of the ROESY and HMBC spectra (Table 6, Figures S29 and S30), indicating the presence of linear carbohydrate moiety with the glucose as second unit and sulfated glucose and 3-*O*-methyl-glucose residues as third and terminal monosaccharides, correspondingly. The comparison of the ^{13}C NMR spectrum of sugar part of psolusoside G (**4**) with that of earlier known okhotoside B_3, isolated from *Cucumaria okhotensis* [19] showed the coincidence of their signals, suggesting the identity of the linear disulfated carbohydrate moieties.

Table 6. ^{13}C and ^{1}H NMR chemical shifts and HMBC and ROESY correlations of carbohydrate moiety of psolusoside G (**4**). [a] Recorded at 176.04 MHz in C_5D_5N/D_2O (4/1). [b] Bold is interglycosidic positions. [c] Italic is sulphate position. [d] Recorded at 700.13 MHz in C_5D_5N. Multiplicity by 1D TOCSY.

Atom	δ_C mult. [a,b,c]	δ_H mult. [d] (*J* in Hz)	HMBC	ROESY
Xyl1 (1→C-3)				
1	105.0 CH	4.72 d (7.5)	C: 3	H-3; H-3, 5 Xyl1
2	**81.8** CH	4.06 t (7.5)	C: 1, 3 Xyl1; C: 1 Glc2	H-1 Glc2
3	77.1 CH	4.16 t (8.8)	C: 2, 4 Xyl1	H-1, 5 Xyl1
4	70.1 CH	4.09 m		
5	66.0 CH_2	4.23 dd (4.6; 10.9)		
		3.62 t (11.3)		H-1, 3 Xyl1
Glc2 (1→2Xyl1)				
1	104.2 CH	5.16 d (7.2)	C: 2 Xyl1	H-2 Xyl1; H-3, 5 Glc2
2	75.2 CH	3.93 t (9.5)	C: 1, 3 Glc2	
3	75.2 CH	4.02 t (9.5)	C: 2, 4 Glc2	H-1, 5 Glc2
4	**81.8** CH	3.95 t (9.5)	C: 3 Glc2, 1 Glc3	H-1 Glc3; H-6 Glc2
5	76.0 CH	3.72 m		H-1 Glc2
6	61.2 CH_2	4.29 m		
Glc3 (1→4Glc2)				
1	103.8 CH	4.86 d (8.4)	C: 4 Glc2	H-4 Glc2; H-3 Glc3
2	73.3 CH	3.80 t (8.4)	C: 1 Glc3	
3	**86.5** CH	4.05 t (9.5)	C: 2, 4 Glc3; 1 MeGlc4	H-1 MeGlc4; H-1 Glc3
4	69.1 CH	3.76 t (9.5)	C: 5, 6 Glc3	
5	74.8 CH	4.04 m		
6	*67.4* CH_2	5.95 dd (1.6; 10.3)		
		4.57 dd (5.4; 10.3)		
MeGlc4 (1→3Glc3)				
1	104.6 CH	5.07 d (7.6)	C: 3 Glc3	H-3 Glc3; H-3, 5 MeGlc4
2	74.2 CH	3.76 t (8.7)	C: 1 MeGlc4	
3	86.4 CH	3.61 t (8.2)	C: 2, 4 MeGlc4, OMe	H-1, 5 MeGlc4
4	69.8 CH	3.96 t (8.2)		
5	75.4 CH	3.99 m		H-1 MeGlc4
6	*67.1* CH_2	4.93 d (10.8)		
		4.71 dd (5.4; 10.8)		
OMe	60.5 CH_3	3.76 s	C: 3 MeGlc4	

The (−)ESI-MS/MS of **4** demonstrated the fragmentation of $[M_{2Na} - Na]^-$ ion at *m/z* 1281.4. The peaks of fragment ions were observed at *m/z*: 1161.5 $[M_{2Na} - Na - HSO_4Na]^-$, 1003.4 $[M_{2Na} - Na - HSO_4Na - C_7H_{12}O_8SNa$ $(MeGlcSO_3Na)]^-$, 813.2 $[M_{2Na} - Na - C_{30}H_{43}O_4$ (Agl) $- H]^-$, 681.1 $[M_{2Na} - Na - C_{30}H_{43}O_4$ (Agl) $-$ Xyl$]^-$, 519.0 $[M_{2Na} - Na - C_{30}H_{43}O_4$ (Agl) $-$ Xyl $-$ Glc$]^-$, 255.0 $[M_{2Na} - Na - C_{30}H_{43}O_4$ (Agl) $-$ Xyl $-$ Glc $- C_6H_9O_8SNa$ $(GlcSO_3Na)]^-$, corroborating the aglycone structure and consequence of monosaccharides in psolusoside G (**4**).

All these data indicate that psolusoside G (**4**) is 3β-*O*-[6-*O*-sodium-sulfate-3-*O*-methyl-β-D-glucopyranosyl-(1→3)-6-*O*-sodium-sulfate-β-D-glucopyranosyl-(1→4)-β-D-glucopyranosyl-(1→2)-β-D-xylopyranosyl]-16-ketoholosta-9(11),25-diene.

The sulfation of third or/and fourth monosaccharide residues in the carbohydrate chain when C(4) position of the first xylose residue is not sulfated as in psolusosides E (**2**), F (**3**) and G (**4**) is probably characteristic structural feature of the glycosides of *Psolus fabricii*. It was also observed in a disulfated tetraoside psolusoside A, with sulfate groups at C-6 of the third (glucose) and fourth (3-*O*-methyl-glucose) residues. Monosulfated colochiroside D, isolated first from the sea cucumber *Colochirus robustus* [16] and later from *Psolus fabricii* as well as the disulfated okhotoside B_3 from *Cucumaria okhotensis* [19], are the glycosides found in the sea cucumbers belonging to other genera, sharing the same structural peculiarity. However, the majority of known sulfated glycosides contain a sulfate group at C-4 of the first xylose residue.

The ^{1}H and ^{13}C NMR spectra of carbohydrate parts of psolusosides H (**5**) and H_1 (**6**) were coincident to each other indicating the identity of carbohydrate chains of these glycosides. The presence of three characteristic doublets at δ(H) 4.73 (*J* = 7.5 Hz), 5.19 (*J* = 6.8 Hz), and 4.88 (*J* = 7.9 Hz) in the ^{1}H NMR spectra of the carbohydrate chains of **5**, **6** correlated with the HSQC spectra with the signals of anomeric carbons at δ(C) 104.9, 104.4, and 104.5, correspondingly, were indicative of a trisaccharide chain and β-configurations of glycosidic bonds (Table 7, Figures S33, S34, S36, S40 and S41, S43). The ^{1}H,^{1}H-COSY, and 1D TOCSY spectra of **5** and **6** showed the signals of three isolated spin systems assigned to two glucose and one xylose residues (Figures S35, S39 and S42). The positions of interglycosidic linkages established by the ROESY and HMBC spectra of **5** and **6** (Table 7, Figures S37, S38, S44 and S45) demonstrated cross-peaks between H(1) of the xylose and H(3) (C(3)) of an aglycone, H(1) of the glucose and H(2), (C(2)) of the xylose and H(1) of the terminal unit (glucose), and H(4) (C(4)) of the second unit (glucose). The terminal glucose unit was sulfated by C(6), which was deduced from character signal at δ(C) 67.2 in comparison with the analogous signal of C(6) of the glucose in the second position of the carbohydrate chain, which was observed at δ(C) 61.6. Therefore, the carbohydrate chain of psolusosides H (**5**) and H_1 (**6**) differed from that of psolusoside G (**4**) in the loss of terminal 3-O-methyl-glucose residue. Actually, the signals in their ^{13}C NMR spectra assigning to xylose and glucose (the second unit) residues were coincident. The signal of C(3) of terminal glucose in the spectra of **5**, **6** was shifted up-field to δ(C) 76.9 due to the absence of the glycosylation effect that was observed in the spectrum of **4** (δ(C) 86.4 C(3) of terminal glucose). The carbohydrate chain of psolusosides H (**5**) and H_1 (**6**) has never been found earlier in the glycosides from sea cucumbers.

Table 7. ^{13}C and ^{1}H NMR chemical shifts and HMBC and ROESY correlations of carbohydrate moieties of psolusosides H (**5**) and H_1 (**6**). [a] Recorded at 176.04 MHz in C_5D_5N. [b] Bold is interglycosidic positions. [c] Italic is sulphate position. [d] Recorded at 700.13 MHz in C_5D_5N. Multiplicity by 1D TOCSY.

Atom	δ_C mult. [a,b,c]	δ_H mult. [d] (*J* in Hz)	HMBC	ROESY
Xyl1 (1→C-3)				
1	104.9 CH	4.73 d (7.5)	C-3	H-3; H-3, 5 Xyl1
2	**82.0** CH	4.07 t (8.0)	C: 1 Glc2; 1, 3 Xyl1	H-1 Glc2
3	77.2 CH	4.18 t (8.9)	C: 2, 4 Xyl1	H-1, 5 Xyl1
4	70.1 CH	4.11 m		
5	66.0 CH_2	4.25 dd (5.6; 11.3)	C: 3, 4 Xyl1	
		3.63 dd (2.0; 11.2)		H-1, 3 Xyl1
Glc2 (1→2Xyl1)				
1	104.4 CH	5.19 d (6.8)	C: 2 Xyl1	H-2 Xyl1; H-3, 5 Glc2
2	75.2 CH	3.95 t (8.2)	C: 1 Glc2	
3	75.5 CH	4.05 t (8.7)	C: 2, 4 Glc2	H-1, 5 Glc2
4	**82.2** CH	3.97 t (8.7)	C: 1 Glc3; 3 Glc2	H-1 Glc3; H-6 Glc2
5	75.9 CH	3.77 m		H-1, 3 Glc2
6	61.6 CH_2	4.33 m		H-4 Glc2
Glc3 (1→4Glc2)				
1	104.5 CH	4.88 d (7.9)	C: 4 Glc2	H-4 Glc2; H-3,5 Glc3
2	74.2 CH	3.83 t (9.0)	C: 1, 3 Glc3	H-4 Glc3
3	76.9 CH	4.10 t (9.0)	C: 2, 4 Glc3	H-1 Glc3
4	70.6 CH	3.97 t (9.0)	C: 5, 6 Glc3	H-2, 6 Glc3
5	75.7 CH	4.06 m	C: 4 Glc3	H-1 Glc3
6	*67.2* CH_2	5.07 dd (2.8; 11.3)		
		4.73 dd (6.8; 11.3)	C: 5 Glc3	

The molecular formula of psolusoside H (**5**) was determined to be $C_{47}H_{71}O_{21}SNa$ from the $[M_{Na}–Na]^-$ ion peak at *m/z* 1003.4213 (calc. 1003.4214) in the (−)HR-ESI-MS. The ^1H and ^{13}C NMR spectra of aglycone part of psolusoside H (**5**) demonstrated the signals characteristic of the holostane-type aglycone (the signals of 18(20)-lactone at δ(C) 179.0 (C(18)) and 83.6 (C(20))) with 16-keto-group (the signals of C(16) at δ(C) 213.8, C(15) at δ(C) 51.8, and C(17) at δ(C) 63.3 with corresponding proton signals at δ(H) 2.65 (d, *J* = 16.0 Hz, H(15)), and 2.32 (d, *J* = 16.0 Hz, H(15)), as well as 2.87 (s, H(17)) (Table 8, Figures S33 and S34). The characteristic signals at δ(C) 121.7 (C(7)), 143.9 (C(8)), and at δ(H) 5.63 (m, H(7)) in the ^{13}C and ^1H NMR spectra of **5** were assigned to 7(8)-double bond in the polycyclic system. The availability of terminal double bond in the side chain of **5** was deduced from the signals at δ(C) 145.4 (C(25)) and 110.3 (C(26)) observed in the ^{13}C NMR and two broad singlets at δ(H) 4.70 and 4.69 (H_2-26) in the ^1H NMR spectra of psolusoside H (**5**). So, the aglycone of psolusoside H (**5**) is a positional isomer (by the double bond position in polycyclic nucleus) of the aglycone comprising psolusosides E (**2**), F (**3**), and G (**4**). This aglycone was found earlier in the glycosides of sea cucumbers belonging to different orders: *Cucumaria japonica* [20,21], *Pseudocolochirus violaceus* [22] (Cucumariidae, Dendrochirotida), and *Australostichopus mollis* [23] (Stichopodidae, Synallactida) [24].

Table 8. ^{13}C and ^1H NMR chemical shifts and HMBC and ROESY correlations of aglycone moiety of psolusoside H (**5**). [a] Recorded at 176.04 MHz in C_5D_5N. [b] Recorded at 700.13 MHz in C_5D_5N/D_2O (4/1).

Position	δ_C mult. [a]	δ_H mult. [b] (*J* in Hz)	HMBC	ROESY
1	35.4 CH_2	1.35 m		H-3, H-5, H-11, H-19
2	26.7 CH_2	2.06 m		
		1.89 m		
3	89.2 CH	3.24 dd (3.8; 11.8)	C: 30, 1 Xyl1	H-1, H-5, H-31, H-1 Xyl1
4	39.2 C			
5	48.1 CH	0.92 dd (4.3; 11.6)		H-1, H-3, H-31
6	23.1 CH_2	1.91 m		H-19, H-31
7	121.7 CH	5.63 m		H-15, H-32
8	143.9 C			
9	46.9 CH	3.54 brd (15.2)		H-19
10	35.7 C			
11	22.2 CH_2	1.80 m		H-1
		1.53 m		H-32
12	29.4 CH_2	2.19 brdd (5.8; 8.8)	C: 13, 18	H-17, H-21, H-32
13	56.6 C			
14	45.6 C			
15	51.8 CH_2	2.65 d (15.9)	C: 13, 16, 32	H-7, H-32
		2.32 d (16.1)	C: 14, 16, 32	H-7
16	213.8 C			
17	63.3 CH	2.87 s	C: 12, 13, 16, 18, 20, 21	H-12, H-21, H-22, H-32
18	179.0 C			
19	23.8 CH_3	1.10 s	C: 1, 9, 10	H-1, H-2, H-6, H-9
20	83.6 C			
21	26.0 CH_3	1.45 s	C: 17, 20, 22	H-12, H-17, H-22
22	38.1 CH_2	1.71 m		H-17, H-21
		1.56 m		
23	22.0 CH_2	1.71 m		
		1.43 m		
24	37.7 CH_2	1.90 m	C: 25, 26	H-26
25	145.4 C			
26	110.3 CH_2	4.70 brs	C: 24, 27	H-27
		4.69 brs	C: 24, 27	H-27
27	22.0 CH_3	1.63 s	C: 24, 25, 26	
30	17.1 CH_3	1.07 s	C: 3, 4, 5, 31	H-2, H-6, H-6 Glc2
31	28.5 CH_3	1.20 s	C: 3, 4, 5, 30	H-3, H-5, H-6, H-1 Xyl1
32	31.7 CH_3	1.16 s	C: 8, 13, 14, 15	H-7, H-11, H-12, H-15, H-17

The (−)ESI-MS/MS of **5** demonstrated the fragmentation of $[M_{Na} – Na]^-$ ion at *m/z* 1003.4. The peaks of fragment ions were observed at *m/z*: 535.1 $[M_{Na} – Na – C_{30}H_{43}O_4$ (Agl) $– H]^-$, 403.1 $[M_{Na}$

– Na – $C_{30}H_{43}O_4$ (Agl) – Xyl]$^-$, 241.0 [M_{Na} – Na – $C_{30}H_{43}O_4$ (Agl) – Xyl – Glc]$^-$ corroborating the structure of psolusoside H (**5**).

All these data indicate that psolusoside H (**5**) is 3β-*O*-[6-*O*-sodium-sulfate-β-D-glucopyranosyl-(1→4)-β-D-glucopyranosyl-(1→2)-β-D-xylopyranosyl]-16-ketoholosta-7,25-diene.

The molecular formula of psolusoside H_1 (**6**) was determined to be $C_{47}H_{73}O_{20}SNa$ from the [M_{Na} – Na]$^-$ ion peak at *m/z* 989.4432 (calc. 989.4421) in the (–)HR-ESI-MS and [M_{Na} + Na]$^+$ at *m/z* 1035.4205 (calc. 1035.4206) in the (+)HR-ESI-MS. In the ^{1}H and ^{13}C NMR spectra of the aglycone part of psolusoside H_1 (**6**), the signals characteristic of the holostane-type aglycone (the signals of 18(20)-lactone at δ(C) 181.0 (C(18)) and 84.7 (C(20))) with 7(8)-double bond in the polycyclic system (the signals at δ(C) 119.8 (C(7)), 146.5 (C(8)) in the ^{13}C NMR, and at δ(H) 5.62 (m, H(7)) in the ^{1}H NMR) and terminal double bond in the side chain (the signals at δ(C) 145.5 (C(25)) and 110.6 (C(26)) in the ^{13}C NMR and two broad singlets at δ(H) 4.78 and 4.74 (H_2-26) in the ^{1}H NMR) were observed (Table 9, Figures S40 and S41). The analysis of ^{1}H,^{1}H-COSY spectrum of **6** showed the protons H_2(15)/H_2(16)/H(17) form the isolated spin system (Figure S42). The signals of C(15), C(16), and C(17) in the ^{13}C NMR spectrum of **6** were observed at δ(C) 34.1, 24.5, and 52.9, correspondingly, and were shielded when compared with the signals C(15)–C(17) in the spectrum of psolusoside H (**5**) due to the absence of 16-keto-group in psolusoside H_1 (**6**). So, the aglycone of psolusoside H_1 (**6**) differed from that of psolusoside H (**5**) only in the lack of 16-keto-group. Such aglycone was earlier found in the glycosides from sea cucumbers of the order Dendrochirotida: *Colochirus robustus* [16] and *Cucumaria japonica* [20].

The (–)ESI-MS/MS of **6** demonstrated the fragmentation of [M_{Na} – Na]$^-$ ion at *m/z* 989.4. The peaks of fragment ions analogous to those for **5** were observed at *m/z*: 535.1 [M_{Na} – Na – $C_{30}H_{45}O_3$ (Agl) – H]$^-$, 403.1 [M_{Na} – Na – $C_{30}H_{45}O_3$ (Agl) – Xyl]$^-$, 241.0 [M_{Na} – Na – $C_{30}H_{45}O_3$ (Agl) – Xyl – Glc]$^-$ corroborating the structure of psolusoside H_1 (**6**).

All these data indicate that psolusoside H_1 (**6**) is 3β-*O*-[6-*O*-sodium-sulfate-β-D-glucopyranosyl-(1→4)-β-D-glucopyranosyl-(1→2)-β-D-xylopyranosyl]-holosta-7,25-diene.

The molecular formula of psolusoside I (**7**) was determined to be $C_{54}H_{82}O_{29}S_2Na_2$ from the [M_{2Na} – Na]$^-$ ion peak at *m/z* 1281.4267 (calc. 1281.4286) in the (–)HR-ESI-MS and [M_{2Na} + Na]$^+$ at *m/z* 1327.4065 (calc. 1327.4071) in the (+)HR-ESI-MS indicating the presence of two sulfate groups. In the ^{1}H and ^{13}C NMR spectra (Table 10, Figures S46 and S47) of the carbohydrate part of **7** four characteristic doublets at δ(H) 4.63–4.82 (J = 7.3–8.1 Hz) and corresponding to them four signals of anomeric carbons at δ(C) 103.9–105.6 were indicative of a tetrasaccharide chain and *β*-configurations of glycosidic bonds (Figure S50). The ^{1}H,^{1}H-COSY and 1D TOCSY spectra of **7** showed the signals of four isolated spin systems assigned to two xylose and two glucose residues (Figures S49, S53 and S54). The positions of interglycosidic linkages established by the ROESY and HMBC spectra of **7** (Table 10, Figures S51 and S52) indicated the branched architecture of tetrasaccharide chain when the fourth glucose residue is attached to C(4) of the first (xylose) residue.

The second sugar unit in the chain of psolusoside I (**7**) is a xylose connected to the first xylose residue by β-(1→2)-glycosidic bond. This feature is very rare occurred in the holothurians glycoside's carbohydrate moieties [25]. The third monosaccharide in the chain is a glucose attached to C(4) of the second (xylose) unit, the fourth residue (glucose) is attached to C-4 of the first xylose unit. Both glucose residues are sulfated by C(6) that was deduced from two signals—at δ(C) 67.6 and 67.9 in the ^{13}C NMR spectrum of **7**—demonstrating α-shifting effect of a sulfate group. The tetrasaccharide branched disulfated carbohydrate moiety of psolusoside I (**7**) with the xylose as the second unit has never been found among the sea cucumber glycosides.

Table 9. ^{13}C and ^{1}H NMR chemical shifts and HMBC and ROESY correlations of aglycone moiety of psolusoside H_1 (**6**). [a] Recorded at 176.04 MHz in C_5D_5N. [b] Recorded at 700.13 MHz in C_5D_5N/D_2O (4/1).

Position	δ_C mult. [a]	δ_H mult. [b] (J in Hz)	HMBC	ROESY
1	36.0 CH_2	1.34 m		H-3, H-11
2	26.8 CH_2	2.03 m		
		1.88 m		H-30
3	89.5 CH	3.24 dd (4.0; 11.7)	C: 30, 31, 1 Xyl1	H-1, H-5, H-31, H-1 Xyl1
4	39.3 C			
5	48.0 CH	0.93 dd (4.4; 10.8)	C: 4, 10, 19, 30, 31	H-3, H-31
6	23.1 CH_2	1.91 m		H-31
7	119.8 CH	5.62 m		H-15, H-32
8	146.5 C			
9	47.2 CH	3.37 brd (14.3)		H-19
10	35.4 C			
11	22.7 CH_2	1.70 m		H-1
		1.49 m		
12	30.3 CH_2	2.00 m	C: 13, 18	H-32
13	58.6 C			
14	51.6 C			
15	34.1 CH_2	1.76 m	C: 13	H-7
		1.50 m		H-32
16	24.5 CH_2	1.97 m		H-32
		1.84 m		
17	52.9 CH	2.29 dd (4.0; 10.5)	C: 13, 18	H-12, H-32
18	181.0 C			
19	23.8 CH_3	1.08 s	C: 1, 5, 9, 10	H-2, H-9
20	84.7 C			
21	22.7 CH_3	1.28 s	C: 17, 20, 22	H-23
22	40.7 CH_2	1.71 m		
		1.55 m		H-24
23	22.0 CH_2	1.47 m	C: 22, 24	H-21
24	37.7 CH_2	1.99 m	C: 22, 23, 25, 26	H-22
		1.88 m		
25	145.5 C			
26	110.6 CH_2	4.78 brs	C: 24, 27	H-27
		4.74 brs	C: 24, 27	H-24, H-27
27	22.1 CH_3	1.66 s	C: 24, 25, 26	
30	17.2 CH_3	1.06 s	C: 3, 4, 5, 31	H-2
31	28.6 CH_3	1.19 s	C: 3, 4, 5, 30	H-3, H-5, H-6, H-1 Xyl1
32	30.7 CH_3	1.11 s	C: 8, 13, 14, 15	H-7, H-11, H-12, H-15, H-17

The aglycone of psolusoside I (**7**) shared some structural features with the aglycones of psolusosides H (**5**) and H_1 (**6**). In the ^{1}H and ^{13}C NMR spectra of the aglycone part of **7**, the signals of holostane-type aglycone (C(18) at δ(C) 180.2 and C(20) at δ(C) 85.5) with 7(8)-double bond in the polycyclic system (the signals at δ(C) 120.2 (C(7)), 145.6 (C(8)), and at δ(H) 5.60 (m, H(7)) and terminal double bond in the side chain (the signals at δ(C) 145.4 (C(25)) and 110.8 (C(26)) in the ^{13}C NMR and two broad singlets at δ(H) 4.72 and 4.73 (H_2-26) in the ^{1}H NMR spectra) were observed (Table 11, Figures S46 and S47). An isolated spin system formed by the protons H_2(15)/H(16)/H(17) was deduced from the ^{1}H,^{1}H-COSY spectrum (Figure S49). The signal of H(16) was observed at δ(H) 5.82 (brq, J = 8.6 Hz) and the corresponding signal of C(16) at δ(C) 75.2 indicated the presence of β-*O*-acetic group. Actually, the additional signals corresponding to this group were observed in the ^{13}C NMR spectrum of **7** at δ(C) 170.7 (carboxyl carbon) and 21.2 (methyl carbon). The holostane aglycone with 7(8)- and 25(26)-double bonds and 16β-acetoxy group frequently occurred in the glycosides of sea cucumbers [2,16,19].

Table 10. ^{13}C and ^{1}H NMR chemical shifts and HMBC and ROESY correlations of carbohydrate moiety of psolusoside I (7). [a] Recorded at 176.04 MHz in C_5D_5N/D_2O (4/1). [b] Bold is interglycosidic positions. [c] Italic is sulphate position. [d] Recorded at 700.13 MHz in C_5D_5N. Multiplicity by 1D TOCSY.

Atom	δ_C mult. [a,b,c]	δ_H mult. [d] (*J* in Hz)	HMBC	ROESY
Xyl1 (1→C-3)				
1	104.4 CH	4.63 d (7.3)	C: 3	H-3; H-3, 5 Xyl1
2	**83.2** CH	3.75 t (7.3)	C: 3 Xyl1; C: 1 Xyl2	H-1 Xyl2; H-4 Xyl1
3	75.2 CH	3.99 t (6.7)	C: 2, 4 Xyl1	H-1, 5 Xyl1
4	**80.2** CH	3.99 t (6.7)	C: 3 Xyl1	
5	63.5 CH_2	4.35 dd (4.9; 11.0)	C: 1, 3, 4 Xyl1	
		3.60 t (9.2)		H-1, 3 Xyl1
Xyl2 (1→2Xyl1)				
1	105.6 CH	4.79 d (7.6)	C: 2 Xyl1	H-2 Xyl1; H-3, 5 Xyl2
2	75.0 CH	3.91 t (9.3)	C: 1, 3 Xyl2	
3	75.0 CH	4.06 t (9.3)	C: 2, 4 Xyl2	H-1, 5 Xyl2
4	**79.9** CH	3.88 m	C: 1 Glc3	H-1 Glc3
5	64.5 CH_2	4.24 dd (5.1; 11.0)	C: 1, 3, 4 Xyl2	
		3.45 t (9.3)		H-1, 3 Xyl2
Glc3 (1→4Xyl2)				
1	103.9 CH	4.71 d (8.0)	C: 4 Xyl2	H-4 Xyl2; H-3, 5 Glc3
2	73.8 CH	3.75 t (8.0)	C: 1, 3 Glc3	H-4 Glc3
3	76.8 CH	4.08 t (9.3)	C: 2, 4 Glc3	H-1 Glc3
4	70.6 CH	3.84 t (9.3)	C: 3, 5, 6 Glc3	
5	75.1 CH	4.03 t (8.7)		
6	*67.6* CH_2	4.97 d (10.7)		
		4.63 dd (6.7; 10.7)	C: 5 Glc3	
Glc4 (1→4Xyl1)				
1	103.9 CH	4.82 d (8.1)	C: 4 Xyl1	H-4 Xyl1; H-3, 5 Glc4
2	73.9 CH	3.82 t (8.1)	C: 1, 3 Glc4	
3	76.8 CH	4.10 t (9.0)	C: 2, 4 Glc4	H-1 Glc4
4	71.0 CH	3.86 t (9.0)	C: 3, 5, 6 Glc4	
5	75.1 CH	4.11 m		H-1 Glc4
6	*67.9* CH_2	5.02 d (9.0)		
		4.57 dd (7.8; 11.2)	C: 5 Glc4	

The (−)ESI-MS/MS of **7** demonstrated the fragmentation of $[M_{2Na} - Na]^-$ ion at *m/z* 1281.4. The peaks of fragment ions were observed at *m/z*: 769.1 $[M_{2Na} - Na - C_{32}H_{47}O_5$ (Agl) $- H]^-$, 505.2 $[M_{2Na} - Na - C_{32}H_{47}O_5$ (Agl) $- C_6H_{10}O_8SNa$ $(GlcSO_3Na)]^-$, 373.0 2 $[M_{2Na} - Na - C_{32}H_{47}O_5$ (Agl) $- GlcSO_3Na - Xyl]^-$, 241.0 2 $[M_{2Na} - Na - C_{32}H_{47}O_5$ (Agl) $- GlcSO_3Na - 2Xyl]^-$. The (+)ESI-MS/MS of **7** demonstrated the fragmentation of $[M_{2Na} + Na]^+$ ion at *m/z* 1327.4. The peaks of fragment ions were observed at *m/z*: 1207.5 $[M_{2Na} + Na - NaHSO_4]^+$, 1147.4 $[M_{2Na} + Na - NaHSO_4 - CH_3COOH]^+$, 1063.4 $[M_{2Na} + Na - C_6H_{10}O_8SNa$ $(GlcSO_3Na)]^+$, 1003.4 $[M_{2Na} + Na - GlcSO_3Na - CH_3COO]^+$, 931.4 $[M_{2Na} + Na - GlcSO_3Na - Xyl + H]^+$, 833.1 $[M_{2Na} + Na - C_{32}H_{47}O_5$ (Agl) $+H]^+$. All these data exhaustively confirmed the structure psolusoside I (**7**) deduced by analyses of NMR data.

All these data indicate that psolusoside I (**7**) is 3β-*O*-{6-*O*-sodium sulfate-β-D-glucopyranosyl-(1→4)-β-D-xylopyranosyl-(1→2)-[6-*O*-sodium sulfate-β-D-glucopyranosyl-(1→4)]-β-D-xylopyranosyl}-16β-acetoxyholosta-7,25-diene.

Table 11. ^{13}C and ^{1}H NMR chemical shifts and HMBC and ROESY correlations of aglycone moiety of psolusoside I (**7**). [a] Recorded at 176.04 MHz in C_5D_5N/D_2O (4/1). [b] Recorded at 700.13 MHz in C_5D_5N/D_2O (4/1).

Position	δ_C mult. [a]	δ_H mult. [b] (J in Hz)	HMBC	ROESY
1	35.8 CH_2	1.32 m		H-3, H-11, H-31
2	26.8 CH_2	1.98 m		
		1.82 m		H-19, H-30
3	88.9 CH	3.19 dd (3.9; 11.6)	C: 4, 30, 31	H-1, H-5, H-31, H-1 Xyl1
4	39.2 C			
5	47.9 CH	0.90 dd (4.5; 11.4)	C: 4, 10, 19, 30, 31	H-3, H-31
6	23.1 CH_2	1.94 m		H-19, H-30
7	120.2 CH	5.60 m		H-15, H-32
8	145.6 C			
9	47.0 CH	3.32 brd (13.8)		H-19
10	35.4 C			
11	22.4 CH_2	1.71 m		H-1
		1.46 m		H-32
12	31.2 CH_2	2.10 m	C: 13, 14, 18	H-12, H-17, H-32
13	59.3 C			
14	47.3 C			
15	43.5 CH_2	2.55 dd (7.3; 12.0)	C: 13, 14, 17, 32	H-7, H-32
		1.62 dd (9.0; 12.0)	C: 14, 16, 32	
16	75.2 CH	5.82 brq (8.6)	OAc	H-32
17	54.5 CH	2.67 d (9.1)	C: 12, 13, 18, 21	H-12, H-21, H-32
18	180.2 C			
19	23.8 CH_3	1.11 s	C: 1, 5, 9, 10	H-1, H-2, H-6, H-9
20	85.5 C			
21	28.0 CH_3	1.52 s	C: 17, 20, 22	H-12, H-17, H-22
22	38.1 CH_2	1.82 dt (3.6; 12.7)	C: 20, 23	H-21
		1.25 dt (4.5; 12.7)	C: 20, 23	
23	22.9 CH_2	1.47 m	C: 22, 24	
		1.36 m		
24	38.1 CH_2	1.92 m	C: 22, 23, 25, 26	H-26
		1.82 m		
25	145.4 C			
26	110.8 CH_2	4.73 brs	C: 24, 27	H-27
		4.72 brs		
27	22.0 CH_3	1.65 s	C: 24, 25, 26	
30	16.7 CH_3	0.94 s	C: 3, 4, 5, 31	H-2, H-6
31	28.2 CH_3	1.12 s	C: 3, 4, 5, 30	H-3, H-5, H-6, H-1 Xyl1
32	32.1 CH_3	1.15 s	C: 8, 13, 14, 15	H-7, H-11, H-12, H-15, H-17
OAc	170.7 C			
	21.2 CH_3	2.02 s	OAc	

2.2. Bioactivity of the Glycosides

The cytotoxic activities of compounds **2**–**7** as well as earlier known psolusoside A (was used as positive control [26]) against mouse erythrocytes (hemolytic activity), the ascite form of mouse Ehrlich carcinoma cells, neuroblastoma Neuro 2A cells, and normal epithelial JB-6 cells are presented in Table 12. The investigated substances, except psolusoside I (**7**) containing a xylose in the second position of carbohydrate chain, demonstrate high hemolytic action, but the majority of them are not active or slightly active against mouse Ehrlich carcinoma cells (ascite form) (except psolusoside A which has a moderate cytotoxic action).

Psolusosides A and E (**2**) are the strongest cytotoxins in this series. The cytotoxicities of psolusosides F (**3**), H (**5**), H_1 (**6**), and I (**7**) against normal JB-6 cells are comparable with those against neuroblastoma Neuro 2A cells. However, there is one interesting exception: Psolusoside G (**4**) (disulfated linear tetraoside with a glucose as second sugar in the chain) is not cytotoxic against normal JB-6 cells but demonstrates high activity against Neuro 2A cells. It opens a possibility to study this compound on models of neurodegenerative diseases. Noteworthy, the cytotoxicity of psolusosides H (**5**) and H_1 (**6**) (the glycosides having trisaccharide chains) and psolusoside I (**7**) (the compound with tetrasaccharide branched carbohydrate chain and a xylose as the second unit) are similar to that of linear tetraosides—psolusosides A, E (**2**), F (**3**), and G (**4**)—whereas it is known that the presence of linear tetrasaccharide chain is one of the necessary conditions for the displaying of membranolytic activity [4,6].

Table 12. Hemolytic activities of glycosides **2–7** and psolusoside A against mouse erythrocytes, cytotoxic activity against the ascite form of mouse Ehrlich carcinoma cells, mouse neuroblasoma Neuro 2A cells, and normal epithelial JB-6 cells.

Glycoside	**Cytotoxicity EC_{50}, μM**			
	Erhythrocytes	**Ehrlich carcinoma**	**Neuro-2A**	**JB-6**
Psolusoside E (**2**)	0.23	55.75	3.96	-
Psolusoside F (**3**)	2.8	96.2	10.8	23.8
Psolusoside G (**4**)	4.2	98.3	7.3	>100.0
Psolusoside H (**5**)	2.5	>100.0	47.5	43.2
Psolusoside H_1 (**6**)	2.7	92.5	10.7	32.3
Psolusoside I (**7**)	18.3	87.4	26.8	37.2
Psolusoside A	1.4	30.9	2.9	7.5

The influence of the psolusosides A, B (**1**), E (**2**), F (**3**), H (**5**), H_1 (**6**), and I (**7**) on cell viability, formation, and growth of colonies of human colorectal adenocarcinoma HT-29 cells was checked. HT-29 cells were treated with various concentrations (0–20 μM) of compounds **1–3**, **5–7** and earlier known psolusoside A for 24 h, and then cell viability was assessed by the MTS assay. It was shown that all investigated compounds are not cytotoxic against HT-29 cells at the dose of 20 μM. The concentrations 10 μM were chosen for the investigation of the glycosides influence on the colony formation of HT-29 cells in soft agar assay. The data concerning inhibitory activity of psolusosides A, B (**1**), E (**2**), F (**3**), H (**5**), H_1 (**6**), and I (**7**) on colony formation of HT-29 cells are presented in Table 13. The highest inhibition of colony formation and growth of HT-29 cells demonstrate psolusosides E (**2**) and F (**3**) ($ICCF_{50}$ 0.1 μM and 0.5 μM, respectively) having the holostane aglycones and linear tetrasaccharide monosulfated carbohydrate chains with the quinovose residue as the second sugar unit. The inhibitory action of compounds **6** and **7** is observed only at the doses 9 and 10 μM, respectively. While psolusosides A, B (**1**), and H (**5**) did not inhibit the colony formation and growth of HT-29 cells for 50% under concentration of 10 μM. Thus, the presence of trisaccharide (as in psolusosides H (**5**) and H_1 (**6**)) or branched tetrasaccharide (as in psolusosides B (**1**) and I (**7**)) chains, even in combination with the holostane aglycones (as in **5–7**), cause the loss of this type of bioactivity.

Table 13. Cytotoxic activities of psolusosides A, B (**1**), E (**2**), F (**3**), H (**5**), H_1 (**6**), and I (**7**) against the HT-29 cells and ability for inhibition of their colony.

Glycoside	**MTS Assay**	**Soft Agar Assay**
	IC_{50}	**$ICCF_{50}$, μM**
psolusoside B (**1**)	>20 μM	>10
psolusoside E (**2**)	>20 μM	0.1±0.03
psolusoside F (**3**)	>20 μM	0.5±0.03
psolusoside H (**5**)	>20 μM	>10
psolusoside H_1 (**6**)	>20 μM	9±0.3
psolusoside I (**7**)	>20 μM	10±0.4
psolusoside A	>20 μM	>10

We also try to study synergic effects of these compounds (0.05 μM) and radioactive irradiation (2 Gy) of HT-29 cells. The number of colonies of HT-29 cells was found to be decreased after radiation exposure at a dose of 2 Gy, but synergic effects of the glycosides and radioactive irradiation (2 Gy) decreasing the number of colonies was not observed.

3. Materials and Methods

3.1. General Experimental Procedures

Specific rotation, Perkin-Elmer 343 Polarimeter; NMR, Bruker Avance III 500 (Bruker BioSpin GmbH, Rheinstetten, Germany) (500.13/125.77 MHz) or Avance III 700 Bruker FT-NMR (Bruker BioSpin GmbH, Rheinstetten, Germany) (700.13/176.04 MHz) ($^1H/^{13}C$) spectrometers; ESI MS (positive and negative ion modes), Agilent 6510 Q-TOF (Agilent Technologies, Santa Clara, CA, USA) apparatus, sample concentration 0.01 mg/mL; HPLC, Agilent 1100 apparatus with a differential refractometer; columns Supelco Ascentis RP-Amide (10×250 mm, 5 μm), Kromasil Cellucoat RP (4.6×150 mm, 5 μm), and Supelcosil LC-Si (4.6×150 mm, 5 μm).

3.2. Animal Material

Specimens of the sea cucumber *Psolus fabricii* (family Psolidae; order Dendrochirotida) were collected in the Sea of Okhotsk near Onekotan Island (Kurile Islands). Sampling was performed with a scallop dredge in August to September,1982 at a depth of 100 m during expedition works on fishing seiners "Mekhanik Zhukov" and "Dalarik". Sea cucumbers were identified by Prof. V.S. Levin; voucher specimens are preserved in A.V. Zhirmunsky National Scientific Center of Marine Biology, Vladivostok, Russia.

3.3. Extraction and Isolation

The sea cucumbers were minced and extracted twice with refluxing 60% EtOH. The extract was evaporated to water residuum and lyophilized followed by the extraction with $CHCl_3$/MeOH (1:1). The obtained extract was evaporated and submitted to the subsequent extraction by EtOAc/H_2O to remove the lipid fraction. The water layer remaining after this extraction was chromatographed on a Polychrom-1 column (powdered Teflon, Biolar, Latvia). The glycosides were eluted with 50% EtOH, evaporated, and subsequently chromatographed on Si gel columns with $CHCl_3$/EtOH/H_2O (100:75:10), (100:100:17), (100:125:25) as mobile phase to give subfractions III–VIII containing different groups of glycosides. These subfractions kept at the temperature −18 °C, then were submitted to HPLC on silica-based (Supelcosil LC-Si) and reversed phase (Supelco Ascentis RP-Amide, Kromasil Cellucoat RP) columns with different solvent systems as mobile phase. The subfraction III was chromatographed on Supelco Ascentis RP-Amide column with 35% acetonitrile (CH_3CN) and then on Supelcosil LC-Si with $CHCl_3$/MeOH/H_2O (85/20/2) as mobile phases to give 10 mg of psolusoside E (**2**). The subfraction IV was submitted to HPLC on Supelco Ascentis RP-Amide column with CH_3CN/H_2O/NH_4OAc (1 M water solution) (45/54/1) to give 1.4 mg of psolusoside H_1 (**6**) as well as five other fractions. Fraction 3 was chromatographed on the column Kromasil Cellucoat-RP with 23% CH_3CN that resulted in isolation of colochiroside D (2.5 mg). The subsequent HPLC of the fraction 4 on Supelco Ascentis RP-Amide column with CH_3CN/H_2O/NH_4OAc (40/59/1) gave 1.4 of psolusoside H (**5**). Fraction 5 was submitted to HPLC on Supelcosil LC-Si column with $CHCl_3$/MeOH/H_2O (65/20/2) as mobile phase to give pure psolusoside F (**3**) (1.4 mg). The subfraction V obtained after Si-gel column chromatography was submitted to HPLC on Supelco Ascentis RP-Amide column with CH_3CN/H_2O/NH_4OAc (40/59/1) resulted in isolation of known psolusoside A (37 mg). HPLC of subfraction VI on the same column with other ratio of the solvents used before—(43/55/2)—gave 46.5 mg of psolusoside G (**4**). Subfraction VII was submitted to HPLC on Kromasil Cellucoat-RP column with 14% CH_3CN followed by HPLC on Supelcosil LC-Si column with $CHCl_3$/MeOH/H_2O (65/20/2) as mobile phase to give 1.1 mg of psolusoside I (**7**). Psolusoside B (**1**) (67 mg) was isolated from subfraction VIII as result of HPLC on Supelco Ascentis RP-Amide column with CH_3CN/H_2O/NH_4OAc (35/64/1) as mobile phase.

3.3.1. Psolusoside B (1)

Colorless powder; $[\alpha]_D^{20}$ −60 (*c* 0.1, 50% MeOH). NMR: See Tables 1 and 2. (+)HR-ESI-MS *m/z*: 1357.4169 (calc. 1357.4176) $[M_{2Na} + Na]^+$, 690.2039 (calc. 690.2034) $[M_{2Na} + 2Na]^{2+}$; (+)ESI-MS/MS *m/z*:

1297.4 [M_{2Na} + Na – $CH_3COOH]^+$, 1237.4 [M_{2Na} + Na – $NaHSO_4]^+$, 1177.4 [M_{2Na} + Na – CH_3COOH – $NaHSO_4]^+$, 1117.4 [M_{2Na} + Na – $2NaHSO_4]^+$, 1075.5 [M_{2Na} + Na – $C_6H_{10}O_9SNa$ ($GlcSO_3Na$)]$^+$, 913.4 [M_{2Na} + Na – $GlcSO_3Na$ – Glc]$^+$, 863.1 [M_{2Na} + Na – $C_{32}H_{47}O_4$ (Agl) + H]$^+$, 743.1 [M_{2Na} + Na – $C_{32}H_{47}O_4$ (Agl) – $NaHSO_4]^+$, 581.1 [M_{2Na} + Na – $C_{32}H_{47}O_4$ (Agl) – $C_6H_{10}O_9SNa$ ($GlcSO_3Na$)]$^+$, 449.0 [M_{2Na} + Na – $C_{32}H_{47}O_4$ (Agl) – $C_6H_{10}O_9SNa$ ($GlcSO_3Na$) – Xyl]$^+$, 287.0 [M_{2Na} + Na – $C_{32}H_{47}O_4$ (Agl) – $C_6H_{10}O_9SNa$ ($GlcSO_3Na$) – Xyl – Glc]$^+$.

3.3.2. Psolusoside E (2)

Colorless powder; $[\alpha]_D^{20}$ −25 (*c* 0.1, 50% MeOH). NMR: See Tables 3 and 4. (−)HR-ESI-MS *m/z*: 1163.4945 (calc. 1163.4950) [M_{Na} – Na]$^-$; (−)ESI-MS/MS *m/z*: 1163.5 [M_{Na} – Na]$^-$, 987.4 [M_{Na} – Na – MeGlc + H]$^-$, 695.2 [M_{Na} – Na – $C_{30}H_{43}O_4$ (Agl) – H]$^-$, 563.1 [M_{Na} – Na – $C_{30}H_{43}O_4$ (Agl) – Xyl]$^-$, 417.1 [M_{Na} – Na – $C_{30}H_{43}O_4$ (Agl) – Xyl – Qui]$^-$, 241.0 [M_{Na} – Na – $C_{30}H_{43}O_4$ (Agl) – Xyl – Qui – MeGlc]$^-$.

3.3.3. Psolusoside F (3)

Colorless powder; $[\alpha]_D^{20}$ −50 (*c* 0.1, 50% MeOH). NMR: See Tables 3 and 5. (−)HR-ESI-MS *m/z*: 1163.4952 (calc. 1163.4950) [M_{Na} – Na]$^-$; (−)ESI-MS/MS *m/z*: 1163.5 [M_{Na} – Na]$^-$, 695.2 [M_{Na} – Na – $C_{30}H_{43}O_4$ (Agl) – H]$^-$, 563.1 [M_{Na} – Na – $C_{30}H_{43}O_4$ (Agl) – Xyl]$^-$, 417.1 [M_{Na} – Na – $C_{30}H_{43}O_4$ (Agl) – Xyl – Qui]$^-$, 255.0 [M_{Na} – Na – $C_{30}H_{43}O_4$ (Agl) – Xyl – Qui – Glc]$^-$.

3.3.4. Psolusoside G (4)

Colorless powder; $[\alpha]_D^{20}$ −49 (*c* 0.1, 50% MeOH). NMR: See Tables 3 and 6. (−)HR-ESI-MS *m/z*: 1281.4313 (calc. 1281.4286) [M_{2Na} – Na]$^-$; (−)ESI-MS/MS *m/z*: 1281.4 [M_{2Na} – Na]$^-$, 1161.5 [M_{2Na} – Na – $HSO_4Na]^-$, 1003.4 [M_{2Na} – Na – HSO_4Na – $C_7H_{12}O_8SNa$ ($MeGlcSO_3Na$)]$^-$, 813.2 [M_{2Na} – Na – $C_{30}H_{43}O_4$ (Agl) – H]$^-$, 681.1 [M_{2Na} – Na – $C_{30}H_{43}O_4$ (Agl) – Xyl]$^-$, 519.0 [M_{2Na} –Na – $C_{30}H_{43}O_4$ (Agl) – Xyl – Glc]$^-$, 255.0 [M_{2Na} – Na – $C_{30}H_{43}O_4$ (Agl) – Xyl – Glc – $C_6H_9O_8SNa$ ($GlcSO_3Na$)]$^-$.

3.3.5. Psolusoside H (5)

Colorless powder; $[\alpha]_D^{20}$ −35 (*c* 0.1, 50% MeOH). NMR: See Tables 7 and 8. (−)HR-ESI-MS *m/z*: 1003.4213 (calc. 12003.4214) [M_{Na} – Na]$^-$; (−)ESI-MS/MS *m/z*: 1003.4 [M_{Na} – Na]$^-$, 535.1 [M_{Na} – Na – $C_{30}H_{43}O_4$ (Agl) – H]$^-$, 403.1 [M_{Na} – Na – $C_{30}H_{43}O_4$ (Agl) – Xyl]$^-$, 241.0 [M_{Na} – Na – $C_{30}H_{43}O_4$ (Agl) – Xyl – Glc]$^-$.

3.3.6. Psolusoside H_1 (6)

Colorless powder; $[\alpha]_D^{20}$ −23 (*c* 0.1, 50% MeOH). NMR: See Tables 7 and 9. (−)HR-ESI-MS *m/z*: 989.4432 (calc. 989.4421) [M_{Na} – Na]$^-$; (+)HR-ESI-MS *m/z*: 1035.4205 (calc. 1035.4206) [M_{Na} + Na]$^+$; (−)ESI-MS/MS *m/z*: 989.4 [M_{Na} – Na]$^-$, 535.1 [M_{Na} – Na – $C_{30}H_{45}O_3$ (Agl) – H]$^-$, 403.1 [M_{Na} – Na – $C_{30}H_{45}O_3$ (Agl) – Xyl]$^-$, 241.0 [M_{Na} – Na – $C_{30}H_{45}O_3$ (Agl) – Xyl – Glc]$^-$.

3.3.7. Psolusoside I (7)

Colorless powder; $[\alpha]_D^{20}$ −17 (*c* 0.1, 50% MeOH). NMR: See Tables 10 and 11. (−)HR-ESI-MS *m/z*: 1281.4267 (calc. 1281.4286) [M_{2Na} – Na]$^-$; (+)HR-ESI-MS *m/z*: 1327.4065 (calc. 1327.4071) [M_{2Na} + Na]$^+$; (−)ESI-MS/MS *m/z*: 1281.4 [M_{2Na} – Na]$^-$, 769.1 [M_{2Na} – Na – $C_{32}H_{47}O_5$ (Agl) – H]$^-$, 505.2 [M_{2Na} – Na – $C_{32}H_{47}O_5$ (Agl) – $C_6H_{10}O_8SNa$ ($GlcSO_3Na$)]$^-$, 373.0 2 [M_{2Na} – Na – $C_{32}H_{47}O_5$ (Agl) – $GlcSO_3Na$ – Xyl]$^-$, 241.0 2 [M_{2Na} – Na – $C_{32}H_{47}O_5$ (Agl) – $GlcSO_3Na$ – 2Xyl]$^-$; (+)ESI-MS/MS *m/z*: 1327.4 [M_{2Na} + Na]$^+$, 1207.5 [M_{2Na} + Na –$NaHSO_4]^+$, 1147.4 [M_{2Na} + Na – $NaHSO_4$ – $CH_3COOH]^+$, 1063.4 [M_{2Na} + Na – $C_6H_{10}O_8SNa$ ($GlcSO_3Na$)]$^+$, 1003.4 [M_{2Na} + Na – $GlcSO_3Na$ – $CH_3COO]^+$, 931.4 [M_{2Na} + Na – $GlcSO_3Na$ – Xyl + H]$^+$, 833.1 [M_{2Na} + Na – $C_{32}H_{47}O_5$ (Agl) +H]$^+$.

3.4. Cell Culture

The museum tetraploid strain of murine ascite Ehrlich carcinoma cells from the All-Russian Oncology Center (Moscow, Russia) was used. The cells were separated from the ascites, which were collected on day 7 after inoculation in mouse CD-1 line. The cells were washed of the ascites triply and resuspended in RPMI-1640 medium containing 8 μg/mL gentamicin (BioloT, St-Petersburg, Russia). Neuroblastoma Neuro 2A cells were cultured in DMEM medium containing 10% fetal bovine serum (FBS), (BioloT, St-Petersburg, Russia) and normal epithelial JB-6 cells were cultured in DMEM medium containing 5% fetal bovine serum (FBS), (BioloT, St-Petersburg, Russia) and 1% penicillin/streptomycine (Invitrogen). The HT-29 (ATCC # HTB-38) human colon cancer cell line was grown in monolayer in McCoy's 5A medium supplemented with 10% (v/v) heat-inactivated FBS, 2 mM L-glutamine, and 1% penicillin-streptomycin in a humidified atmosphere containing 5% CO_2. Cells were maintained in a sterile environment and kept in an incubator at 5% CO_2 and 37 °C to promote growth. HT-29 cells were sub-cultured every 3–4 days by rinsing with phosphate buffered saline (PBS), adding trypsin to detach the cells from the tissue culture flask, and transferring 10%–20% of the harvested cells to a new flask containing fresh growth media.

3.5. Cytotoxic Activity

3.5.1. Nonspecific Esterase Activity Assay

Cytotoxic activity against ascite form of mouse Ehrlich carcinoma cells was investigated by nonspecific esterase activity assay. In total, 10μL of the test substance solution and 100 μL of the cell suspension were placed into each well of a 96-well microplate. The plate was incubated in a CO_2 incubator at 37 °C for 1 or 24 h. A stock solution of the probe fluorescein diacetate (FDA; Sigma, St. Louis, MO, USA) in DMSO (1 mg/mL) was prepared. After incubation of the cells with test compounds, 10 μL of FDA solution (50 μg/mL) was added to each well and the plate was incubated at 37 °C for 15 min. Cells were washed with PBS, and fluorescence was measured with a Fluoroskan Ascent plate reader (Thermo Labsystems, Helsinky, Finland) at λex = 485 nm and λem = 518 nm. All experiments were repeated in triplicate. Cytotoxic activity was expressed as the percent of cell viability. Nonspecific esterase activity assay has been used for determination of cytotoxicity against neuroblastoma Neuro 2A and normal epithelial JB-6 cells.

3.5.2. MTT Assay

The solutions of tested substances in different concentrations (20 μL) and cell suspension (200 μL) were added in wells of 96-well plate and incubated over night at 37 °C and 5% CO_2. After incubation, the cells were sedimented by centrifugation, 200 μL of medium from each well were collected, and 100 μL of pure medium were added. Then, 10 μL of MTT solution 5 μg/mL (Sigma, St. Louis, MO, USA) were added in each well. The plate was incubated for 4 h, after which 100 μL SDS-HCl were added to each well, and the plate was incubated at 37 °C for 4–18 h. Optical density was measured at 570 nm and 630–690 nm. Cytotoxic activity of the substances was calculated as the concentration that caused 50% metabolic cell activity inhibition (IC_{50}). The MTT assay has been used for determination of cytotoxic activity against Ehrlich carcinoma cells.

3.5.3. MTS Assay

The HT-29 cells (1.0×10^4/well) were seeded in 96-well plates for 24 h at 37 °C in 5% CO_2 incubator. The cells were treated with the tested substances at concentrations range from 0 to 20 μM for an additional 24 h. Subsequently, cells were incubated with 15 μL MTS reagent for 3 h and the absorbance in each well was measured at 490/630 nm using microplate reader "Power Wave XS" (Bio Tek, Winooski, VT, USA). All the experiments were repeated three times, and the mean absorbance values were calculated. The results are expressed as the percentage of inhibition that produced a reduction in absorbance by compound's treatment compared to the non-treated cells (control).

3.6. *Hemolytic Activity*

Blood was taken from CD-1 mice (18–20 g). The mice were anesthetized with diethyl ether, their chests were rapidly opened, and blood was collected in cold (4 °C) 10 mM phosphate-buffered saline, pH 7.4 (PBS) without anticoagulant. Erythrocytes were washed 3 times in PBS using at least 10 vol. of washing solution by centrifugation (2000 rpm) for 5 min. Erythrocytes were used at a concentration that provided an optical density of 1.0 at 700 nm for a non-hemolyzed sample. Then, 20 µL of a water solution of test substance with a fixed concentration was added to a well of a 96-well plate containing 180 µL of the erythrocyte suspension. Erythrocyte suspension was incubated with substances for 24 h at 37 °C. After that, the optical density of the obtained solutions was measured and ED_{50} for hemolytic activity of each compound was calculated.

3.7. *Soft Agar Assay*

The HT-29 cells (8.0×10^3) were seeded in 6-well plate and treated with the tested compounds at concentrations range 0–10 µM in 1 mL of 0.3% Basal Medium Eagle (BME) agar containing 10% FBS, 2mM L-glutamine, and 25 µg/mL gentamicin. The cultures were maintained at 37 °C in a 5% CO_2 incubator for 14 days, and the cell's colonies were scored using a microscope "Motic AE 20" (Scientific Instrument Company, Campbell, CA, USA) and the Motic Image Plus version 2.0 (Scientific Instrument Company, Campbell, CA, USA) computer program.

3.8. *Radiation Exposure*

Irradiation was delivered at room temperature using single doses of X-ray system XPERT 80 (KUB Technologies, Inc, Milford, CT, USA). The dose 2 Gy was used for colony formation assay. The absorber dose was measured using X-ray radiation clinical dosimeter DRK-1 (Moscow, Russia).

3.9. *Cell Irradiation*

The HT-29 cells (6.0×10^5) were plated at 60 mm dishes and incubated for 24 h. After the incubation, the cells were cultured in the presence or absence of 0.05 µM of tested compounds for additional 24 h before irradiation at the dose of 2 Gy. Immediately after irradiation, the cells were returned to the incubator for recovery. Three hours later, the cells were harvested and used for soft agar assay to establish the synergism of radioactive irradiation and investigated compounds effects on colony formation of tested cells.

4. Conclusions

Seven individual compounds **1**–**7**, including mono- and disulfated, linear, and branched tetraosides, as well as monosulfated triosides, were isolated from the sea cucumber *Psolus fabricii* in addition to recently obtained non-sulfated hexaosides [14,15]. The structural analysis of the glycosides **1**–**7** demonstrated the variability of their aglycones and carbohydrate chains. Five aglycones and six different carbohydrate chains were found in these compounds. Although all the aglycones were earlier known, five types of sugar chains in these glycosides were new. Three linear tetraosides—psolusosides E (**2**), F (**3**), and G (**4**)—are biogenetically interrelated. These compounds share the same aglycone and differ from each other in positions and numbers of sulfate groups and in the nature of the second monosaccharide residue in the carbohydrate chain (quinovose or glucose). The compounds **2**–**4** and two disulfated branched tetraosides —psolusosides B (**1**) and I (**7**)—altogether are a good illustration of the mosaicism of carbohydrate chain biosynthesis. Firstly, diverse monosaccharide residues (quinovose, glucose, or xylose) can glycosylate C(2) of the first xylose unit; secondly, the fourth (terminal) monosaccharide residue can bind to C(3) of the third monosaccharide unit (glucose), which resulted in the formation of linear chains of psolusosides E (**2**), F (**3**), and G (**4**), or to C(4) of the first (xylose) residue resulted in the formation of branched chains of the glycosides **1** and **7**. At that, carbohydrate chain of psolusosides H (**5**) and H_1 (**6**) can be biosynthetic precursor for the both type of tetrasaccharide chains—linear chain of psolusoside G (**4**) and branched chain of psolusoside B (**1**) (Figure 2).

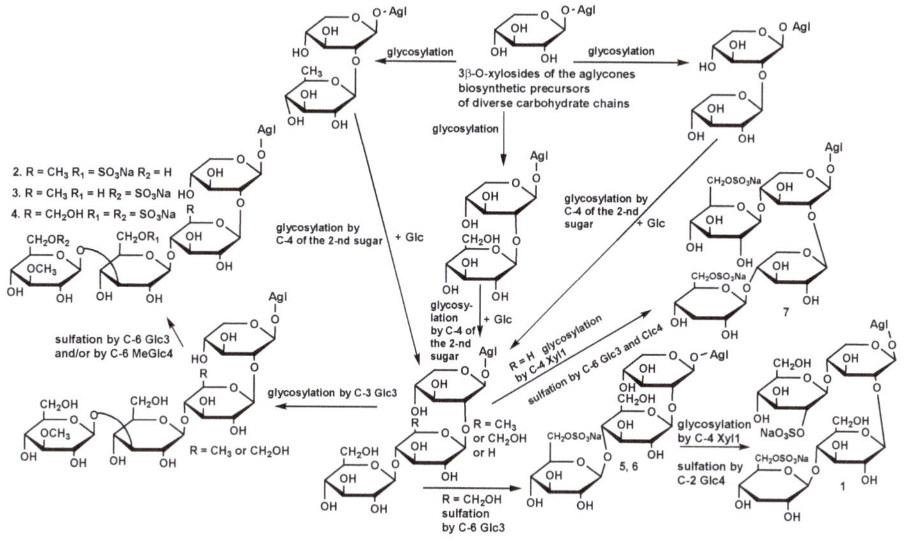

Figure 2. The hypothetic scheme of the carbohydrate chains biosynthesis of the glycosides of *P. fabricii*.

The aglycones of triosides, psolusosides H (**5**) and H_1 (**6**), and branched tetraosides, psolusosides B (**1**) and I (**7**), are structurally diverse. These aglycones belong either to non-holostane (with 18(16)-lactone—as in **1**) or to holostane (with 18(20)-lactone as in **5–7**) types and share the presence of 7(8)- and 25(26)-double bonds. It is known that holostane-type aglycones are biosynthesized via the hydroxylation of C(20) in triterpene precursor followed by C(18) oxidation, resulting in the formation of 18(20)-lactone. When the hydroxyl groups are simultaneously present at C(16) and C(20) of 18-carboxylated derivative, the formation of 18(16)-lactone occurred [25]. This situation is obviously realized in the aglycone of psolusoside B (**1**). The acetylation of C(16) (as in psolusoside I (**7**)) or oxidation of the corresponding hydroxyl group at C(16) to a carbonyl (as in psolusoside H (**5**)) prevent the formation of 18(16)-lactone and lead to the synthesis of holostane derivatives. However, the incorporation of this type of functionalities to C(16) could be also realized after the 18(20)-lactonization due to the mosaic type of biosynthesis of triterpene glycosides of sea cucumbers. Hence, unsubstituted at C(16) precursors of holostane aglycone as well as their acetates and 16-keto derivatives may give holostane glycosides after previous 18(20)-lactonization, as it realized at biosynthesis of **2–4** and **5–7** (Figure 3).

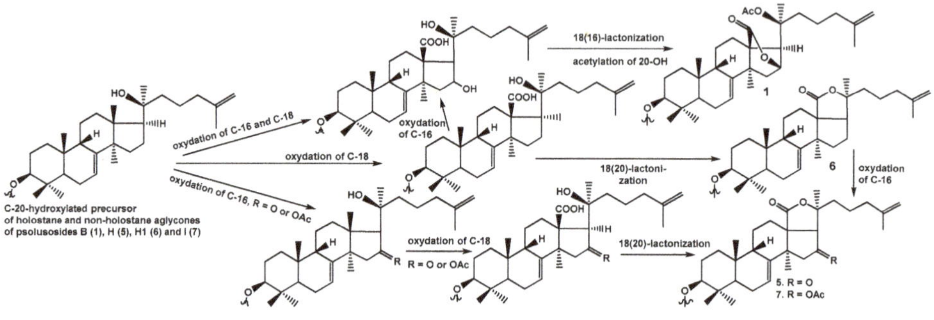

Figure 3. The hypothetic scheme of biosynthesis of holostane and non-holostane aglycones of the glycosides of *P. fabricii*.

Supplementary Materials: The following are available online at http://www.mdpi.com/1660-3397/17/6/358/s1.

Author Contributions: Conceptualization, A.S.S., S.A.A., and V.I.K.; methodology, A.S.S., S.A.A., and S.P.E.; investigation, A.S.S., A.I.K., S.A.A., V.I.K., P.S.D., R.S.P., E.A.C., O.S.M., S.P.E., and P.V.A.; writing—original draft preparation, A.S.S. and V.I.K.; writing—review and editing, V.I.K.

Funding: The chemical structure and part of bioassay were carried out at partial financial support of the Grant of the Russian Foundation for Basic Research No. 19-04-000-14. The studies of cytotoxic activities on a series of human cancer cell lines and investigation of the radiosensitizing effects were supported by the Grant of the Russian Science Foundation No. 16-14-10131.

Acknowledgments: Authors thank Valentin A. Stonik for reading the manuscript and critical remarks.

Conflicts of Interest: The authors declare no conflict of interest.

References

1. Kalinin, V.I.; Silchenko, A.S.; Avilov, S.A.; Stonik, V.A.; Smirnov, A.V. Sea cucumbers triterpene glycosides, the recent progress in structural elucidation and chemotaxonomy. *Phytochem. Rev.* **2005**, *4*, 221–236. [CrossRef]
2. Bahrami, Y.; Franko, C.M.M. Acetylated triterpene glycosides and their biological activity from Holothuroidea reported in the past six decades. *Mar. Drugs* **2016**, *14*, 147. [CrossRef] [PubMed]
3. Mondol, M.A.M.; Shin, H.J.; Rahman, M.A. Sea cucumber glycosides: Chemical structures, producing species and important biological properties. *Mar. Drugs* **2017**, *15*, 317. [CrossRef] [PubMed]
4. Kalinin, V.I.; Aminin, D.L.; Avilov, S.A.; Silchenko, A.S.; Stonik, V.A. Triterpene glycosides from sea cucumbers (Holothuroidea, Echinodermata), biological activities and functions. In *Studies in Natural Product Chemistry (Bioactive Natural Products)*; Atta-ur-Rahman, Ed.; Elsevier Science Publisher: Amsterdam, The Netherlands, 2008; Volume 35, pp. 135–196.
5. Janakiram, N.B.; Mohammed, A.; Rao, C. Sea cucumbers metabolites as potent anti-cancer agents. *Mar. Drugs* **2015**, *13*, 2909–2923. [CrossRef] [PubMed]
6. Kim, S.K.; Himaya, S.W.A. Triterpene glycosides from sea cucumbers and their biological activities. *Adv. Food Nutr. Res.* **2012**, *63*, 297–319.
7. Aminin, D.L.; Pislyagin, E.A.; Menchinskaya, E.S.; Silchenko, A.S.; Avilov, S.A.; Kalinin, V.I. Immunomodulatory and anticancer activity of sea cucumber triterpene glycosides. In *Studies in Natural Products Chemistry (Bioactive Natural Products)*; Atta-ur-Rahman, Ed.; Elsevier Science Publisher: Amsterdam, The Netherlands, 2014; Volume 41, pp. 75–94.
8. Careaga, V.P.; Maier, M.S. Cytotoxic triterpene glycosides from sea cucumbers. In *Handbook of Anticancer Drugs from Marine Origin*; Kim, S.-K., Ed.; Springer: Cham, Switzerland, 2015; pp. 515–528.
9. Aminin, D.L.; Menchinskaya, E.S.; Pisliagin, E.A.; Silchenko, A.S.; Avilov, S.A.; Kalinin, V.I. Sea cucumber triterpene glycosides as anticancer agents. In *Studies in Natural Products Chemistry*; Atta-ur-Rahman, Ed.; Elsevier Science Publisher: Amsterdam, The Netherlands, 2016; Volume 49, pp. 55–105.

10. Kalinin, V.I.; Kalinovskii, A.I.; Stonik, V.A. Psolusoside A—A new triterpene glycoside from the holothurian *Psolus fabricii*. *Chem. Nat. Compd.* **1983**, *19*, 753–754. [CrossRef]
11. Kalinin, V.I.; Kalinovskii, A.I.; Stonik, V.A. Structure of psolusoside A—The main triterpene glycoside from the holothurian *Psolus fabricii*. *Chem. Nat. Compd.* **1985**, *21*, 197–202. [CrossRef]
12. Kalinin, V.I.; Kalinovskii, A.I.; Stonik, V.A. Onekotanogenin—A new triterpene genine from the holothurian *Psolus fabricii*. *Chem. Nat. Compd.* **1987**, *23*, 560–563. [CrossRef]
13. Kalinin, V.I.; Kalinovskii, A.I.; Stonik, V.A.; Dmitrenok, P.S.; El'kin, Y.N. Structure of psolusoside B—A nonholostane triterpene glycoside from the holothurian genus *Psolus*. *Chem. Nat. Compd.* **1989**, *25*, 311–317. [CrossRef]
14. Silchenko, A.S.; Avilov, S.A.; Kalinovsky, A.I.; Kalinin, V.I.; Andrijaschenko, P.V.; Dmitrenok, P.S. Psolusosides C_1, C_2, and D_1, novel triterpene hexaosides from the sea cucumber *Psolus fabricii* (Psolidae, Dendrochirotida). *Nat. Prod. Commun.* **2018**, *13*, 1623–1628.
15. Silchenko, A.S.; Avilov, S.A.; Kalinovsky, A.I.; Kalinin, V.I.; Andrijaschenko, P.V.; Dmitrenok, P.S.; Popov, R.S.; Chingizova, E.A.; Kasakin, M.F. Psolusosides C_3 and D_2–D_5, Five novel triterpene hexaosides from the sea cucumber *Psolus fabricii* (Psolidae, Dendrochirotida): Chemical structures and bioactivities. *Nat. Prod. Commun.* **2019**, in press.
16. Silchenko, A.S.; Kalinovsky, A.I.; Avilov, S.A.; Andryjaschenko, P.V.; Dmitrenok, P.S.; Kalinin, V.I.; Yurchenko, E.A.; Dolmatov, I.Y. Colochirosides A_1, A_2, A_3 and D, four novel sulfated triterpene glycosides from the sea cucumber *Colochirus robustus* (Cucumariidae, Dendrochirotida). *Nat. Prod. Commun.* **2016**, *11*, 381–387. [CrossRef] [PubMed]
17. Silchenko, A.S.; Kalinovsky, A.I.; Avilov, S.A.; Andryjaschenko, P.V.; Dmitrenok, P.S.; Yurchenko, E.A.; Dolmatov, I.Y.; Dautov, S.S.; Stonik, V.A.; Kalinin, V.I. Colochiroside E, unusual non-holostane triterpene sulfated trioside from the sea cucumber *Colochirus robustus* and evidence of the impossibility of a 7(8)-double bond migration in lanostane derivatives having an 18(16)-lactone. *Nat. Prod. Commun.* **2016**, *11*, 741–746. [CrossRef] [PubMed]
18. Maltsev, I.I.; Stonik, V.A.; Kalinovsky, A.I.; Elyakov, G.B. Triterpene glycosides from sea cucumber *Stichopus japonicus* Selenka. *Comp. Biochem. Physiol. Part B.* **1984**, *78*, 421–426. [CrossRef]
19. Silchenko, A.S.; Avilov, S.A.; Kalinin, V.I.; Kalinovsky, A.I.; Dmitrenok, P.S.; Fedorov, S.N.; Stepanov, V.G.; Dong, Z.; Stonik, V.A. Constituents of the Sea Cucumber *Cucumaria okhotensis*. Structures of okhotosides B_1–B_3 and cytotoxic activities of some glycosides from this species. *J. Nat. Prod.* **2008**, *71*, 351–356. [CrossRef] [PubMed]
20. Avilov, S.A.; Kalinin, V.I.; Smirnov, A.V. Use of triterpene glycosides for resolving taxonomic problems in the sea cucumber genus *Cucumaria* (Holothuroidea, Echinodermata). *Biochem. Syst. Ecol.* **2004**, *32*, 715–733. [CrossRef]
21. Silchenko, A.S.; Kalinovskym, A.I.; Dmitrenok, P.S.; Kalinin, V.I.; Mazeika, A.N.; Vorobieva, N.S.; Sanina, N.M.; Kostetsky, E.Y. Cucumarioside E from the Far Eastern sea cucumber *Cucumaria japonica* (Cucumariidae, Dendrochirotida), new minor monosulfated holostane triterpene pentaoside with glucose as the second monosaccharide residue. *Nat. Prod. Commun.* **2015**, *10*, 877–880. [CrossRef] [PubMed]
22. Silchenko, A.S.; Kalinovsky, A.I.; Avilov, S.A.; Andryjaschenko, P.V.; Dmitrenok, P.S.; Kalinin, V.I.; Yurchenko, E.A.; Dautov, S.S. Structures of violaceusosides C–E and G, sulfated triterpene glycosides from the sea cucumber *Pseudocolochirus violaceus* (Cucumariidae, Denrochirotida). *Nat. Prod. Commun.* **2014**, *9*, 391–399.
23. Moraes, G.; Northcote, P.T.; Silchenko, A.S.; Antonov, A.S.; Kalinovsky, A.I.; Dmitrenok, P.S.; Avilov, S.A.; Kalinin, V.I.; Stonik, V.A. Mollisosides A, B_1 and B_2: Minor triterpene glycosides from the New Zealand and South Australian sea cucumber *Australostichopus mollis*. *J. Nat. Prod.* **2005**, *68*, 842–847. [CrossRef]
24. Miller, A.K.; Kerr, A.M.; Paulay, G.; Reich, M.; Wilson, N.G.; Carvajal, J.I.; Rouse, G.W. Molecular phylogeny of extant Holothuroidea (Echinodermata). *Mol. Phylogenet. Evol.* **2017**, *111*, 110–131. [CrossRef]

25. Silchenko, A.S.; Kalinovsky, A.I.; Avilov, S.A.; Andryjaschenko, P.V.; Dmitrenok, P.S.; Yurchenko, E.A.; Ermakova, S.P.; Malyarenko, O.S.; Dolmatov, I.Y.; Kalinin, V.I. Cladolosides C_4, D_1, D_2, M, M_1, M_2, N and Q, new triterpene glycosides with diverse carbohydrate chains from sea cucumber *Cladolabes schmeltzii*. An uncommon 20,21,22,23,24,25,26,27-okta-*nor*-lanostane aglycone. The synergism of inhibitory action of non-toxic dose of the glycosides and radioactive irradiation on colony formation of HT-29 cells. *Carb. Res.* **2018**, *468*, 36–44.
26. Fedorov, S.N.; Dyshlovoy, S.A.; Kuzmich, A.S.; Shubina, L.K.; Avilov, S.A.; Silchenko, A.S.; Bode, A.M.; Dong, Z.; Stonik, V.A. In Vitro anticancer activities of some triterpene glycosides from holothurians of Cucumariidae, Stichopodidae, Psolidae, Holothuriidae and Synaptidae families. *Nat. Prod. Commun.* **2016**, *11*, 1239–1242. [PubMed]

marine drugs

MDPI

Article

Structures and Bioactivities of Psolusosides B_1, B_2, J, K, L, M, N, O, P, and Q from the Sea Cucumber *Psolus fabricii*. The First Finding of Tetrasulfated Marine Low Molecular Weight Metabolites

Alexandra S. Silchenko, Anatoly I. Kalinovsky, Sergey A. Avilov, Vladimir I. Kalinin *, Pelageya V. Andrijaschenko, Pavel S. Dmitrenok, Roman S. Popov and Ekaterina A. Chingizova

G.B. Elyakov Pacific Institute of Bioorganic Chemistry, Far Eastern Branch of the Russian Academy of Sciences, Pr. 100-letya Vladivostoka 159, Vladivostok 690022, Russia; sialexandra@mail.ru (A.S.S.); kaaniv@piboc.dvo.ru (A.I.K.); avilov-1957@mail.ru (S.A.A.); pandryashchenko@mail.ru (P.V.A.); paveldmt@piboc.dvo.ru (P.S.D.); rs.popov@outlook.com (R.S.P.); martyyas@mail.ru (E.A.C.)

* Correspondence: kalininv@hotmail.com; Tel./Fax: +7-(423)2-31-40-50

Received: 13 October 2019; Accepted: 2 November 2019; Published: 6 November 2019

Abstract: Ten new di-, tri- and tetrasulfated triterpene glycosides, psolusosides B_1 (**1**), B_2 (**2**), J (**3**), K (**4**), L (**5**), M (**6**), N (**7**), O (**8**), P (**9**), and Q (**10**), were isolated from the sea cucumber *Psolus fabricii* collected in the Sea of Okhotsk near the Kurile Islands. Structures of these glycosides were established by two-dimensional (2D) NMR spectroscopy and HR-ESI mass-spectrometry. It is particularly interesting that highly polar compounds **9** and **10** contain four sulfate groups in their carbohydrate moieties, including two sulfates in the same terminal glucose residue. Glycoside **2** has an unusual non-holostane aglycone with 18(16)-lactone and a unique 7,8-epoxy fragment. Cytotoxic activities of compounds **1–10** against several mouse cell lines such as Ehrlich ascites carcinoma cells, neuroblastoma Neuro 2A, normal epithelial JB-6 cells, and erythrocytes were quite different depending both on structural peculiarities of these glycosides and the type of cells subjected to their actions. Psolusoside L (**5**), pentaoside, with three sulfate groups at C-6 of two glucose and one 3-*O*-methylglucose residue and holostane aglycone, is the most active compound in the series. The presence of a sulfate group at C-2 of the terminal glucose residue attached to C-4 of the first (xylose) residue significantly decreases activities of the corresponding glycosides. Psolusosides of group B (**1**, **2**, and known psolusoside B) are inactive in all tests due to the presence of non-holostane aglycones and tetrasaccharide-branched sugar chains sulfated by C-2 of Glc4.

Keywords: *Psolus fabricii*; triterpene glycosides; psolusosides; sea cucumber; cytotoxic activity

1. Introduction

Triterpene glycosides of sea cucumbers are well known by their structural diversity and promising biological effects [1–3], including cytotoxicity against cancer cells and antitumor activity [4–6]. Therefore, the search for new representatives of this class of marine natural products and studies of their biological activities seem to be relevant. Moreover, structural analysis of diverse glycosides of sea cucumbers allows us to understand the peculiarities of biosynthesis of these complicated and numerous marine metabolites.

As a continuation of our investigation of triterpene glycoside composition of the sea cucumber *Psolus fabricii* (Psolidae, Dendrochirotida) [7–12] we report herein the isolation of ten new glycosides, psolusosides B_1 (**1**), B_2 (**2**), J–Q (**3–10**), and their structural elucidation based on the ^{1}H, ^{13}C NMR, one-dimensional (1D) TOCSY and 2D NMR (^{1}H,^{1}H-COSY, HMBC, HSQC, ROESY), and HR-ESI mass spectrometry. The hemolytic activities against mouse erythrocytes and cytotoxic activities of **1–10** against mouse Ehrlich ascites carcinoma cells, neuroblastoma Neuro 2A, and normal epithelial JB-6 cells have been studied.

2. Results and Discussion

2.1. Structural Elucidation of the Glycosides

The initial stages of isolation of compounds **1–10** were the same as for other glycosides from *P. fabricii* and were described earlier [10–12]. The individual glycosides were isolated by HPLC on reversed-phase columns to give psolusosides: B_1 (**1**) (7,3 mg), B_2 (**2**) (3.4 mg), J (**3**) (4.8 mg), K (**4**) (3.4 mg), L (**5**) (60 mg), M (**6**) (1.0 mg), N (**7**) (8.8 mg), O (**8**) (0.6 mg), P (**9**) (8.5 mg), and Q (**10**) (1.4 mg) (Figure 1).

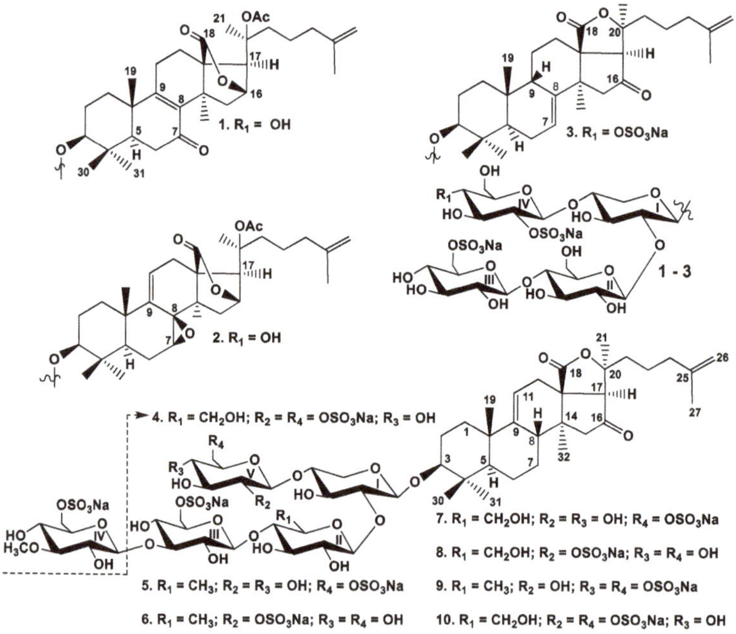

Figure 1. Chemical structures of glycosides isolated from *Psolus fabricii:* **1**—psolusoside B_1; **2**—psolusoside B_2; **3**—psolusoside J; **4**—psolusoside K; **5**—psolusoside L; **6**—psolusoside M; **7**—psolusoside N; **8**—psolusoside O; **9**—psolusoside P; **10**—psolusoside Q.

The ^{1}H and ^{13}C NMR spectra corresponding to the carbohydrate chains of psolusosides B_1 (**1**) and B_2 (**2**) were coincident to each other and to those of known psolusoside B [12] showing the identity of their tetrasaccharide carbohydrate moieties branched by C-4 of the xylose unit and having two sulfate groups (Table S1).

The molecular formula of psolusoside B_1 (**1**) was determined to be $C_{55}H_{82}O_{31}S_2Na_2$ from the $[M_{2Na} - Na]^-$ ion peak at *m/z* 1325.4164 (calc. 1325.4185) and $[M_{2Na} - 2Na]^{2-}$ ion peak at *m/z* 651.2157 (calc. 651.2146) in the (−)HR-ESI-MS. The signal of H-16 was observed as a broad singlet at δ_H 4.89 and the signal of H-17 was observed as a singlet at δ_H 2.97 in the ^{1}H NMR spectrum of **1**. These data as well as corresponding signals of carbons at δ_C 79.9 (C-16) and δ_C 58.8 (C-17) (Table 1) were indicative for 18(16)-lactone moiety (Table 1). *O*-acetyl group (δ_C 170.9 ($CH_3\underline{C}OO$) and 21.6 ($\underline{C}H_3COO$) in the ^{13}C NMR spectrum), attached to C-20, caused the deshielding of its signal to δ_C 83.8 in the same manner as in the spectrum of psolusoside B [12]. The side chain of **1** was identical to that of psolusoside B due to the coincidence of those signals in the ^{1}H and ^{13}C NMR spectra. The signal at δ_C 199.3 corresponded to a keto-group adjacent to a double bond (the signals of olefinic carbons at δ_C 135.3 (C-8) and 169.0 (C-9)). The position of the keto-group was deduced as C-7 based on the correlations between H_2-6 (δ_H 2.42 and δ_H 2.29) and C-7 (δ_C 199.3) in the HMBC spectrum of **1**. This was also corroborated by an isolated spin system between the doublet of doublets at δ_H 1.54 (H-5) and another doublet of doublets at δ_H 2.42 (H-6a) and the triplet at δ_H 2.29 (H-6b) observed in the ^{1}H,^{1}H-COSY spectrum. The 8(9)-position of double bond was confirmed by the HMBC correlations H_3-32/C-8 and H_3-19/C-9. So, the aglycone of psolusoside B_1 (**1**) is characterized by the unique combination of such structural features as 7-keto-8(9)-ene fragment and 18(16)-lactone.

Table 1. ^{13}C and ^{1}H NMR chemical shifts and HMBC and ROESY correlations of aglycone moiety of psolusoside B_1 (**1**). [a] Recorded at 176.04 MHz in C_5D_5N/D_2O (4/1). [b] Recorded at 700.13 MHz in C_5D_5N/D_2O (4/1).

Position	δ_C Mult. [a]	δ_H Mult. (*J* in Hz) [b]	HMBC	ROESY
1	34.3 CH_2	1.71 m		H-11, H-19
		1.24 m		H-3, H-5, H-11
2	26.3 CH_2	2.00 m		
		1.81 m		H-19, H-30
3	87.7 CH	3.08 dd (4.5; 11.6)	C: 30, C-1 Xyl1	H-5, H-31, H-1 Xyl1
4	39.4 C			
5	50.8 CH	1.54 dd (2.9; 14.5)		H-1, H-3, H-31
6	36.5 CH_2	2.42 dd (2.6; 15.4)	C: 7, 10	H-31
		2.29 t (15.0)	C: 5, 7	H-19, H-30
7	199.3 C			
8	135.3 C			
9	169.0 C			
10	40.0 C			
11	23.5 CH_2	2.96 m		H-19
		2.50 dd (9.7; 20.8)		
12	19.1 CH_2	2.27 m		
		2.14 m		H-32
13	54.9 C			
14	40.1 C			
15	42.5 CH_2	2.56 d (14.7)	C: 8, 14, 16, 17, 32	
		2.20 d (14.7)	C: 14, 32	H-17, H-32
16	79.9 CH	4.89 brs	C: 13, 14, 18	H-21, H-22, H-23
17	58.8 CH	2.97 s	C: 13, 14, 18, 20, 21, 22	H-15, H-21, H-32
18	179.2 C			
19	18.1 CH_3	1.14 s	C: 1, 5, 9, 10	H-1, H-2, H-6, H-11, H-30
20	83.8 C			

Table 1. *Cont.*

Position	δ_C Mult. [a]	δ_H Mult. (*J* in Hz) [b]	HMBC	ROESY
21	23.3 CH_3	1.64 s	C: 17, 20, 22	H-16, H-17, H-22
22	37.6 CH_2	2.23 m		H-17, H-21
		1.83 m		H-16
23	21.1 CH_2	1.52 m		
		1.45 m		H-17, H-21
24	37.7 CH_2	1.96 brdd (8.9; 16.2)	C: 22, 23, 25, 26	H-22, H-26, H-27
25	145.4 C			
26	110.7 CH_2	4.74 brs	C: 24, 27	H-27
27	22.1 CH_3	1.66 s	C: 24, 25, 26	H-26
30	15.8 CH_3	0.94 s	C: 3, 4, 5, 31	H-2, H-6, H-19, H-31
31	26.8 CH_3	1.01 s	C: 3, 4, 5, 30	H-3, H-5, H-6, H-30, H-1 Xyl1
32	27.8 CH_3	1.41 s	C: 8, 13, 14, 15	H-12, H-15, H-17
OAc	170.9 C			
	21.6 CH_3	2.09 s	OAc	

The (−)ESI-MS/MS of **1** demonstrated the fragmentation of $[M_{2Na} - Na]^-$ ion at *m/z* 1325.4. The peaks of fragment ions were observed at *m/z* 1265.4 $[M_{2Na} - Na - CH_3COOH]^-$, 1145.4 $[M_{2Na} - Na - CH_3COOH - NaHSO_4]^-$, 1001.4 $[M_{2Na} - Na - CH_3COOH - C_6H_{10}O_8SNa$ ($GlcSO_3Na$) + $H]^-$, and 839.3 $[M_{2Na} - Na - CH_3COOH - GlcSO_3Na - Glc + H]^-$ corroborating the structure of psolusoside B_1 (**1**).

All these data indicate that psolusoside B_1 (**1**) is 3*β*-*O*-{6-*O*-sodium-sulfate-*β*-D-glucopyranosyl-(1→4)-*β*-D-glucopyranosyl-(1→2)-[2-*O*-sodium-sulfate-*β*-D-glucopyranosyl-(1→4)]-*β*-D-xylopyranosyl}-7-keto-20(*S*)-acetoxylanosta-8,25-diene-18(16)-lactone.

The molecular formula of psolusoside B_2 (**2**) ($C_{55}H_{82}O_{31}S_2Na_2$) was determined to be the same as of **1** from the $[M_{2Na} - Na]^-$ ion peak at *m/z* 1325.4163 (calc. 1325.4185) and $[M_{2Na} - 2Na]^{2-}$ ion peak at *m/z* 651.2159 (calc. 651.2146) in the (−)HR-ESI-MS. In the ^1H and ^{13}C NMR spectra of the aglycone part of **2** the signals characteristic of 18(16)-lactone (δ_H 4.94 (brs, H-16), δ_H 3.01 (s, H-17), δ_C 79.3 (C-16), and δ_C 60.2 (C-17) as well as *O*-acetylated C-20 (δ_H 2.05 (s, $\underline{CH_3}COO$), δ_C 21.8 ($\underline{C}H_3COO$), δ_C 170.9($CH_3\underline{C}OO$), and δ_C 83.9 (C-20)) were observed (Table 2). The side chains in aglycones of **1** and **2** were identical to each other. The signal at δ_H 3.10 (d, 6.6, H-7) was assigned by the $^1H,^1H$-COSY spectrum where the protons H-5/H-6/H-7 formed an isolated spin system. The corresponding signal of C-7 at δ_C 56.2 was deduced by the HSQC spectrum of **2**. The signal of quaternary C-8 assigned by the HMBC correlations H_3-32/C-8, H_2-6/C-8, and H-7/C-8 was deshielded to δ_C 59.6 in the ^{13}C NMR spectrum. These data indicated the presence of an oxygen-bearing substituent at C-7 and C-8, which was supposed to be an 7,8-epoxide [13] that correlated with the MS data. The olefinic broad doublet of doublets at δ_H 6.00 was assigned to H-11 due to its correlation with H_2-12 (δ_H 2.81 (dd, 5.2; 17.6, H-12a) and 2.60 (brdd, 2.3; 17.6, H-12b)) in the $^1H,^1H$-COSY spectrum. The signal at δ_C 122.7 corresponded to olefinic C-11 and was deduced by the HSQC spectrum. So, the double bond could occupy the 9(11)-position only. The signal of C-9 at δ_C 143.2 correlated in the HMBC spectrum with both δ_H 2.81 (H-12a) and 2.60 (H-12b) and the methyl singlet δ_H 1.13 (H_3-19).

Table 2. ^{13}C and ^{1}H NMR chemical shifts and HMBC and ROESY correlations of aglycone moiety of psolusoside B_2 (**2**). [a] Recorded at 176.04 MHz in C_5D_5N/D_2O (4/1). [b] Recorded at 700.13 MHz in C_5D_5N/D_2O (4/1).

Position	δ_C Mult. [a]	δ_H Mult. (J in Hz) [b]	HMBC	ROESY
1	37.1 CH_2	1.84 m		H-11, H-19
		1.32 m		H-11
2	26.3 CH_2	1.99 m		
		1.78 m		H-30
3	88.5 CH	3.04 dd (4.3; 11.8)	C: 30, 31, C-1 Xyl1	H-1, H-5, H-31, H-1 Xyl1
4	39.4 C			
5	48.9 CH	1.00 brd (12.1)	C: 4, 6, 10, 19	H-1, H-3, H-31
6	21.0 CH_2	1.94 m		H-31
		1.73 m	C: 7, 8	H-19, H-30
7	56.2 CH	3.10 d (6.6)	C: 5, 6, 8, 14	H-15, H-32
8	59.6 C			
9	143.2 C			
10	36.7 C			
11	122.7 CH	6.00 brdd (2.6; 5.2)	C: 10, 13	H-1
12	25.2 CH_2	2.81 dd (5.2; 17.6)	C: 9, 11, 13, 14, 18	H-21
		2.60 brdd (2.3; 17.6)	C: 9, 11, 13, 18	H-21, H-32
13	53.3 C			
14	42.0 C			
15	35.1 CH_2	1.93 m	C: 14, 16, 17, 32	
		1.50 brdd (2.3; 13.6)	C: 32	H-17, H-32
16	79.3 CH	4.94 brs	C: 13, 14, 18	H-22
17	60.2 CH	3.01 s	C: 13, 14, 18, 20, 21	H-12, H-15, H-21, H-32
18	177.9 C			
19	22.4 CH_3	1.13 s	C: 1, 5, 9, 10	H-2, H-6, H-30
20	83.9 C			
21	23.4 CH_3	1.64 s	C: 17, 20, 22	H-12, H-17, H-22
22	37.6 CH_2	2.24 td (4.6; 13.3)		
		1.85 dd (4.6; 13.7)		H-16
23	21.6 CH_2	1.52 m	C: 22, 24	
		1.46 m	C: 22, 24	
24	37.7 CH_2	1.96 m	C: 22, 23, 25	H-22, H-26
25	145.4 C			
26	110.7 CH_2	4.73 brs	C: 24, 25, 27	H-24, H-27
27	22.1 CH_3	1.65 s	C: 25, 26	H-26
30	15.8 CH_3	0.90 s	C: 3, 4, 5, 31	H-2, H-6, H-19
31	27.4 CH_3	1.09 s	C: 3, 4, 5, 30	H-3, H-5, H-6, H-30, H-1 Xyl1
32	23.9 CH_3	1.33 s	C: 8, 13, 14, 15	H-7, H-12, H-15, H-17
OAc	170.9 C			
	21.8 CH_3	2.05 s	OAc	

The configuration of C-7 was established as (S) by the ROE-correlation H-7/H_3-32 and was confirmed by the coupling pattern of H-7 (δ_H 3.10 (d, 6.6)), that coincided with the calculated coupling constant based on dihedral angle values in the optimized MM2 model of aglycone of psolusoside B_2 (**2**) having H-7α-orientation and 8(R)-configuration. Thus, the aglycone of psolusoside B_2 (**2**) has unprecedented 7(S),8(R)-epoxy-20(S)-acetoxylanosta-9(11),25-diene-18(16)-lactone structure.

The (−)ESI-MS/MS of **2** showed the fragmentation of $[M_{2Na} - Na]^-$ ion at m/z 1325.4. The peaks of fragment ions were observed at the same m/z values of 1265.4, 1145.4, 1001.4, and 839.3 as in the spectrum of **1**, corroborating the identity of the carbohydrate chains of **1** and **2**. Additionally, the fragment ion-peaks at m/z 535.1 $[M_{2Na} - Na - C_{32}H_{45}O_6$ (Agl) $- C_6H_{10}O_8SNa$ $(GlcSO_3Na)]^-$ and 403 $[M_{2Na} - Na - C_{32}H_{45}O_6$ (Agl) $- C_6H_{10}O_8SNa$ $(GlcSO_3Na)$ − Xyl $(C_5H_8O_4)]^-$ corresponding to the tri- and disaccharide fragments, were observed in the MS/MS spectrum of **2**.

All these data indicate that psolusoside B_2 (**2**) is 3β-O-{6-O-sodium-sulfate-β-D-glucopyranosyl-(1→4)-β-D-glucopyranosyl-(1→2)-[2-O-sodium-sulfate-β-D-glucopyranosyl-(1→4)]-β-D-xylopyranosyl}-7(S),8(R)-epoxy-20(S)-acetoxylanosta-9(11),25-diene-18(16)-lactone.

The molecular formula of psolusoside J (**3**) ($C_{53}H_{79}O_{32}S_3Na_3$) was determined from the $[M_{3Na} - Na]^-$ ion peak at m/z 1369.3485 (calc. 1369.3517), $[M_{3Na} - 2Na]^{2-}$ ion peak at m/z 673.1812 (calc. 673.1813), and $[M_{3Na} - 3Na]^{3-}$ ion peak at m/z 441.1248 (calc. 441.1244) in the (−)HR-ESI-MS. The ^1H and ^{13}C NMR spectra of the aglycone part of psolusoside J (**3**) coincided with those of psolusoside H isolated earlier from *P. farbricii* [12] (Table S2) indicating the identity of their holostane-type aglycones having 7(8)- and 25(26)-double bonds and 16-keto-group. This aglycone is common for the glycosides of sea cucumbers belonging to the orders Dendrochirotida and Aspidochirotida [2,12].

In the ^1H and ^{13}C NMR spectra of the carbohydrate part of psolusoside J (**3**) four characteristic doublets at δ_H 4.60–5.12 (J = 7.3 − 8.1 Hz) and, corresponding to them, signals of anomeric carbons at δ_C 101.7–105.5 were indicative of a tetrasaccharide chain and β-configurations of glycosidic bonds. The ^{13}C NMR spectra of tetrasaccharide carbohydrate chain of **3** and those of **1** and **2** were quite different, while the $^1H,^1H$-COSY and 1D TOCSY spectra of **3** showed the signals of four isolated spin systems assigned to one xylose and three glucose residues as in psolusosides B [12], B_1 (**1**), and B_2 (**2**). The positions of interglycosidic linkages were elucidated by the ROESY and HMBC spectra of **3** (Table 3), where the correlations between H-1 of the xylose (Xyl1) and H-3 (C-3) of the aglycone, H-1 of the second residue (glucose, Glc2) and H-2 (C-2) of the xylose (Xyl1), H-1 of the third residue (glucose, Glc3) and H-4 (C-4) of the second residue (glucose, Glc2), H-1 of the fourth residue (glucose, Glc4) and H-4 (C-4) of the first residue (xylose, Xyl1) were observed, indicating the same architecture of sugar chains in **3** and **1** and **2**. The comparison of the NMR spectra of **1** and **3** showed the coincidence of the signals of three monosaccharide residues corresponding to the linear part of the carbohydrate chain (residues I–III). The signals of terminal monosaccharide unit attached to C-4 of the first (Xyl1) unit, assigned by the $^1H,^1H$-COSY and 1D TOCSY spectra of **3** were indicative of a sulfated by C-2 glucose residue due to characteristic shifting effects observed in the ^{13}C NMR spectrum: the signal of C-2 Glc4 was deshielded to δ_C 81.2 and the signal of C-1 Glc4 was shielded to δ_C 101.7 in comparison with the corresponding signals of the same sugar unit in the ^{13}C NMR spectrum of psolusoside I isolated by us earlier [12].

The δ_C of the signals of C-2 and C-1 of the fourth monosaccharide unit (Glc4) in the ^{13}C NMR spectrum of psolusoside J (**3**) were very close to those in the ^{13}C NMR spectrum of **1**, corroborating the presence of a sulfate group at C-2 of this residue (Glc4). The correlations between H-2/H-3/H-4 in this monosaccharide residue, deduced by the $^1H,^1H$-COSY spectrum of **3**, indicated the signal of H-4 Glc4 at δ_H 4.90. The signal of the corresponding carbon (C-4 Glc4), deduced by the HSQC spectrum, was downshifted to δ_C 77.3 as compared with the same signal (C-4 Glc4) at δ_C 70.7 in the ^{13}C NMR spectrum of **1**. Actually, the signals at δ_C ~70.4–70.8 were absent and the signals of C-3 Glc4 and C-5 Glc4 were upshifted to δ_C 75.6 and 76.6, correspondingly, in the ^{13}C NMR spectrum of **3** due to β-shifting effect of sulfate group, when compared with the corresponding signals in the ^{13}C NMR spectrum of **1**. Considering that (−)HR-ESI-MS indicated the presence of three sulfate groups as well as the NMR data, the attachment of the third sulfate group to C-4 of Glc4 was supposed. The signal at δ_C 62.4 (C-6 Glc4) was characteristic for carbons of non-sulfated hydroxy-methylene groups of glucopyranose residues and excluded the positioning of the third sulfate group at C-6 Glc4 that confirmed our supposition. Hence psolusoside J (**3**) is a trisulfated tetraoside with two sulfate groups attached to the same glucose residue. To the best our knowledge, this structural feature is first found in the glycosides.

Table 3. ^{13}C and ^{1}H NMR chemical shifts and HMBC and ROESY correlations of carbohydrate moiety of psolusoside J (**3**). [a] Recorded at 176.04 MHz in C_5D_5N/D_2O (4/1). [b] Bold = interglycosidic positions. [c] Italic = sulphate position. [d] Recorded at 700.13 MHz in C_5D_5N/D_2O (4/1). [e] Recorded at 500.13 MHz in C_5D_5N/D_2O (4/1). Multiplicity by one-dimensional (1D) TOCSY.

Atom	δ_C Mult. [a, b, c]	δ_H Mult. [d] (*J* in Hz)	HMBC	ROESY [e]
		Xyl1 (1→C-3)		
1	105.5 CH	4.60 d (7.3)	C: 3	H-3; H-3, 5 Xyl1
2	**81.6 CH**	4.03 t (8.3)	C: 1 Glc2; C: 1, 3 Xyl1	H-1 Glc2
3	75.8 CH	4.24 t (8.8)	C: 2 Xyl1	H-1, 5 Xyl1
4	**79.3 CH**	4.11 m	C: 1 Clc4	H-1 Glc4
5	64.3 CH_2	4.50 dd (4.9; 11.9)		
		3.81 t (11.2)	C: 1 Xyl1	H-1, 3 Xyl1
		Glc2 (1→2Xyl1)		
1	104.8 CH	5.12 d (8.1)	C: 2 Xyl1	H-2 Xyl1; H-3, 5 Glc2
2	75.9 CH	3.84 t (8.1)	C: 1, 3 Glc2	
3	76.0 CH	3.98 t (9.5)	C: 2, 4 Glc2	H-1, 5 Glc2
4	**82.8 CH**	3.89 t (9.5)	C: 1 Glc3; C: 5, 6 Glc2	H-1 Glc3
5	76.7 CH	3.71 brd (9.5)		H-1, 3 Glc2
6	62.2 CH_2	4.31 dd (2.2; 12.0)		
		4.26 dd (7.4; 12.0)		
		Glc3 (1→4Glc2)		
1	105.3 CH	4.82 d (8.1)	C: 4 Glc2	H-4 Glc2; H-3, 5 Glc3
2	74.9 CH	3.80 t (8.1)	C: 1, 3 Glc3	
3	77.6 CH	4.08 t (9.5)	C: 2, 4 Glc3	
4	71.4 CH	3.92 t (9.5)	C: 3, 6 Glc3	H-6 Glc3
5	76.3 CH	4.03 dd (4,7; 10.1)		H-1, 3 Glc3
6	*68.2* CH_2	5.01 brd (10.1)		
		4.66 dd (6.1; 10.1)	C: 5 Glc3	
		Glc4 (1→4Xyl1)		
1	101.7 CH	4.99 d (7.4)	C: 4 Xyl1	H-4 Xyl1; H-3, 5 Glc4
2	*81.2* CH	4.87 t (8.8)	C: 1, 3 Glc4	
3	75.6 CH	4.40 t (8.8)	C: 2, 4 Glc4	H-1, 5 Glc4
4	*77.3* CH	4.90 t (8.8)	C: 3, 5, 6 Glc4	H-6 Glc4
5	76.6 CH	3.84 t (8.8)		H-1, 3 Glc4
6	62.4 CH_2	4.41 brd (10.2)		
		4.24 dd (5.4; 12.2)		

The (−)ESI-MS/MS of **3** demonstrated the fragmentation of $[M_{3Na} - Na]^-$ ion at *m/z* 1369.3. The peaks of fragment ions were observed at *m/z* 1249.4 $[M_{3Na} - Na - NaHSO_4]^-$, 1105.4 $[M_{3Na} - Na - C_6H_9O_8SNa\ (GlcSO_3Na)]^-$, 1003.4 $[M_{3Na} - Na - C_6H_9O_8SNa\ (GlcSO_3Na) - NaSO_3 + H]^-$, 841.4 $[M_{3Na} - Na - NaSO_3 - GlcSO_3Na - Glc + H]^-$, 403.0 $[M_{3Na} - Na - C_{30}H_{43}O_4\ (Agl) - C_6H_9O_{11}S_2Na_2\ (Glc(SO_3Na)_2) - Xyl\ (C_5H_8O_4)]^-$, and 241.0 $[M_{3Na} - Na - C_{30}H_{43}O_4\ (Agl) - C_6H_9O_{11}S_2Na_2\ (Glc(SO_3Na)_2) - Xyl\ (C_5H_8O_4) - Glc\ (C_6H_{10}O_5)]^-$, corroborating the structure of psolusoside J (**3**).

All these data indicate that psolusoside J (**3**) is 3*β*-*O*-{6-*O*-sodium-sulfate-*β*-D-glucopyranosyl-(1→4)-*β*-D-glucopyranosyl-(1→2)-[2,4-*O*-sodium-disulfate-*β*-D-glucopyranosyl-(1→4)]-*β*-D-xylopyranosyl}-16-ketoholosta-7,25-diene.

The ^{13}C NMR spectra of the aglycone moieties of the glycosides **4–10** were identical to each other (Table S3) and to those of psolusosides E, F, and G containing 16-ketoholosta-9(11),25-dien-3β-ol as an aglycone, known earlier and frequently occurring in the glycosides of sea cucumbers [12].

The molecular formula of psolusoside K (**4**) was determined to be $C_{53}H_{79}O_{32}S_3Na_3$ from the $[M_{3Na} - Na]^-$ ion peak at *m/z* 1369.3485 (calc. 1369.3517), $[M_{3Na} - 2Na]^{2-}$ ion peak at *m/z* 673.1821 (calc. 673.1813), and $[M_{3Na} - 3Na]^{3-}$ ion peak at *m/z* 441.1255 (calc. 441.1244) in the (−)HR-ESI-MS and was coincident with the formula of psolusoside J (**3**). In the ^1H and ^{13}C NMR spectra of the carbohydrate moiety of psolusoside K (**4**) four characteristic doublets at δ_H 4.61–5.07 (J = 7.2–8.4 Hz) and corresponding signals of anomeric carbons at δ_C 101.4–104.7 were indicative of a tetrasaccharide chain and *β*-configurations of glycosidic bonds. The positions of interglycosidic linkages were elucidated by the ROESY and HMBC spectra of **4** (Table 4) as described above indicating the presence of a tetrasaccharide carbohydrate chain branched by C-4 of the xylose residue (Xyl1). The monosaccharide composition of **4**, deduced from the $^1H,^1H$-COSY and 1D TOCSY spectra, was the same as in glycosides **1–3**. The comparison of the ^{13}C NMR spectra of trisulfated compounds **3** and **4** showed the coincidence of the signals corresponding to three monosaccharide residues (residues I–III in the formula) forming the linear part of the sugar chain. The signals of C-2 Glc4 at δ_C 80.3 and C-1 Glc4 at δ_C 101.4 in the ^{13}C NMR spectrum of **4** were very close to those in the spectrum of **3** that indicated the attachment of a sulfate group to C-2 Glc4 in psolusoside K (**4**). All of the signals of this monosaccharide residue were assigned using the $^1H,^1H$-COSY and 1D TOCSY spectra. The doublet at δ_H 5.00 and the doublet of doublets at δ_H 4.63 corresponded to the protons of the hydroxy-methylene group of the terminal glucose unit (H_2-6 Glc4) and were deshielded as compared with the corresponding signals in the ^1H NMR spectrum of **3**. The signal at δ_C 67.4 (C-6 Glc4) also indicated the presence of a sulfate group at C-6 of Glc4 in addition to another sulfate group at C-2 of Glc4. So, psolusoside K (**4**) is an isomer of psolusoside J (**3**) by the sulfate position and is the second glycoside from sea cucumbers that contains two sulfate groups bonded to the same monosaccharide residue.

The (−)ESI-MS/MS of **4** demonstrated the fragmentation of $[M_{3Na} - Na]^-$ ion at *m/z* 1369.3. The peaks of fragment ions were observed at the same *m/z*: 1249.4 $[M_{3Na} - Na - NaHSO_4]^-$, 1105.4 $[M_{3Na} - Na - C_6H_9O_8SNa\ (GlcSO_3Na)]^-$, 1003.4 $[M_{3Na} - Na - C_6H_9O_8SNa\ (GlcSO_3Na) - NaSO_3 + H]^-$, 403.0 $[M_{3Na} - Na - C_{30}H_{43}O_4\ (Agl) - C_6H_9O_{11}S_2Na_2\ (Glc(SO_3Na)_2) - Xyl\ (C_5H_8O_4)]^-$, and 241.0 $[M_{3Na} - Na - C_{30}H_{43}O_4\ (Agl) - C_6H_9O_{11}S_2Na_2\ (Glc(SO_3Na)_2) - Xyl\ (C_5H_8O_4) - Glc\ (C_6H_{10}O_5)]^-$ as in the MS/MS of psolusoside J (**3**) corroborating their isomerism.

All these data indicate that psolusoside K (**4**) is 3*β*-*O*-{6-*O*-sodium-sulfate-*β*-D-glucopyranosyl-(1→4)-*β*-D-glucopyranosyl-(1→2)-[2,6-*O*-sodium-disulfate-*β*-D-glucopyranosyl-(1→4)]-*β*-D-xylopyranosyl}-16-ketoholosta-9,25-diene.

The molecular formula of psolusoside L (**5**) ($C_{60}H_{91}O_{36}S_3Na_3$) was determined from the $[M_{3Na} - Na]^-$ ion peak at *m/z* 1529.4222 (calc. 1529.4253), $[M_{3Na} - 2Na]^{2-}$ ion peak at *m/z* 753.2190 (calc. 753.2180), and $[M_{3Na} - 3Na]^{3-}$ ion peak at *m/z* 494.4835 (calc. 494.4823) in the (−)HR-ESI-MS, indicating the presence of three sulfate groups. In the ^1H and ^{13}C NMR spectra of the carbohydrate part of psolusoside L (**5**) five characteristic doublets at δ_H 4.65–5.16 (J = 6.9–8.1 Hz) and, corresponding to them, signals of anomeric carbons at δ_C 103.4–104.8 were indicative of a pentasaccharide chain and *β*-configurations of glycosidic bonds (Table 5). Analysis of the ^{1}H,^{1}H-COSY and 1D TOCSY spectra of psolusoside L (**5**) showed the presence of one xylose, one quinovose, two glucose, and one 3-*O*-methylglucose residues. The presence of a quinovose residue was confirmed by the ^{1}H and ^{13}C NMR spectra demonstrating the characteristic doublet at δ_H 1.59 (H-6 Qui2) and the signal at δ_C 17.7 (C-6 Qui2). The positions of interglycosidic linkages and the consequence of monosaccharides in the chain of **5** were established by analysis of the ROESY and HMBC spectra (Table 5) indicating the presence of branched pentasaccharide moiety with glucose, attached to C-4 Xyl1, and 3-*O*-methylglucose, attached to C-3 Glc3, as terminal residues. The ^{13}C NMR spectrum of **5** demonstrated three signals at δ_C 67.0, 67.5, and 67.6, corresponding to sulfated hydroxy-methylene groups of glucopyranose residues that indicated the sulfation of two glucose and 3-O-methylglucose units in the carbohydrate chain of **5**.

Table 4. ^{13}C and ^{1}H NMR chemical shifts and HMBC and ROESY correlations of carbohydrate moiety of psolusoside K (**4**). [a] Recorded at 176.04 MHz in C_5D_5N/D_2O (4/1). [b] Bold = interglycosidic positions. [c] Italic = sulphate position. [d] Recorded at 700.13 MHz in C_5D_5N/D_2O (4/1). [e] Recorded at 500.13 MHz in C_5D_5N/D_2O (4/1). Multiplicity by 1D TOCSY.

Atom	δ_C Mult. [a, b, c]	δ_H Mult. [d] (*J* in Hz)	HMBC	ROESY [e]
		Xyl1 (1→C-3)		
1	104.7 CH	4.61 d (7.2)	C: 3	H-3; H-3, 5 Xyl1
2	**81.0 CH**	4.00 t (8.7)	C: 1 Glc2; C: 1, 3 Xyl1	H-1 Glc2
3	74.9 CH	4.20 t (8.7)	C: 2, 4 Xyl1	H-1, 5 Xyl1
4	**79.7 CH**	4.01 m	C: 1 Clc4	H-1 Glc4
5	63.6 CH_2	4.48 dd (5.9; 11.9)	C: 1, 3, 4 Xyl1	
		3.77 t (10.9)		H-1, 3 Xyl1
		Glc2 (1→2Xyl1)		
1	104.1 CH	5.07 d (7.9)	C: 2 Xyl1	H-2 Xyl1; H-3, 5 Glc2
2	75.2 CH	3.84 t (7.9)	C: 1, 3 Glc2	
3	74.8 CH	4.00 t (8.9)	C: 2, 4 Glc2	H-5 Glc2
4	**82.1 CH**	3.90 t (8.9)	C: 1 Glc3; C: 3, 5, 6 Glc2	H-1 Glc3
5	75.9 CH	3.72 d (9.9)		H-1, 3 Glc2
6	61.2 CH_2	4.29 dd (4.1; 11.6)		
		4.27 dd (8.4; 11.9)		
		Glc3 (1→4Glc2)		
1	104.5 CH	4.80 d (8.4)	C: 4 Glc2	H-4 Glc2; H-3, 5 Glc3
2	74.1 CH	3.80 t (8.4)	C: 1, 3 Glc3	
3	76.8 CH	4.08 t (9.2)	C: 2, 4 Glc3	H-1 Glc3
4	70.8 CH	3.88 t (9.2)	C: 3, 5, 6 Glc3	H-6 Glc3
5	75.3 CH	4.03 dd (5.0; 10.1)		H-1 Glc3
6	*67.5* CH_2	5.00 d (10.9)		
		4.62 dd (7.6; 10.9)	C: 5 Glc3	
		Glc4 (1→4Xyl1)		
1	101.4 CH	4.90 d (7.6)	C: 4 Xyl1	H-4 Xyl1; H-3, 5 Glc4
2	*80.3* CH	4.72 t (8.4)	C: 1, 3 Glc4	H-4 Glc4
3	76.4 CH	4.26 t (9.2)	C: 2, 4 Glc4	H-1, 5 Glc4
4	70.5 CH	3.92 t (9.2)	C: 3, 5, 6 Glc4	H-2, 6 Glc4
5	75.2 CH	4.04 dd (5.0; 10.9)		H-1, 3 Glc4
6	*67.4* CH_2	5.00 d (10.1)		
		4.63 dd (6.7; 11.8)	C: 5 Glc4	

Table 5. ^{13}C and ^{1}H NMR chemical shifts and HMBC and ROESY correlations of carbohydrate moiety of psolusoside L (**5**). [a] Recorded at 176.04 MHz in C_5D_5N/D_2O (4/1). [b] Bold = interglycosidic positions. [c] Italic = sulphate position. [d] Recorded at 700.13 MHz in C_5D_5N/D_2O (4/1). [e] Recorded at 500.13 MHz in C_5D_5N/D_2O (4/1). Multiplicity by 1D TOCSY.

Atom	δ_C Mult. [a, b, c]	δ_H Mult. [d] (*J* in Hz)	HMBC	ROESY [e]
		Xyl1 (1→C-3)		
1	104.6 CH	4.65 d (7.1)	C: 3; C: 5 Xyl1	H-3; H-5 Xyl1
2	**82.3 CH**	3.89 t (7.9)	C: 1, 3 Xyl1; 1 Qui2	H-1 Qui2; H-4 Xyl1
3	75.0 CH	4.09 t (7.9)	C: 2, 4 Xyl1	H-1, 5 Xyl1
4	**79.2 CH**	4.04 m	C: 3 Xyl1; 1 Glc5	
5	63.5 CH_2	4.35 dd (5.5; 11.1)	C: 1, 3, 4 Xyl1	
		3.61 dd (9.4; 11.1)	C: 1 Xyl1	H-1 Xyl1
		Qui2 (1→2Xyl1)		
1	104.7 CH	4.89 d (7.5)	C: 2 Xyl1	H-2 Xyl1; H-5 Qui2
2	75.4 CH	3.84 t (9.0)	C: 1, 3 Qui2	H-4 Qui2
3	74.6 CH	3.91 t (9.0)	C: 2, 4 Qui2	H-1, 5 Qui2
4	**87.1 CH**	3.37 t (8.7)	C: 3, 5 Qui2, 1 Glc3	H-1 Glc3; H-2 Qui2
5	71.4 CH	3.63 dd (6.4; 9.5)		H-1 Qui2
6	17.7 CH_3	1.59 d (6.4)	C: 4, 5 Qui2	H-4, 5 Qui2
		Glc3 (1→4Qui2)		
1	104.2 CH	4.73 d (8.1)	C: 4 Qui2	H-4 Qui2; H-3 Glc3
2	73.4 CH	3.84 t (8.1)	C: 1, 3 Glc3	H-4 Glc3
3	**86.5 CH**	4.15 t (8.1)	C: 2, 4 Glc3; 1 MeGlc4	H-1 MeGlc4; H-1 Glc3
4	69.4 CH	3.75 t (9.1)	C: 3, 5, 6 Glc3	H-6 Glc3
5	74.6 CH	4.12 t (9.1)		
6	*67.5* CH_2	4.98 dd (2.0; 11.0)		
		4.57 dd (7.7; 11.0)	C: 5 Glc3	H-4 Glc3
		MeGlc4 (1→3Glc3)		
1	104.8 CH	5.16 d (6.9)	C: 3 Glc3	H-3 Glc3; H-3, 5 MeGlc4
2	74.3 CH	3.79 t (8.8)	C: 1, 3 MeGlc4	H-4 MeGlc4
3	86.3 CH	3.64 t (8.8)	C: 2, 4 MeGlc4, OMe	H-1, 5 MeGlc4, OMe
4	69.8 CH	4.01 m	C: 3, 5 MeGlc4	H-2, 6 MeGlc4
5	75.5 CH	4.01 m	C: 4, 6 MeGlc4	H-1, 3 MeGlc4
6	*67.0* CH_2	4.92 d (10.8)	C: 4, 5 MeGlc4	
		4.75 dd (3.0; 10.8)	C: 5 MeGlc4	
OMe	60.4 CH_3	3.75 s	C: 3 MeGlc4	
		Glc5 (1→4Xyl1)		
1	103.4 CH	4.81 d (7.8)	C: 4 Xyl1	H-4 Xyl1; H-3 Glc5
2	73.8 CH	3.81 t (7.8)	C: 1, 3 Glc5	H-4 Glc5
3	76.8 CH	4.10 t (8.8)	C: 2, 4 Glc5	H-1 Glc5
4	70.7 CH	3.92 t (8.8)	C: 3, 5, 6 Glc5	H-2, 6 Glc5
5	75.6 CH	4.06 dd (4.9; 9.8)		H-1 Glc5
6	*67.6* CH_2	5.02 d (9.8)	C: 4 Glc5	
		4.65 dd (6.9; 11.8)	C: 5 Glc5	H-4 Glc5

The comparison of the ^{13}C NMR spectrum of the sugar part of psolusoside L (**5**) with those of known achlioniceosides A_1, A_2, and A_3, with identical carbohydrate chains, isolated earlier from the sea cucumber *Rhipidothuria racowitzai* [14] showed the coincidence of the signals of four monosaccharide residues in their spectra. The signals of terminal 3-*O*-methylglucose residues of the novel and known compounds were different due to the absence of a sulfate group in this residue of known compounds. All these data indicated that psolusoside L (**5**) is a pentaoside with a new trisulfated carbohydrate chain branched by C-4 Xyl1.

The (−)ESI-MS/MS of **5** demonstrated the fragmentation of $[M_{3Na} - Na]^-$ ion at *m/z* 1529.4. The peaks of fragment ions were observed at *m/z*: 1409.5 $[M_{3Na} - Na - NaHSO_4]^-$, 1265.4 $[M_{3Na} - Na - C_6H_9O_8SNa\ (GlcSO_3Na)]^-$, 1131.5 $[M_{3Na} - Na - C_7H_{12}O_9SNa\ (MeGlcSO_3Na) - NaSO_3]^-$, 665.1 $[M_{3Na} - Na - C_{30}H_{43}O_4\ (Agl) - C_7H_{12}O_9SNa\ (MeGlcSO_3Na) - NaSO_3]^-$, and 519.0 $[M_{3Na} - Na - C_{30}H_{43}O_4\ (Agl) - C_7H_{12}O_9SNa\ (MeGlcSO_3Na) - C_6H_9O_7SNa\ (GlcSO_3Na)]^-$, confirming the structure of psolusoside L (**5**).

All these data indicate that psolusoside L (**5**) is 3*β*-*O*-{6-*O*-sodium-sulfate-3-*O*-methyl-*β*-D-glucopyranosyl-(1→3)-6-*O*-sodium-sulfate-*β*-D-glucopyranosyl-(1→4)-*β*-D-quinovopyranosyl-(1→2)-[6-*O*-sodium-sulfate-*β*-D-glucopyranosyl-(1→4)]-*β*-D-xylopyranosyl}-16-ketoholosta-9(11),25-diene.

The molecular formula of psolusoside M (**6**) was determined to be $C_{60}H_{91}O_{36}S_3Na_3$ from the ion peaks at *m/z* 1529.4273 (calc. 1529.4253) $[M_{3Na} - Na]^-$, 753.2202 (calc. 753.2180) $[M_{3Na} - 2Na]^{2-}$, and 494.4844 (calc. 494.4823) $[M_{3Na} - 3Na]^{3-}$ in the (−)HR-ESI-MS, indicating this glycoside to be an isomer of psolusoside L (**5**). In the ^1H and ^{13}C NMR spectra of the carbohydrate part of psolusoside M (**6**) five characteristic doublets at δ_H 4.58–5.15 (J = 7.1–8.5 Hz) and, corresponding to them, signals of anomeric carbons at δ_C 100.9–104.8, were indicative of a pentasaccharide chain and *β*-configurations of glycosidic bonds (Table 6). Analysis of the $^1H,^1H$-COSY and 1D TOCSY, ROESY, and HMBC spectra of psolusoside M (**6**) showed the same monosaccharide composition and architecture of the carbohydrate chain as in **5**. Actually, the comparison of their ^{13}C NMR spectra showed the closeness of the signals corresponding to the monosaccharides from the first to the fourth. The differences of the ^{13}C NMR spectra of compounds **6** and **5** were concerned with the terminal glucose residue (Glc5) connected to C-4 Xyl1. The characteristic signals at δ_C 100.9 (C-1 Glc5) and at δ_C 80.6 (C-2 Glc5) in the ^{13}C NMR spectrum of **6** were very close to the corresponding signals in the spectra of the compounds **1**–**4** indicating the presence of a sulfate group at C-2 Glc5 in the psolusoside M (**6**). At the same time, the hydroxy-methylene group of this sugar was free from sulfation, since the signal of C-6 Glc5 was observed at δ_C 61.8. Two signals of sulfated hydroxy-methylene groups of the glucose (Glc3) and 3-*O*-methylglucose (MeGlc4) residues were observed at δ_C 67.5 and 67.0 in the ^{13}C NMR spectrum of **6**. Therefore, psolusoside M (**6**) is an isomer of psolusoside L (**5**) by the sulfate group position.

The (−)ESI-MS/MS of **6** demonstrated the fragmentation of $[M_{3Na} - Na]^-$ ion at *m/z* 1529.4. The peaks of fragment ions were observed at the same *m/z*: 1409.5 $[M_{3Na} - Na - NaHSO_4]^-$, 1265.4 $[M_{3Na} - Na - C_6H_9O_8SNa\ (GlcSO_3Na)]^-$, 1131.5 $[M_{3Na} - Na - C_7H_{12}O_9SNa\ (MeGlcSO_3Na) - NaSO_3]^-$, 665.1 $[M_{3Na} - Na - C_{30}H_{43}O_4\ (Agl) - C_7H_{12}O_9SNa\ (MeGlcSO_3Na) - NaSO_3]^-$, and 519.0 $[M_{3Na} - Na - C_{30}H_{43}O_4\ (Agl) - C_7H_{12}O_9SNa\ (MeGlcSO_3Na) - C_6H_9O_7SNa\ (GlcSO_3Na)]^-$ as in the MS/MS spectrum of glycoside **5**.

Table 6. ^{13}C and ^{1}H NMR chemical shifts and HMBC and ROESY correlations of carbohydrate moiety of psolusoside M (**6**). [a] Recorded at 176.04 MHz in C_5D_5N/D_2O (4/1). [b] Bold = interglycosidic positions. [c] Italic = sulphate position. [d] Recorded at 700.13 MHz in C_5D_5N/D_2O (4/1). [e] Recorded at 500.13 MHz in C_5D_5N/D_2O (4/1). Multiplicity by 1D TOCSY.

Atom	δ_C Mult. [a, b, c]	δ_H Mult. [d] (*J* in Hz)	HMBC	ROESY [e]
		Xyl1 (1→C-3)		
1	104.8 CH	4.58 d (7.1)	C: 3	H-3, H-3, 5 Xyl1
2	**82.5 CH**	3.88 t (7.1)	C: 1, 3 Xyl1	H-1 Qui2
3	75.0 CH	4.20 t (8.7)	C: 2, 4 Xyl1	H-1 Xyl1
4	**78.6 CH**	4.13 m		H-1 Glc5
5	63.6 CH_2	4.48 dd (4.7; 11.9)	C: 3 Xyl1	
		3.77 t (11.9)		H-1 Xyl1
		Qui2 (1→2Xyl1)		
1	104.5 CH	4.92 d (7.9)	C: 2 Xyl1	H-2 Xyl1; H-5 Qui2
2	75.4 CH	3.85 t (8.7)		H-4 Qui2
3	74.9 CH	3.91 t (8.7)		H-1 Qui2
4	**86.9 CH**	3.39 t (9.6)	C: 1 Glc3; 3, 5 Qui2	H-1 Glc3
5	71.3 CH	3.59 dd (6.0; 9.6)		H-1 Qui2
6	17.7 CH_3	1.58 d (6.0)		
		Glc3 (1→4Glc2)		
1	104.1 CH	4.74 d (8.5)	C: 4 Qui2	H-4 Qui2; H-5 Glc3
2	73.4 CH	3.82 t (8.5)		
3	**86.5 CH**	4.13 t (9.3)	C: 4 Glc3; 1 MeGlc4	H-1 MeGlc4; H-1 Glc3
4	69.3 CH	3.75 t (9.3)		
5	74.7 CH	4.11 t (10.1)		H-1 Glc3
6	*67.5* CH_2	4.99 brd (10.1)		
		4.57 m		
		MeGlc4 (1→3Glc3)		
1	104.7 CH	5.15 d (7.8)	C: 3 Glc3	H-3 Glc3; H-3, 5 MeGlc4
2	74.3 CH	3.78 t (8.5)	C: 1, 3 MeGlc4	H-4 MeGlc4
3	86.3 CH	3.64 m	C: 4 MeGlc4	H-1, 5 MeGlc4
4	69.8 CH	4.01 m	C: 5 MeGlc4	H-2, 6 MeGlc4
5	75.5 CH	4.01 m		H-1, 3 MeGlc4
6	*67.0* CH_2	4.92 brd (11.6)		
		4.75 brd (11.6)		H-4 MeGlc4
OMe	60.5 CH_3	3.76 s	C: 3 MeGlc4	
		Glc5 (1→4Xyl1)		
1	100.9 CH	4.96 d (7.8)	C: 4 Xyl1	H-4 Xyl1; H-3, 5 Glc5
2	*80.6* CH	4.76 t (7.8)	C: 1 Glc5	H-4 Glc5
3	76.9 CH	4.29 t (8.5)	C: 2, 4 Glc5	H-1, 5 Glc5
4	70.8 CH	3.90 t (8.5)	C: 5 Glc5	
5	77.4 CH	3.87 m		
6	61.8 CH_2	4.34 brd (10.1)		
		4.01 dd (6.2; 12.4)		

All these data indicate that psolusoside M (**6**) is 3*β*-*O*-{6-*O*-sodium-sulfate-3-*O*-methyl-*β*-D-glucopyranosyl-(1→3)-6-*O*-sodium-sulfate-*β*-D-glucopyranosyl-(1→4)-*β*-D-quinovopyranosyl-(1→2)-[2-*O*-sodium-sulfate-*β*-D-glucopyranosyl-(1→4)]-*β*-D-xylopyranosyl}-16-ketoholosta-9(11),25-diene.

The molecular formula of psolusoside N (**7**) was determined to be $C_{60}H_{91}O_{37}S_3Na_3$ from the $[M_{3Na} - Na]^-$ ion peak at *m/z* 1545.4171 (calc. 1545.4202), $[M_{3Na} - 2Na]^{2-}$ ion peak at *m/z* 761.2164 (calc. 761.2155), and $[M_{3Na} - 3Na]^{3-}$ ion peak at *m/z* 499.8151 (calc. 499.8139) in the (−)HR-ESI-MS. In the ^1H and ^{13}C NMR spectra of the carbohydrate part of psolusoside N (**7**) five characteristic doublets at δ_H 4.67–5.12 (J = 6.8–8.3 Hz) and, corresponding to them, signals of anomeric carbons at δ_C 103.3–104.7 were indicative of a pentasaccharide chain and *β*-configurations of glycosidic bonds (Table 7). Analysis of the ^1H,^1H-COSY and 1D TOCSY spectra of psolusoside N (**7**) showed the presence of one xylose, three glucose, and one 3-*O*-methylglucose residues. The positions of interglycosidic linkages and the consequence of monosaccharides in the carbohydrate chain of **7** were established in the same manner as for **1**–**6** (Table 7) indicating the presence of branched pentasaccharide moiety having the same architecture as in compounds **5** and **6**. The comparison of the ^{13}C NMR spectra of **7** and **5** showed the closeness of the signals of all the monosaccharide residues except for the signals assigned to the second sugar units in their chains. Actually, in the ^1H and ^{13}C NMR spectra of **7**, the signals characteristic of quinovose residue were absent but two doublets of doublets at δ_H 4.95 (H-6a Glc2) and at δ_H 4.75 (H-6b Glc2) and the signal at δ_C 61.0 (C-6 Glc2), assigned to hydroxy-methylene group of glucopyranose moiety, were detected. These data indicated the replacement of quinovose by the glucose residue in the second position of a carbohydrate chain in psolusoside N (**7**) as compared with psolusoside L (**5**). Three sulfate groups were supposed to attach the C-6 of two glucose and 3-*O*-methylglucose residues due to the signals at δ_C 67.4, 67.5, and 66.9 observed in the spectrum of **7**. The carbohydrate chain of psolusoside N (**7**) is the first found in the glycosides from holothurians.

The (−)ESI-MS/MS of **7** demonstrated the fragmentation of $[M_{3Na} - Na]^-$ ion at *m/z* 1545.4. The peaks of fragment ions were observed at *m/z*: 1425.5 $[M_{3Na} - Na - NaHSO_4]^-$, 1281.4 $[M_{3Na} - Na - C_6H_9O_8SNa\ (GlcSO_3Na)]^-$, 1147.5 $[M_{3Na} - Na - C_7H_{12}O_9SNa\ (MeGlcSO_3Na) - NaSO_3]^-$, 1003.4 $[M_{3Na} - Na - C_6H_9O_8SNa\ (GlcSO_3Na) - C_7H_{12}O_8SNa\ (MeGlcSO_3Na) + H]^-$, 681.1 $[M_{3Na} - Na - C_{30}H_{43}O_4\ (Agl) - C_7H_{12}O_9SNa\ (MeGlcSO_3Na) - NaSO_3 + H]^-$, and 519.0 $[M_{3Na} - Na - C_{30}H_{43}O_4\ (Agl) - C_7H_{12}O_9SNa\ (MeGlcSO_3Na) - C_6H_9O_7SNa\ (GlcSO_3Na)]^-$, corroborating the structure of psolusoside N (**7**).

All these data indicate that psolusoside N (**7**) is 3*β*-*O*-{6-*O*-sodium-sulfate-3-*O*-methyl-*β*-D-glucopyranosyl-(1→3)-6-*O*-sodium-sulfate-*β*-D-glucopyranosyl-(1→4)-*β*-D-glucopyranosyl-(1→2)-[6-*O*-sodium-sulfate-*β*-D-glucopyranosyl-(1→4)]-*β*-D-xylopyranosyl}-16-ketoholosta-9(11),25-diene.

The molecular formula of psolusoside O (**8**) was established as the same ($C_{60}H_{91}O_{37}S_3Na_3$) as compound **7** from the $[M_{3Na} - Na]^-$ ion peak at *m/z* 1545.4197 (calc. 1545.4202), $[M_{3Na} - 2Na]^{2-}$ ion peak at *m/z* 761.2171 (calc. 761.2155), and $[M_{3Na} - 3Na]^{3-}$ ion peak at *m/z* 499.8155 (calc. 499.8139) in the (−)HR-ESI-MS.

In the ^1H and ^{13}C NMR spectra of the carbohydrate part of psolusoside O (**8**), five characteristic doublets at δ_H 4.60–5.12 (J = 7.0–8.6 Hz) and, corresponding to them, signals of anomeric carbons at δ_C 101.0–104.8 indicated a pentasaccharide carbohydrate chain and *β*-configurations of glycosidic bonds (Table 8).

Table 7. ^{13}C and ^{1}H NMR chemical shifts and HMBC and ROESY correlations of carbohydrate moiety of psolusoside N (**7**). [a] Recorded at 176.04 MHz in C_5D_5N/D_2O (4/1). [b] Bold = interglycosidic positions. [c] Italic = sulphate position. [d] Recorded at 700.13 MHz in C_5D_5N/D_2O (4/1). [e] Recorded at 500.13 MHz in C_5D_5N/D_2O (4/1). Multiplicity by 1D TOCSY.

Atom	δ_C Mult.[a, b, c]	δ_H Mult.[d] (*J* in Hz)	HMBC	ROESY [e]
		Xyl1 (1→C-3)		
1	104.7 CH	4.67 d (6.8)	C: 3	H-3; H-3, 5 Xyl1
2	**81.3 CH**	4.01 t (9.0)	C: 1 Xyl1; 1 Glc2	H-1 Glc2
3	75.0 CH	4.13 t (9.0)	C: 4 Xyl1	H-1, 5 Xyl1
4	**78.8 CH**	4.04 dd (4.5; 9.8)	C: 1 Glc5	H-1 Glc5
5	63.5 CH_2	4.36 dd (4.5; 10.5)		
		3.63 dd (9.8; 12.0)		H-1 Xyl1
		Glc2 (1→2Xyl1)		
1	104.3 CH	5.06 d (8.3)	C: 2 Xyl1	H-2 Xyl1; H-3, 5 Glc2
2	75.2 CH	3.87 t (9.0)	C: 1, 3 Glc2	
3	75.2 CH	4.00 t (9.0)	C: 4 Glc2	H-1 Glc2
4	**81.8 CH**	3.95 t (9.0)	C: 3 Glc2, 1 Glc3	H-1 Glc3
5	75.9 CH	3.71 d (9.8)		H-1 Glc2
6	61.0 CH_2	4.31 dd (3.0; 11.3)		
		4.26 brd (11.3)		
		Glc3 (1→4Glc2)		
1	103.8 CH	4.84 d (8.3)	C: 4 Glc2	H-4 Glc2; H-3, 5 Glc3
2	73.4 CH	3.83 t (8.3)	C: 1, 3 Glc3	
3	**86.3 CH**	4.10 t (9.0)	C: 4 Glc3; 1 MeGlc4	H-1 MeGlc4; H-1 Glc3
4	69.3 CH	3.75 t (9.0)	C: 5, 6 Glc3	
5	74.8 CH	4.04 dd (6.8; 10.0)		
6	*67.4* CH_2	4.95 dd (2.3; 10.5)		
		4.57 dd (7.5; 10.5)		
		MeGlc4 (1→3Glc3)		
1	104.6 CH	5.12 d (8.3)	C: 3 Glc3	H-3 Glc3; H-3, 5 MeGlc4
2	74.3 CH	3.78 t (9.0)	C: 1, 3 MeGlc4	
3	86.3 CH	3.62 t (9.0)	C: 2, 4 MeGlc4, OMe	H-1 MeGlc4, OMe
4	69.7 CH	4.00 t (9.0)	C: 3, 5 MeGlc4	
5	75.5 CH	3.98 m		H-1 MeGlc4
6	*66.9* CH_2	4.92 dd (2.3; 11.3)		
		4.75 dd (4.5; 11.3)		
OMe	60.4 CH_3	3.75 s	C: 3 MeGlc4	
		Glc5 (1→4Xyl1)		
1	103.3 CH	4.81 d (8.3)	C: 4 Xyl1	H-4 Xyl1; H-3 Glc5
2	73.8 CH	3.82 t (9.0)	C: 1, 3 Glc5	
3	76.8 CH	4.10 t (9.0)	C: 2, 4 Glc5	H-1 Glc5
4	70.7 CH	3.94 t (9.0)	C: 3, 6 Glc5	
5	75.6 CH	4.05 dd (3.8; 9.8)		
6	*67.5* CH_2	5.02 d (9.0)		
		4.67 dd (6.8; 11.3)		

Table 8. ^{13}C and ^{1}H NMR chemical shifts and HMBC and ROESY correlations of carbohydrate moiety of psolusoside O (**8**). [a] Recorded at 176.04 MHz in C_5D_5N/D_2O (4/1). [b] Bold = interglycosidic positions. [c] Italic = sulphate position. [d] Recorded at 700.13 MHz in C_5D_5N/D_2O (4/1). [e] Recorded at 500.13 MHz in C_5D_5N/D_2O (4/1). Multiplicity by 1D TOCSY.

Atom	δ_C Mult. [a, b, c]	δ_H Mult. [d] (*J* in Hz)	HMBC	ROESY [e]
		Xyl1 (1→C-3)		
1	104.8 CH	4.60 d (7.1)	C: 3	H-3, H-3, 5 Xyl1
2	**81.1 CH**	4.04 t (8.8)		H-1 Glc2
3	75.1 CH	4.23 t (8.8)	C: 2 Xyl1	
4	**78.7 CH**	4.11 m		H-1 Glc5
5	63.7 CH_2	4.48 m		
		3.78 t (11.2)		H-1 Xyl1
		Glc2 (1→2Xyl1)		
1	104.2 CH	5.12 d (8.6)	C: 2 Xyl1	H-2 Xyl1; H-3, 5 Glc2
2	75.2 CH	3.86 t (8.6)	C: 1 Glc2	
3	75.5 CH	3.98 t (8.6)		
4	**81.9 CH**	3.93 t (9.4)		H-1 Glc3
5	75.9 CH	3.70 brd (11.7)		
6	61.2 CH_2	4.31 brd (11.7)		
		4.27 dd (5.5; 11.7)		
		Glc3 (1→4Glc2)		
1	103.8 CH	4.86 d (7.8)	C: 4 Glc2	H-4 Glc2; H-3 Glc3
2	73.4 CH	3.81 t (8.6)	C: 1, 3 Glc3	
3	**86.3 CH**	4.08 t (8.6)	C: 2, 4 Glc3; 1 MeGlc4	H-1 MeGlc4; H-1 Glc3
4	69.2 CH	3.77 t (9.4)		
5	74.8 CH	4.04 m		H-1 Glc3
6	*67.3* CH_2	4.97 brd (9.4)		
		4.58 dd (7.8; 11.7)		
		MeGlc4 (1→3Glc3)		
1	104.6 CH	5.11 d (7.8)	C: 3 Glc3	H-3 Glc3; H-3, 5 MeGlc4
2	74.3 CH	3.77 t (8.6)	C: 1, 3 MeGlc4	
3	86.4 CH	3.63 t (8.6)	C: 2, 4 MeGlc4, OMe	H-1 MeGlc4
4	69.7 CH	4.02 t (8.6)	C: 5 MeGlc4	
5	75.2 CH	3.99 m		H-1 MeGlc4
6	*66.9* CH_2	4.92 brd (10.1)		
		4.76 dd (4.7; 12.5)		
OMe	60.4 CH_3	3.76 s	C: 3 MeGlc4	
		Glc5 (1→4Xyl1)		
1	101.0 CH	4.94 d (7.0)	C: 4 Xyl1	H-4 Xyl1; H-3, 5 Glc5
2	*80.6* CH	4.76 t (8.6)	C: 1, 3 Glc5	H-4 Glc5
3	76.9 CH	4.30 t (8.6)	C: 2, 4 Glc5	H-1, 5 Glc5
4	70.8 CH	3.91 t (8.6)	C: 5 Glc5	
5	77.4 CH	3.87 m		H-1 Glc5
6	61.8 CH_2	4.34 dd (2.3; 12.5)		
		4.02 dd (7.0; 12.5)		

Analysis of the $^{1}H,^{1}H$-COSY and 1D TOCSY spectra of psolusoside O (**8**) showed the same monosaccharide composition and positions of interglycosidic linkages as in the carbohydrate chain of compound **7** (Table 8). The coincidence of the molecular formulae of **8** and **7** and the presence of three-charged ions in the (−)HR-ESI-MS of **8** indicated their difference in the position of a sulfate group. Really, the signals of monosaccharide residues from the first to the fourth were almost coincident in their ^{13}C NMR spectra. The characteristic signals at δ_C 101.0 and δ_C 80.6 indicated the bonding of a sulfate group to C-2 of a terminal residue which glycosylates C-4 Xyl1. Analysis of the $^{1}H,^{1}H$-COSY and 1D TOCSY spectra of **8** showed this unit is a glucose (Glc5). Indeed, the comparison of the ^{13}C NMR spectra of **8** and **6** revealed their difference only in the signals of the second monosaccharide unit and the coincidence of the signals of the remaining ones. All these data indicate psolusoside O (**8**) has new trisulfated carbohydrate chain with the sulfate groups attached to C-6 of the third (Glc3), to C-6 of the fourth (MeGlc), and to C-2 of the fifth (Glc5) monosaccharide residues.

The (−)ESI-MS/MS of **8** demonstrated the fragmentation of $[M_{3Na} - Na]^-$ with peaks of fragment ions, observed at *m/z* 1425.5 $[M_{3Na} - Na - NaHSO_4]^-$, 1281.4 $[M_{3Na} - Na - C_6H_9O_8SNa$ $(GlcSO_3Na)]^-$, 1161.5 $[M_{3Na} - Na - C_6H_9O_8SNa$ $(GlcSO_3Na) - NaHSO_4]^-$, 1147.5 $[M_{3Na} - Na - C_7H_{12}O_9SNa$ $(MeGlcSO_3Na) - NaSO_3]^-$, 1003.4 $[M_{3Na} - Na - C_6H_9O_8SNa$ $(GlcSO_3Na) - C_7H_{12}O_8SNa$ $(MeGlcSO_3Na) + H]^-$, and 681.1 $[M_{3Na} - Na - C_{30}H_{43}O_4$ (Agl) $- C_7H_{12}O_9SNa$ $(MeGlcSO_3Na) - NaSO_3 + H]^-$, 519.0 $[M_{3Na} - Na - C_{30}H_{43}O_4$ (Agl) $- C_7H_{12}O_9SNa$ $(MeGlcSO_3Na) - C_6H_9O_7SNa$ $(GlcSO_3Na)]^-$, corroborating the isomerism of psolusosides O (**8**) and N (**7**).

All these data indicate that psolusoside O (**8**) is 3*β*-*O*-{6-*O*-sodium-sulfate-3-*O*-methyl-*β*-D-glucopyranosyl-(1→3)-6-*O*-sodium-sulfate-*β*-D-glucopyranosyl-(1→4)-*β*-D-glucopyranosyl-(1→2)-[2-*O*-sodium-sulfate-*β*-D-glucopyranosyl-(1→4)]-*β*-D-xylopyranosyl}-16-ketoholosta-9(11),25-diene.

The molecular formula of psolusoside P (**9**) was determined to be $C_{60}H_{90}O_{39}S_4Na_4$ from the $[M_{4Na} - Na]^-$ ion peak at *m/z* 1631.3598 (calc. 1631.3641), $[M_{4Na} - 2Na]^{2-}$ ion peak at *m/z* 804.1879 (calc. 804.1874), $[M_{4Na} - 3Na]^{3-}$ ion peak at *m/z* 528.4628 (calc. 528.4619), and $[M_{4Na} - 4Na]^{4-}$ ion peak at *m/z* 390.6001 (calc. 390.5991) in the (−)HR-ESI-MS indicating the presence of four sulfate groups. In the ^{1}H and ^{13}C NMR spectra of the carbohydrate part of psolusoside P (**9**), five characteristic doublets at δ_H 4.66–5.16 (J = 7.2–8.3 Hz) and corresponding signals of anomeric carbons at δ_C 103.1–104.8 were indicative of a pentasaccharide chain and *β*-configurations of glycosidic bonds (Table 9). Analysis of the $^{1}H,^{1}H$-COSY and 1D TOCSY spectra of psolusoside P (**9**) showed the presence of one xylose, one quinovose, two glucose, and one 3-*O*-methylglucose residues. The positions of interglycosidic linkages and the consequence of monosaccharides in the chain of **9** established by the ROESY and HMBC spectra were the same as in the glycosides **5** and **6** (Table 9). The comparison of the ^{13}C NMR spectra of the compounds **9** and **5** showed the coincidence of the signals corresponding to the monosaccharides from the first to the fourth indicating their identity in these glycosides. The signals of the fifth terminal sugar residue assigned by the $^{1}H,^{1}H$-COSY and 1D TOCSY spectra corresponded to the glucose residue sulfated by C-6 (the signal at δ_C 67.9 (C-6 Glc5)). Thus, three sulfate groups were positioned at C-6 of 3-*O*-methylglucose (MeGlc4) and C-6 of two glucose residues (Glc3 and Glc5) in the carbohydrate chain of psolusoside P (**9**). The position of the fourth sulfate group at C-4 Glc5 was established by the comparison of the ^{13}C NMR spectra of psolusosides P (**9**) and L (**5**). The signal of C-4 Glc5, deduced by the $^{1}H,^{1}H$-COSY spectrum of **9**, was deshielded to δ_C 77.1 due to α-shifting effect of the sulfate group, as compared with the corresponding signal in the ^{13}C NMR spectrum of **5** observed at δ_C 70.7. Oppositely, the signal of C-5 Glc5 was shielded to δ_C 73.7 in the spectrum of **9** due to the β-shifting effect of the sulfate group as compared with the spectrum of **5** (δ_C 75.65 (C-5 Glc5)). So, psolusoside P (**9**) is the first case of triterpene glycoside having four sulfate groups, in that two of them were connected to one monosaccharide residue.

Table 9. ^{13}C and ^{1}H NMR chemical shifts and HMBC and ROESY correlations of carbohydrate moiety of psolusoside P (**9**). [a] Recorded at 176.04 MHz in C_5D_5N/D_2O (4/1). [b] Bold = interglycosidic positions. [c] Italic = sulphate position. [d] Recorded at 700.13 MHz in C_5D_5N/D_2O (4/1). [e] Recorded at 500.13 MHz in C_5D_5N/D_2O (4/1). Multiplicity by 1D TOCSY.

Atom	δ_C Mult. [a, b, c]	δ_H Mult. [d] (*J* in Hz)	HMBC	ROESY [e]
		Xyl1 (1→C-3)		
1	104.8 CH	4.66 d (7.2)	C: 3	H-3
2	**82.2 CH**	3.87 t (8.8)	C: 1 Xyl1; 1 Qui2	H-1 Qui2
3	75.0 CH	4.08 t (8.8)	C: 2, 4 Xyl1	H-1, 5 Xyl1
4	**79.7 CH**	4.05 m	C: 1 Glc5	H-1 Glc5
5	63.4 CH_2	4.34 dd (5.6; 11.2)	C: 3 Xyl1	
		3.61 dd (9.6; 12.0)		H-1, 3 Xyl1
		Qui2 (1→2Xyl1)		
1	104.6 CH	4.87 d (7.8)	C: 2 Xyl1	H-2 Xyl1; H-3, 5 Qui2
2	75.4 CH	3.83 t (7.8)	C: 1, 3 Qui2	H-4 Qui2
3	74.6 CH	3.90 t (8.6)	C: 2, 4 Qui2	H-1, 5 Qui2
4	**87.2 CH**	3.36 t (8.6)	C: 3, 5 Qui2, 1 Glc3	H-1 Glc3, H-2 Qui2
5	71.4 CH	3.62 dd (6.3; 9.4)		H-1 Qui2
6	17.7 CH_3	1.57 d (5.7)		
		Glc3 (1→4Qui2)		
1	104.2 CH	4.72 d (8.0)	C: 4 Qui2	H-4 Qui2; H-5 Glc3
2	73.7 CH	3.83 t (8.8)	C: 1, 3 Glc3	H-4 Glc3
3	**86.4 CH**	4.14 t (8.8)	C: 2, 4 Glc3; 1 MeGlc4	H-1 MeGlc4; H-1 Glc3
4	69.4 CH	3.74 t (9.6)	C: 3, 5, 6 Glc3	H-6 Glc3
5	74.5 CH	4.11 t (9.6)		H-1 Glc3
6	*67.5* CH_2	4.95 dd (2.4; 11.2)		
		4.55 dd (8.0; 11.2)	C: 5 Glc3	H-4 Glc3
		MeGlc4 (1→3Glc3)		
1	104.7 CH	5.16 d (8.3)	C: 3 Glc3	H-3 Glc3; H-3, 5 MeGlc4
2	74.3 CH	3.78 t (8.3)	C: 1, 3 MeGlc4	H-4 MeGlc4
3	86.5 CH	3.63 t (8.3)	C: 2, 4 MeGlc4, OMe	H-1, 5 MeGlc4, OMe
4	69.8 CH	4.00 m	C: 3, 5 MeGlc4	H-2, 6 MeGlc4
5	75.5 CH	4.01 m	C: 4 MeGlc4	H-1, 3 MeGlc4
6	*67.0* CH_2	4.92 d (10.6)	C: 4 MeGlc4	
		4.74 dd (4.5; 11.3)		H-4 MeGlc4
OMe	60.5 CH_3	3.76 s	C: 3 MeGlc4	
		Glc5 (1→4Xyl1)		
1	103.1 CH	4.79 d (7.7)	C: 4 Xyl1	H-4 Xyl1; H-3, 5 Glc5
2	73.4 CH	3.84 t (7.7)	C: 1, 3 Glc5	H-4 Glc5
3	76.1 CH	4.21 t (8.8)	C: 2, 4 Glc5	H-1 Glc5
4	*77.1* CH	4.68 t (8.8)	C: 3, 5, 6 Glc5	
5	73.7 CH	4.15 dt (9.6; 12.0)		H-1 Glc5
6	*67.9* CH_2	5.29 dd (2.2; 11.9)		
		4.65 dd (8.8; 11.7)	C: 5 Glc5	

The (−)ESI-MS/MS of **9** demonstrated the fragmentation of $[M_{4Na} - Na]^-$ ion at *m/z* 1631.4. The peaks of fragment ions were observed at *m/z*: 1265.4 $[M_{4Na} - Na - C_6H_8O_{11}S_2Na_2$ $(Glc(SO_3Na)_2)]^-$, 1233.4 $[M_{4Na} - Na - C_7H_{12}O_9SNa$ $(MeGlcSO_3Na) - NaSO_3]^-$, 1145.5 $[M_{4Na} - Na - C_6H_9O_{11}S_2Na_2$ $(Glc(SO_3Na)_2 - NaSO_4]^-$, 1089.4 $[M_{4Na} - Na - C_7H_{12}O_9SNa$ $(MeGlcSO_3Na) - C_6H_8O_7SNa$ $(GlcSO_3Na)]^-$, 969.4 $[M_{4Na} - Na - C_7H_{12}O_9SNa$ $(MeGlcSO_3Na) - C_6H_8O_7SNa$ $(GlcSO_3Na) - NaHSO_4]^-$, and 943.3 $[M_{4Na} - Na - C_7H_{12}O_9SNa$ $(MeGlcSO_3Na) - C_6H_8O_7SNa$ $(GlcSO_3Na) - C_6H_{10}O_4$ $(Qui)]^-$ corroborating the structure of carbohydrate chain of psolusoside P (**9**).

All these data indicate that psolusoside P (**9**) is 3*β*-*O*-{6-*O*-sodium-sulfate-3-*O*-methyl-*β*-D-glucopyranosyl-(1→3)-6-*O*-sodium-sulfate-*β*-D-glucopyranosyl-(1→4)-*β*-D-quinovopyranosyl-(1→2)-[4,6-*O*-sodium-disulfate-*β*-D-glucopyranosyl-(1→4)]-*β*-D-xylopyranosyl}-16-ketoholosta-9(11),25-diene.

The molecular formula of psolusoside Q (**10**) was determined to be $C_{60}H_{90}O_{40}S_4Na_4$ from the $[M_{4Na} - Na]^-$ ion peak at *m/z* 1647.3544 (calc. 1647.3590), $[M_{4Na} - 2Na]^{2-}$ ion peak at *m/z* 812.1854 (calc. 812.1849), $[M_{4Na} - 3Na]^{3-}$ ion peak at *m/z* 533.7944 (calc. 533.7935), and $[M_{4Na} - 4Na]^{4-}$ ion peak at *m/z* 394.5989 (calc. 394.5978) in the (−)HR-ESI-MS demonstrating the presence of four sulfate groups. In the ^1H and ^{13}C NMR spectra of the carbohydrate part of psolusoside Q (**10**), five characteristic doublets at δ_H 4.61–5.12 (J = 6.7–8.4 Hz) and, corresponding to them, signals of anomeric carbons at δ_C 101.5–104.8, were indicative of a pentasaccharide chain and *β*-configurations of glycosidic bonds (Table 10). The molecular weights of tetrasulfated psolusosides P (**9**) and Q (**10**) differed by 16 *amu* in HR-ESI-MS that along with the absence of the signals corresponding to the quinovose residue in the NMR spectra of **10** indicated the presence of a glucose residue in the second position of its carbohydrate chain. Actually, the coincidence of the signals of monosaccharide residues from the first to the fourth the ^{13}C NMR spectra of psolusosides Q (**10**), N (**7**), and O (**8**) confirmed this supposition. Analysis of the $^1H,^1H$-COSY, 1D TOCSY, ROESY, and HMBC spectra of psolusoside Q (**10**) showed the same monosaccharide composition and the consequence of monosaccharides in the chain of **10** as in psolusosides N (**7**) and O (**8**) (Table 10). The characteristic signals at δ_C 101.5 (C-1 Glc5) and δ_C 80.3 (C-2 Glc5) indicated attachment of a sulfate group to C-2 of the fifth residue (Glc5) in the sugar part of **10**. The signal of C-6 Glc5 was assigned by the HSQC spectrum of **10**, demonstrating the correlation of the both doublet at δ_H 5.02 (H-6a Glc5) and doublet of doublets at δ_H 4.64 (H-6b Glc5) with the corresponding resonance at δ_C 67.5 that indicated the presence of an additional sulfate group at C-6 Glc5 in psolusoside Q (**10**). All these data show that psolusoside Q (**10**) has a new carbohydrate chain with four sulfate groups, in that two of them are attached to C-2 and C-6 of the same (Glc5) residue.

The (−)ESI-MS/MS of **10** demonstrated the fragmentation of $[M_{4Na} - Na]^-$ ion at *m/z* 1647.4. The peaks of fragment ions were observed at *m/z* 1527.4 $[M_{4Na} - Na - NaHSO_4]^-$, 1281.4 $[M_{4Na} - Na - C_6H_8O_{11}S_2Na_2$ $(Glc(SO_3Na)_2)]^-$, 1161.5 $[M_{4Na} - Na - C_6H_8O_{11}S_2Na_2$ $(Glc(SO_3Na)_2) - NaHSO_4]^-$, 1003.4 $[M_{4Na} - Na - C_6H_8O_{11}S_2Na_2$ $(Glc(SO_3Na)_2) - C_7H_{11}O_8SNa$ $(MeGlcSO_3Na)]^-$, 681.1 $[M_{4Na} - Na - C_{30}H_{43}O_4$ $(Agl) - C_6H_9O_{11}S_2Na_2$ $(Glc(SO_3Na)_2) - C_5H_8O_4$ $(Xyl)]^-$, and 519.0 $[M_{4Na} - Na - C_{30}H_{43}O_4$ $(Agl) - C_6H_9O_{11}S_2Na_2$ $(Glc(SO_3Na)_2) - C_5H_8O_4$ $(Xyl) - C_6H_{10}O_5$ $(Glc)]^-$ corroborating the sequence of monosaccharide residues in psolusoside Q (**10**).

All these data indicate that psolusoside Q (**10**) is 3*β*-*O*-{6-*O*-sodium-sulfate-3-*O*-methyl-*β*-D-glucopyranosyl-(1→3)-6-*O*-sodium-sulfate-*β*-D-glucopyranosyl-(1→4)-*β*-D-glucopyranosyl-(1→2)-[2,6-*O*-sodium-disulfate-*β*-D-glucopyranosyl-(1→4)]-*β*-D-xylopyranosyl}-16-ketoholosta-9(11),25-diene.

Thus, highly polar tetrasulfated glycosides are first discovered in sea cucumbers. Although polysulfated polysaccharides are common biopolymers of marine macrophytes and invertebrates, low molecular weight metabolites, containing several sulfate groups are extremely rare. So far, trisulfated natural compounds such as steroid glycosides were found only in sponges [15–17] and trisulfated triterpene glycosides, in some representatives of the class Holothuroidea [18,19].

Table 10. ^{13}C and ^{1}H NMR chemical shifts and HMBC and ROESY correlations of carbohydrate moiety of psolusoside Q (**10**). [a] Recorded at 176.04 MHz in C_5D_5N/D_2O (4/1). [b] Bold = interglycosidic positions. [c] Italic = sulphate position. [d] Recorded at 700.13 MHz in C_5D_5N/D_2O (4/1). [e] Recorded at 500.13 MHz in C_5D_5N/D_2O (4/1). Multiplicity by 1D TOCSY.

Atom	δ_C Mult. [a, b, c]	δ_H Mult. [d] (*J* in Hz)	HMBC	ROESY [e]
		Xyl1 (1→C-3)		
1	104.8 CH	4.61 d (6.7)	C: 3	H-3, H-3, 5 Xyl1
2	**81.2 CH**	4.01 t (8.9)		H-1 Glc2
3	74.9 CH	4.19 t (8.9)	C: 2 Xyl1	H-1, 5 Xyl1
4	**79.8 CH**	4.02 m	C: 3 Xyl1	H-1 Glc5
5	63.6 CH_2	4.49 dd (5.2; 11.2)		
		3.77 dd (9.4; 11.2)		H-1 Xyl1
		Glc2 (1→2Xyl1)		
1	104.2 CH	5.06 d (8.4)	C: 2 Xyl1	H-2 Xyl1; H-3, 5 Glc2
2	75.1 CH	3.84 t (9.2)	C: 3 Glc2	
3	74.6 CH	3.99 t (9.2)	C: 2, 4 Glc2	
4	**81.8 CH**	3.94 t (8.4)	C: 3 Glc2	H-1 Glc3, H-2, 6 Glc2
5	75.9 CH	3.71 d (10.2)		H-1 Glc2
6	60.9 CH_2	4.29 m		
		4.26 m		
		Glc3 (1→4Glc2)		
1	103.9 CH	4.84 d (7.4)	C: 4 Glc2	H-4 Glc2; H-3, 5 Glc3
2	73.4 CH	3.82 t (9.2)	C: 1, 3 Glc3	
3	**86.4 CH**	4.10 t (9.2)	C: 2, 4 Glc3; 1 MeGlc4	H-1 MeGlc4; H-1 Glc3
4	69.3 CH	3.76 t (9.2)		
5	74.7 CH	4.04 m		H-1 Glc3
6	*67.4* CH_2	4.94 d (9.2)		
		4.58 dd (7.4; 12.0)		H-4 Glc3
		MeGlc4 (1→3Glc3)		
1	104.7 CH	5.12 d (8.3)	C: 3 Glc3	H-3 Glc3; H-3, 5 MeGlc4
2	74.3 CH	3.77 t (8.3)	C: 1, 3 MeGlc4	
3	86.3 CH	3.63 t (8.3)	C: 2, 4 MeGlc4, OMe	H-1, 5 MeGlc4, OMe
4	69.7 CH	4.03 t (9.2)	C: 3, 5, 6 MeGlc4	
5	75.5 CH	3.99 m		H-1, 3 MeGlc4
6	*66.9* CH_2	4.91 dd (2.1; 11.5)		
		4.77 dd (4.4; 11.1)		
OMe	60.4 CH_3	3.75 s	C: 3 MeGlc4	
		Glc5 (1→4Xyl1)		
1	101.5 CH	4.90 d (8.2)	C: 4 Xyl1	H-4 Xyl1; H-3, 5 Glc5
2	*80.3* CH	4.72 t (8.2)	C: 1, 3 Glc5	
3	76.5 CH	4.27 t (8.2)	C: 2, 4 Glc5	H-1, 5 Glc5
4	70.5 CH	3.93 t (9.1)	C: 6 Glc5	
5	75.2 CH	4.05 t (9.1)		H-1 Glc5
6	*67.5* CH_2	5.02 d (10.4)		
		4.64 dd (6.7; 11.4)		

2.2. Bioactivity of the Glycosides

The cytotoxic activities of the compounds **1–10** as well as known earlier psolusosides G (used as a positive control) and B [12] against mouse erythrocytes (hemolytic activity), the ascite form of mouse Ehrlich ascites carcinoma cells, neuroblastoma Neuro 2A cells, and normal epithelial JB-6 cells are presented in Table 11. The biological effects of the investigated substances were quite different due to the diverse structures of their aglycones and carbohydrate chains. Moreover, hemolytic effects of these compounds were higher than their cytotoxicity against other cells, especially against the Ehrlich ascites carcinoma cells. For instance, psolusoside P (**9**) demonstrated high hemolytic action, but moderate cytotoxicity against Neuro 2A and JB-6 cells and was not active against mouse Ehrlich carcinoma cells (ascite form). The analogic dependency was observed for psolusosides M (**6**) and O (**8**), which were not cytotoxic against all the cell lines except erythrocytes.

Table 11. The cytotoxic activities of glycosides **1–10** and psolusosides B and G (positive control) against mouse erythrocytes, Ehrlich ascites carcinoma cells, mouse neuroblastoma Neuro 2A cells, and normal epithelial JB-6 cells.

Glycoside	Cytotoxicity EC_{50}, μM			
	Erythrocytes	**Ehrlich Carcinoma**	**Neuro-2A**	**JB-6**
Psolusoside B	>100.0	>100.0	>100.0	>100.0
Psolusoside B_1 (**1**)	>100.0	>100.0	>100.0	>100.0
Psolusoside B_2 (**2**)	>100.0	>100.0	>100.0	>100.0
Psolusoside J (**3**)	>100.0	>100.0	>100.0	>100.0
Psolusoside K (**4**)	>100.0	>100.0	>100.0	>100.0
Psolusoside L (**5**)	2.42	9.73	10.60	7.37
Psolusoside M (**6**)	67.83	>100.0	>100.0	>100.0
Psolusoside N (**7**)	12.37	57.32	13.52	19.94
Psolusoside O (**8**)	34.82	>100.0	>100.0	>100.0
Psolusoside P (**9**)	10.92	>100.0	59.96	56.40
Psolusoside Q (**10**)	>100.0	>100.0	>100.0	>100.0
Psolusoside G	8.86	82.16	35.14	>100.0

Psolusoside L (**5**) was shown to be the most active substance in the series. It has a holostane-type aglycone and pentasaccharide chain with three sulfate groups at C-6 of two glucose and 3-*O*-methylglucose residues. It is very unusual for a glycoside with three sulfate groups to demonstrate high cytotoxic properties, because it is known that sulfate groups attached to the C-6 position of the terminal glucose and 3-*O*-methylglucose residues greatly decrease the activity of pentaosides branched by the second monosaccharide unit (quinovose) sugar chains [3]. Probably, the peculiarities of architecture of a carbohydrate chain of **5** (the branching at C-4 Xyl1) compensate the negative influence of the three sulfate groups.

The activity of psolusoside N (**7**) was slightly lower than that of **5**, due to the presence of a glucose residue as the second unit in the sugar chain instead of the quinovose (in **5**) that is in good accordance with the earlier observations of the glycoside's SAR [3]. The alteration of the sulfate position attached to the terminal (glucose) residue from C-6 Glc5 to C-2 Glc5 caused the extreme decrease in the activity. This was illustrated by the effects of psolusoside M (**6**) differing from the compound **5** in this character only and demonstrating much lower hemolytic action than **5** and the absence of the activity against other tested cells. The same relationship was observed for psolusosides N (**7**) and O (**8**) differing from each other in the position of the sulfate group in the fifth (Glc5) residue.

The tetrasulfated (at C-6 Glc3, C-6 MeGlc4, C-6 Glc5, and C-4 Glc5) psolusoside P (**9**) demonstrated high hemolytic and moderate cytotoxic action against Neuro-2A and JB-6 cells and was not active against ascites of Ehrlich carcinoma. However, it was much more active than trisulfated psolusoside M (**6**) containing sulfate group at C-2 Glc5. The activity of tetrasulfated psolusoside Q (**10**) was also strongly reduced by the sulfate group attached to C-2 Glc5 as well as by the presence of glucose in the second position of its carbohydrate chain.

Psolusosides B [12], B_1 (**1**), and B_2 (**2**) were not active in all the tests due to the presence of non-holostane aglycones in combination with the tetrasaccharide-branched carbohydrate chain sulfated by C-2 of terminal residue (Glc4) attached to C-4 Xyl1. Moreover, psolusosides J (**3**) and K (**4**) with carbohydrate chains with the same architecture and sulfate group at C-2 of the terminal residue (Glc4) were also inactivated despite the presence of holostane aglycones.

3. Materials and Methods

3.1. General Experimental Procedures

Specific rotation, Perkin-Elmer 343 Polarimeter; NMR, Bruker Avance III 500 (Bruker BioSpin GmbH, Rheinstetten, Germany) (500.13/125.77 MHz) or Avance III 700 Bruker FT-NMR (Bruker BioSpin GmbH, Rheinstetten, Germany) (700.13/176.04 MHz) ($^1H/^{13}C$) spectrometers were used with tetramethylsilane as the internal standard. ESI MS (positive and negative ion modes), Agilent 6510 Q-TOF apparatus was used with a sample concentration of 0.01 mg/mL. HPLC, Agilent 1100 apparatus with a differential refractometer was used with columns Supelco Ascentis RP-Amide (10 × 250 mm, 5 μm) and Supelco Discovery HS F5-5 (10 × 250 mm, 5 μm).

3.2. Animals and Cells

Specimens of the sea cucumber *Psolus fabricii* (family Psolidae; order Dendrochirotida) were collected in the Sea of Okhotsk near Onekotan Island (Kurile Islands). Sampling was performed with a scallop dredge in August–September 1982 at a depth of 100 m during expedition works on fishing seiners "Mekhanik Zhukov" and "Dalarik". Sea cucumbers were identified by V.S. Levin. Voucher specimens were preserved in the A.V. Zhirmunsky National Scientific Center of Marine Biology, Vladivostok, Russia.

CD-1 mice weighing 18–20 g were purchased from RAMS 'Stolbovaya' nursery (Russia) and kept at the animal facility in standard conditions. All experiments were conducted in compliance with all of the rules and international recommendations of the European Convention for the Protection of Vertebrate Animals used for Experimental Studies.

The museum tetraploid strain of murine ascite Ehrlich carcinoma (EAC) cells from the All-Russian Oncology Center (Moscow, Russia) was used. EAC cells were injected into the peritoneal cavity of CD-1 mice. Cells for experimentation were collected 7 days after inoculation. For this purpose, mice were killed by cervical dislocation, and the ascitic fluid containing tumor cells was collected with a syringe. The cells were washed triply by centrifugation at 2000 rpm (450 g) for 10 min in PBS (pH 7.4) followed by resuspension in RPMI-1640 medium containing 8 μg/mL gentamicin (BioloT, Saint Peterburg, Russia). Neuroblastoma Neuro 2A cells were cultured in DMEM medium containing 10% fetal bovine serum (FBS; BioloT, Saint Petersburg, Russia), normal epithelial JB-6 cells were cultured in DMEM medium containing 5% fetal bovine serum (BioloT, Saint Petersburg, Russia), and 1% penicillin/streptomycine (Termo Fisher Scientific (Invitrogen), Waltham, Massachusetts, USA).

3.3. Extraction and Isolation

The sea cucumbers (about 800 specimens, average weight of one specimen is about 100 g) were minced and extracted twice with refluxing 60% EtOH. The extract was evaporated to water residuum and lyophilized followed by extraction with $CHCl_3$/MeOH (1:1). The obtained extract was evaporated and submitted to the subsequent extraction by EtOAc/H_2O to remove the lipid fraction.

The water layer remaining after this extraction was chromatographed on a Polychrom-1 column (powdered Teflon, Biolar, Olaine, Latvia). The glycosides were eluted with 50% EtOH, evaporated, and subsequently chromatographed on Si gel columns with $CHCl_3$/EtOH/H_2O (100:75:10), (100:100:17), and (100:125:25) as the mobile phase to give subfractions III–VIII containing different groups of glycosides. The continued chromatography on Si gel column of glycosidic sum with $CHCl_3$/EtOH/H_2O (100:125:25) as the mobile phase also gave subfractions IX (602 mg) and X (405 mg). The total weight of all the glycosidic fractions was about 2 g. HPLC of the subfraction VIII on Supelco Ascentis RP-Amide column with CH_3CN/H_2O/NH_4OAc (35/64/1) as the mobile phase gave psolusoside B [6] and other fractions: Ps-B(2) and Ps-B(3). The pure psolusoside B_1 (**1**) (7.3 mg) was isolated as a result of recromatography of the Ps-B(3) fraction on Discovery HS F5-5 column with MeOH/H_2O/NH_4OAc (1 M water solution) (50/49/1) as the mobile phase. Psolusoside B_2 (**2**) (3.4 mg) was isolated by HPLC of the Ps-B(2) fraction on the same column with MeOH/H_2O/NH_4OAc (1 M water solution) (60/38/2) as the mobile phase. The subfraction IX was chromatographed on Supelco Ascentis RP-Amide column with MeOH/H_2O/NH_4OAc (1 M water solution) (60/39/1) as the mobile phase to give psolusoside L (**5**) (60 mg) and another subfraction (IXa), that was rechromatographed on Discovery HS F5-5 column with the same solvents in ratio (50/49/1) as the mobile phase to obtain 3.4 mg of psolusoside J (**3**) and 4.8 mg of psolusoside K (**4**). The subfraction X was subjected to HPLC on Supelco Discovery HS F5-5 column with MeOH/H_2O/NH_4OAc (1 M water solution) (55/44/1) as mobile phase to give several subsubfractions, rechromatography of which was carried out using different ratios of MeOH/H_2O/NH_4OAc (1 M water solution) as mobile phases. The use of the chromatographic system MeOH/H_2O/NH_4OAc (60/39/1) resulted in psolusosides M (**6**) (1 mg), N (**7**) (8.8 mg), and P (**9**) (8.5 mg) isolation, the system (52/47/1) gave pure psolusoside O (**8**) (0.6 mg), and the system (50/48.5/1.5) gave 1.4 mg of pure psolusoside Q (**10**).

3.3.1. Psolusoside B_1 (1)

Colorless powder; $[\alpha]_D^{20}$ −23 (*c* 0.1, 50% MeOH). NMR: See Table 1 and Table S1. (−)HR-ESI-MS *m/z*: 1325.4164 (calc. 1325.4185) $[M_{2Na} - Na]^-$, 651.2157 (calc. 651.2146) $[M_{2Na} - 2Na]^{2-}$; (−)ESI-MS/MS *m/z*: 1265.4 $[M_{2Na} - Na - CH_3COOH]^-$, 1145.4 $[M_{2Na} - Na - CH_3COOH - NaHSO_4]^-$, 1001.4 $[M_{2Na} - Na - CH_3COOH - C_6H_{10}O_8SNa$ $(GlcSO_3Na) + H]^-$, 839.3 $[M_{2Na} - Na - CH_3COOH - GlcSO_3Na - Glc + H]^-$.

3.3.2. Psolusoside B_2 (2)

Colorless powder; $[\alpha]_D^{20}$ −18 (*c* 0.1, 50% MeOH). NMR: See Table 2 and Table S1. (−)HR-ESI-MS *m/z*: 1325.4163 (calc. 1325.4185) $[M_{Na} - Na]^-$, 651.2159 (calc. 651.2146) $[M_{2Na} - 2Na]^{2-}$; (−)ESI-MS/MS *m/z*: 1265.4 $[M_{2Na} - Na - CH_3COOH]^-$, 1145.4 $[M_{2Na} - Na - CH_3COOH - NaHSO_4]^-$, 1001.4 $[M_{2Na} - Na - CH_3COOH - C_6H_{10}O_8SNa$ $(GlcSO_3Na) + H]^-$, 839.3 $[M_{2Na} - Na - CH_3COOH - GlcSO_3Na - Glc + H]^-$, 535.1 $[M_{2Na} - Na - C_{32}H_{45}O_6$ (Agl) $- C_6H_{10}O_8SNa$ $(GlcSO_3Na)]^-$, 403 $[M_{2Na} - Na - C_{32}H_{45}O_6$ (Agl) $- C_6H_{10}O_8SNa$ $(GlcSO_3Na)$ $-$ Xyl $(C_5H_8O_4)]^-$.

3.3.3. Psolusoside J (3)

Colorless powder; $[\alpha]_D^{20}$ −17 (*c* 0.1, 50% MeOH). NMR: See Table 3 and Table S2. (−)HR-ESI-MS *m/z*: 1369.3485 (calc. 1369.3517) $[M_{3Na} - Na]^-$, 673.1812 (calc. 673.1813) $[M_{3Na} - 2Na]^{2-}$, 441.1248 (calc. 441.1244) $[M_{3Na} - 3Na]^{3-}$; (−)ESI-MS/MS *m/z*: 1249.4 $[M_{3Na} - Na - NaHSO_4]^-$, 1105.4 $[M_{3Na} - Na - C_6H_9O_8SNa$ $(GlcSO_3Na)]^-$, 1003.4 $[M_{3Na} - Na - C_6H_9O_8SNa$ $(GlcSO_3Na) - NaSO_3 + H]^-$, 841.4 $[M_{3Na} - Na - NaSO_3 - GlcSO_3Na - Glc + H]^-$, 403.0 $[M_{3Na} - Na - C_{30}H_{43}O_4$ (Agl) $- C_6H_9O_{11}S_2Na_2$ $(Glc(SO_3Na)_2)$ $-$ Xyl $(C_5H_8O_4)]^-$ and 241.0 $[M_{3Na} - Na - C_{30}H_{43}O_4$ (Agl) $- C_6H_9O_{11}S_2Na_2$ $(Glc(SO_3Na)_2)$ $-$ Xyl $(C_5H_8O_4)$ $-$ Glc $(C_6H_{10}O_5)]^-$.

3.3.4. Psolusoside K (4)

Colorless powder; $[\alpha]_D^{20}$ −16 (*c* 0.1, 50% MeOH). NMR: See Table 4 and Table S3. (−)HR-ESI-MS *m/z*: 1369.3485 (calc. 1369.3517) $[M_{3Na} - Na]^-$, 673.1821 (calc. 673.1813) $[M_{3Na} - 2Na]^{2-}$, 441.1255 (calc. 441.1244) $[M_{3Na} - 3Na]^{3-}$; (−)ESI-MS/MS *m/z*: 1249.4 $[M_{3Na} - Na - NaHSO_4]^-$, 1105.4 $[M_{3Na} - Na - C_6H_9O_8SNa$ (GlcSO$_3$Na)$]^-$, 1003.4 $[M_{3Na} - Na - C_6H_9O_8SNa$ (GlcSO$_3$Na) $- NaSO_3 + H]^-$, 403.0 $[M_{3Na} - Na - C_{30}H_{43}O_4$ (Agl) $- C_6H_9O_{11}S_2Na_2$ (Glc(SO$_3$Na)$_2$) − Xyl ($C_5H_8O_4$)$]^-$, 241.0 $[M_{3Na} - Na - C_{30}H_{43}O_4$ (Agl) $- C_6H_9O_{11}S_2Na_2$ (Glc(SO$_3$Na)$_2$) − Xyl ($C_5H_8O_4$) − Glc ($C_6H_{10}O_5$)$]^-$.

3.3.5. Psolusoside L (5)

Colorless powder; $[\alpha]_D^{20}$ −35 (*c* 0.1, 50% MeOH). NMR: See Table 5 and Table S3. (−)HR-ESI-MS *m/z*: 1529.4222 (calc. 1529.4253) $[M_{3Na} - Na]^-$, 753.2190 (calc. 753.2180) $[M_{3Na} - 2Na]^{2-}$, 494.4835 (calc. 494.4823) $[M_{3Na} - 3Na]^{3-}$; (−)ESI-MS/MS *m/z*: 1409.5 $[M_{3Na} - Na - NaHSO_4]^-$, 1265.4 $[M_{3Na} - Na - C_6H_9O_8SNa$ (GlcSO$_3$Na)$]^-$, 1131.5 $[M_{3Na} - Na - C_7H_{12}O_9SNa$ (MeGlcSO$_3$Na) $- NaSO_3]^-$, 665.1 $[M_{3Na} - Na - C_{30}H_{43}O_4$ (Agl) $- C_7H_{12}O_9SNa$ (MeGlcSO$_3$Na) $- NaSO_3]^-$, 519.0 $[M_{3Na} - Na - C_{30}H_{43}O_4$ (Agl) $- C_7H_{12}O_9SNa$ (MeGlcSO$_3$Na) $- C_6H_9O_7SNa$ (GlcSO$_3$Na)$]^-$.

3.3.6. Psolusoside M (6)

Colorless powder; $[\alpha]_D^{20}$ −20 (*c* 0.1, 50% MeOH). NMR: See Table 6 and Table S3. (−)HR-ESI-MS *m/z*: 1529.4273 (calc. 1529.4253) $[M_{3Na} - Na]^-$, 753.2202 (calc. 753.2180) $[M_{3Na} - 2Na]^{2-}$, 494.4844 (calc. 494.4823) $[M_{3Na} - 3Na]^{3-}$; (−)ESI-MS/MS *m/z*: 1409.5 $[M_{3Na} - Na - NaHSO_4]^-$, 1265.4 $[M_{3Na} - Na - C_6H_9O_8SNa$ (GlcSO$_3$Na)$]^-$, 1131.5 $[M_{3Na} - Na - C_7H_{12}O_9SNa$ (MeGlcSO$_3$Na) $- NaSO_3]^-$, 665.1 $[M_{3Na} - Na - C_{30}H_{43}O_4$ (Agl) $- C_7H_{12}O_9SNa$ (MeGlcSO$_3$Na) $- NaSO_3]^-$, 519.0 $[M_{3Na} - Na - C_{30}H_{43}O_4$ (Agl) $- C_7H_{12}O_9SNa$ (MeGlcSO$_3$Na) $- C_6H_9O_7SNa$ (GlcSO$_3$Na)$]^-$.

3.3.7. Psolusoside N (**7**)

Colorless powder; $[\alpha]_D^{20}$ −12 (*c* 0.1, 50% MeOH). NMR: See Table 7 and Table S3. (−)HR-ESI-MS *m/z*: 1545.4171 (calc. 1545.4202) $[M_{3Na} - Na]^-$, 761.2164 (calc. 761.2155) $[M_{3Na} - 2Na]^{2-}$, 499.8151 (calc. 499.8139) $[M_{3Na} - 3Na]^{3-}$; (−)ESI-MS/MS *m/z*: 1425.5 $[M_{3Na} - Na - NaHSO_4]^-$, 1281.4 $[M_{3Na} - Na - C_6H_9O_8SNa$ (GlcSO$_3$Na)$]^-$, 1147.5 $[M_{3Na} - Na - C_7H_{12}O_9SNa$ (MeGlcSO$_3$Na) $- NaSO_3]^-$, 1003.4 $[M_{3Na} - Na - C_6H_9O_8SNa$ (GlcSO$_3$Na) $- C_7H_{12}O_8SNa$ (MeGlcSO$_3$Na) $+ H]^-$, 681.1 $[M_{3Na} - Na - C_{30}H_{43}O_4$ (Agl) $- C_7H_{12}O_9SNa$ (MeGlcSO$_3$Na) $- NaSO_3 + H]^-$, 519.0 $[M_{3Na} - Na - C_{30}H_{43}O_4$ (Agl) $- C_7H_{12}O_9SNa$ (MeGlcSO$_3$Na) $- C_6H_9O_7SNa$ (GlcSO$_3$Na)$]^-$.

3.3.8. Psolusoside O (**8**)

Colorless powder; $[\alpha]_D^{20}$ −60 (*c* 0.1, 50% MeOH). NMR: See Table 8 and Table S3. (−)HR-ESI-MS *m/z*: 1545.4197 (calc. 1545.4202) $[M_{3Na} - Na]^-$, 761.2171 (calc. 761.2155) $[M_{3Na} - 2Na]^{2-}$, 499.8155 (calc. 499.8139) $[M_{3Na} - 3Na]^{3-}$; (−)ESI-MS/MS *m/z*: 1425.5 $[M_{3Na} - Na - NaHSO_4]^-$, 1281.4 $[M_{3Na} - Na - C_6H_9O_8SNa$ (GlcSO$_3$Na)$]^-$, 1161.5 $[M_{3Na} - Na - C_6H_9O_8SNa$ (GlcSO$_3$Na) $- NaHSO_4]^-$, 1147.5 $[M_{3Na} - Na - C_7H_{12}O_9SNa$ (MeGlcSO$_3$Na) $- NaSO_3]^-$, 1003.4 $[M_{3Na} - Na - C_6H_9O_8SNa$ (GlcSO$_3$Na) $- C_7H_{12}O_8SNa$ (MeGlcSO$_3$Na) $+ H]^-$, 681.1 $[M_{3Na} - Na - C_{30}H_{43}O_4$ (Agl) $- C_7H_{12}O_9SNa$ (MeGlcSO$_3$Na) $- NaSO_3 + H]^-$, 519.0 $[M_{3Na} - Na - C_{30}H_{43}O_4$ (Agl) $- C_7H_{12}O_9SNa$ (MeGlcSO$_3$Na) $- C_6H_9O_7SNa$ (GlcSO$_3$Na)$]^-$.

3.3.9. Psolusoside P (**9**)

Colorless powder; $[\alpha]_D^{20}$ −26 (*c* 0.1, 50% MeOH). NMR: See Table 9 and Table S3. (−)HR-ESI-MS *m/z*: 1631.3598 (calc. 1631.3641) $[M_{4Na} - Na]^-$, 804.1879 (calc. 804.1874) $[M_{4Na} - 2Na]^{2-}$, 528.4628 (calc. 528.4619) $[M_{4Na} - 3Na]^{3-}$, 390.6001 (calc. 390.5991) $[M_{4Na} - 4Na]^{4-}$; (−)ESI-MS/MS *m/z*: 1265.4 $[M_{4Na} - Na - C_6H_8O_{11}S_2Na_2$ (Glc(SO$_3$Na)$_2$)$]^-$, 1233.4 $[M_{4Na} - Na - C_7H_{12}O_9SNa$ (MeGlcSO$_3$Na) $- NaSO_3]^-$, 1145.5 $[M_{4Na} - Na - C_6H_9O_{11}S_2Na_2$ (Glc(SO$_3$Na)$_2$ $- NaSO_4]^-$, 1089.4 $[M_{4Na} - Na - C_7H_{12}O_9SNa$

(MeGlcSO$_3$Na) – $C_6H_8O_7SNa$ (GlcSO$_3$Na)]$^-$, 969.4 [M_{4Na} – Na – $C_7H_{12}O_9SNa$ (MeGlcSO$_3$Na) – $C_6H_8O_7SNa$ (GlcSO$_3$Na) – $NaHSO_4$]$^-$, 943.3 [M_{4Na} – Na – $C_7H_{12}O_9SNa$ (MeGlcSO$_3$Na) – $C_6H_8O_7SNa$ (GlcSO$_3$Na) – $C_6H_{10}O_4$ (Qui)]$^-$.

3.3.10. Psolusoside Q (**10**)

Colorless powder; $[\alpha]_D{}^{20}$ −10 (*c* 0.1, 50% MeOH). NMR: See Table 10 and Table S3. (−)HR-ESI-MS *m/z*: 1647.3544 (calc. 1647.3590) [M_{4Na} – Na]$^-$, 812.1854 (calc. 812.1849) [M_{4Na} – 2Na]$^{2-}$, 533.7944 (calc. 533.7935) [M_{4Na} – 3Na]$^{3-}$, 394.5989 (calc. 394.5978) [M_{4Na} – 4Na]$^{4-}$; (−)ESI-MS/MS *m/z*: 1527.4 [M_{4Na} – Na – $NaHSO_4$]$^-$, 1281.4 [M_{4Na} – Na – $C_6H_8O_{11}S_2Na_2$ (Glc(SO$_3$Na)$_2$)]$^-$, 1161.5 [M_{4Na} – Na – $C_6H_8O_{11}S_2Na_2$ (Glc(SO$_3$Na)$_2$) – $NaHSO_4$]$^-$, 1003.4 [M_{4Na} – Na – $C_6H_8O_{11}S_2Na_2$ (Glc(SO$_3$Na)$_2$) – $C_7H_{11}O_8SNa$ (MeGlcSO$_3$Na)]$^-$, 681.1 [M_{4Na} – Na – $C_{30}H_{43}O_4$ (Agl) – $C_6H_9O_{11}S_2Na_2$ (Glc(SO$_3$Na)$_2$) – $C_5H_8O_4$ (Xyl)]$^-$, 519.0 [M_{4Na} – Na – $C_{30}H_{43}O_4$ (Agl) – $C_6H_9O_{11}S_2Na_2$ (Glc(SO$_3$Na)$_2$) – $C_5H_8O_4$ (Xyl) – $C_6H_{10}O_5$ (Glc)]$^-$.

3.4. *Cytotoxic Activity (MTT Assay)*

The solutions (20 µL) of tested substances in different concentrations and cell suspension (200 µL) were added in wells of 96-well plates and incubated over night at 37 °C and 5% CO_2. After incubation the cells were precipitated by centrifugation, 200 µL of medium from each well were collected and 100 µL of pure medium were added. Then 10 µL of MTT (3-(4,5-dimethylthiazol-2-yl)-2,5-diphenyltetrazolium bromide) solution 5 µg/mL (Sigma, St. Louis, MO, USA) were added in each well. The plate was incubated for 4 h, after that 100 µL SDS-HCl were added to each well and the plate was incubated at 37 °C for 4–18 h. Optical density was measured at 570 nm and 630–690 nm. Cytotoxic activity of the substances was calculated as the concentration that caused 50% metabolic cell activity inhibition (IC_{50}).

3.5. *Hemolytic Activity*

Blood was taken from CD-1 mice (18–20 g). The mice were anesthetized with diethyl ether, their chests were rapidly opened, and blood was collected in cold (4 °C) 10 mM phosphate-buffered saline, pH 7.4 (PBS) without an anticoagulant. Erythrocytes were washed by centrifugation (2000 rpm) for 5 min, 3 times in PBS using at least 10 vol. of washing solution. Erythrocytes were used at a concentration that provided an optical density of 1.0 at 700 nm for a non-hemolyzed sample. In addition, 20 µL of a water solution of test substance with a fixed concentration was added to a well of a 96-well plate containing 180 µL of the erythrocyte suspension. Erythrocyte suspension was incubated with substances for 24 h at 37 °C. After that, the optical density of the obtained solutions was measured and EC_{50} for hemolytic activity of each compound was calculated.

4. Conclusions

Ten individual compounds **1–10**, including di-, tri-, and unprecedented tetrasulfated glycosides were isolated from the sea cucumber *Psolus fabricii*. Psolusosides B_1 (**1**) and B_2 (**2**) have disulfated branched by C-4 Xyl1 tetrasaccharide chains identical to that of psolusoside B [12] and non-holostane aglycones with 18(16)-lactone moiety. They differ from other glycosides by their unique structural feature such as 7,8-epoxy-fragment in **2** or by their combination of unusual features such as 7-keto-8,9-ene fragment and 18(16)-lactone in **1**. The compounds **3–10** contain common for the sea cucumbers glycosides aglycones, but unique carbohydrate chains. Psolusosides J (**3**) and K (**4**) are characterized by new trisulfated tetrasaccharide-branched chains with the terminal glucose unit sulfated by two positions: by C-2 and C-4 in the compound **3** and by C-2 and C-6 in the compound **4**. Psolusosides L (**5**), M (**6**), and P (**9**) have branched by C-4 Xyl1 pentasaccharide chains with the quinovose as the second sugar unit. These compounds differ from each other by the quantity and positions of sulfate groups in the fifth (Glc5) residue. Psolusoside P (**9**) is tetrasulfated glycoside, containing two sulfate groups at C-4 and C-6 of the same terminal (Glc5) residue. Psolusosides N (**7**), O (**8**), and Q (**10**) have carbohydrate chains with the same architecture, as the glycosides **5**, **6**, and **9** and differ from those by the second monosaccharide residue, which is a glucose instead of a quinovose. Psolusosides N (**7**) and O (**8**) are the structural analogs of psolusosides L (**5**) and M (**6**), correspondingly, having identical sulfate groups positions. Tetrasulfated psolusoside Q (**10**) differs from psolusoside P (**9**) by the positions of sulfation—at C-2 and C-6 of terminal (Glc5) residue. Tetrasulfated glycosides have not ever been found in any natural objects.

The present investigation is conclusive in a series of research concerning the glycosides of the sea cucumber *Psolus fabricii*. Generally, 27 new and 5 known earlier triterpene glycosides have been isolated from this animal. These compounds contain six previously unknown aglycones and 13 novel carbohydrate chains.

The sulfated oligosaccharide moieties predominate in the glycosides of *P. fabricii*. Monosulfated trisaccharide (psolusosides H and H_1) and linear tetrasaccharide (psolusosides E and F) moieties, disulfated branched tetrasaccharide (psolusosides B, B_1 (**1**), B_2 (**2**), and I) or linear tetrasaccharide (psolusosides A and G) carbohydrate chains, trisulfated branched tetrasaccharide (psolusosides J (**3**) and K (**4**)) carbohydrate chains, and finally pentasaccharide trisulfated (psolusosides L (**5**), M (**6**), N (**7**), O (**8**)) and tetrasulfated (psolusosides P (**9**) and Q (**10**)) carbohydrate chains were found in glycosides of *P. fabricii*. The sugar chains also differ from each other by the second monosaccharide unit (quinovose, glucose, or xylose). The most variable structural feature of the carbohydrate chains of the glycosides **1–10** is the quantity (one or two) and positions of the sulfate groups in terminal glucose unit, attached to C-4 Xyl1. There are three combinations of such positions of sulfate groups in these residues: C-2 and C-4, C-2 and C-6, or C-4 and C-6. Whereas the single sulfate group bonds only to C-2 or C-6 of terminal glucose unit.

It is interesting to note that diverse groups of psolusosides (A–Q) (a certain group of glycosides consists of substances with the same carbohydrate chain and diverse aglycones) characterized by different structural variability of aglycones. All psolusosides belonging to groups C and D (both containing hexasaccharide non-sulfated sugar chains) have the holostane-type aglycones with 9(11)-double bond, 16-keto-group, and different side chains (5 variants). Psolusosides of the group B contain exclusively non-holostane aglycones with 18(16)-lactone and 7(8)-double bond, completely different from the aglycones of the other groups of psolusosides. These could be explained by their special biological functions in the organism-producer. Four holostane aglycones with 7(8)-, or 9(11)-double bond were found in five glycosides having trisaccharide (psolusosides H and H_1) or tetrasaccharide-branched carbohydrate chains (psolusosides I, J, K).

Pentaosides (psolusosides L–Q) and tetraosides with linear sugar chains (psolusosides A, E, F, G) contain the same holostane-type aglycone with 9(11)-double bond. It suggests, that linear tetraosides are biosynthetic precursors of pentaosides—psolusosides L (**5**), M (**6**) and P (**9**)—which are biosynthesized via glycosylation and sulfation of psolusosides A, E, and F, correspondingly. Psolusosides N (**7**), O (**8**), and Q (**10**) are formed from psolusoside G through the same processes.

Hence, the biogenetic analysis of the structures of glycosides found in *P. fabricii* showed that carbohydrate chains and aglycones biosynthesis possesses a mosaic (combinatoric) character, which also has some trends.

Supplementary Materials: The following are available online at http://www.mdpi.com/1660-3397/17/11/631/s1, Table S1: NMR data for the carbohydrate moiety of psolusosides B1 (1) and B2 (2), Table S2: NMR data for the aglycone moiety of psolusoside J (3), Table S3: NMR data for the aglycone moiety of psolusoside K (4), L (5), M (6), N (7), O (8), P (9), Q (10), Figures S1–S56: original 2D and 1D NMR specra of glycosides 1–10, Figures S57–S66: original mass-spectra of glycosides 1–10.

Author Contributions: Conceptualization, A.S.S. and V.I.K.; methodology, A.S.S. and S.A.A.; investigation, A.S.S., A.I.K., S.A.A., V.I.K., P.S.D., R.S.P., E.A.C., and P.V.A.; writing—original draft preparation, A.S.S. and V.I.K.; writing—review and editing, V.I.K.

Funding: The chemical structure and part of the bioassay were carried out with partial financial support of the Grant of the Russian Foundation for Basic Research No. 19-04-000-14.

Acknowledgments: The authors are very appreciative to Valentin A. Stonik (G.B. Elyakov Pacific Institute of Bioorganic Chemistry, Vladivostok, Russia) for the reading and discussion of the manuscript.

Conflicts of Interest: The authors declare no conflicts of interest.

References

1. Stonik, V.A.; Kalinin, V.I.; Avilov, S.A. Toxins from sea cucumbers (holothuroids): Chemical structures, properties, taxonomic distribution, biosynthesis and evolution. *J. Nat. Toxins* **1999**, *8*, 235–248. [PubMed]
2. Mondol, M.A.M.; Shin, H.J.; Rahman, M.A. Sea cucumber glycosides: Chemical structures, producing species and important biological properties. *Mar. Drugs* **2017**, *15*, 317. [CrossRef] [PubMed]
3. Kalinin, V.I.; Aminin, D.L.; Avilov, S.A.; Silchenko, A.S.; Stonik, V.A. Triterpene glycosides from sea cucumbers (Holothurioidea, Echinodermata). Biological activities and functions. In *Studies in Natural Product Chemistry (Bioactive Natural Products)*; Atta-ur-Rahman, Ed.; Elsevier Science Publisher: Amsterdam, The Netherlands, 2008; Volume 35, pp. 135–196.
4. Dyshlovoy, S.A.; Menchinskaya, E.S.; Venz, S.; Rast, S.; Amann, K.; Hauschild, J.; Otte, K.; Kalinin, V.I.; Silchenko, A.S.; Avilov, S.A.; et al. The marine triterpene glycoside frondoside A exhibits activity in vitro and in vivo in prostate cancer. *Int. J. Cancer* **2016**, *28*, 2450–2465. [CrossRef] [PubMed]
5. Yun, S.H.; Park, E.S.; Shin, S.W.; Na, Y.W.; Han, J.Y.; Jeong, J.S.; Shastina, V.V.; Stonik, V.A.; Park, J.I.; Kwak, J.Y. Stichoposide C induces apoptosis through the generation of ceramide in leukemia and colorectal cancer cells and shows in vivo antitumor activity. *Clin. Cancer Res.* **2012**, *18*, 5934–5948. [CrossRef] [PubMed]
6. Aminin, D.L.; Chaykina, E.L.; Agafonova, I.G.; Avilov, S.A.; Kalinin, V.I.; Stonik, V.A. Antitumor activity of the immunomodulatory lead Cumaside. *Int. Immunopharmacol.* **2010**, *10*, 648–654. [CrossRef] [PubMed]
7. Kalinin, V.I.; Kalinovskii, A.I.; Stonik, V.A. Psolusoside A—A new triterpene glycoside from the holothurian *Psolus fabricii*. *Chem. Nat. Compd.* **1983**, *19*, 753–754. [CrossRef]
8. Kalinin, V.I.; Kalinovskii, A.I.; Stonik, V.A. Structure of psolusoside A—The main triterpene glycoside from the holothurian *Psolus fabricii*. *Chem. Nat. Compd.* **1985**, *21*, 197–202. [CrossRef]
9. Kalinin, V.I.; Kalinovskii, A.I.; Stonik, V.A.; Dmitrenok, P.S.; El'kin, Y.N. Structure of psolusoside B—A nonholostane triterpene glycoside from the holothurian genus *Psolus*. *Chem. Nat. Compd.* **1989**, *25*, 311–317. [CrossRef]
10. Silchenko, A.S.; Avilov, S.A.; Kalinovsky, A.I.; Kalinin, V.I.; Andrijaschenko, P.V.; Dmitrenok, P.S. Psolusosides C_1, C_2, and D_1, novel triterpene hexaosides from the sea cucumber *Psolus fabricii* (Psolidae, Dendrochirotida). *Nat. Prod. Commun.* **2018**, *13*, 1623–1628.
11. Silchenko, A.S.; Avilov, S.A.; Kalinovsky, A.I.; Kalinin, V.I.; Andrijaschenko, P.V.; Dmitrenok, P.S.; Popov, R.S.; Chingizova, E.A.; Kasakin, M.F. Psolusosides C_3 and D_2–D_5, Five novel triterpene hexaosides from the sea cucumber *Psolus fabricii* (Psolidae, Dendrochirotida): Chemical structures and bioactivities. *Nat. Prod. Commun.* **2019**, *14*. [CrossRef]
12. Silchenko, A.S.; Kalinovsky, A.I.; Avilov, S.A.; Kalinin, V.I.; Andrijaschenko, P.V.; Dmitrenok, P.S.; Popov, R.S.; Chingizova, E.A.; Ermakova, S.P.; Malyarenko, O.S. Structures and bioactivities of six new triterpene glycosides, psolusosides E, F, G, H, H_1 and I and the corrected structure of psolusoside B from the sea cucumber *Psolus fabricii*. *Mar. Drugs* **2019**, *17*, 358. [CrossRef] [PubMed]
13. Liu, H.; Heilmann, J.; Rali, T.; Sticher, O. New Tirucallane-Type Triterpenes from *Dysoxylum variabile*. *J. Nat. Prod.* **2001**, *64*, 159–163. [CrossRef] [PubMed]
14. Antonov, A.S.; Avilov, S.A.; Kalinovsky, A.I.; Anastyuk, S.D.; Dmitrenok, P.S.; Kalinin, V.I.; Taboada, S.; Bosh, A.; Avila, C.; Stoni, V.A. Triterpene glycoasides from Antarctic sea cucumbers. 2. Structure of achlioniceosides A_1, A_2 and A_3 from the sea cucumber *Achlionice violaescupidata* (=*Rhipidothuria racowitzai*). *J. Nat. Prod.* **2009**, *72*, 33–38. [CrossRef] [PubMed]
15. Fusetani, N.; Matsunaga, S.; Konosu, S. Bioactive marine metabolites II. Halistanol sulfate, an antimicrobial novel steroid sulfate from the marine sponge *Halichondria* cf. moorei Bergquist. *Tetrahedron Lett.* **1981**, *2*, 1985–1988. [CrossRef]
16. Makarieva, T.N.; Shubina, L.K.; Kalinovsky, A.I.; Stonik, V.A.; Elyakov, G.B. Steroids in porifera. II. Steroid derivatives from two sponges of the family *Halichondriidae*. Sokotrasterol sulfate, a marine steroid with a new pattern of side chain alkylation. *Steroids* **1983**, *42*, 267–281. [CrossRef]
17. Kanazawa, S.; Fusetani, N.; Matsunaga, S. Halistanol sulfates A-E, new steroid sulfates, from a marine sponge, *Epipolasis* sp. *Tetrahedron* **1992**, *48*, 5467–5472. [CrossRef]

18. Avilov, S.A.; Kalinin, V.I.; Smirnov, A.V. Use of triterpene glycosides for resolving taxonomic problems in the sea cucumber genus *Cucumaria* (Holothorioidea, Echinodermata). *Biochem. Syst. Ecol.* **2004**, *32*, 715–733. [CrossRef]
19. Silchenko, A.S.; Kalinovsky, A.I.; Avilov, S.A.; Andryjaschenko, P.V.; Dmitrenok, P.S.; Kalinin, V.I.; Chingizova, E.A.; Minin, K.V.; Stonik, V.A. Structures and biogenesis of fallaxosides D_4, D_5, D_6 and D_7, trisulfated non-holostane triterpene glycosides from the sea cucumber *Cucumaria fallax*. *Molecules* **2016**, *21*, 939. [CrossRef] [PubMed]

marine drugs

MDPI

Article

Glycosaminoglycan from *Apostichopus japonicus* Improves Glucose Metabolism in the Liver of Insulin Resistant Mice

Yunmei Chen [1], Yuanhong Wang [1,2], Shuang Yang [1,2], Mingming Yu [1,2], Tingfu Jiang [1,2] and Zhihua Lv [1,2,*]

1 Key Laboratory of Marine Drugs, Ministry of Education of China, Key Laboratory of Glycoscience & Glycotechnology of Shandong Province, School of Medicine and Pharmacy, Ocean University of China, Qingdao 266003, China; yunmchen@hotmail.com (Y.C.); yhwang@ouc.edu.cn (Y.W.); yangshuang@ouc.edu.cn (S.Y.); yumingming@ouc.edu.cn (Y.M.); jiangtingfu@ouc.edu.cn (T.J.)

2 Laboratory for Marine Drugs and Bioproducts of Qingdao National Laboratory for Marine Science and Technology, Qingdao 266003, China

* Correspondence: lvzhihua@ouc.edu.cn; Tel./Fax: +86-532-8203-2064

Received: 11 November 2019; Accepted: 16 December 2019; Published: 18 December 2019

Abstract: Holothurian glycosaminoglycan isolated from *Apostichopus japonicus* (named AHG) can suppress hepatic glucose production in insulin resistant hepatocytes, but its effects on glucose metabolism in vivo are unknown. The present study was conducted to investigate the effects of AHG on hyperglycemia in the liver of insulin resistant mice induced by a high-fat diet (HFD) for 12 weeks. The results demonstrated that AHG supplementation apparently reduced body weight, blood glucose level, and serum insulin content in a dose-dependent manner in HFD-fed mice. The protein levels and gene expression of gluconeogenesis rate-limiting enzymes G6Pase and PEPCK were remarkedly suppressed in the insulin resistant liver. In addition, although the total expression of IRS1, Akt, and AMPK in the insulin resistant liver was not affected by AHG supplementation, the phosphorylation of IRS1, Akt, and AMPK were clearly elevated by AHG treatment. These results suggest that AHG could be a promising natural marine product for the development of an antihyperglycemic agent.

Keywords: glycosaminoglycan; *Apostichopus japonicus*; glucose metabolism; Akt; AMPK

---

1. Introduction

Type 2 diabetes mellitus(T2DM) is characterized by long-term persistent hyperglycemia with several complications such as eye injury, renal failure, cardiovascular disease, and nervous system damage [1]. According to a report from the International Diabetes Federation (IDF) published in 2017, the number of people with diabetes has risen to 425 million in the world (2017, IDF, http://diabetesatlas.org) and T2DM makes up about 90% of cases of diabetes [2]. Thus, T2DM has become a chronic disease globally. Insulin resistance is the hallmark of T2DM and describes a condition whereby the ability of insulin to trigger glucose uptake, metabolism, or storage is impaired [3,4]. As the major site of glucose utilization during the post-prandial period and the main human tissue of glucose synthetization, the liver is the vital target organ of insulin resistance, and liver insulin resistance is the main reason for the development of T2DM [5,6]. When insulin resistance develops in the liver, increased hepatic glucose production and decreased glucose utilization contribute to the elevated level of blood glucose [7]. Therefore, an increasing number of studies are examining the liver, and specifically the inappropriate glucose metabolism in insulin resistant individuals.

Up to now, due to the huge species biodiversity and extremely harsh living environment, marine organisms have been found to have biologically active agents which show antidiabetic effects, and the sea cucumber is one of the most widely used marine traditional medicines in Asia [8,9]. Sea cucumbers are echinoderms from the class Holothuroidea with a wide spectrum of biological and pharmacological activities, such as anti-cancer [10], anticoagulant [11], antifungal [12], anti-hypertension [13], antioxidant [14], anti-inflammatory [15], anti-diabetic [16] and antiviral [17] activities. Recent studies demonstrate that the sea cucumber exerts various activities on account of the high content of bioactive constituents, including saponins, polypeptides, gangliosides, and polysaccharides [18,19]. Polysaccharides from sea cucumber are divided into two groups, namely holothurian fucan and holothurian glycosaminoglycan. Holothurian glycosaminoglycan is a kind of highly sulfated polysaccharide isolated from the sea cucumber body wall, possessing a chondroitin sulfate backbone with a high percentage of sulfated α-L-fucose (Fuc) branches [20]. Among the various biological macromolecules, holothurian glycosaminoglycan is considered as the major bioactive constituent of sea cucumber [21].

Apostichopus japonicus is an economically important sea cucumber which is widely distributed in Russia, China, Japan and Korea [22]. AHG (the structure shown in Figure 1) is a novel glycosaminoglycan from *Apostichopus japonicus* with a chondroitin sulfate-like backbone structure of →4GlcA(Fuc2S,4Sα1→3)β1→3GalNAc4S6Sβ1→[23]. In our previous study, AHG exhibits anti-diabetic activity by suppressing hepatic glucose production in insulin resistant hepatocytes [24]. However, the physiological effects of AHG in vivo are unknown. In this study, we investigated the protective ability of AHG on dysregulated glucose homeostasis in insulin resistant mice induced by a high-fat diet (HFD). Also, we further explored the possible biochemical regulator involving in the effects of AHG in the liver glucose metabolism of HFD-fed insulin resistant mice.

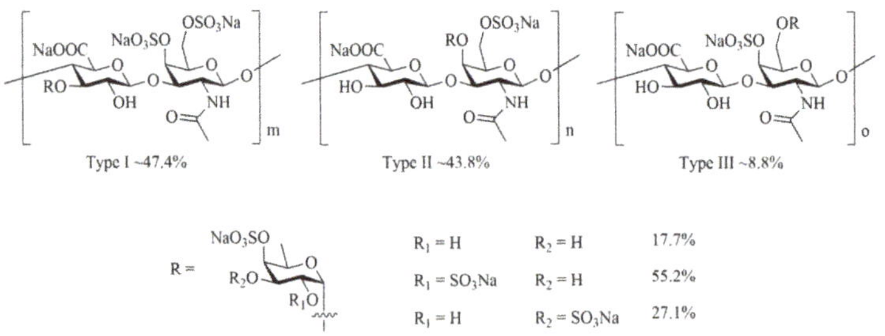

Figure 1. Chemical Structure of AHG from Sea Cucumber *Apostichopus japonicus.*

2. Results

2.1. AHG Decreased Body Weight in Mice Fed with HFD

As shown in Figure 2A, after 12 weeks of HFD treatment, mice exerted two-fold increase in body weight gain compared to mice of LFD group. AHG (20 mg/kg/day) supplementation showed no significant change in body weight gain compared to HFD group, however, treatment of AHG (50 mg/kg/day) and AHG (100 mg/kg/day) alleviated the enhancement by showing 12.1% and 18.9% inhibition in the body weight gain of the HFD group, indicating that AHG affected an HFD-induced insulin resistant model. As a positive control, the body weight gain of mice fed metformin was 24% lower than that of mice fed HFD. To further detect whether the decreasing body weight was caused by the change of energy intake, the food intake was measured. The results showed (Figure 2B) that food intake was significantly lower in the HFD group compared to the LFD group. However, AHG supplementation showed no effect on food intake in HFD treatment. Metformin showed a markedly suppression on food intake compared to LFD group, which was consistently in agreement with the literature [25].

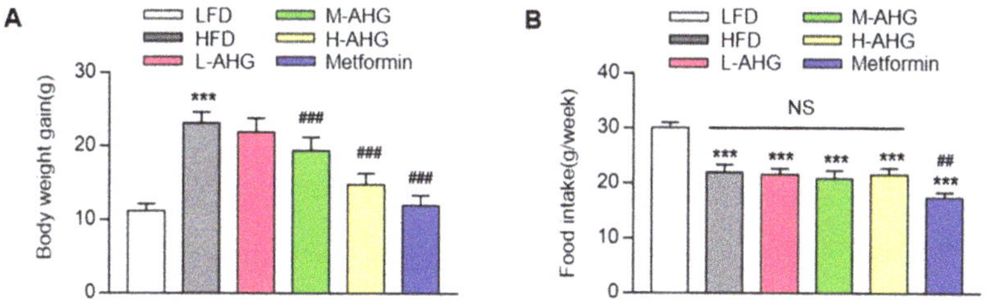

Figure 2. Effects of AHG supplementation on body weight in insulin resistant mice induced with HFD. C57BL/6J mice were fed with HFD for 12 weeks and treated with low, medium and high doses (20, 50, and 100 mg/kg/day, respectively) of AHG for eight weeks. (**A**) Body weight gain; (**B**) food intake per week. Data are showed as mean ± SD, $n = 8$; $^{*}p < 0.5$, $^{**}p < 0.1$, $^{***}p < 0.01$ vs. LFD group, $^{\#}p < 0.5$, $^{\#\#}p < 0.1$, $^{\#\#\#}p < 0.01$ vs. HFD group.

2.2. AHG Improved Glucose Metabolism in Mice Fed with HFD

As shown in Figure 3A, compared to LFD group, an obvious increase in fasting blood glucose levels was observed in HFD group ($p < 0.01$), however, the increasement was attenuated by AHG in a dose-dependent manner. Also, fasting plasma glucose in H-AHG group was similar to Metformin group, indicating that AHG supplementation (100 mg/kg/day) had a similar hypoglycemic effect as metformin. HFD impaired glucose tolerance sharply, which was attenuated by the supplementation of AHG in a concentration-dependent pattern (Figure 3C–D). Consistent with this result, insulin injection failed to decline blood glucose in HFD mice, whereas blood glucose decreased normally in H-AHG mice compared to HFD mice, which was reflected in the area under the curve for ITT (Figure 3E–F). Moreover, no significant difference of blood glucose level was found between H-AHG group and Metformin group in OGTT ($p = 0.48$) and ITT ($p = 0.25$) (Figure 3C–E). The consumption of the HFD mice also caused high level of serum insulin, which was increased four-fold when compared with the basal level of insulin content in LFD mice (Figure 3B). However, this effect was abolished in the H-AHG group. There was no notable difference in the serum insulin content between the H-AHG and Metformin groups. Overall results confirmed that the treatment of AHG improved insulin resistance induced by HFD in C57BL/6J mice.

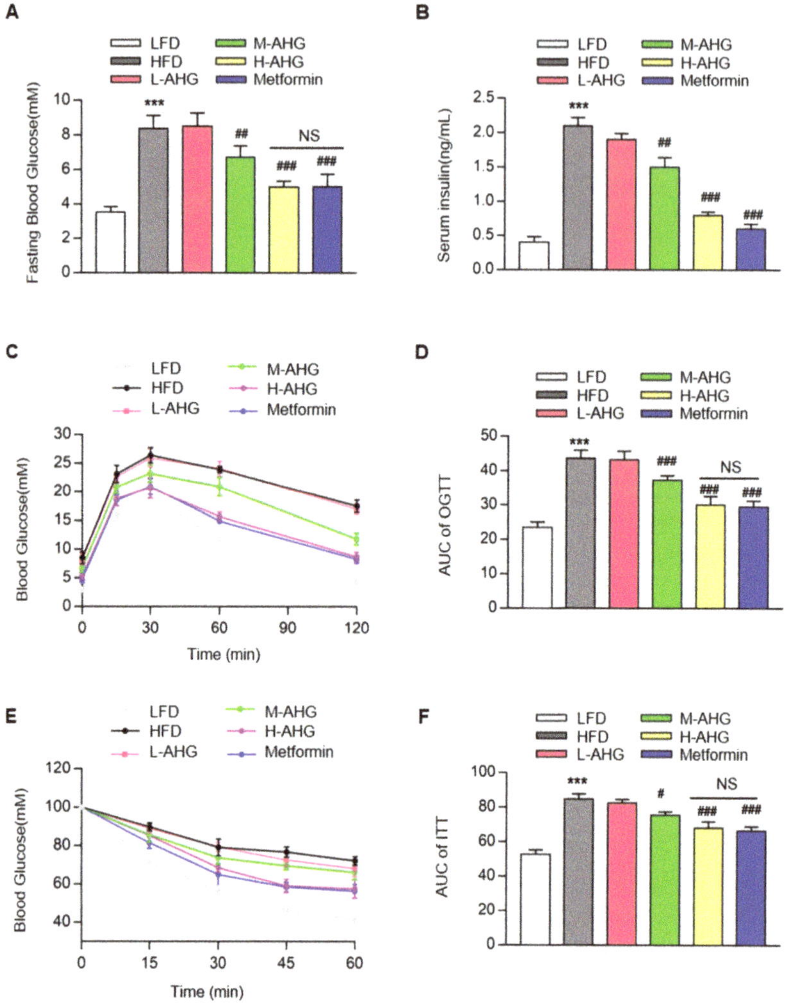

Figure 3. Effects of AHG supplementation on glucose metabolism in insulin resistant mice induced with HFD. C57BL/6J mice were fed with HFD for 12 weeks and treated with low, medium and high doses (20, 50 and 100 mg/kg/day, respectively) of AHG for eight weeks. (**A**) Fasting blood glucose; (**B**) serum insulin content; (**C**) Oral glucose tolerance test (OGTT); (**D**) The values of AUC for OGTT; (**E**) Insulin tolerance test (ITT); (**F**) The values of AUC for ITT. Data are showed as mean ± SD, $n = 8$; $^{*}p < 0.5$, $^{**}p < 0.1$, $^{***}p < 0.01$ vs. LFD group, $^{\#}p < 0.5$, $^{\#\#}p < 0.1$, $^{\#\#\#}p < 0.01$ vs. HFD group.

2.3. AHG alleviated liver injury in mice fed with HFD

Considering that the liver is the main target tissue of insulin resistance, we next explored whether AHG affected liver tissue in HFD-induced insulin resistance mice. The liver tissue weight, ALT level and AST level were measured. As shown in Figure 4A, a high dose of AHG significantly decreased the liver/body weight ratio in HFD mice ($p < 0.001$). Additionally, the value of ALT and AST showed the similar trend, indicating that AHG alleviated liver injury caused by HFD (Figure 4B,C). Gene expression analysis indicated that HFD stimulated inflammatory cytokines transcriptional levels of TNF-α, IL-6 and IL-1β in control mice by 5.7, 4.0, and 11.5-fold, respectively. The elevation of the gene expression of TNF-α, IL-6 and IL-1β was reduced when mice fed with high dose AHG (Figure 4D). These results indicate that AHG attenuated spontaneous liver injury and inflammation caused by HFD.

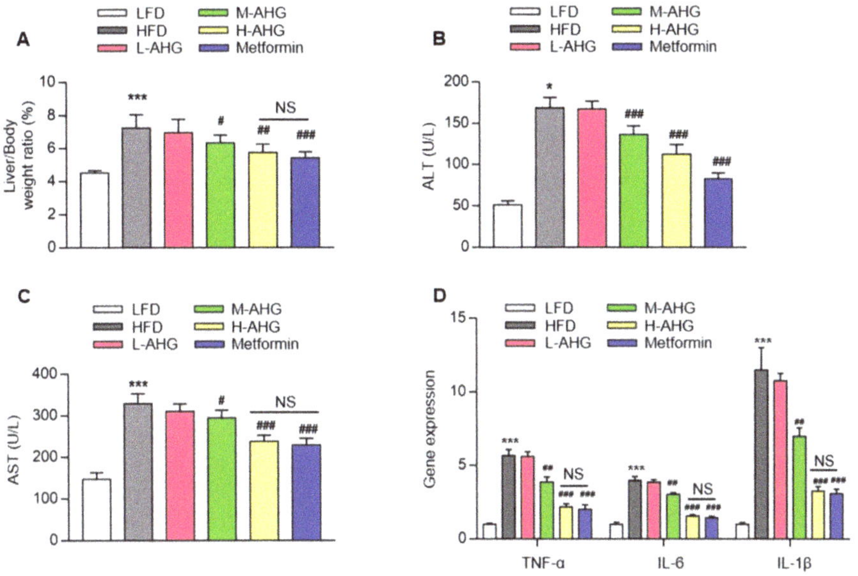

Figure 4. Effects of AHG supplementation on liver injury and inflammation induced by HFD. C57BL/6J mice were fed with HFD for 12 weeks and treated with low, medium and high doses (20, 50 and 100 mg/kg/day, respectively) of AHG for eight weeks. (**A**) Liver/body weight ratio; (**B**) The values of ALT; (**C**) The values of AST; (**D**) The mRNA expression analysis of TNF-α, IL-6 and IL-1β were measured by RT-PCR and normalized by Cyclophilin. Data are showed as mean ± SD, $n = 8$; $^{*}p < 0.5$, $^{**}p < 0.1$, $^{***}p < 0.01$ vs. LFD group, $^{\#}p < 0.5$, $^{\#\#}p < 0.1$, $^{\#\#\#}p < 0.01$ vs. HFD group

2.4. AHG Suppressed Hepatic Gluconeogenesis in Fasting Mice Fed A High Fat Diet

Our previous study showed that AHG suppressed hepatic gluconeogenesis in insulin resistant hepatocytes [24]. Next, we determined whether AHG affected hepatic gluconeogenesis in an insulin resistant liver. As shown in Figure 5, the protein level and gene expression of gluconeogenesis rate-limiting enzymes G6Pase and PEPCK were measured. In the livers, HFD significantly increased the protein level of G6Pase and PEPCK by 50% and 20%, respectively, but the increased potential in mice with the administration of high dose AHG was, respectively, 12% and 5%, with a significant reduction compared to HFD mice (Figure 5A,B). The Q-PCR result showed the same trend. As a positive control, metformin repressed the transcriptional and translational level of G6Pase and PEPCK as reported in the literature [26]. This confirms that AHG suppressed hepatic gluconeogenesis in insulin resistant mice fed with a HFD.

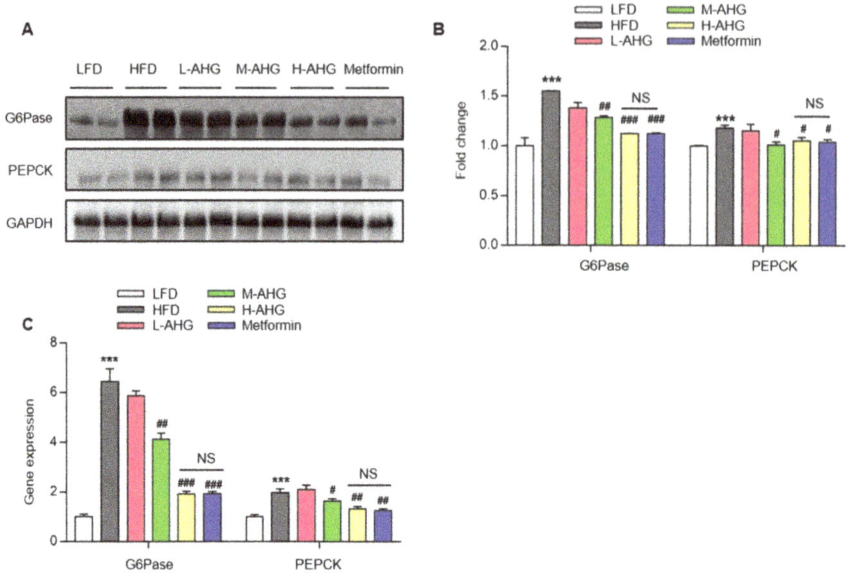

Figure 5. Effects of AHG supplementation on gluconeogenesis in insulin resistant mice induced with HFD. C57BL/6J mice were fed with HFD for 12 weeks and treated with low, medium and high doses (20, 50 and 100 mg/kg/day, respectively) of AHG for eightweeks. (**A**) The protein levels of G6Pase and PEPCK in livers were measured by Western blot. (**B**) Quantification of protein levels of G6Pase and PEPCK was performed by Image J. GAPDH as a control to normalize the expression of the protein. Data are showed as mean ± SD, $n = 3$; (**C**) Gene expression of G6Pase and PEPCK in livers were quantified by RT-PCR. Data are shown as mean ± SD, $n = 4$; $^{*}p < 0.5$, $^{**}p < 0.1$, $^{***}p < 0.01$ vs. LFD group, $^{\#}p < 0.5$, $^{\#\#}p < 0.1$, $^{\#\#\#}p < 0.01$ vs. HFD group.

2.5. AHG Improved Insulin Signaling Pathway in Liver of Mice Fed with HFD

To explore the effect of AHG on insulin signaling pathway in liver, mice were administered with or without insulin by intraperitoneal injection 2 min before euthanasia. The protein level of major cascades of insulin signaling pathway were assessed by Western Blot. HFD consumption suppressed the basal and insulin-stimulated protein level of p-IRS1(S302) and p-Akt (T308), supplementation of AHG resulted in a dose-dependent increase of IRS1 and Akt protein level in basal and insulin-stimulated state (Figure 6A,B). Metformin also activated the insulin signaling cascade pathway, which was consistent with the previously reported work [27]. The above results suggest that the observed alterations in glucose metabolism caused by HFD are associated with an impaired IRS1/Akt pathway in liver, and this effect is mitigated by AHG supplementation.

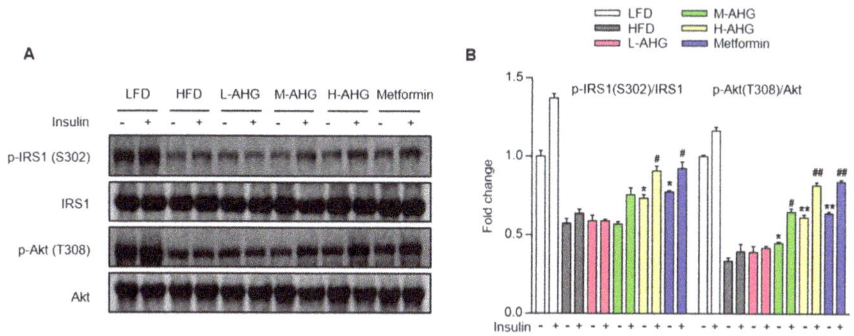

Figure 6. Effects of AHG supplementation on insulin signaling pathway in the livers of insulin resistant mice induced with HFD. C57BL/6J mice were fed with HFD for 12 weeks and treated with low, medium and high doses (20, 50 and 100 mg/kg/day, respectively) of AHG for eight weeks. At the end of treatment, mice were fasted for 12 h and intraperitoneally injected with either saline or insulin (20 units/kg). (**A**) The protein expression of p-IRS (S302), IRS1, p-Akt (T308) and Akt were measured by Western blot. (**B**) Quantification of p-IRS (S302)/IRS1 and p-Akt (T308)/Akt was performed by Image J. Data are showed as mean ± SD, n=3; *$p < 0.5$, **$p < 0.1$, ***$p < 0.01$ vs. insulin untreated HFD group, #$p < 0.5$, ##$p < 0.1$, ###$p < 0.01$ vs. insulin treated HFD group.

2.6. AHG Activated AMPK in Liver of Mice Fed a High Fat Diet

AMPK and Akt are the two primary effectors in response to glucose metabolism [28]. AMPK maintains blood glucose levels and reduces hepatic gluconeogenesis gene expression-PEPCK and G6Pase. It is currently unknown that whether AHG activates AMPK in the insulin resistant liver. To address whether AMPK was involved in the effect of AHG on glucose homeostasis in liver, the protein levels of phosphorylated AMPK and an AMPK substrate, acetyl-CoA carboxylase (ACC) were measured. As illustrated in Figure 7, in the liver of insulin resistance mice induced by HFD, AHG elevated the p-AMPK and p-ACC in a dose-dependent manner. Also, there was no difference in activating AMPK between H-AHG and Metformin groups. These results state clearly that the possible role of AHG on repressing hepatic gluconeogenesis via AMPK signaling pathway.

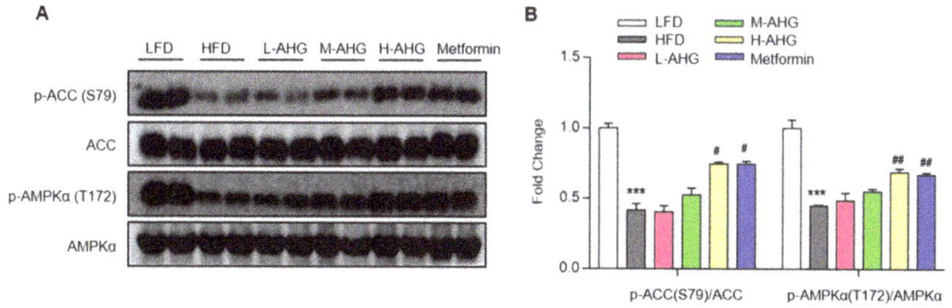

Figure 7. Effects of AHG supplementation on AMPK signaling pathway in the livers of insulin resistant mice induced with HFD. C57BL/6J mice were fed with HFD for 12 weeks and treated with low, medium and high doses (20, 50 and 100 mg/kg/day, respectively) of AHG for eight weeks. (**A**) The protein expression of p-ACC (S79), ACC, p-AMPKα (T172) and AMPKα were measured by Western blot. (**B**) Quantification of p-ACC (S79)/ACC and p-AMPKα (T172)/ AMPKαwas performed by Image J. Data are showed as mean ± SD, $n = 3$; * $p < 0.5$, **$p < 0.1$, ***$p < 0.01$ vs. LFD group, # $p < 0.5$, ## $p < 0.1$, ### $p < 0.01$ vs. HFD group.

3. Discussion

Having high nutritional and medicinal qualities, sea cucumber has been drawn notable attention [9]. There is culminating evidence that polysaccharides from sea cucumber have an effective anti-diabetic activity [16,29–31]. In our previous study, AHG was shown to have a therapeutic effect on hepatic glucose metabolism in vitro, however, there is no report target the effect of AHG on glucose metabolism in vivo. In the current study, we provided evidence that AHG treatment significantly alleviated fasting blood glucose and improved insulin resistance by suppressing gluconeogenesis in HFD-fed insulin resistant mice. Also, AHG supplementation elevated IRS1/Akt activation and AMPK phosphorylation in the liver of insulin resistant mice induced by HFD. All of these data suggest that AHG improved glucose metabolism in vivo.

The liver plays a crucial role in organismic energy metabolism, especially glucose metabolism, and dysfunction of the liver can result in acute metabolic abnormalities [32,33]. Due to the primary site of insulin clearance, liver is associated with the development of insulin resistance. In insulin resistance, the increased gluconeogenesis is the main mechanism involving in elevated blood glucose [34]. In the present study, the effect of AHG on gluconeogenesis was detected. HFD supplementation significantly enlarged the gene expression and protein levels of G6Pase and PEPCK in the liver, and that administration with high dose AHG remarkably inhibited this trend. These results implied that AHG maintained glucose metabolism in the liver via suppressing gluconeogenesis.

Insulin is the crucial hormone regulating gluconeogenesis and insulin represses gluconeogenesis by activating the insulin signaling pathway [35]. Insulin binds to insulin receptor and then activates phosphorylation of insulin receptor substrate (p-IRS), followed by recruitment and phosphorylation of Akt. These are key protein kinases involving in restraining gluconeogenesis related genes G6Pase and PEPCK [36,37]. Therefore, the activation of insulin signaling pathway in liver may be a promising strategy for improving insulin sensitivity. In the current study, HFD impaired the phosphorylation of IRS1 and Akt induced by insulin in the liver. However, AHG supplementation alleviated these alterations. Also, AHG improved insulin tolerance, glucose tolerance and alleviated serum insulin content in HFD group mice, suggesting that AHG improved insulin sensitivity induced by HFD, these provided insight into the anti-insulin resistance actions of AHG.

As an intracellular energy sensor, AMPK is another well-known suppressor of gluconeogenesis in addition to insulin pathway and the suppression of AMPK accompanies insulin resistance [38]. Thus, AMPK is considered as a potential target for diabetes prevention and insulin resistance treatment. So far, cumulative research has reported that various natural products regulated glucose metabolism through AMPK, such as epigallocatechin-3-gallate (EGCG) [39], ginsenoside Rg1 [40], rosmarinic acid [41], and astragalus polysaccharide [42]. To the best of our knowledge, it is barely published to target the polysaccharide isolated from marine organism especially sea cucumber on the effect of glucose metabolism through AMPK. Here, we explored how AHG activated AMPK in an insulin resistant liver induced by HFD. Thus, one of the mechanisms of the effect of AHG on glucose homeostasis in insulin resistance mice might be related to AMPK pathway.

Chronic inflammation plays a vital role in the process of liver injury induced by insulin resistance and associated with disease risk [43]. It is reported that proinflammatory cytokines TNF-α and IL-6 produced by HFD were able to induce insulin resistance [44]. In inflammation, proinflammatory cytokine is reported to activate various serine kinases such as c-Jun N-terminal kinase (JNK) [45], S6 kinase (S6K) [46], double-stranded RNA-dependent protein kinase mammalian [47], and inhibitor of nuclear factor kappa-B kinase subunit β (IKKβ) [48]. These serine kinases can alleviate the activation of IRS1, resulting in the impairment of insulin receptor-mediated signaling and the occurrence of insulin resistance. In our HFD-induced insulin resistant model, AHG attenuated HFD-induced liver inflammation by inhibiting the expression of inflammatory genes. The anti-inflammatory effect of AHG on the liver may be one potential explanation for the beneficial effects of AHG on insulin resistance.

The high molecular weight is thought to be the main factor that suppress the transepithelial transport of glycosaminoglycan. It is reported that a fucosyl chondroitin sulfate chains (Mw=12.00 KDa) isolated from sea cucumber can be detected in plasma and urine after oral administration (50 mg/kg) [49]. Additionally, a glycosaminoglycan from Cucumaria frondosa (Mw=21.53 kDa) can improve glucose metabolism in the liver of the insulin resistant mice after oral administration [30]; a glycosaminoglycan from Acaudina molpadioides (Mw=21.53 kDa) was found to improve glucose metabolism in the liver and skeletal muscle of the insulin resistant mice after oral administration [16,29]. A sulfated polysaccharide from sea cucumber Stichopus japonicus (SCSP, Mw=179.4 kDa) can prevent obesity in association with modification of gut microbiota in HFD-fed mice [31]. These studies indicated that glycosaminoglycans from sea cucumber have the possibility to be absorbed in liver and muscle skeletal of insulin resistant mice. Based on our previously study (data unpublished), AHG is a moderate absorption drug in an endocytosis manner using Caco-2 and M cell models. In Caco-2 and M cell models, the Papp (AP-BL) values of AHG were about 2 $\times 10^{-6}$ cm/s and 8 $\times 10^{-6}$ cm/s, respectively. This indicates AHG is absorbed in the intestinal tracts. The pharmacokinetic study of AHG in animals will be of interest for further study.

Cumulative evidence supports a beneficial effect of AHG on glucose homeostasis as we observed in the model of HFD-induced insulin resistant mice. Oral treatment with AHG significantly reduced body weight and fasting blood glucose level. Furthermore, AHG supplementation improved insulin resistance by repressing gluconeogenesis related genes. In addition, the activation of IRS1/Akt and AMPK signaling pathway are the possible mechanisms underlying the effect of AHG on glucose metabolism in insulin resistant liver. In summary, these findings provide evidence that AHG has the potential to become a novel marine natural product to provide a therapy for the prevention and treatment of insulin resistance and type 2 diabetes. To have a better understanding of the effect of AHG on glucose metabolism in vivo, further investigation is necessary to verify the effect of AHG on regulation of glucose homeostasis in other insulin target tissues, such as skeletal muscle and adipose tissues.

4. Materials and Methods

4.1. Chemical Reagents

Insulin and metformin were obtained from Aladdin (Shanghai, China). Antibodies PEPCK, G6Pase, p-IRS1(S302), p-Akt (T308), IRS1, Akt, p-AMPKα (T172), p-ACC (S79), AMPKα and ACC were all purchased from Cell Signaling Technology (Danvers, MA, United States). Insulin was obtained from Millipore Sigma (Danvers, MA, United States). Radioimmunoprecipitation assay buffer (RIPA buffer), SDS-PAGE, PVDF membrane and ECL were obtained from Beyotime (Shanghai, China).

4.2. Preparation of AHG

AHG was prepared as previously described [23]. Specifically, AHG was extracted from the body wall of the sea cucumber Apostichopus japonicus, purchased from the Nanshan market of Qingdao, China. AHG was prepared as previously described. The body wall of fresh sea cucumber Apostichopus japonicus was grinded into homogenate and extracted with KOH (1 mol/L) at 60 °C for 60 min. After neutralization with cold HCl, diastase vera (EC 3.3.21.4) was added to hydrolyze the protein. The crude polysaccharide was precipitated with 60% ethanol. The crude AHG was further fractionated by a Q Sepharose Fast Flow column (300 mm × 30 mm) with elution by a step-wise gradient of 0.75 and 1.5 M NaCl. The fractions eluted with 1.5 M NaCl were further purified on a Sephadex 25 column (100 × 2.6 cm) with deionized water at a flow rate of 0.3 mL/min. The purified AHG were pooled and lyophilized. The yield of AHG isolated from the fresh sea cucumber Apostichopus japonicus was 0.51% by weight. The average molecular weight of AHG was 98.07 kDa and the purity of AHG was over 99% by using gel filtration chromatography. Monosaccharide composition analysis based on pre-column derivatization reversed-phase HPLC showed that AHG consisted of glucuronic acid, galactosamine and fucose in the molar ratio of 1/1.03/1.16. The sulfate content was 33.20% determined by high performance capillary electrophoresis.

4.3. Animals and Animal Care

Male C57BL/6J mice (18-22 g) were obtained from Jinan Pengyue Experimental Animal Breeding Co. Ltd. (License Number: SCXK (lu) 2014-0007) and housed in a controlled condition (25 °C, 50 ± 5 % humidity and 12 h dark-light cycles). All mice were randomly divided into six groups for 12 weeks (8 mice/group): (1) LFD group, given low fat diet (LFD) for 12 weeks; (2) HFD group, given High Fat Diet (HFD) for 12 weeks; (3) L-AHG group, given HFD for 4 weeks, then given HFD simultaneously with oral administration of low dose AHG (20 mg/kg/day) for another 8 weeks; (4) M-AHG group, given HFD for 4 weeks, then given HFD simultaneously with oral administration of medium dose AHG (50 mg/kg/day) for another 8 weeks; (5) H-AHG group, given HFD for 4 weeks, then given HFD simultaneously with high dose oral administration of AHG (100 mg/kg/day) for another 8 weeks; (6) Metformin group, given HFD for 4 weeks, then given HFD simultaneously with oral administration of metformin (200 mg/kg/day) for another 8 weeks. LFD (TP23523) and HFD (60% fat, TP23520) were purchased from Nantong Teluofei Feed Technology Co., Ltd. (Nantong, China). At the end of treatment, mice were fasted for 12 h and intraperitoneally injected with either saline or insulin (20 units/kg). Mice were euthanized 2 min after injection. Blood were collected and centrifuged at 3000 rpm for 15min at 4 °C to obtain serum. Tissues were weighed and flash frozen in liquid nitrogen. The anti-hyperglycemic action of metformin is mainly a consequence of suppressed glucose output owing to inhibition of liver gluconeogenesis [50]. The centre of mechanism of metformin on liver gluconeogenesis is the alteration of the cell energy metabolism. metformin inhibits mitochondrial complex I, resulting a drop in ATP production. the inhibition of ATP stimulates the activation of 5′-AMP-activated protein kinase (AMPK) [51–53]. All procedures applied to animals were performed in accordance with the guidelines of the Laboratory Animal Center of Ocean University of China and were approved by Animal Ethics Committee of School of Medicine and Pharmacy, Ocean University of China (Qingdao, China).

4.4. Glucose Tolerance Test and Insulin Tolerance Test

For the glucose tolerance test (GTT), mice were fasted for 16 h and administered orally with glucose (2 g/kg). Blood glucose was collected before and at 15, 30, 60 and 120 min. For insulin tolerance test (ITT), mice were injected with insulin (2 units/kg) after 4 h fasting. Blood glucose was collected before and at 15, 30, 45, and 60 min. Blood glucose was measured with a glucometer (Roche, Penzberg, Germany).

4.5. Serum Insulin Level Assay

For serum insulin level, blood was collected from mice in fed state. The concentration of serum insulin was measured by Invitrogen insulin mouse ELISA kit (Thermo Fisher, Shanghai, China).

4.6. Serum Biochemical Analysis

Serum samples were obtained from blood after 30 min centrifugation at 4 °C. The levels of Alanine aminotransferase (ALT) and aspartate aminotransferase (AST) were assessed by ALT and AST assay kits (Nanjing Jiancheng, Nanjing, China).

4.7. Western Blot analysis

Proteins from the liver tissues were measured by BCA Protein Assay Kit (Thermo Fisher, Shanghai, China) and separated by SDS-PAGE. After separated, proteins were transferred to PVDF membranes and then membranes were blocked in TBST containing 5% BSA and incubated with indicated antibodies overnight at 4 °C, followed by secondary antibody for 1 hour at room temperature. Finally, the protein bands were detected by using ECL (Nanjing Jiancheng, Nanjing, China) and were quantified by Image J.

4.8. Quantitative Real-time PCR

Total RNA was extracted from the liver tissues by using TRIzol reagent (Invitrogen, Carlsbad, USA) and was reversed to cDNA by using iScript cDNA Synthesis Kit (Bio-Rad, USA). Gene expression was measured with SYBR Green Supermix (Bio-Rad, Hercules, California, USA) in Bio-Rad CFX384 system. The mRNA expression was calculated by $\Delta\Delta$CT methods and Cyclophilin was used as a load control. The primers used for this study were as follows, G6Pase (forward-5′- TGG TAG CCC TGT CTT TCT TTG-3′; reverse-5′- TTC CAG CAT TCA CAC TTT CCT-3′), PEPCK (forward-5′-ACA CAC ACA CAT GCT CAC AC-3′; reverse-5′- ATC ACC GCA TAG TCT CTG AA-3′), TNF-α (forward-5′-CCC GAG TGA CAA GCC TGT AG-3′; reverse-5′-GAT GGC AGA GAG GAG GTT GAC-3′), IL-6 (forward-5′-ACA GCC ACT CAC CTC TTC AG -3′; reverse-5′-CCA TCT TTT TCA GCC ATC TTT-3′), IL-1β (forward-5′-AGA TGA TAA GCC CAC TCT ACA G-3′; reverse-5′-ACA TTC AGC ACA GGA CTC TC-3′),Cyclophilin (forward-5′- AGC TAG ACT TGA AGG GGA ATG-3′; reverse-5′-ATT TCT TTT GAC TTG CGG GC-3′).

4.9. Statistical Analysis

All results were expressed as the mean ± SD. Statistical significance between the two groups was calculated using unpaired two-tailed t test. In Figure 2B, ANOVA tests were used to compare two groups among multiple groups. The value of $p < 0.05$ was considered statistically significantly. Data were considered no statistically significant at NS.

Author Contributions: Conceptualization, Z.L., Y.W. and Y.C.; data curation, Y.C., M.Y. and T.J.; writing—original draft preparation, Y.C.; writing—review and editing, S.Y.; visualization, M.Y.; supervision, Z.L.; project administration, Z.L.; funding acquisition, Z.L. and S.Y. All the authors read and approved the final manuscript. All authors have read and agreed to the published version of the manuscript.

Funding: "This research was funded by NSFC-Shandong Joint Fund for Marine Science Research Centers (grant No. U1606403), the National Natural Science Foundation of China (grant No. 31700704), Shandong Provincial Natural Science Foundation (grant No. ZR2017BC092) and the Fundamental Research Funds for the Central Universities (grant No. 201964019, Ocean University of China).

Acknowledgments: The authors gratefully acknowledged to the NSFC-Shandong Joint Fund for Marine Science Research Centers (grant No. U1606403), the National Natural Science Foundation of China (grant No. 31700704), Shandong Provincial Natural Science Foundation (grant No. ZR2017BC092) and the Fundamental Research Funds for the Central Universities (grant No. 201964019, Ocean University of China) for partially funding this work.

Conflicts of Interest: The authors declare no conflict of interest.

References

1. Forbes, J.M.; Cooper, M.E. Mechanisms of diabetic complications. *Physiol. Rev.* **2013**, *93*, 137–188. [CrossRef] [PubMed]
2. Petersmann, A.; Nauck, M.; Muller-Wieland, D.; Kerner, W.; Muller, U.A.; Landgraf, R.; Freckmann, G.; Heinemann, L. Definition, Classification and Diagnosis of Diabetes Mellitus. *Exp. Clin. Endocrinol. Diabetes Off. J. Ger. Soc. Endocrinol. Ger. Diabetes Assoc.* **2018**, *126*, 406–410. [CrossRef] [PubMed]
3. Kahn, B.B.; Flier, J.S. Obesity and insulin resistance. *J. Clin. Invest.* **2000**, *106*, 473–481. [CrossRef] [PubMed]
4. Czech, M.P. Insulin action and resistance in obesity and type 2 diabetes. *Nat. Med.* **2017**, *23*, 804–814. [CrossRef]
5. Adeva-Andany, M.M.; Perez-Felpete, N.; Fernandez-Fernandez, C.; Donapetry-Garcia, C.; Pazos-Garcia, C. Liver glucose metabolism in humans. *Biosci Rep.* **2016**, *36*, e00416. [CrossRef]
6. Biddinger, S.B.; Kahn, C.R. From mice to men: Insights into the insulin resistance syndromes. *Annu. Rev. Physiol.* **2006**, *68*, 123–158. [CrossRef]
7. Rines, A.K.; Sharabi, K.; Tavares, C.D.; Puigserver, P. Targeting hepatic glucose metabolism in the treatment of type 2 diabetes. *Nat. Rev. Drug Discov.* **2016**, *15*, 786–804. [CrossRef]
8. Lauritano, C.; Ianora, A. Marine Organisms with Anti-Diabetes Properties. *Mar. Drugs* **2016**, *14*. [CrossRef]
9. Khotimchenko, Y. Pharmacological Potential of Sea Cucumbers. *Int. J. Mol. Sci.* **2018**, *19*, 1342. [CrossRef]

10. Janakiram, N.B.; Mohammed, A.; Rao, C.V. Sea Cucumbers Metabolites as Potent Anti-Cancer Agents. *Mar. Drugs* **2015**, *13*, 2909–2923. [CrossRef]
11. Mourao, P.A.; Pereira, M.S.; Pavao, M.S.; Mulloy, B.; Tollefsen, D.M.; Mowinckel, M.C.; Abildgaard, U. Structure and anticoagulant activity of a fucosylated chondroitin sulfate from echinoderm. Sulfated fucose branches on the polysaccharide account for its high anticoagulant action. *J. Biol. Chem.* **1996**, *271*, 23973–23984. [CrossRef] [PubMed]
12. Wang, Z.; Zhang, H.; Yuan, W.; Gong, W.; Tang, H.; Liu, B.; Krohn, K.; Li, L.; Yi, Y.; Zhang, W. Antifungal nortriterpene and triterpene glycosides from the sea cucumber Apostichopus japonicus Selenka. *Food Chem.* **2012**, *132*, 295–300. [CrossRef] [PubMed]
13. Zhao, Y.; Li, B.; Liu, Z.; Dong, S.; Zhao, X.; Zeng, M. Antihypertensive effect and purification of an ACE inhibitory peptide from sea cucumber gelatin hydrolysate. *Process. Biochem.* **2007**, *42*, 1586–1591. [CrossRef]
14. Mamelona, J.; Pelletier, É.; Girard-Lalancette, K.; Legault, J.; Karboune, S.; Kermasha, S. Quantification of phenolic contents and antioxidant capacity of Atlantic sea cucumber, Cucumaria frondosa. *Food Chem.* **2007**, *104*, 1040–1047. [CrossRef]
15. Herencia, F.; Ubeda, A.; Ferrandiz, M.L.; Terencio, M.C.; Alcaraz, M.J.; Garcia-Carrascosa, M.; Capaccioni, R.; Paya, M. Anti-inflammatory activity in mice of extracts from Mediterranean marine invertebrates. *Life Sci.* **1998**, *62*, P115–P120. [CrossRef]
16. Hu, S.; Chang, Y.; Wang, J.; Xue, C.; Shi, D.; Xu, H.; Wang, Y. Fucosylated chondroitin sulfate from Acaudina molpadioides improves hyperglycemia via activation of PKB/GLUT4 signaling in skeletal muscle of insulin resistant mice. *Food Funct.* **2013**, *4*, 1639–1646. [CrossRef]
17. Farshadpour, F.; Gharibi, S.; Taherzadeh, M.; Amirinejad, R.; Taherkhani, R.; Habibian, A.; Zandi, K. Antiviral activity of Holothuria sp. a sea cucumber against herpes simplex virus type 1 (HSV-1). *Eur. Rev. Med. Pharmacol. Sci.* **2014**, *18*, 333–337.
18. Bordbar, S.; Anwar, F.; Saari, N. High-value components and bioactives from sea cucumbers for functional foods—A review. *Mar. Drugs* **2011**, *9*, 1761–1805. [CrossRef]
19. Mondol, M.A.M.; Shin, H.J.; Rahman, M.A.; Islam, M.T. Sea Cucumber Glycosides: Chemical Structures, Producing Species and Important Biological Properties. *Mar. Drugs* **2017**, *15*. [CrossRef]
20. Pomin, V.H. Holothurian fucosylated chondroitin sulfate. *Mar. Drugs* **2014**, *12*, 232–254. [CrossRef]
21. Valcarcel, J.; Novoa-Carballal, R.; Perez-Martin, R.I.; Reis, R.L.; Vazquez, J.A. Glycosaminoglycans from marine sources as therapeutic agents. *Biotechnol. Adv.* **2017**, *35*, 711–725. [CrossRef] [PubMed]
22. Yamana, Y.; Hamano, T.; Goshima, S. Seasonal distribution pattern of adult sea cucumber Apostichopus japonicus (Stichopodidae) in Yoshimi Bay, western Yamaguchi Prefecture, Japan. *Fish. Sci.* **2009**, *75*, 585–591. [CrossRef]
23. Yang, J.; Wang, Y.; Jiang, T.; Lv, Z. Novel branch patterns and anticoagulant activity of glycosaminoglycan from sea cucumber Apostichopus japonicus. In *International Journal of Biological Macromolecules*; Elsevier: Amsterdam, The Netherlands, 2015; Volume 72, pp. 911–918.
24. Chen, Y.; Liu, H.; Wang, Y.; Yang, S.; Yu, M.; Jiang, T.-F.; Lv, Z. Glycosaminoglycan from Apostichopus japonicus inhibits hepatic glucose production via activating Akt/FoxO1 and inhibiting PKA/CREB signaling pathways in insulin resistance hepatocytes. *Food Funct.* **2019**. [CrossRef] [PubMed]
25. Matsui, Y.; Hirasawa, Y.; Sugiura, T.; Toyoshi, T.; Kyuki, K.; Ito, M. Metformin Reduces Body Weight Gain and Improves Glucose Intolerance in High-Fat Diet-Fed C57BL/6J Mice. *Biol. Pharm. Bull.* **2010**, *33*, 963–970. [CrossRef] [PubMed]
26. Kim, Y.D.; Park, K.G.; Lee, Y.S.; Park, Y.Y.; Kim, D.K.; Nedumaran, B.; Jang, W.G.; Cho, W.J.; Ha, J.; Lee, I.K.; et al. Metformin inhibits hepatic gluconeogenesis through AMP-activated protein kinase-dependent regulation of the orphan nuclear receptor SHP. *Diabetes* **2008**, *57*, 306–314. [CrossRef]
27. Xu, H.; Zhou, Y.; Liu, Y.; Ping, J.; Shou, Q.; Chen, F.; Ruo, R. Metformin improves hepatic IRS2/PI3K/Akt signaling in insulin-resistant rats of NASH and cirrhosis. *J. Endocrinol.* **2016**, *229*, 133–144. [CrossRef]
28. Zhao, Y.; Hu, X.; Liu, Y.; Dong, S.; Wen, Z.; He, W.; Zhang, S.; Huang, Q.; Shi, M. ROS signaling under metabolic stress: Cross-talk between AMPK and AKT pathway. *Mol. Cancer* **2017**, *16*, 79. [CrossRef]
29. Hu, S.W.; Tian, Y.Y.; Chang, Y.G.; Li, Z.J.; Xue, C.H.; Wang, Y.M. Fucosylated chondroitin sulfate from sea cucumber improves glucose metabolism and activates insulin signaling in the liver of insulin-resistant mice. *J. Med. Food* **2014**, *17*, 749–757. [CrossRef]

30. Hu, S.; Chang, Y.; Wang, J.; Xue, C.; Li, Z.; Wang, Y. Fucosylated chondroitin sulfate from sea cucumber in combination with rosiglitazone improved glucose metabolism in the liver of the insulin-resistant mice. *Biosci. Biotechnol. Biochem.* **2013**, *77*, 2263–2268. [CrossRef]
31. Zhu, Z.; Zhu, B.; Sun, Y.; Ai, C.; Wang, L.; Wen, C.; Yang, J.; Song, S.; Liu, X. Sulfated Polysaccharide from Sea Cucumber and its Depolymerized Derivative Prevent Obesity in Association with Modification of Gut Microbiota in High-Fat Diet-Fed Mice. *Mol. Nutr. Food Res.* **2018**, *62*, e1800446. [CrossRef]
32. Han, H.S.; Kang, G.; Kim, J.S.; Choi, B.H.; Koo, S.H. Regulation of glucose metabolism from a liver-centric perspective. *Exp. Mol. Med.* **2016**, *48*, e218. [CrossRef] [PubMed]
33. Petersen, M.C.; Vatner, D.F.; Shulman, G.I. Regulation of hepatic glucose metabolism in health and disease. *Nat. Rev. Endocrinol.* **2017**, *13*, 572–587. [CrossRef] [PubMed]
34. Meshkani, R.; Adeli, K. Hepatic insulin resistance, metabolic syndrome and cardiovascular disease. *Clin. Biochem.* **2009**, *42*, 1331–1346. [CrossRef] [PubMed]
35. Lin, H.V.; Accili, D. Hormonal regulation of hepatic glucose production in health and disease. *Cell. Metab.* **2011**, *14*, 9–19. [CrossRef] [PubMed]
36. Manning, B.D.; Toker, A. AKT/PKB Signaling: Navigating the Network. *Cell* **2017**, *169*, 381–405. [CrossRef] [PubMed]
37. Dong, X.C.; Copps, K.D.; Guo, S.; Li, Y.; Kollipara, R.; DePinho, R.A.; White, M.F. Inactivation of hepatic Foxo1 by insulin signaling is required for adaptive nutrient homeostasis and endocrine growth regulation. *Cell. Metab.* **2008**, *8*, 65–76. [CrossRef]
38. Lin, S.C.; Hardie, D.G. AMPK: Sensing Glucose as well as Cellular Energy Status. *Cell. Metab.* **2018**, *27*, 299–313. [CrossRef]
39. Collins, Q.F.; Liu, H.Y.; Pi, J.; Liu, Z.; Quon, M.J.; Cao, W. Epigallocatechin-3-gallate (EGCG), a green tea polyphenol, suppresses hepatic gluconeogenesis through 5′-AMP-activated protein kinase. *J. Biol. Chem.* **2007**, *282*, 30143–30149. [CrossRef]
40. Kim, S.J.; Yuan, H.D.; Chung, S.H. Ginsenoside Rg1 suppresses hepatic glucose production via AMP-activated protein kinase in HepG2 cells. *Biol. Pharm. Bull.* **2010**, *33*, 325–328. [CrossRef]
41. Vlavcheski, F.; Naimi, M.; Murphy, B.; Hudlicky, T.; Tsiani, E. Rosmarinic Acid, a Rosemary Extract Polyphenol, Increases Skeletal Muscle Cell Glucose Uptake and Activates AMPK. *Molecules* **2017**, *22*. [CrossRef]
42. Zhang, R.; Qin, X.; Zhang, T.; Li, Q.; Zhang, J.; Zhao, J. Astragalus Polysaccharide Improves Insulin Sensitivity via AMPK Activation in 3T3-L1 Adipocytes. *Molecules* **2018**, *23*. [CrossRef] [PubMed]
43. Shoelson, S.E.; Lee, J.; Goldfine, A.B. Inflammation and insulin resistance. *J. Clin. Invest.* **2006**, *116*, 1793–1801. [CrossRef] [PubMed]
44. Cai, D.; Yuan, M.; Frantz, D.F.; Melendez, P.A.; Hansen, L.; Lee, J.; Shoelson, S.E. Local and systemic insulin resistance resulting from hepatic activation of IKK-beta and NF-kappaB. *Nat. Med.* **2005**, *11*, 183–190. [CrossRef] [PubMed]
45. Hirosumi, J.; Tuncman, G.; Chang, L.; Gorgun, C.Z.; Uysal, K.T.; Maeda, K.; Karin, M.; Hotamisligil, G.S. A central role for JNK in obesity and insulin resistance. *Nature* **2002**, *420*, 333–336. [CrossRef] [PubMed]
46. Gao, Z.; Zuberi, A.; Quon, M.J.; Dong, Z.; Ye, J. Aspirin inhibits serine phosphorylation of insulin receptor substrate 1 in tumor necrosis factor-treated cells through targeting multiple serine kinases. *J. Biol. Chem.* **2003**, *278*, 24944–24950. [CrossRef] [PubMed]
47. Nakamura, T.; Furuhashi, M.; Li, P.; Cao, H.; Tuncman, G.; Sonenberg, N.; Gorgun, C.Z.; Hotamisligil, G.S. Double-stranded RNA-dependent protein kinase links pathogen sensing with stress and metabolic homeostasis. *Cell* **2010**, *140*, 338–348. [CrossRef] [PubMed]
48. Yuan, M.; Konstantopoulos, N.; Lee, J.; Hansen, L.; Li, Z.W.; Karin, M.; Shoelson, S.E. Reversal of obesity- and diet-induced insulin resistance with salicylates or targeted disruption of Ikkbeta. *Sci. (New York, N.Y.)* **2001**, *293*, 1673–1677. [CrossRef]
49. Imanari, T.; Washio, Y.; Huang, Y.; Toyoda, H.; Suzuki, A.; Toida, T. Oral absorption and clearance of partially depolymerized fucosyl chondroitin sulfate from sea cucumber. *Thromb. Res.* **1999**, *93*, 129–135. [CrossRef]
50. Pernicova, I.; Korbonits, M. Metformin–mode of action and clinical implications for diabetes and cancer. *Nat. Rev. Endocrinol.* **2014**, *10*, 143–156. [CrossRef]

51. El-Mir, M.Y.; Nogueira, V.; Fontaine, E.; Avéret, N.; Rigoulet, M.; Leverve, X. Dimethylbiguanide inhibits cell respiration via an indirect effect targeted on the respiratory chain complex I. *J. Biol. Chem.* **2000**, *275*, 223–228. [CrossRef]
52. Miller, R.A.; Birnbaum, M.J. An energetic tale of AMPK-independent effects of metformin. *J. Clin. Investig.* **2010**, *120*, 2267–2270. [CrossRef] [PubMed]
53. Rena, G.; Hardie, D.G.; Pearson, E.R. The mechanisms of action of metformin. *Diabetologia* **2017**, *60*, 1577–1585. [CrossRef] [PubMed]

marine drugs

Article

Virescenosides from the Holothurian-Associated Fungus *Acremonium striatisporum* Kmm 4401

Olesya I. Zhuravleva [1,2], Alexandr S. Antonov [1], Galina K. Oleinikova [1], Yuliya V. Khudyakova [1], Roman S. Popov [1], Vladimir A. Denisenko [1], Evgeny A. Pislyagin [1], Ekaterina A. Chingizova [1] and Shamil Sh. Afiyatullov [1,*]

1 G.B. Elyakov Pacific Institute of Bioorganic Chemistry, Far Eastern Branch of the Russian Academy of Sciences, Prospect 100-letiya Vladivostoka, 159, Vladivostok 690022, Russia; zhuravleva.oi@dvfu.ru (O.I.Z.); pibocfebras@gmail.com (A.S.A.); oleingk@mail.ru (G.K.O.); 161070@rambler.ru (Y.V.K.); prs_90@mail.ru (R.S.P.); vladenis@pidoc.dvo.ru (V.A.D.); pislyagin@hotmail.com (E.A.P.); martyyas@mail.ru (E.A.C.)

2 School of Natural Science, Far Eastern Federal University, Sukhanova St., 8, Vladivostok 690000, Russia

* Correspondence: afiyat@piboc.dvo.ru; Tel.: +7-423-231-1168

Received: 10 October 2019; Accepted: 25 October 2019; Published: 29 October 2019

Abstract: Ten new diterpene glycosides virescenosides Z_9-Z_{18} (**1–10**) together with three known analogues (**11–13**) and aglycon of virescenoside A (**14**) were isolated from the marine-derived fungus *Acremonium striatisporum* KMM 4401. These compounds were obtained by cultivating fungus on wort agar medium with the addition of potassium bromide. Structures of the isolated metabolites were established based on spectroscopic methods. The effects of some isolated glycosides and aglycons **15–18** on urease activity and regulation of Reactive Oxygen Species (ROS) and Nitric Oxide (NO) production in macrophages stimulated with lipopolysaccharide (LPC) were evaluated.

Keywords: *Acremonium striatisporum*; secondary metabolites; marine fungi; diterpene glycosides; urease activity

1. Introduction

Marine fungi are promising and prolific sources of new biologically active compounds. At the same time, glycosylated secondary metabolites of marine fungi such as ribofuranosides, containing as aglycon moieties anthraquinones [1–3], diphenyl ethers [4,5], isocoumarin [6] and naphthyl derivatives [7] are relatively rare. Recently, two steroid glycosides with β-D-mannose as sugar part were isolated from ascomycete *Dichotomomyces cejpii* [8] and new triterpene glycoside auxarthonoside bearing rare sugar N-acetyl-6-methoxy-glucosamine was described from sponge-derived fungus *Auxarthron reticulatum* [9]. Some of these glycosides exhibited cytotoxic [5], radical scavenging [3,4], and neurotropic [8] activities.

During our ongoing search for new natural compounds from marine-derived fungi, we have investigated the strain *Acremonium striatisporum* KMM 4401 associated with the holothurian *Eupentacta fraudatrix*. Twenty-one new diterpene glycosides, virescenosides have previously been isolated from this strain under cultivation on solid rice medium and wort agar medium [10–12]. Virescenosides Z_5 and Z_7 exhibited an unusual 16-chloro-15-hydroxyethyl group as their side chains in aglycones [12]. So, we attempted directed biosynthesis for the production of other halogenated compounds by culturing the fungus *Acremonium striatisporum* KMM 4401 in media containing potassium bromide. Unfortunately, we were unable to obtain glycoside derivatives with the incorporation of a bromine atom in a molecule structure. Chromatographic separation of the $CHCl_3$-EtOH extract of the culture of fungus has now led to the isolation of ten undescribed diterpene glycosides virescenosides Z_9-Z_{18} (**1-10**) (Figure 1) together with known virescenosides F (**11**) and G (**12**), lactone of virescenoside G (**13**) and aglycon of virescenoside A (**14**) (Figure S1).

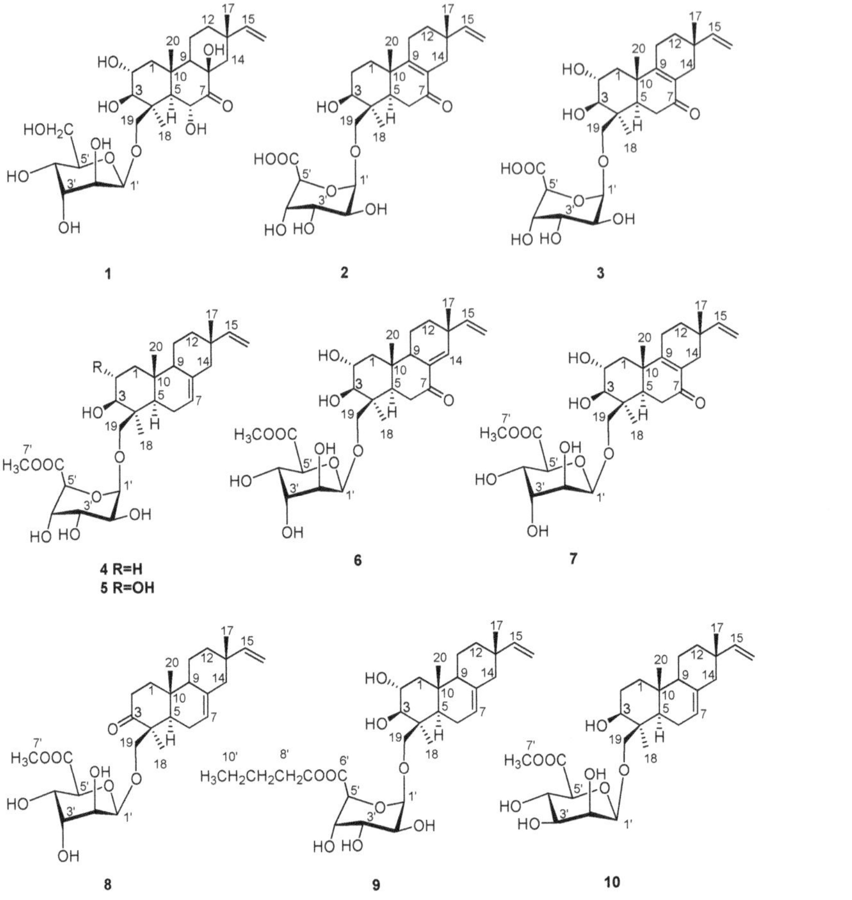

Figure 1. Chemical structures of **1–10**.

2. Results and Discussion

The $CHCl_3$-EtOH (2:1, v/v) extract of the culture of *A. striatisporum* was separated by low-pressure reversed-phase column chromatography on Teflon powder Polycrome-1 followed by Si gel flash column chromatography and then by RP HPLC to yield individual compounds **1-14** as colorless, amorphous solids.

The molecular formula of virescenoside Z_9 (**1**) was determined as $C_{26}H_{42}O_{11}$ based on the analysis of HRESIMS (*m/z* 529.2656 [M−H]$^-$, calcd for $C_{26}H_{41}O_{11}$, 529.2654) and NMR data. A close inspection of the ^1H and ^{13}C NMR data (Tables 1 and 2; Figures S3-S8) of **1** by DEPT and HSQC revealed the presence of three quaternary methyls (δ_H 0,95, 1.28, 1.81; δ_C 28.5, 17.7, 25.8), six methylenes (δ_C 18.4, 34.3, 46.9, 49.8, 64.0 and 74.0), including two oxygen-bearing, eight oxygenated methines (δ_H 3.61, 3.70, 4.28, 4.50, 4.56, 4.69, 4.74, 5.43; δ_C 84.7, 57.3, 69.1, 75.7, 72.7, 67.2, 72.3, 101.2) including one methine linked to an anomeric carbon, two tertiary (δ_H 1.93, 2.41; δ_C 60.5, 55.9), four saturated quaternary carbons (δ_C 35.8, 43.8 (2C) and 80.2), including one oxygen-bearing, one monosubstituted double bond (δ_C 108.4, 151.0) and one carbonyl or carboxyl carbon (δ_C 178.0). HMBC correlations from H_3-20 (δ_H 1.28) to C-1 (δ_C 46.9), C-5 (δ_C 55.9), C-9 (δ_C 60.5) and C-10 (δ_C 43.8), from H_3-18 (δ_H 1.81) to C-3 (δ_C 84.7), C-4 (δ_C 43.8), C-5 (δ_C 55.9) and C-19 (δ_C 74.0), from H-3 (δ_H 3.61) to C-2 (δ_C 69.1), C-4 and C-19, from H-1β (δ_H 2.34) to C-3, from H-6 (δ_H 3.70) to C-4, C-5, C-7 (δ_C 178.0) and C-8 (δ_C 80.2), from H-9 (δ_H 1.93) to C-8 and C-10 established the structures of the A and B rings and the location of hydroxy groups at C-2, C-3, C-6, C-8 and carbonyl function at C-7. The correlations observed in the COSY and HSQC spectra of **1** indicated the presence of the isolated spin system: >CH-CH_2-CH_2- (C-9-C-11-C-12). These data and HMBC correlations from H_3-17 (δ_H 0.95) to C-12 (δ_C 34.3), C-13 (δ_C 35.8), C-14 (δ_C 49.8), C-15 (δ_C 151.0) and from H-14β (δ_H 1.48) to C-8, C-9 and C-12 established the structure of the C ring in **1**.

The proton signals of a typical ABX system of a vinyl group at δ_H 6.64 (1H, dd, 10.8, 17.6 Hz), 4.96 (1H, dd, 1.8, 17.6) and 4.85 (1H, dd, 1.8, 10.8) indicated the C-15, C-16 position of this double bond [13–16]. NOE correlations (Figure 2) H_3-20 (δ_H 1.28)/H-2 (δ_H 4.28), H-6 (δ_H 3.70), H-19b (δ_H 4.96) and H-5 (δ_H 2.41)/H-3 (δ_H 3.62), H_3-18 (δ_H 1.81) indicated a *trans*-ring fusion of the A and B rings, as well as the stereochemistry of the methyl and hydroxymethyl groups at C-4, methyl group at C-10 and hydroxy groups at C-2, C-3 and C-6. NOE cross-peaks H-9 (δ_H 1.93)/H-5 and H-14β (δ_H 1.48)/H_3-20, H_3-17 (δ_H 0.95), H-6 showed the stereochemistry of the methyl group at C-13 and suggested the β-orientation of hydroxy group at C-8.

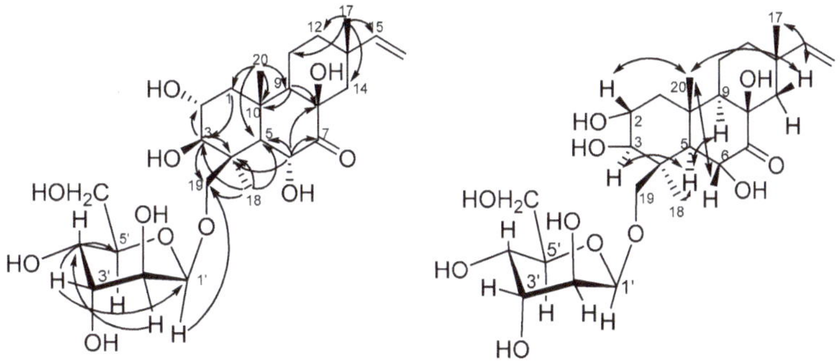

Figure 2. Key HMBC and NOESY correlations of **1**.

Interpretation of the COSY data gave rise to spin systems for monosaccharide involving one anomeric proton, four oxymethines and protons of a hydroxymethyl group. A comparison of the ^{13}C NMR spectrum of **1** with the data published for α-D-altropyranoses and β-D-altropyranoses as well as a good coincidence of carbon signals due to the glycosidic moiety with those of virescenosides O, T, W [10] together with magnitudes of ^1H-^1H spin coupling constants in ^1H NMR spectra of **1** elucidated the presence of a β-D-altropyranoside unit of $^4C_1$ form in **1**. A long-range correlation H-1′ (δ_H 5.43)/C-19 as well as the NOESY cross-peak between H-1′ and H-19a and downfield chemical shift of C-19 (δ_C 74.0) revealed a linkage between the altrose and aglycon. Thus, the structure of virescenoside Z_9 (**1**) was represented as 19-*O*-β-D-altropyranosyl-7-oxo-isopimara-15-en-2α,3β,6α,8β-tetraol.

In HRESIMS virescenoside Z_{10} (**2**) gave a quasimolecular ion at *m/z* 493.2446 $[M–H]^-$. These data, coupled with ^{13}C NMR spectral data (DEPT), established the molecular formula of **2** as $C_{26}H_{38}O_9$. ^1H and ^{13}C NMR spectra of **2** (Tables 1 and 2; Figures S9-S13) indicated the presence of a Δ^{15}-pimarene-type aglycon possessing primary alcohol on a quaternary carbon (AB system, coupling at 3.73 d, 10.2 Hz and 4.17 d, 10.2 Hz) and one secondary alcohol function at δ_C 80.0. The remaining functionality, corresponding to the carbon signals at δ 202.9 (C), 168.7 (C) and 130.3 (C), suggested the presence of the tetrasubstituted enone chromophore. The structure of the aglycon part of **2** was found by extensive NMR spectroscopy to be the same as that of virescenoside P [17].

Table 1. ^{13}C NMR data (δ in ppm) for virescenosides Z_9-Z_{18} (**1–10**).

Position	1[a]	2[b]	3[b]	4[c]	5[b]	6[c]	7[b]	8[c]	9[b]	10[d]
1	46.9, CH_2	35.9, CH_2	44.1, CH_2	40.0, CH_2	48.0, CH_2	46.6, CH_2	44.1, CH_2	40.3, CH_2	48.1, CH_2	38.5, CH_2
2	69.1, CH	29.0, CH_2	69.6, CH	29.4, CH_2	69.5, CH	69.3, CH	69.4, CH	37.3, CH_2	69.5, CH	28.6, CH_2
3	84.7, CH	80.0, CH	84.6, CH	81.5, CH	85.6, CH	84.7, CH	84.4, CH	218.4, C	85.6, CH	79.0, CH
4	43.8, C	41.3, C	44.7, C	43.8, C	44.6, C	44.9, C	44.7, C	54.1, C	44.6, C	43.0, C
5	55.9, CH	51.8, CH	51.7, CH	53.5, CH	53.2, CH	51.9 CH	51.5, CH	55.5, CH	53.2, CH	51.5 CH
6	57.3, CH	37.3, CH_2	37.6, CH_2	25.2, CH_2	25.3, CH_2	39.3, CH_2	37.8, CH_2	25.8, CH_2	25.3, CH_2	24.6, CH_2
7	178.0, C	202.9, C	202.7, C	123.1, CH	123.2, CH	203.1, C	202.8, C	123.1, CH	123.2, CH	122.8, CH
8	80.2, C	130.3, C	130.3, C	137.2, C	137.1, C	136.8, C	130.2, C	137.7, C	136.9, C	134.9, C
9	60.5, CH	168.7, C	167.8, C	54.1, CH	54.1, CH	52.6, CH	167.9, C	53.1, CH	54.1, CH	52.3, CH
10	43.8, C	43.9, C	42.2, C	36.8, C	37.9, C	38.4, C	42.2, C	37.0, C	37.9, C	35.5, C
11	18.4, CH_2	24.9, CH_2	25.0, CH_2	22.1, CH_2	22.2, CH_2	21.0, CH_2	25.0, CH_2	22.1, CH_2	22.2, CH_2	20.5, CH_2
12	34.3, CH_2	35.3, CH_2	35.2, CH_2	37.9, CH_2	37.8, CH_2	35.6, CH_2	35.2, CH_2	37.8, CH_2	37.8, CH_2	36.4, CH_2
13	35.8, C	36.0, C	36.1, C	38.4, C	38.4, C	40.4, C	36.1, C	38.4, C	38.4, C	37.1, C
14	49.8, CH_2	34.9, CH_2	34.9, CH_2	47.7, CH_2	47.6, CH_2	146.5, CH	34.9, CH_2	47.6, CH_2	47.6, CH_2	46.3, CH_2
15	151.0, CH	147.0, CH	147.2, CH	152.0, CH	151.9, CH	148.6, CH	146.9, CH	151.9, CH	151.9, CH	150.6, CH
16	108.4, CH_2	112.7, CH_2	112.7, CH_2	110.4, CH_2	110.4, CH_2	112.9, CH_2	112.7, CH_2	110.5, CH_2	110.4, CH_2	109.6, CH_2
17	28.5, CH_3	29.2, CH_3	29.1, CH_3	22.6, CH_3	22.6, CH_3	26.8, CH_3	29,2, CH_3	22.6, CH_3	22.6, CH_3	21.7, CH_3
18	25.8, CH_3	22.8, CH_3	23.7, CH_3	23.9, CH_3	24.6, CH_3	23.6, CH_3	23.7, CH_3	22.0, CH_3	24.6, CH_3	23.9, CH_3
19	74.0, CH_2	73.3, CH_2	73.8, CH_2	73.9, CH_2	74.1, CH_2	73.1, CH_2	73,6, CH_2	75.1, CH_2	73.9, CH_2	72.1, CH_2
20	17.7, CH_3	18.7, CH_3	19.7, CH_3	16.9, CH_3	17.6, CH_3	16.1, CH_3	19,7, CH_3	16.7, CH_3	17.6, CH_3	15.7, CH_3
1′	101.2, CH	103.5, CH	103.3, CH	103.7, CH	103.3, CH	102.8, CH	102.8, CH	101.9, CH	103.3, CH	103.5, CH
2′	72.7, CH	71.1, CH	71.3, CH	70.9, CH	71.2, CH	71.6 CH	71.7, CH	72.1, CH	71.3, CH	71.6 CH
3′	72.3, CH	70.8, CH	71.1, CH	70.6, CH	71.1, CH	71.8, CH	72.0, CH	72.0, CH	71.2, CH	75.1, CH
4′	67.2, CH	69.9, CH	69.7 CH	69.9, CH	69.6, CH	69.1, CH	69.0, CH	68.7, CH	69.7, CH	70.0, CH
5′	75.7, CH	76.5, CH	76.6, CH	76.6, CH	76.4, CH	76.4, CH	76.1, CH	76.0, CH	76.4, CH	77.8, CH
6′	64.0, CH_2	174.0, C	174.0, C	172.7, C	172.9, C	173.2, C	172.8, C	172.6, C	172.9, C	170.7, C
7′				53.3, CH_3	53.3, CH_3	53.4 CH_3	53.3, CH_3	53.3, CH_3	66.8, CH_2	51.8 CH_3
8′									32.3, CH_2	
9′									20.7, CH_2	
10′									14.7, CH_3	

[a] Chemical shifts were measured at 176.04 in Pyr-d_5. [b] Chemical shifts were measured at 176.04 in CD_3OD. [c] Chemical shifts were measured at 125.77 in CD_3OD. [d] Chemical shifts were measured at 125.77 in Pyr-d_5.

Table 2. ^{1}H NMR data (δ in ppm, *J* in Hz) for virescenosides Z_9-Z_{13} (**1–5**).

Position	1[a]	2[b]	3[b]	4[c]	5[b]
1	α: 1.54 t (11.5) β: 2.34 dd (4.6, 12.2)	α: 1.35 m β: 1.94 m	α: 1.23 m β: 2.17 dd (4.5, 12.8)	α: 1.22 dt (4.6, 13.5) β: 1.90 dd (3.5, 13.5)	α: 1.11 β: 2.11 dd (4.2, 12.6)
2	4.28 ddd (4.5, 9.3, 11.5)	α: 1.82 dd (3.5, 11.9) β: 1.75 dd (4.0, 13.6)	3.82 m	α: 1.74 dd (3.4, 11.8) β: 1.65 dd (3.0, 13.4)	3.76 m
3	3.61 d (9.3)	3.26 dd (4.0, 11.9)	2.99 d (9.8)	3.24 dd (4.1, 11.8)	2.98 d (9.8)
5	2.41 d (13.2)	1. 67 dd (3.6, 14.4)	1. 76 dd (3.5, 14.7)	1. 26 t (8.2)	1.34 dd (3.9, 11.4)
6	3.70 d (13.2)	α: 2.54 dd (3.6, 18.0) β: 2.64 dd (14.4, 18.0)	α: 2.56 dd (3.3, 18.2) β: 2.70 dd (14.7, 18.2)	2.03 m	2.03 m
7				5.38 brs	5.39 brs
9	1.93 t (7.5)			1.66 dd (3.9, 7.8)	1.74 m
11	α: 1.38 m β: 1.69 m	α: 2.23 m β: 2.27 m	α: 2.26 m β: 2.32 m	α: 1.38 m β: 1.58 m	α: 1.41 m β: 1.61 m
12	α: 1.87 β: 1.36	α: 1.35 m β: 1.62 m	α: 1.64 m β: 1.36 m	α: 1.37 m β: 1.48 dt (2.7, 8.9)	α: 1.50 dd (2.8, 12.1) β: 1.39 td (2.8, 11.5)
14	α: 2.37 d (14.0) β: 1.48 d (14.0)	α: 2.30 d (17.5) β: 1.93 d (17.5)	α: 2.31 m β: 1.94 dt (2.6, 17.9)	α: 1.97 brd (14.1) β: 1.91 dd (2.6, 14.1)	α: 1.99 m β: 1.92 dd (2.6, 14.1)
15	6.64 dd (10.8, 17.6)	5.70 dd (10.6, 17.5)	5.70 dd (10.8, 17.5)	5.80 dd (10.7, 17.5)	5.81 dd (10.8, 17.6)
16	a: 4.85 dd (1.8, 10.8) b: 4.96 dd (1.8, 17.6)	a: 4.82 dd (1.4, 17.5) b: 4.93 dd (1.4, 10.6)	a: 4.82 dd (1.5, 17.6) b: 4.93 dd (1.5, 10.8)	a: 4.84 dd (1.3, 10.7) b: 4.92 dd (1.3, 17.5)	a: 4.85 dd (1.4, 10.8) b: 4.93 dd (1.4, 17.6)
17	0.95 s	1.00 s	1.01 s	0.86 s	0.86 s
18	1.81 s	1.13 s	1.16 s	1.10 s	1.11 s
19	a: 4.23 d (9.9) b: 4.98 d (9.9)	a: 3.73 d (10.2) b: 4.17 d (10.2)	a: 3.67 d (10.4) b: 4.14 d (10.4)	a: 3.83 d (10.2) b: 4.04 d (10.2)	a: 3.72 d (10.3) b: 4.03 d (10.3)
20	1.28 s	1.14 s	1.21 s	0.87 s	0.95 s
1	5.43 d (1.2)	4.84 d (2.9)	4.82 d (2.5)	4.85 d (2.5)	4.85 d (2.7)
2′	4.56 dd (1.2, 3.9)	3.77 dd (2.7, 7.8)	3.77 dd (2.5, 7.4)	3.77 dd (2.8, 7.9)	3.77 m
3′	4.74 t (3.7)	3.93 dd (2.9, 7.7)	3.94 dd (3.0, 7.4)	3.89 dd (3.0, 7.9)	3.90 dd (3.3, 7.3)
4′	4.50 m	4.23 t (4.8)	4.20 dd (3.0, 5.7)	4.26 dd (3.0, 4.9)	4.23 dd (3.3, 5.6)
5′	4.69 d (3.2, 12.2)	4.24 d (4.8)	4.22 d (5.7)	4.28 d (4.9)	4.28 d (5.6)
6′	a: 4.40 dd (6.5, 12.3) b: 4.51 m				
7′				3.78 s	3.78 s

[a] Chemical shifts were measured at 700.13 in Pyr-d_5. [b] Chemical shifts were measured at 700.13 in CD_3OD. [c] Chemical shifts were measured at 500.13 in CD_3OD.

The HRESIMS of virescenosides Z_{11} (**3**) showed the quasimolecular ion at *m/z* 509.2408 $[M-H]^-$. These data, coupled with ^{13}C NMR spectral data (DEPT), established the molecular formula of **3** as $C_{26}H_{38}O_{10}$. The structure of the aglycon moiety of **3** was found by extensive NMR spectroscopy (^1H, ^{13}C, HSQC, HMBC and NOESY) (Tables 1 and 2; Figures S14-S18) to be the same as those of virescenoside M [18].

The ^{13}C and ^1H NMR spectra of the sugar moieties of virescenoside Z_{10} (**2**) and Z_{11} (**3**) showed a close similarity of all proton and carbon chemical shifts with those of virescenosides Z_7 and Z_8 [12]. The 7.7-,7.4-Hz splitting between H-2 and H-3 indicated that both were axial, whereas the 4.8-, 5.7-Hz splitting between H-4 and H-5 showed that these protons in equatorial position. These data and HMBC correlations between anomeric protons and C-5′-methine groups and between H-5′ (δ_H 4.24, 4.22) and C-6′ (δ_C 174.0) suggested the presence of a *β*-altruronopyranoside unit of $^1C_4$ conformation in **2** and **3**. The long-range correlations H-1′ (δ_H 4.84, 4.82)/C-19 as well as the NOESY cross-peak between H-1′ and H-19a and downfield chemical shifts of C-19 (δ_C 73.3, 73.8) indicated that sugar moieties were linked at C-19. Earlier in result of reduction of the sum of virescenosides Z_4-Z_8 with $LiAlH_4$ and the acid hydrolysis of obtained products was isolated D-altrose as the only sugar that was identified by GLC of the corresponding acetylated (+)- and (-)-2-octyl glycosides using authentic samples prepared from D-altrose [12]. Thus, the structure of virescenoside Z_{10} (**2**) was determined as 19-*O*-β-D-altruronopyranosyl-7-oxo-isopimara-8(9),15-dien-3β-ol, and the structure of virescenoside Z_{11} (**3**) was established as 19-*O*-β-D-altruronopyranosyl-7-oxo-isopimara-8(9),15-dien-2α,3β-diol.

The HRESIMS of virescenosides Z_{12} (**4**) and Z_{13} (**5**) showed the quasimolecular ions at *m/z* 517.2770 $[M + Na]^+$ and *m/z* 533.2718 $[M + Na]^+$, respectively. These data, coupled with ^{13}C NMR spectral data (DEPT), established the molecular formula of **4** and **5** as $C_{27}H_{42}O_8$ and $C_{27}H_{42}O_9$, respectively. A close inspection of the ^1H and ^{13}C NMR data of **4** (Tables 1 and 2; Figures S19-S23) revealed that virescenoside Z_{12} (**4**) was structurally identical to virescenosides B [13] and G [19] (See Extraction and Isolation) with respect to the aglycon. The structure of the aglycon moiety of **5** was found by extensive NMR spectroscopy (Tables 1 and 2; Figures S24-S28) to be the same as that of virescenosides A [13,20] and F [19] (See Extraction and Isolation).

The NMR spectra of glycosides **4** and **5** indicated that both compounds contained closed carbohydrate moieties (Tables 1 and 2). Initial examination of the 1-D proton and one bond correlation NMR data suggested the presence of one sugar (anomeric signals at δ_H 4.85, δ_C 103.7 for **4** and δ_H 4.85, δ_C 103.3 for **5**). The ^1H and ^{13}C NMR spectra of the sugar parts of **4** and **5** indicated the presence of the methoxy groups (both, δ_H 3.78, δ_C 53.3). HMBC correlations from anomeric protons to C5′-methine groups and from H-5′ (δ_H 4.28) to C-6′ (δ_C 172.7, 172.9) and from H_3-7′ (δ_H 3.78) to C-6′ together with magnitudes of ^1H-^1H spin coupling constants suggested the presence of the methyl ester of a *β*-altruronopyranoside unit of $^1C_4$ form in **4** and **5**. A long-range correlations H-1′ (δ_H 4.85)/C-19 (δ_C 73.9, 74.1) as well as the NOESY cross-peaks between H-1′ and H-19a (δ_H 3.83, 3.72) and downfield chemical shifts of C-19 indicated that sugar moieties were linked at C-19. Thus, the structure of virescenoside Z_{12} (**4**) was determined as 19-*O*-[(methyl-β-D-altruronopyranosyl)-uronat]-isopimara-7,15-dien-3β-ol, and the structure of virescenoside Z_{13} (**5**) was established as 19-*O*-[(methyl-β-D-altruronopyranosyl)-uronat]-isopimara-7,15-dien-2α,3β-diol.

The NMR data (Tables 1 and 3) of virescenosides Z_{14} (**6**), Z_{15} (**7**) and Z_{16} (**8**) suggested the presence of one sugar (anomeric signals at δ_H 4.78, δ_C 102.8, δ_H 4.79, δ_C 102.8, δ_H 4.75, δ_C 101.9). The ^1H and ^{13}C NMR spectra of the sugar moieties of **6**, **7** and **8** showed a close similarity of all proton and carbon chemical shifts and proton multiplicities. These data and HMBC correlations from anomeric protons to C-5′ methine groups and from H-5′ (δ_H 4.24, 4.24, 4.23) to C-6′ (δ_C 173.2, 172.8, 172.6) and from H_3-7′ (δ_H 3.76, 3.76, 3.77) to C-6′ suggested the presence of the methyl ester of a *β*-altruronopyranoside unit in **6**, **7** and **8**. The 7.0-, 7.3-, 8.0-Hz splitting between H-4 and H-5 indicated that both were axial and conformation of sugar parts in **6**, **7** and **8** is $^4C_1$.

Table 3. ^1H NMR data (δ in ppm, J in Hz) for virescenosides Z_{14}-Z_{18} (**6–10**)

Position	6[c]	7[b]	8[c]	9[b]	10[d]
1	α: 1.22 m β: 2.07 dd (4.3, 12.7)	α: 1.23 m β: 2.18 dd (4.5, 12.8)	α: 1.50 m β: 2.19 m	α: 1.11 m β: 2.11 dd (4.2, 12.5)	α: 1.15 dt β: 1.78 brd (3.9, 13.1)
2	3.79 m	3.80 dd (9.8, 13.9)	α: 2.84 dt (5.4, 14.2) β: 2.23 m	3.76 m	α: 1.85 m β: 1.92 m
3	3.04 d (9.9)	3.00 d (9.8)		2.98 d (9.8)	3.55 dd (4.0, 11.9)
5	1.73 dd (5.0, 13.8)	1. 75 dd (3.4, 14.7)	1.63 dd (4.1, 12.3)	1.34 dd (4.5, 11.8)	1.27 m
6	α: 2.59 dd (5.0, 19.0)	α: 2.53 dd (3.4, 18.2) β: 2.80 dd (14.7, 18.2)	α: 2.04 m β: 2.11 m	α: 2.01 m β: 2.06 m	α: 2.06 m β: 2.40 m
7			5.41 brs	5.38 m	5.30 m
9	2.13 m		1.76 m	1.74 m	1.60 m
11	α: 1.79 m β: 1.54 m	α: 2.26 m β: 2.31 m	α: 1.64 m β: 1.47 m	α: 1.41 m β: 1.61 m	α: 1.46 m β: 1.32 m
12	α: 1.54 m β: 1.67 m	α: 1.64 m β: 1.36 m	α: 1.51 β: 1.44	α: 1.50 m β: 1.39 m	α: 1.32 m β: 1.45 m
14	6.68 t (2.1)	α: 2.32 m β: 1.95 d (17.8)	α: 2.00 m β: 1.94 d (2.6, 14.0)	α: 1.99 m β: 1.92 dd (2.6, 14.1)	α: 2.03 brd (14.0) β: 1.94 brd (14.0)
15	5.83 dd (10.7, 17.5)	5.71 dd (10.8, 17.5)	5.81 dd (10.7, 17.5)	5.81 dd (10.8, 17.4)	5.87 dd (10.6, 17.4)
16	5.00 m	a: 4.83 dd (1.2, 17.5) b: 4.93 dd (1.2, 10.8)	a: 4.86 dd (1.4, 10.7) b: 4.94 dd (1.4, 17.5)	a: 4.85 dd (1.4, 10.8) b: 4.93 dd (1.4, 17.4)	a: 4.95 d (10.6) b: 5.02 d (17.4)
17	1.12 s	1,01 s	0.89 s	0.86 s	0.90 s
18	1.13 s	1.15 s	1.11 s	1.11 s	1.41 s
19	a: 3.68 d (10.3) b: 4.09 d (10.3)	a: 3.65 d (10,4) b: 4.11 d (10,4)	a: 3.90 d (9.8) b: 3.96 d (9.8)	a: 3.71 d (10.5) b: 4.04 d (1053)	a: 4.26 d (10.3) b: 4.59 d (10.3)
20	0.95 s	1.22 s	1.17 s	0.95 s	0.93 s
1′	4.78 brs	4.79 d (2.0)	4.75 d (1.9)	4.84 d (2.1)	4.97 brs
2′	3.76 m	3.76 m	3.66 dd (1.9, 5.6)	3.77 dd (2.5, 7.4)	4.55 d (3.2)
3′	3.91 dd (3.3, 6.3)	3.92 dd (3.3, 6.0)	3.89 dd (3.0, 5.6)	3.92 dd (3.0, 7.4)	4.14 dd (3.3, 9.4)
4′	4.15 dd (3.3, 7.0)	4.14 dd (3.3, 7.3)	4.08 dd (3.2, 8.0)	4.22 m	4.87 t (9.3)
5′	4.24 d (7.0)	4.24 d (7.3)	4.23 d (8.0)	4.25 d (5.6)	4.40 d (9.3)
7′	3.76 s	3.76 s	3.77 s	a: 4.15 dt (6.6, 10.7) b: 4.19 m	3.64 s
8′				a,b: 1.68 m	
9′				a,b: 1.45 m	
10′				0.96 t (7.5)	

[a] Chemical shifts were measured at 700.13 in Pyr-d_5. [b] Chemical shifts were measured at 700.13 in CD_3OD. [c] Chemical shifts were measured at 500.13 in CD_3OD. [d] Chemical shifts were measured at 500.13 in Pyr-d_5.

The HRESIMS of virescenosides Z_{14} (**6**) showed the quasimolecular ion at *m/z* 547.2508 [M + Na]$^+$. These data, coupled with ^{13}C NMR spectral data (DEPT), established the molecular formula of **6** as $C_{27}H_{40}O_{10}$. The structure of the aglycon moiety of **6** was found by extensive NMR spectroscopy (^1H, ^{13}C, HSQC, HMBC and NOESY) (Tables 1 and 3; Figures S29-S33) to be the same as those of virescenoside V [21].

The molecular formula of virescenoside Z_{15} (**7**) was determined as $C_{27}H_{40}O_{10}$ based on the analysis of HRESIMS (*m/z* 523.2550 [M-H]$^-$, calcd for $C_{27}H_{39}O_{10}$, 523.2549) and NMR data. The ^1H and ^{13}C NMR data (Tables 1 and 3; Figures S34-S38) observed for the aglycon part of **7** closely resembled those obtained for virescenoside Z_{10} (**2**) with the exception of the C-1-C-4 carbon and proton signals of ring A. The HMBC correlations from H-5 (δ_H 1.75) to C-3 (δ_C 84.4), H-3 (δ_H 3.00) and from H_2-1 (δ_H 1.23, 2.18) to C-2 (δ_C 69.2) and downfield chemical shifts of C-2 placed an additional hydroxy group at C-2 of ring A. The relative stereochemistry of protons on C-2 and C-3 was defined based on the ^1H-^1H coupling constant (J=9.8) and assigned as axial. Previously, a similar aglycon has been described for virescenoside M [10].

The HRESIMS of virescenoside Z_{16} (**8**) showed the quasimolecular at *m/z* 515.2617 [M + Na]$^+$. These data, coupled with ^{13}C NMR spectral data (DEPT), established the molecular formula of **8** as $C_{27}H_{40}O_8$ (Tables 1 and 3). The structure of the aglycon moiety of **8** was found by 2D NMR experiments (Figures S39-S43) to be the same as that of virescenoside Z_4 [12].

The attachment of a carbohydrate chains at C-19 of aglycon moieties of **6**, **7** and **8** was confirmed by cross-peaks H-1′ (δ_H 4.78, 4.79, 4.75)/H-19a (δ_H 3.68, 3.65, 3.90) and H-1′/C-19 (δ_C 73.1, 73.6, 75.1) in the NOESY and HMBC spectra, respectively. From all these data, virescenoside Z_{14} (**6**) was structurally identified as 19-*O*-[(methyl-β-D-altruronopyranosyl)-uronat]-7-oxo-isopimara-8(14),15-dien-2α,3β-diol, virescenoside Z_{15} (**7**) as 19-*O*-[(methyl-β-D-altruronopyranosyl)-uronat]-7-oxo-isopimara-8(9),15-dien-2α,3β-diol and virescenoside Z_{16} (**8**) as 19-*O*-[(methyl-β-D-altruronopyranosyl)-uronat]-3-oxo-isopimara-7,15-dien.

The HRESIMS of virescenoside Z_{17} (**9**) showed the quasimolecular ion at *m/z* 575.3194 [M + Na]$^+$. These data, coupled with ^{13}C NMR spectral data (DEPT), established the molecular formula of **9** as $C_{30}H_{48}O_9$. The ^1H and ^{13}C NMR data observed for aglycon and sugar (C-1′-C-6′) parts of **9** (Tables 1 and 3; Figures S44-S48) matched those reported for virescenoside Z_{13} (**5**). The correlations observed in the COSY and HSQC spectra of **9** indicated the presence of the isolated spin system: $-CH_2-CH_2-CH_2-CH_3$ (C-7′-C-10′). These data and HMBC correlations from H_3-10′ (δ_H 0.96) to C-8′ (δ_C 32.3), C-9′ (δ_C 20.7) and from Ha-7′ (δ_H 4.15) to C-6′ (δ_C 172.9), C-8′ and C-9′ suggested the presence of the butyl ester of a *β*-altruronopyranoside unit of $^1C_4$ form in **9**. On the basis of all the data above, the structure of virescenosides Z_{17} (**9**) was established as 19-*O*-[(butyl-β-D-altruronopyranosyl)-uronat]-isopimara-7,15-dien-2α,3β-diol.

The HRESIMS of virescenoside Z_{18} (**10**) showed the quasimolecular at *m/z* 517.2773 [M + Na]$^+$. These data, coupled with ^{13}C NMR spectral data (DEPT), established the molecular formula of **10** as $C_{27}H_{42}O_8$. The ^1H and ^{13}C NMR data observed for the aglycon part of **10** (Tables 1 and 3; Figures S49-S54) matched those reported for virescenoside Q [17]. Initial examination of the 1-D proton and one bond correlation NMR data suggested the presence of one sugar (anomeric signal at δ_H 4.97, δ_C 103.5). The ^1H and ^{13}C NMR spectra of the sugar part of **10** indicated the presence of the methoxycarbonyl group (δ_H 3.64, δ_C 51.8, 170.7). A comparison of the ^{13}C NMR spectrum with the data published for *α*- and *β*-D-mannopyranoses as well as a good coincidence of carbon signals C-1′-C-4′ with those of virescenoside Q together with magnitudes of ^1H-^1H spin coupling constants in ^1H NMR spectrum of **10** elucidated the presence of *β*-D-mannouronopyranoside unit of $^4C_1$ form in **10** [17,22,23]. A long-range correlation H-1′ (δ_H 4.97)/C-19 (δ_C 72.1) as well as the NOESY cross-peak between H-1′ and H-19a (δ_H 4.26) and downfield chemical shifts of C-19 indicated that sugar moiety was linked at C-19. Thus, the structure of virescenoside Z_{18} (**10**) was determined as 19-*O*-[(methyl-β-D-mannopyrananosyl)-uronat]-isopimara-7,15-dien-3β-ol.

Since methanol is used in the isolation procedure of virescenosides, it is possible that the methyl esters of the sugar units may be obtained during the course of isolation. Therefore, we separated the part of subfraction II by RP-HPLC using acetonitrile instead of methanol and obtain virescenosides Z_{12} (**4**) and Z_{13} (**5**) which were characterized by ^{1}H and ^{13}C NMR spectra. Furthermore, we observed compounds **4-8** and **10** in subfraction II by HPLC-MS method (See Supplementary Figure S2).

The structures of known compounds virescenosides F (**11**) and G (**12**), lactone of virescenoside G (**13**) [19] and aglycon of virescenoside A (**14**) [13] (See Supplementary Figure S1) were determined based on HRESIMS and NMR data and comparison with literature. The aglycons of virescenosides B (**15, 16**), C (**17**) and M (**18**) (See Supplementary Figure S1, Experimental Section) were prepared as a result of acid hydrolysis of the corresponding glycosides for examination of their biological activity.

Next, we investigated the effects of some isolated compounds and aglycones **15-18** on urease activity and regulation of ROS and NO production in macrophages stimulated with lipopolysaccharide (LPS).

The development of urease inhibitors, usually considered as antiulcer agents, carries a significant interest for medicinal chemists. Urease is an enzyme that is clinically used as diagnostic to determine the presence of pathogens in the gastrointestinal and urinary tracts. It has been described that the bacterial urease causes many clinically harmful infections, like stomach cancer, infectious stones and peptic ulcer formation in human and animal health [24]. Urease is also involved in the pathogenesis of hepatic coma, urolithiasis, urinary catheter encrustation and oral cavity infections by hydrolyzing the salivary urea [25].

Aglycons **14** and **15** inhibit urease activity with an IC_{50} of 138.8 and 125.0 μM, respectively. Thiourea used as positive control inhibited urease activity with IC_{50} of 23.0 μM.

Compounds **1, 2, 5, 15-18** at a concentration of 10 μM induced a significant down-regulation of ROS production in macrophages stimulated with lipopolysaccharide (LPS) (Figure 3). Virescenoside Z_{10} (**2**) decreased the ROS content in macrophages by 45%.

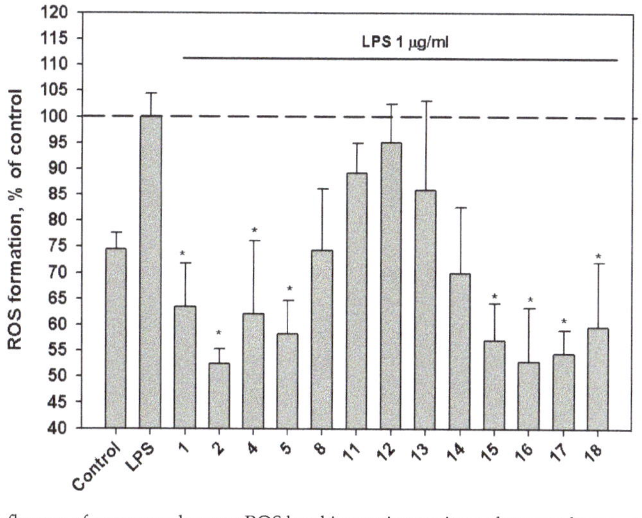

Figure 3. Influence of compounds upon ROS level in murine peritoneal macrophages, co-incubated with LPS from *E. coli*. The compounds were tested at a concentration of 10 μM. Time of cell incubation with compounds was 1 h at 37 °C. * $p < 0.05$.

Compounds **2, 5, 16** and **17** induced a moderate down-regulation of NO production in LPS-stimulated macrophages at concentration of 1 μM (Figure 4).

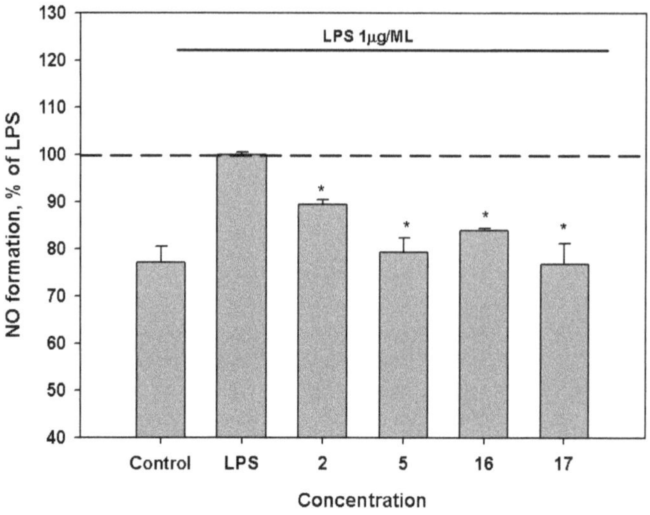

Figure 4. Influence of compounds upon RNS level in murine peritoneal macrophages, co-incubated with LPS from *E. coli*. The compounds were tested at a concentration of 1 μM. Time of cell incubation with compounds was 1 h at 37 °C. * $p < 0.05$.

3. Materials and Methods

3.1. General Experimental Procedures

Optical rotations were measured on a Perkin-Elmer 343 polarimeter (Perkin Elmer, Waltham, MA, USA). UV spectra were recorded on a Shimadzu UV-1601PC spectrometer (Shimadzu Corporation, Kyoto, Japan) in methanol. NMR spectra were recorded in CD_3OD, $CDCl_3$, DMSO-d_6 and C_5D_5N on a Bruker DPX-500 (Bruker BioSpin GmbH, Rheinstetten, Germany) and a Bruker DRX-700 (Bruker BioSpin GmbH, Rheinstetten, Germany) spectrometer, using TMS as an internal standard. The Bruker Impact II Q-TOF mass spectrometer (Bruker Daltonics, Bremen, Germany) was used to record the MS and MS/MS spectra within m/z range 50–1500. The capillary voltage was set to 1300 V, and the drying gas was heated to 150 °C at the flow rate 3 L/min. Collision-induced dissociation (CID) product ion mass spectra were obtained using nitrogen as the collision gas. The instrument was operated using the program otofControl (ver. 4.0, Bruker Daltonics, Bremen, Germany) and the data were analyzed using the DataAnalysis Software (ver. 4.3, Bruker Daltonics, Bremen, Germany).

Low-pressure liquid column chromatography was performed using silica gel (50/100 μm, Imid, Russia) and Polychrome-1 (powder Teflon, Biolar, Latvia). Plates precoated with silica gel (5–17 μm, 4.5 × 6.0 cm, Imid) and silica gel 60 RP-18 $F_{254}S$ (20 × 20 cm, Merck KGaA, Germany) were used for thin-layer chromatography. Preparative HPLC was carried out on a Agilent 1100 chromatography (Agilent Technologies, USA) using a YMC ODS-AM (YMC Co., Ishikawa, Japan) (5 μm, 10 × 250 mm) and YMC ODS-A (YMC Co., Ishikawa, Japan) (5 μm, 4.6 × 250 mm) columns with a Agilent 1100 refractometer (Agilent Technologies, USA).

3.2. Cultivation of Fungus

The fungus was grown stationary at 22 °C for 14 days on 6 flasks (1 L) (medium: wort-200 mL, sodium tartrate-0.05 g/L, agar-20 g/L, potassium bromide-30 g/L, seawater-800 mL).

3.3. Extraction and Isolation

At the end of the incubation period, the mycelium and medium were homogenized and extracted three times with a mixture of $CHCl_3$–EtOH (2:1, v/v, 2.5 L). The combined extracts (4.5 g) were concentrated to dryness and separated by low pressure RP CC (the column 20 × 8 cm) on Polychrome-1 Teflon powder in H_2O and 50% EtOH. After elution of inorganic salts and highly polar compounds by H_2O, 50% EtOH was used to obtain the fraction of amphiphilic compounds, including the virescenosides. After evaporation of the solvent, the residual material (2.6 g) was subjected to Si gel flash CC (7 × 13 cm) chromatography with a solvent gradient system of increasing polarity from 10 to 60% EtOH in $CHCl_3$ (total volume 8 L). Fractions of 20 mL were collected and combined by TLC examination to obtain two subfractions. Subfraction I ($CHCl_3$–EtOH 5:1, 3:1, 180 mg) was purified and separated by RP HPLC on a YMC ODS-A column eluting with MeOH–H_2O-TFA (85:15:0.1) to yield **8** (2.4 mg), **9** (3.6 mg), **13** (2.4 mg) and **14** (4.0 mg). Subfraction II ($CHCl_3$–EtOH 2:1, 840 mg) was purified by RP HPLC on a YMC ODS-AM column eluting at first with MeOH–H_2O-TFA (80:20:0.1) and then with MeOH–H_2O-TFA (70:30:0.1) to yield **1** (2.5 mg), **2** (2.5 mg), **3** (7.5 mg), **4** (15.5 mg), **5** (71 mg), **6** (1.4 mg), **7** (6.6 mg) **10** (1.4 mg), **11** (98 mg) and **12** (63 mg).

The part of subfraction II (35 mg) was purified by RP HPLC on a YMC ODS-A column eluting with CH_3CN–H_2O-TFA (50:50:0.1) to yield **4** (1.1 mg), **5** (4.5 mg), **11** (6 mg) and **12** (2 mg).

3.4. Spectral Data

Virescenoside Z_9 (**1**): amorphous solids; $[\alpha]_D^{20}$ +1.5 (*c* 0.15, MeOH); ^1H and ^{13}C NMR data, see Tables 1 and 2, Supplementary Figures S3–S8; HRESIMS *m/z* 553.2618 $[M + Na]^+$ (calcd. for $C_{26}H_{42}O_{11}Na$, 553.2619, Δ + 0.2 ppm).

Virescenoside Z_{10} (**2**): amorphous solids; $[\alpha]_D^{20}$ +10.0 (*c* 0.09, MeOH); UV (MeOH) λ_{max} (log ε) 248 (3.91) nm; ^1H and ^{13}C NMR data, see Tables 1 and 2, Supplementary Figures S9–S13; HRESIMS *m/z* 493.2446 $[M–H]^-$ (calcd. for $C_{26}H_{37}O_9$, 493.2443, Δ–0.6 ppm).

Virescenoside Z_{11} (**3**): amorphous solids; $[\alpha]_D^{20}$ + 7.5 (*c* 0.10, MeOH); UV (MeOH) λ_{max} (log ε) 248 (3.64) nm; ^1H and ^{13}C NMR data, see Tables 1 and 2, Supplementary Figures S14–S18; HRESIMS *m/z* 509.2403 $[M–H]^-$ (calcd. for $C_{26}H_{37}O_{10}$, 509.2392, Δ–2.0 ppm).

Virescenoside Z_{12} (**4**): amorphous solids; $[\alpha]_D^{20}$ –50.0 (*c* 0.10, MeOH); ^1H and ^{13}C NMR data, see Tables 1 and 2, Supplementary Figures S19–S23; HRESIMS *m/z* 517.2770 $[M + Na]^+$ (calcd. for $C_{27}H_{42}O_8Na$, 517.2772, Δ + 0.4 ppm).

Virescenoside Z_{13} (**5**): amorphous solids; $[\alpha]_D^{20}$ –69.2 (*c* 0.13, MeOH); ^1H and ^{13}C NMR data, see Tables 1 and 2, Supplementary Figures S24–S28; HRESIMS *m/z* 533.2718 $[M + Na]^+$ (calcd. for $C_{27}H_{42}O_9Na$, 533.2721, Δ + 0.6 ppm).

Virescenoside Z_{14} (**6**): amorphous solids; $[\alpha]_D^{20}$ –44.0 (*c* 0.10, MeOH); UV (MeOH) λ_{max} (log ε) 249 (3.81) nm; ^1H and ^{13}C NMR data, see Tables 1 and 3, Supplementary Figures S29–S33; HRESIMS *m/z* 547.2508 $[M + Na]^+$ (calcd. for $C_{27}H_{40}O_{10}Na$, 547.2514, Δ +1.0 ppm).

Virescenoside Z_{15} (**7**): amorphous solids; $[\alpha]_D^{20}$ + 17.3 (*c* 0.15, MeOH); UV (MeOH) λ_{max} (log ε) 248 (3.99) nm; ^1H and ^{13}C NMR data, see Tables 1 and 3, Supplementary Figures S34–S38; HRESIMS *m/z* 547.2515 $[M + Na]^+$ (calcd. for $C_{27}H_{40}O_{10}Na$, 547.2514, Δ –0.2 ppm).

Virescenoside Z_{16} (**8**): amorphous solids; $[\alpha]_D^{20}$ –78.0 (*c* 0.05, MeOH); ^1H and ^{13}C NMR data, see Tables 1 and 3, Supplementary Figures S39–S43; HRESIMS *m/z* 515.2617 $[M + Na]^+$ (calcd. for $C_{27}H_{40}O_8Na$, 515.2615, Δ–0.4 ppm).

Virescenoside Z_{17} (**9**): amorphous solids; $[\alpha]_D^{20}$ –60.0 (*c* 0.10, MeOH); ^1H and ^{13}C NMR data, see Tables 1 and 3, Supplementary Figures S44–S48; HRESIMS *m/z* 575.3194 $[M + Na]^+$ (calcd. for $C_{29}H_{48}O_9Na$, 575.3191, Δ-0.5 ppm), *m/z* 551.3229 $[M-H]^-$ calcd. for $C_{29}H_{47}O_9$, 551.3226, Δ -0.6 ppm).

Virescenoside Z_{18} (**10**): amorphous solids; $[\alpha]_D^{20}$ -32.5 (*c* 0.12, MeOH); ^1H and ^{13}C NMR data, see Tables 1 and 3, Supplementary Figures S49–S54; HRESIMS *m/z* 517.2773 $[M + Na]^+$ (calcd. for $C_{27}H_{42}O_8Na$, 517.2772, Δ-0.3 ppm).

Virescenoside F (**11**): amorphous solids; ^{1}H NMR (700 MHz, CD_3OD) δ: 5.80 (1H, dd, J = 10.8, 17.4 Hz, H-15), 5.38 (1H, m, H-7), 4.92 (1H, dd, J = 1.4, 17.4 Hz, H-16b), 4.86 (1H, d, J = 3.0 Hz, H-1′), 4.85 (1H, dd, J = 1.4, 10.8 Hz, H-16a), 4.27 (1H, t, J = 3.7 Hz, H-4′),4.24 (1H, d, J = 4.3 Hz, H-5′), 4.12 (1H, d, J = 9.8 Hz, H-19b), 3.93 (1H, dd, J = 3.3, 8.3 Hz, H-3′), 3.78 (1H, dd, J = 3.0, 8.0 Hz, H-2′), 3.76 (1H, m, H-2), 3.73 (1H, d, J = 9.8 Hz, H-19a), 2.98 (1H, d, J = 9.8 Hz, H-3), 2.11 (1H, dd, J = 4.2, 12.5 Hz, H-1β), 2.04 (1H, m, H2-6), 1.99 (1H, m, H-14α), 1.92 (1H, dd, J = 2.8, 14.4 Hz, H-14β), 1.73 (1H, m, H-9), 1.60 (1H, m, H-11α), 1.50 (1H, m, H-12α), 1.39 (1H, m, H-12β), 1.39 (1H, m, H-11β), 1.34 (1H, dd, J = 5.8, 10.7 Hz, H-5), 1.14 (3H, s, Me-18), 1.11 (1H, d, J = 12.3 Hz, H-1α), 0.92 (3H, s, Me-20), 0.86 (3H, s, Me-17). ^{13}C NMR (176 MHz, CD_3OD) δ: 174.0 (C-6′), 151.9 (C-15), 136.9 (C-8), 123.2 (C-7), 110.4 (C-16), 103.5 (C-1′), 85.8 (C-3), 76.6 (C-5′), 74.3 (C-19), 70.9 (C-2′), 70.4 (C-3′), 70.2 (C-4′), 69.7 (C-2), 54.1 (C-9), 53.4 (C-5), 48.0 (C-1), 47.6 (C-14), 44.4 (C-4), 38.4 (C-13), 37.9 (C-10), 37.9 (C-12), 25.1 (C-6), 24.2 (C-18), 22.6 (C-17), 22.1 (C-11), 17.7 (C-20); Supplementary Figures S55–S59; HRESIMS m/z 519.2563 $[M + Na]^+$ (calcd. for $C_{26}H_{40}O_9Na$, 519.2565, Δ + 0.3 ppm).

Virescenoside G (**12**): amorphous solids; ^{1}H NMR (700 MHz, CD_3OD) δ: 5.80 (1H, dd, J = 10.8, 17.5 Hz, H-15), 5.37 (1H, m, H-7), 4.92 (1H, dd, J = 1.6, 17.5 Hz, H-16b), 4.86 (1H, d, J = 3.3 Hz, H-1′), 4.84 (1H, dd, J = 1.6, 10.8 Hz, H-16a), 4.30 (1H, t, J = 3.5 Hz, H-4′),4.24 (1H, d, J = 3.7 Hz, H-5′), 4.10 (1H, d, J = 9.9 Hz, H-19b), 3.93 (1H, dd, J = 3.3, 8.7 Hz, H-3′), 3.84 (1H, d, J = 9.9 Hz, H-19a), 3.78 (1H, dd, J = 3.3, 8.7 Hz, H-2′), 3.24 (1H, dd, J = 4.0, 11.8 Hz, H-3), 2.07 (1H, m, H-6α), 2.01 (1H, m, H-6β), 1.97 (1H, m, H-14α), 1.91 (1H, dd, J = 2.7, 14.0 Hz, H-14β), 1.90 (1H, dd, J = 3.4, 13.4 Hz, H-1β), 1.74 (1H, dd, J = 3.5, 11.8 Hz, H-2β),1.68 (1H, dd, J = 3.5, 7.5 Hz, H-2α),1.66 (1H, dd, J = 3.7, 7.7 Hz, H-9), 1.57 (1H, m, H-11α), 1.47 (1H, td, J = 2.9, 9.1 Hz, H-12β), 1.37 (1H, m, H-12α), 1.38 (1H, m, H-11β), 1.26 (1H, dd, J = 4.5, 12.1 Hz, H-5), 1.23 (1H, m, H-1α), 1.12 (3H, s, Me-18), 0.86 (3H, s, Me-17), 0.85 (3H, s, Me-20). ^{13}C NMR (176 MHz, CD_3OD) δ: 173.7 (C-6′), 152.0 (C-15), 137.1 (C-8), 123.2 (C-7), 110.4 (C-16), 103.9 (C-1′), 81.7 (C-3), 76.7 (C-5′), 73.9 (C-19), 70.6 (C-4′), 70.4 (C-3′), 70.1 (C-2′), 54.1 (C-9), 53.7 (C-5), 47.7 (C-14), 43.7 (C-4), 40.0 (C-1), 38.4 (C-13), 37.9 (C-12), 36.8 (C-10), 29.4 (C-2), 25.0 (C-6), 23.6 (C-18), 22.6 (C-17), 22.1 (C-11), 17.0 (C-20); Supplementary Figures S60–S64; HRESIMS m/z 503.2617 $[M + Na]^+$ (calcd. for $C_{26}H_{40}O_8Na$, 503.2615, Δ–0.3 ppm).

Lactone of virescenoside G (**13**): amorphous solids; ^{1}H NMR (700 MHz, DMSO-d_6) δ: 5.80 (1H, dd, *J* = 10.8, 17.6 Hz, H-15), 5.66 (1H, d, *J* = 7.8 Hz, 5′-OH), 5.56 (1H, d, *J* = 5.6 Hz, 2′-OH), 5.37 (1H, m, H-7), 5.29 (1H, d, *J* = 3.4 Hz, 2′-OH), 4.93 (1H, dd, *J* = 1.7, 17.6 Hz, H-16b), 4.85 (1H, dd, *J* = 1.7, 10.7 Hz, H-16a), 4.68 (1H, d, *J* = 6,8 Hz, H-1′), 4.41 (1H, dd, *J* = 5.6, 7.8 Hz, H-5′), 4.39 (1H, brs, H-3′), 4.14 (1H, d, *J* = 11.0 Hz, H-19b), 4.12 (1H, dd, *J* = 3.4, 5.8 Hz, H-4′), 3.55 (1H, ddd, *J* = 1.5, 5.4, 6.8 Hz, H-2′), 3.49 (1H, dd, *J* = 4.0, 11.9 Hz, H-3), 3.48 (1H, d, *J* = 11.0 Hz, H-19a), 2.22 (1H, dd, *J* = 3.1, 12.4 Hz, H-2b), 1.93 (1H, m, H-6β), 1.91 (1H, m, H-14α), 1.88 (1H, m, H-1β), 1.87 (1H, m, H-14β), 1.71 (1H, m, H-6α), 1.64 (1H, m, H-9), 1.55 (1H, m, H-11b), 1.46 (1H, m, H-2a), 1.44 (1H, m, H-12b), 1.31 (2H, m, H-11a, H-12a), 1.25 (3H, s, Me-18), 1.22 (1H, dd, *J* = 2.5, 9.3 Hz, H-5), 1.12 (1H, dt, *J* = 3.1, 12.5 Hz, H-1α), 0.89 (3H, s, Me-20), 0.82 (3H, s, Me-17). ^{13}C NMR (176 MHz, DMSO-d_6) δ: 174.0 (C-6′), 149.9 (C-15), 135.5 (C-8), 121.1 (C-7), 109.8 (C-16), 93.8 (C-1′), 84.0 (C-3′), 80.1 (C-3), 71.4 (C-2′), 70.1 (C-4′), 68.5 (C-5′), 68.4 (C-19), 50.8 (C-9), 49.9 (C-5), 45.4 (C-14), 36.5 (C-13), 36.1 (C-4), 35.9 (C-1), 35.5 (C-12), 34.7 (C-10), 25.5 (C-18), 21.7 (C-6), 21.3 (C-17), 21.2 (C-2), 19.7 (C-11), 15.7 (C-20); Supplementary Figures S65–S68; HRESIMS *m/z* 485.2508 $[M + Na]^+$ (calcd. for $C_{26}H_{38}O_7Na$, 485.2510, Δ + 0.4 ppm).

Aglycon of virescenoside A (**14**): amorphous solids; ^{1}H NMR (500MHz, CD_3OD) δ: 5.80 (1H, dd, J = 10.8, 17.5 Hz, H-15), 5.37 (1H, brs, H-7), 4.93 (1H, dd, J = 1.4, 17.5 Hz, H-16b), 4.85 (1H, dd, J = 1.4, 10.8 Hz, H-16a), 4.14 (1H, d, J = 11.2 Hz, H-19b), 3.79 (1H, ddd, J = 4.3, 9.8, 11.7 Hz, H-2), 3.50 (1H, d, J = 11.2 Hz, H-19a), 3.09 (1H, d, J = 9.8 Hz, H-3), 2.11 (1H, dd, J = 4.3, 12.6 Hz, H-1β), 1.98 (1H, m, H-6β), 1.97 (1H, m, H-14α), 1.91 (1H, dd, J = 2.2, 13.7 Hz, H-14β), 1.92 (1H, m, H-6α), 1.73 (1H, m, H-9), 1.61 (1H, dt, J = 3.9, 10.0 Hz, H-11β), 1.50 (1H, d, J = 8.7 Hz, H-12α), 1.39 (2H, m, H-11α, H-12β), 1.35 (1H, dd, J = 4.2, 12.0 Hz, H-5), 1.21 (3H, s, Me-18), 1.12 (1H, t, J = 12.3 Hz, H-1α), 0.93 (3H, s, Me-20), 0.86 (3H, s, Me-17). ^{13}C NMR (125 MHz, CD_3OD) δ: 151.9 (C-15), 137.1 (C-8), 123.1 (C-7), 110.4 (C-16), 86.5 (C-3), 69.7 (C-2), 66.6 (C-19), 54.0 (C-9), 53.0 (C-5), 47.7 (C-1), 47.6 (C-14), 44.4 (C-4), 38.4 (C-13), 37.9

(C-10), 37.8 (C-12), 24.8 (C-6), 24.5 (C-18), 22.6 (C-17), 22.1 (C-11), 17.9 (C-20), Supplementary Figures S69–S73; HRESIMS *m/z* 343.2241 $[M + Na]^+$ (calcd. for $C_{20}H_{32}O_3Na$, 343.2244, Δ + 0.8 ppm).

3.5. Urease Inhibition Assay

The reaction mixture consisting of 25 μL enzyme solution (urease from *Canavalia ensiformis*, Sigma, 1U final concentration) and 5 μL of test compounds dissolved in water (10–300.0 μM final concentration) was preincubated at 37 °C for 60 min in 96-well plates. Then 55 μL of phosphate buffer solution with 100 μM urea was added to each well and incubated at 37 °C for 10 min. The urease inhibitory activity was estimated by determining of ammonia production using indophenol method [26]. Briefly, 45 μL of phenol reagent (1% w/v phenol and 0.005% w/v sodium nitroprusside) and 70 μL of alkali reagent (0.5% w/v NaOH and 0.1% active chloride NaOCl) were added to each well. The absorbance was measured after 50 min at 630 nm using a microplate reader Multiskan FC (Thermo Scientific, Canada). All the reactions were performed in triplicate in a final volume of 200 μL. The pH was maintained 7.3–7.5 in all assays. DMSO 5% was used as a positive control.

3.6. Reactive Oxygen Species (ROS) Level Analysis in LPS-Treated Cells

The suspension of macrophages on 96-well plates (2 × 104 cells/well) were washedwith the PBS and treated with 180 μL/well of the tested compounds (10 μM) for 1 h and 20 μL/well LPS from E. coli serotype 055:B5 (Sigma, 1.0 μg/mL), which were both dissolved in PBS and cultured at 37 °C in a CO2-incubator for one hour. For the ROS levels measurement, 200 μL of 2,7-dichlorodihydrofluorescein diacetate (DCF-DA, Sigma, final concentration 10 μM) fresh solution was added to each well, and the plates were incubated for 30 min at 37 °C. The intensity of DCF-DA fluorescence was measured at λex 485 n/λem 518 nm using the plate reader PHERAstar FS (BMG Labtech, Offenburg, Germany) [27].

3.7. Reactive Nitrogen Species (RNS) Level Analysis in LPS-Treated Cells

The suspension of macrophages on 96-well plates (2 × 104 cells/well) were washed withthe PBS and treated with 180 μL/well of the tested compounds (10 μM) for 1 h and 20 μL/well LPS from E. coli serotype 055:B5 (Sigma, 1.0 μg/mL), which were both dissolved in PBS and cultured at 37 °C in a CO2-incubator for one hour. For the RNS levels measurement, 200 μL Diaminofluorescein-FM diacetate (DAF FM-DA, Sigma, final concentration 10 μM) fresh solution was added to each well, and the plates were incubated for 40 min at 37 °C, then replaced with fresh PBS, and then incubated for an additional 30 min to allow complete de-esterification of the intracellular diacetates. The intensity of DAF FM-DA fluorescence was measured at λex 485 n/λem 520 nm using the plate reader PHERAstar FS (BMG Labtech, Offenburg, Germany).

3.8. Peritoneal Macrophage Isolation

Mice BALB/c were sacrificed by cervical dislocation. Peritoneal macrophages were isolated using standard procedures. For this purpose, 3 mL of PBS (pH 7.4) was injected into the peritoneal cavity and the body intensively palpated for 1–2 min. Then the peritoneal fluid was aspirated with a syringe. Mouse peritoneal macrophage suspension was applied to a 96-well plate left at 37 °C in an incubator for 2 h to facilitate attachment of peritoneal macrophages to the plate. Then a cell monolayer was triply flushed with PBS (pH 7.4) for deleting attendant lymphocytes, fibroblasts and erythrocytes and cells were used for further analysis.

All animal experiments were conducted in compliance with all rules and international recommendations of the European Convention for the Protection of Vertebrate Animals used for experimental and other scientific purposes. All procedures were approved by the Animal Ethics Committee at the G. B. Elyakov Pacific Institute of Bioorganic Chemistry, Far Eastern Branch of the Russian Academy of Sciences (Vladivostok, Russia), according to the Laboratory Animal Welfare guidelines.

3.9. Statistical Analysis

Average value, standard error, standard deviation and p-values in all experiments were calculated and plotted on the chart using SigmaPlot 3.02 (Jandel Scientific, San Rafael, CA, USA). Statistical difference was evaluated by t-test, and results were considered as statistically significant at $p < 0.05$.

4. Conclusions

Ten new diterpene glycosides, virescenosides Z_9-Z_{18} (**1-10**) were isolated from a marine strain of *Acremonium striatisporum* KMM 4401 associated with the holothurian *Eupentacta fraudatrix*. Virescenoside Z_9 (**1**) is an altroside of a new 7-oxo-isopimara-15-en-2α,3β,6α,8β-tetraol aglycon. Virescenosides Z_{12}-Z_{16} (**4-8**) were determined as the monosides having unique methyl esters of altruronic acid as their sugar moieties. Carbohydrate chain of virescenoside Z_{18} (**10**) was structurally identified as the methyl ester of mannuronic acid. The effects of some isolated glycosides and aglycons **15-18** on urease activity and regulation of ROS and NO production in macrophages stimulated with lipopolysaccharide (LPC) were evaluated.

Supplementary Materials: ^{1}H, ^{13}C, HSQC, HMBC and NOESY spectra of all compounds are available online at http://www.mdpi.com/1660-3397/17/11/616/s1.

Author Contributions: O.I.Z. supervised research, analyzed of NMR spectra and wrote the manuscript; S.S.A. conceptualization, analyzed of NMR spectra and wrote the manuscript; A.S.A. and G.K.O. investigation; Y.V.K. cultivated the fungus; R.S.P. performed MS experiments; V.A.D. performed NMR experiments; E.A.P. evaluated inhibitory effects on ROS and NO production; E.A.C. examined urease activity.

Funding: The study was supported by Russian Science Foundation (grant No 19-74-10014).

Acknowledgments: The study was carried out on the equipment of the Collective Facilities Center "The Far Eastern Center for Structural Molecular Research (NMR/MS) PIBOC FEB RAS". The study was carried out using the Collective Facilities Center "Collection of Marine Microorganisms PIBOC FEB RAS".

Conflicts of Interest: The authors declare no conflict of interest.

References

1. Du, F.Y.; Li, X.M.; Song, J.Y.; Li, C.S.; Wang, B.G. Anthraquinone derivatives and an orsellinic acid ester from the marine alga-derived endophytic fungus *Eurotium cristatum* EN-220. *Helv. Chim. Acta* **2014**, *97*, 973–978. [CrossRef]
2. Li, D.L.; Li, X.M.; Wang, B.G. Natural anthraquinone derivatives from a marine mangrove plant-derived endophytic fungus *Eurotium rubrum*: Structural elucidation and DPPH radical scavenging activity. *J. Microbiol. Biotechnol.* **2009**, *19*, 675–678.
3. Li, Y.; Li, X.; Lee, U.; Kang, J.S.; Choi, H.D.; Son, B.W. A new radical scavenging anthracene glycoside, asperflavin ribofuranoside, and polyketides from a marine isolate of the fungus *Microsporum*. *Chem. Pharm. Bull.* **2006**, *54*, 882–883. [CrossRef]
4. Li, X.F.; Xia, Z.Y.; Tang, J.Q.; Wu, J.H.; Tong, J.; Li, M.J.; Ju, J.H.; Chen, H.R.; Wang, L.Y. Identification and biological evaluation of secondary metabolites from marine derived fungi—*Aspergillus* sp. SCSIOW3, cultivated in the presence of epigenetic modifying agents. *Molecules* **2017**, *22*, 1302. [CrossRef] [PubMed]
5. Wang, Y.N.; Mou, Y.H.; Dong, Y.; Wu, Y.; Liu, B.Y.; Bai, J.; Yan, D.J.; Zhang, L.; Feng, D.Q.; Pei, Y.H.; et al. Diphenyl ethers from a marine-derived *Aspergillus sydowii*. *Mar. Drugs* **2018**, *16*, 451. [CrossRef] [PubMed]
6. Hu, Z.X.; Xue, Y.B.; Bi, X.B.; Zhang, J.W.; Luo, Z.W.; Li, X.N.; Yao, G.M.; Wang, J.P.; Zhang, Y.H. Five new secondary metabolites produced by a marine-associated fungus, *Daldinia eschscholzii*. *Mar. Drugs* **2014**, *12*, 5563–5575. [CrossRef] [PubMed]
7. Du, L.; Zhu, T.J.; Liu, H.B.; Fang, Y.C.; Zhu, W.M.; Gu, Q.Q. Cytotoxic polyketides from a marine-derived fungus *Aspergillus glaucus*. *J. Nat. Prod.* **2008**, *71*, 1837–1842. [CrossRef]
8. Harms, H.; Kehraus, S.; Nesaei-Mosaferan, D.; Hufendieck, P.; Meijer, L.; Konig, G.M. A beta-42 lowering agents from the marine-derived fungus *Dichotomomyces cejpii*. *Steroids* **2015**, *104*, 182–188. [CrossRef]

9. Nazir, M.; Harms, H.; Loef, I.; Kehraus, S.; El Maddah, F.; Arslan, I.; Rempel, V.; Muller, C.E.; Konig, G.M. GPR18 inhibiting amauromine and the novel triterpene glycoside auxarthonoside from the sponge-derived fungus *Auxarthron reticulatum*. *Planta Med.* **2015**, *81*, 1141–1145. [CrossRef]
10. Afiyatullov, S.S.; Pivkin, M.V.; Kalinovsky, A.I.; Kuznetsova, T.A. New cytotoxic glycosides of the fungus *Acremonium striatisporum* isolated from a sea cucumber. *Nat. Prod. I* **2007**, *15*, 85–114.
11. Afiyatullov, S.S.; Kalinovsky, A.I.; Antonov, A.S.; Zhuravleva, O.I.; Khudyakova, Y.V.; Aminin, D.L.; Yurchenko, A.N.; Pivkin, M.V. Isolation and structures of virescenosides from the marine-derived fungus *Acremonium striatisporum*. *Phytochem. Lett.* **2016**, *15*, 66–71. [CrossRef]
12. Afiyatullov, S.S.; Kalinovsky, A.I.; Antonov, A.S. New virescenosides from the marine-derived fungus *Acremonium striatisporum*. *Nat. Prod. Commun.* **2011**, *6*, 1063–1068. [CrossRef] [PubMed]
13. Polonsky, J.; Baskevitch, Z.; Cagnoli-Bellavita, N.; Ceccherelli, P. Structures des virescenols A et B, metabolites de *Oospora virescens* (link) Wallr. *Tetrahedron* **1970**, *29*, 449–454.
14. Polonsky, J.; Baskevitch, Z.; Cagnoli-Bellavita, N.; Ceccherelli, P.; Buckwalter, B.L.; Wenkert, E. Carbon-13 nuclear magnetic resonance spectroscopy of naturally occurring substances. XI. Biosynthesis of virescenosides. *J. Am. Chem. Soc.* **1972**, *94*, 4369–4370. [CrossRef]
15. Wenkert, E.; Beak, P. The stereochemistry of rumiene. *J. Am. Chem. Soc.* **1961**, *83*, 998–1000. [CrossRef]
16. De Pascual, J.T.; Barrero, A.F.; Muriel, L.; San Feliciano, A.; Grande, M. New natural diterpene acid from *Juniperus communis*. *Phytochemistry* **1980**, *19*, 1153–1156. [CrossRef]
17. Afiyatullov, S.S.; Kalinovsky, A.I.; Kuznetsova, T.A.; Isakov, V.V.; Pivkin, M.V.; Dmitrenok, P.S.; Elyakov, G.B. New diterpene glycosides of the fungus *Acremonium striatisporum* isolated from a sea cucumber. *J. Nat. Prod.* **2002**, *65*, 641–644. [CrossRef]
18. Afiyatullov, S.S.; Kuznetsova, T.A.; Isakov, V.V.; Pivkin, M.V.; Prokof'eva, N.G.; Elyakov, G.B. New diterpenic altrosides of the fungus *Acremonium striatisporum* isolated from a sea cucumber. *J. Nat. Prod.* **2000**, *63*, 848–850. [CrossRef]
19. Ceccherelli, P.; Cagnoli-Bellavita, N.; Polonsky, J.; Baskevitch, Z. Structures des virescenosides F et G, nouveaux metabolites de *Oospora virescens* (link) Wallr. *Tetrahedron* **1973**, *29*, 449–454. [CrossRef]
20. Bellavita, N.; Bernassau, J.-M.; Ceccherelli, P.; Raju, M.S.; Wenkert, E. An unusual solvent dependence of the carbone-13 nuclear magnetic resonance spectral features of some glycosides as studied by relaxation-time measurements. *J. Am. Chem. Soc.* **1980**, *2*, 17–20. [CrossRef]
21. Afiyatullov, S.S.; Kalinovsky, A.I.; Pivkin, M.V.; Dmitrenok, P.S.; Kuznetsova, T.A. New diterpene glycosides of the fungus *Acremonium striatisporum* isolated from a sea cucumber. *Nat. Prod. Res.* **2006**, *20*, 902–908. [CrossRef] [PubMed]
22. King-Morris, M.J.; Serianni, A.S. ^{13}C NMR studies of [1-^{13}C] aldoses: Empirical rules correlating pyranose ring configuration and conformation with ^{13}C chemical shifts and ^{13}C-^{13}C spin couplings. *J. Am. Chem. Soc.* **1987**, *109*, 3501–3508. [CrossRef]
23. Podlasek, C.A.; Wu, J.; Stripe, W.A.; Bondo, P.B.; Serrianni, A.S. [^{13}C]-Enriched methyl aldopyranosides: Structural interpretations of ^{13}C-^{1}H spin-coupling constants and ^{1}H chemical shifts. *J. Am. Chem. Soc.* **1995**, *117*, 8635–8644. [CrossRef]
24. Burne, R.A.; Chen, Y.Y. Bacterial ureases in infectious diseases. *Microbes Infect.* **2000**, *2*, 533–542. [CrossRef]
25. Mobley, H.L.; Hausinger, R.P. Microbial ureases: Significance, regulation, and molecular characterization. *Microbiol. Rev.* **1989**, *53*, 85–108.
26. Weatherburn, M.W. Phenol hypochlorite reaction for determination of ammonia. *Anal. Chem.* **1967**, *39*, 971–974. [CrossRef]
27. Ivanchina, N.V.; Kicha, A.A.; Malyarenko, T.V.; Kalinovsky, A.I.; Menchinskaya, E.S.; Pislyagin, E.A.; Dmitrenok, P.S. The influence on LPS-induced ROS formation in macrophages of capelloside A, a new steroid glycoside from the starfish *Ogmaster capella*. *Nat. Prod. Commun.* **2015**, *10*, 1937–1940. [CrossRef]

marine drugs

Article

Structural and Serological Studies of the O6-Related Antigen of *Aeromonas veronii* bv. *sobria* Strain K557 Isolated from *Cyprinus carpio* on a Polish Fish Farm, which Contains L-perosamine (4-amino-4,6-dideoxy-L-mannose), a Unique Sugar Characteristic for *Aeromonas* Serogroup O6

Katarzyna Dworaczek [1], Dominika Drzewiecka [2], Agnieszka Pękala-Safińska [3] and Anna Turska-Szewczuk [1,*]

1 Department of Genetics and Microbiology, Maria Curie-Skłodowska University in Lublin, Akademicka 19, 20-033 Lublin, Poland

2 Laboratory of General Microbiology, Department of Biology of Bacteria, Faculty of Biology and Environmental Protection, University of Łódź, Banacha 12/16, 90-237 Łódź, Poland

3 Department of Fish Diseases, National Veterinary Research Institute, Partyzantów 57, 24-100 Puławy, Poland

* Correspondence: aturska@hektor.umcs.lublin.pl; Tel.: +48-81-537-50-18; Fax: +48-81-537-59-59

Received: 10 June 2019; Accepted: 3 July 2019; Published: 5 July 2019

Abstract: Amongst *Aeromonas* spp. strains that are pathogenic to fish in Polish aquacultures, serogroup O6 was one of the five most commonly identified immunotypes especially among carp isolates. Here, we report immunochemical studies of the lipopolysaccharide (LPS) including the O-specific polysaccharide (O-antigen) of *A. veronii* bv. *sobria* strain K557, serogroup O6, isolated from a common carp during an outbreak of motile aeromonad septicemia (MAS) on a Polish fish farm. The O-polysaccharide was obtained by mild acid degradation of the LPS and studied by chemical analyses, mass spectrometry, and ^{1}H and ^{13}C NMR spectroscopy. It was revealed that the O-antigen was composed of two O-polysaccharides, both containing a unique sugar 4-amino-4,6-dideoxy-L-mannose (*N*-acetyl-L-perosamine, L-Rha*p*4NAc). The following structures of the O-polysaccharides (O-PS 1 and O-PS 2) were established: O-PS 1: →2)-α-L-Rha*p*4NAc-(1→; O-PS 2: →2)-α-L-Rha*p*4NAc-(1→3)-α-L-Rha*p*4NAc-(1→3)-α-L-Rha*p*4NAc-(1→. Western blotting and an enzyme-linked immunosorbent assay (ELISA) showed that the cross-reactivity between the LPS of *A. veronii* bv. *sobria* K557 and the *A. hydrophila* JCM 3968 O6 antiserum, and *vice versa*, is caused by the occurrence of common α-L-Rha*p*4NAc-(1→2)-α-L-Rha*p*4NAc and α-L-Rha*p*4NAc-(1→3)-α-L-Rha*p*4NAc disaccharides, whereas an additional →4)-α-D-Gal*p*NAc-associated epitope defines the specificity of the O6 reference antiserum. Investigations of the serological and structural similarities and differences in the O-antigens provide knowledge of the immunospecificity of *Aeromonas* bacteria and are relevant in epidemiological studies and for the elucidation of the routes of transmission and relationships with pathogenicity.

Keywords: *Aeromonas*; fish pathogen; lipopolysaccharide (LPS); structure; O-antigen; O-polysaccharide; L-perosamine; immunospecificity; NMR spectroscopy; mass spectrometry

1. Introduction

The genus *Aeromonas*, which belongs to the family *Aeromonadaceae* along with four other genera *Telumonas*, *Oceanimonas*, *Oceanisphaera* and *Zobellella*, is composed of a large number of species classified within 17 DNA-hybridization groups (HG) or genomospecies, and 14 phenospecies [1–5]. These

Gram-negative rods are typically found in aquatic environments, e.g., freshwater, estuarine and coastal water, seawater, drinking water supplies, polluted waters, marine, and freshwater sediment and sand. Aeromonads have also been isolated from animals, food, and various clinical samples. Most frequently, *Aeromonas* spp. strains are pathogenic to poikilothermic animals including amphibians, fish, and reptiles. They also can be associated with infections of birds and mammals [4–6]. In fish, non-motile psychrophilic species of *Aeromonas salmonicida* subsp. salmonicida are pathogenic to Salmonidae, provoking systemic furunculosis [7,8]. In turn, the representatives of *Aeromonas hydrophila*, *Aeromonas bestiarum*, *Aeromonas salmonicida*, *Aeromonas jandaei*, and *Aeromonas veronii* bv. *sobria* have been described as causative agents of diseases in a variety of fish species. These include motile Aeromonas septicemia (MAS), clinical conditions associated with systemic infection resulting in high mortality rates and severe economic losses in aquacultures, and a chronic type of disease called motile Aeromonas infection (MAI) causing erosion of fins, skin lesions, and ulcerations [9–12]. The mesophilic and motile *A. veronii*, *A. hydrophila*, *A. caviae* and *A. schubertii* species are potentially pathogenic to humans. Immunocompromised patients and children are especially vulnerable. Clinical presentations include gastroenteritis, wound infections, biliary tract infections, pneumonia, meningitis, septic arthritis, or septicemia without an obvious focus of infection [4,13–16].

The pathogenicity of *Aeromonas* is determined by their ability to produce extracellularly secreted enzymes and toxins i.e., hemolysins, cytotonic and cytotoxic enterotoxins, proteases, lipases, gelatinases, and leucocidins, which play an important role in both the initial steps and spread of the infection process. Moreover, cell-surface constituents, including outer membrane proteins, S-layers, surface polysaccharides (capsule, lipopolysaccharide, and glucan), flagella, and pili, have been identified as important compounds associated with *Aeromonas* virulence [4,10,17–19].

Lipopolysaccharide (LPS) is the dominant glycolipid in the outer leaflet of the outer membrane of Gram-negative bacteria, which mediates their virulence. The high-molecular-weight S-LPS glycoforms have a tripartite structure comprising lipid A, core oligosaccharide, which together with lipid A contributes to maintenance of the integrity of the outer membrane, and the O-specific polysaccharide (O-PS), which is connected to the core and most frequently consists of a heteropolymer composed of repeating oligosaccharide units [20–23]. The O-specific polysaccharide is a surface antigen called the somatic O-antigen, whose high degree of diversity determines the specificity of each bacterium and gives the basis for their serological classification. Serotyping to identify bacterial strains is invaluable for epidemiological investigations since many O-serotypes are associated with specific disease syndromes [21,24].

Aeromonas strains are serologically heterogeneous and they have been characterized and classified based on O-antigens into 44 serogroups using the NIH scheme (National Institute of Health, Japan) proposed by Sakazaki and Shimada [25], which can be further extended to 97 O serogroups after inclusion of provisional new serogroups [26]. The variants of the O-specific polysaccharide, which represents a specific antigenic fingerprint for bacteria, might be very useful, e.g., for identification of *Aeromonas* strains eliciting infections in farmed fish and for diagnosis of etiological agents of gastrointestinal infections in humans.

Despite the large antigenic diversity of *Aeromonas*, only several serogroups i.e., O11, O16, O18, and O34 (NIH scheme) have been reported, most frequently, as predominating especially in clinical specimens. These O-serotypes of mesophilic *Aeromonas* were associated with most cases of bacteremia, suggesting their role in the pathogenesis of some systemic diseases [4].

As shown in previous reports, the distribution of the serogroups of *Aeromonas* strains may be related to geographic location [27,28]. Such differences in the occurrence of *Aeromonas* species serogroups were associated with outbreaks of septicemia in fish. Strains belonging to serogroup O14 have been identified as pathogens of European eels [29]. As revealed in recently published studies by Kozińska and Pękala [28], the majority of isolates that are pathogenic to carp in Polish aquacultures represented serogroups O3, O6, O41, PGO4, and PGO6, whereas serogroups O11, O16, O18, O33, PGO1, and PGO2 dominated among both carp and trout isolates. In turn, motile aeromonas septicemia

(MAS) incidences in rainbow trout have been related to the strains of serogroups O11, O16, O34, and O14. Moreover, pathogenic isolates of two species: *Aeromonas veronii* bv. *sobria* and *Aeromonas sobria* were mainly classified within the O6 serogroup.

The *A. veronii* species, originally described by Hickman-Brenner et al. [30] as a novel member of the genus, is commonly associated with diarrhea. Nevertheless, this species, which consists of two biovars, *A. veronii* bv. *sobria*, which is negative for aesculin hydrolysis and ornithine decarboxylase, and *A. veronii* bv. *veronii*, which is positive for these reactions [15,31], is commonly known as a fish pathogen associated mainly with ulcerative syndrome [32,33]. It is worth emphasizing that *A. veroni* bv. *sobria* was one of the dominant species among carp isolates collected during 5 years in Polish culture facilities, and 76% of these isolates were classified as virulent [28]. In the light of the increased *Aeromonas* infection incidence rate and the economic importance of these diseases in cultured fish, it is essential to characterize virulence factors associated with the pathogenesis of this species, particularly including LPS, which is the major surface glycoconjugate of Gram-negative bacteria. Recently, two new structures of O-antigens were established for the species *A. veronii* bv. *sobria* and *A. sobria* [34,35].

Here we report immunochemical investigation of LPS, especially the O-specific polysaccharide of *A. veronii* bv. *sobria* strain K557, which was isolated from the common carp (*Cyprinus carpio* L.) during an outbreak of motile aeromonad infection/motile aeromonad septicemia (MAI/MAS) on a Polish fish farm [28,36]. The structural characterization revealed that the O-antigen was composed of two O-polysaccharides, both containing a unique sugar 4-amino-4,6-dideoxy-L-mannose (*N*-acetyl-L-perosamine, L-Rha*p*4NAc). Serological studies using Western blotting and an enzyme-linked immunosorbent assay (ELISA) with intact and adsorbed O-antisera showed that the O-antigen of *A. veronii* bv. *sobria* strain K557 is related but not identical to that of *A. hydrophila* JCM 3968 O6, which is a reference strain for *Aeromonas* serogroup O6 [37].

2. Results

2.1. Bacterial Cultivation, Isolation of LPS, and SDS-PAGE Study

Cells of *Aeromonas veronii* bv. *sobria* strain K557 were extracted with hot aqueous 45% phenol [38], and LPS species were harvested from the phenol phase in a yield of 3.9% of the bacterial cell mass. SDS-PAGE analysis of the LPS followed by silver staining showed a pattern typical for LPS isolated from smooth bacterial cells (Figure 1) with the content of both slow-migrating high-molecular-weight (HMW) S-LPS species and fast-migrating low-molecular-weight R-LPS glycoforms. Moreover, the electrophoregram indicated that the studied S-LPS contained molecules where the core oligosaccharides were substituted with shorter O-chains in comparison to the LPS species of *A. hydrophila* JCM 3968, O6.

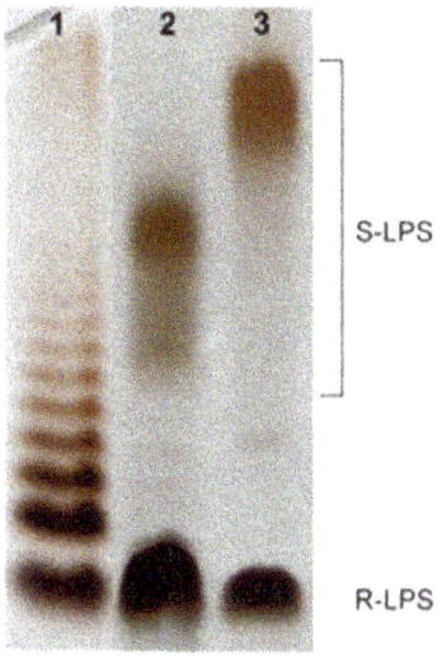

Figure 1. Silver-stained SDS-PAGE of the LPS of *A. veronii* bv. *sobria* strain K557 (lane 2, 2 μg), *A. hydrophila* JCM 3968, O6 (lane 3, 2 μg), and *Salmonella enterica* sv. Typhimurium (Sigma-Aldrich, St. Louis, MO, USA) as a reference (lane 1, 2 μg). R-LPS and S-LPS species are indicated.

2.2. Serological Studies of the A. veronii bv. sobria K557 O-antigen

A. veronii bv. *sobria* strain K557 was serologically typed by agglutination tests using heat-inactivated bacteria and antisera for 44 defined *Aeromonas* O-serogroups (NIH system) and 20 provisional serogroups (PGO1–PGO20) for selected Polish isolates and classified as a representative of the *Aeromonas* serogroup O6 [28,36].

The LPS preparations from *A. veronii* bv. *sobria* strain K557 and *A. hydrophila* JCM 3968, which is the reference strain to serogroup O6, were studied by Western blotting and ELISA with intact and adsorbed polyclonal rabbit O-antisera.

In Western blot, the reference antiserum O6 recognized electrophoretically separated LPS molecules of both strains (Figure 2a). Strong reaction to slow-migrating bands corresponding to the O-PS containing LPS species was observed and suggested that the O-antigens shared common epitopes. Additionally, the stained bands of fast-migrating R-LPS molecules may indicate similarities in the core region of the studied strains. In turn, the other Western blot (Figure 2b) with *A. veronii* bv. *sobria* K557 O-antiserum revealed the recognition of antigenic determinants within LPS molecules of both *A. veronii* bv. *sobria* K557 and *A. hydrophila* JCM 3968 strains; however, the cross-reaction with the heterologous LPS was weaker than in the homologous system.

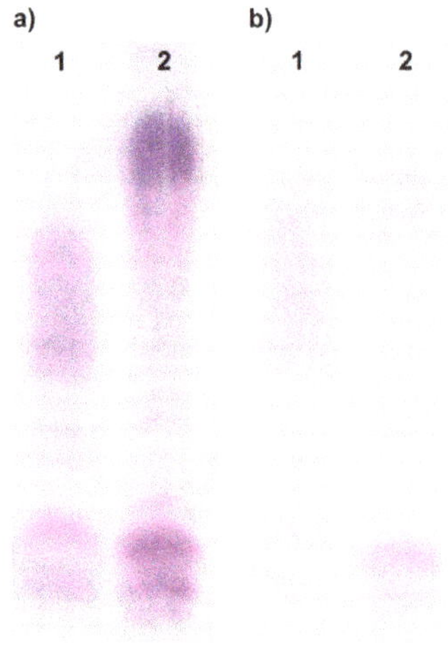

Figure 2. Western blots of lipopolysaccharides after SDS-PAGE with the intact reference O6 (**a**) and anti-K557 O-antisera (**b**). Lanes: 1, LPS of *A. veronii* bv. *sobria* strain K557; 2, LPS of *A. hydrophila* strain JCM 3968.

Accordingly, in ELISA (Table 1), the rabbit polyclonal O antiserum specific for *A. hydrophila* JCM 3968 O6 reacted strongly with the homologous LPS, and cross-reaction was observed with the LPS of *A. veronii* bv. *sobria* strain K557 to the lower titer of 1:64,000. In turn, the polyclonal O-antiserum against *A. veronii* bv. *sobria* strain K557 revealed reactions at the same level as both the homologous and heterologous LPS samples. The O6 reference antiserum revealed stronger reactivity than the K557-specific one, as demonstrated by the Western blotting results.

Adsorption of the antiserum specific for the *A. veronii* bv. *sobria* K557 O-antigen with *A. hydrophila* strain JCM 3968 O6 cells totally abolished the reactivity of this antiserum with both LPS preparations. The opposite reaction, i.e., adsorption of the reference O6 antiserum with the *A. veronii* bv. *sobria* K557 cells, only decreased its reactivity in the homologous system, whereas there was no reaction of the adsorbed O6 antiserum with *A. veronii* bv. *sobria* K557 LPS. The latter findings indicated that the adsorption process was complete and resulted in the removal of anti-K557 antibodies from the reference O6 antiserum. The remaining antibodies, strongly reacting with the homologous O6 LPS, were most probably specific to an additional epitope characteristic for the *A. hydrophila* JCM 3968 O6 antigen.

Table 1. Reactivity (reciprocal titers) of the studied antisera (intact or adsorbed) with the LPS preparations from the *A. hydrophila* JCM 3968 and *A. veronii* bv. *sobria* K557 strains.

Serum Specific to Strain		JCM 3968 LPS	K557 LPS
JCM 3968	intact	1,024,000	64,000
	K557 adsorbed	256,000	<1000
K557	intact	64,000	64,000
	JCM 3968 adsorbed	<1000	<1000

These data indicated that the reference O6 antiserum and the *A. veronii* bv. *sobria* K557 O-antiserum shared common antibodies but the reference one contained additional immunoglobulins recognizing structural determinants that were not present in the *A. veronii* bv. *sobria* K557 O-antigen.

Therefore, detailed chemical analyses were performed to establish the structure of the *A. veronii* bv. *sobria* strain K557 O-PS.

2.3. Chemical and Mass Spectrometry Analyses of LPS

Compositional analysis of the degraded polysaccharide (dgPS) liberated from the phenol-soluble LPS was performed using GLC-MS of alditol acetates. It showed the presence of D-glucose (D-Glc), D-galactose (D-Gal), 2-amino-2-deoxy-D-glucose (D-GlcN), D-*glycero*-D-*manno*-heptose (D,D-Hep), and L-*glycero*-D-*manno*-heptose (L,D-Hep) in a ratio of 1.0 : 1.2 : 1.3 : 1.5 : 3.8 as the core oligosaccharide components. The chemical analysis also revealed 6-deoxymannose (Rha), 2-amino-2,6-dideoxyglucose (QuiN), and 4-amino-4,6-dideoxyhexose (identified as Rha4N, see Section 2.4.). Rha4N was found as the main component of the O-polysaccharide part (see Section 2.4.). Kdo (3-deoxy-D-*manno*-2-octulosonic acid)—the only acidic sugar—was found in the LPS after treatment of the LPS with 48% aqueous HF (hydrofluoric acid), which suggested its phosphorylation [37,39,40]. The GLC-MS analysis of fatty acids as methyl esters and *O*-TMS derivatives revealed that 3-hydroxytetradecanoic [14:0(3-OH)], 3-hydroxy*iso*pentadecanoic [*i*15:0(3-OH)], as well as dodecanoic (12:0) and tetradecanoic (14:0) acids, were the most abundant species in a ratio of 6.3 : 2.4 : 1.5 : 1.0. GlcN was identified as the sugar component of lipid A.

The negative ion matrix-assisted laser desorption/ionization time-of-flight (MALDI-TOF) mass spectrum of the *A. veronii* bv. *sobria* K557 lipopolysaccharide (Figure 3) showed the most intensive signals in the *m*/*z* range 1600–2000, which were attributable to a lipid A and a core oligosaccharide species (Y- and C-type fragment ions) arising from an in-source fragmentation at the glycosidic bond between the Kdo and the lipid A [41]. The ions at *m*/*z* 1768.17, 1796.19, and 1824.22 originated from hexaacylated lipid A species (Y-fragment ions) [37]. The ion at *m*/*z* 1824.22 represented a variant of lipid A, where the diglucosaminyl backbone bisphosphorylated at O-1 and O-4′ was substituted by four 3-hydroxytetradecanoic acids 14:0(3-OH) and two tetradecanoic acids 14:0, instead of two dodecanoic acids 12:0, compared with the ion at *m*/*z* 1768.17. In turn, the ion at *m*/*z* 1796.19 contained a sugar backbone acylated by three, instead of four, 3-hydroxytetradecanoic acids, one 3-hydroxy*iso*pentadecanoic acid, and two dodecanoic acids. The composition of the ions is shown in Table 2.

The ions at *m*/*z* 1954.5 and 1874.5 (C-fragment ions) were assigned to the core oligosaccharide with the following composition: $Hep_6Hex_2HexN_1Kdo$. The mass difference between the ions of 80 amu corresponded to species containing phosphorylated and dephosphorylated variants of the core oligosaccharide, respectively.

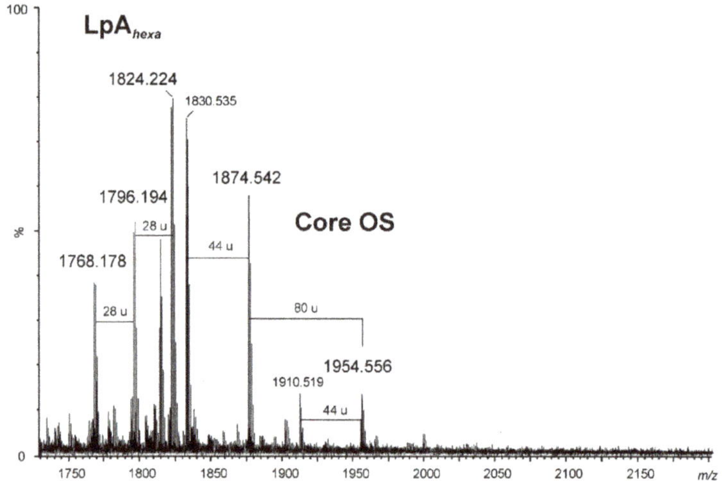

Figure 3. MALDI-TOF mass spectrum (negative ion mode) of the LPS of *A. veronii* bv. *sobria* strain K557. The notations indicate: 28 u—differences in $(CH_2)_2$ in the fatty acid chain length; 44 u—loss of CO_2; 80 u—loss of phosphate; LpA_{hexa} - hexaacylated lipid A; Core OS—core oligosaccharide.

Table 2. Composition of the main species present in the negative ion MALDI-TOF mass spectrum of the LPS of *A. veronii* bv. *sobria* strain K557.

$[M-H]^-$ Observed	$[M-H]^-$ Calculated	M Monoisotopic	Assigned Composition
1768.178	1768.181	1769.188	$HexN_2P_2$[14:0(3-OH)]$_4$(12:0)$_2$
1796.194	1796.139	1797.146	$HexN_2P_2$[14:0(3-OH)]$_3$[*i*15:0(3-OH)](12:0)$_2$
1824.224	1824.243	1825.250	$HexN_2P_2$[14:0(3-OH)]$_4$(14:0)$_2$
1830.535	1830.626	1831.633	$Hep_6Hex_2HexN_1$Kdo-COO
1874.542	1874.616	1875.624	$Hep_6Hex_2HexN_1$Kdo
1910.519	1910.593	1911.600	[$Hep_6Hex_2HexN_1$KdoP]-COO
1954.556	1954.583	1955.599	$Hep_6Hex_2HexN_1$KdoP

2.4. Structural Studies of O-polysaccharide (O-PS)

The O-PS was released from phenol-soluble LPS by mild-acid degradation followed by gel-permeation-chromatography (GPC) on Sephadex G-50 fine to give a high-molecular-weight O-polysaccharide with the yield of 33% of the LPS mass. GLC-MS sugar analysis of alditol acetates obtained after full acid hydrolysis of the O-PS showed the presence of rhamnose and 4-amino-4,6-dideoxymannose (Rha4N) in a peak area ratio ~1: 28.4. Other compounds detected in a small amount (below 10 %) in the GLC chromatogram of the *A. veronii* bv. *sobria* K557 O-PS, i.e., Glc, Gal and two heptose isomers (D,D-Hep and L,D-Hep), represented the core oligosaccharide sugars

Determination of the absolute configurations of the monosaccharides by GLC of the acetylated (*S*)- and (*SR*)-2-octyl glycosides [42] showed that Rha4N had the L configuration. Samples from the O-polysaccharide of *Citrobacter gillenii* O9a,9b [43] and the O-PS of *A. hydrophila* JCM 3968 [37] were used as a reference standard of D-Rha4N and L-Rha4N, respectively.

The methylation analysis of the O-PS by GLC-MS of partially methylated alditol acetates resulted in the identification of 4,6-dideoxy-3-*O*-methyl-4-(*N*-methyl)acetamidohexose (derived from 2-substituted Rha4N), and 4,6-dideoxy-2-*O*-methyl-4-(*N*-methyl)acetamidohexose (derived from 3-substituted Rha4N). The *EI* (electron impact) mass spectrum (Figure 4) of 4,6-dideoxy-2-*O*-methyl-4-(*N*-methyl)acetamidomannose was characterized by the presence of intense ion peaks at *m/z* 118 (C-1 ÷ C2 fragment), 275 (C-1 ÷ C4 fragment), and 172 (C-4 ÷ C6 fragment), and allowed distinguishing this derivative from 4,6-dideoxy-3-*O*-methyl-4-(*N*-methyl)acetamidomannose, whose *EI* mass spectrum contained ions at *m/z* 190, 275, and 172 characteristic of the C-1 ÷ C3, C-1 ÷ C4, and C-4 ÷ C6 primary fragments, respectively.

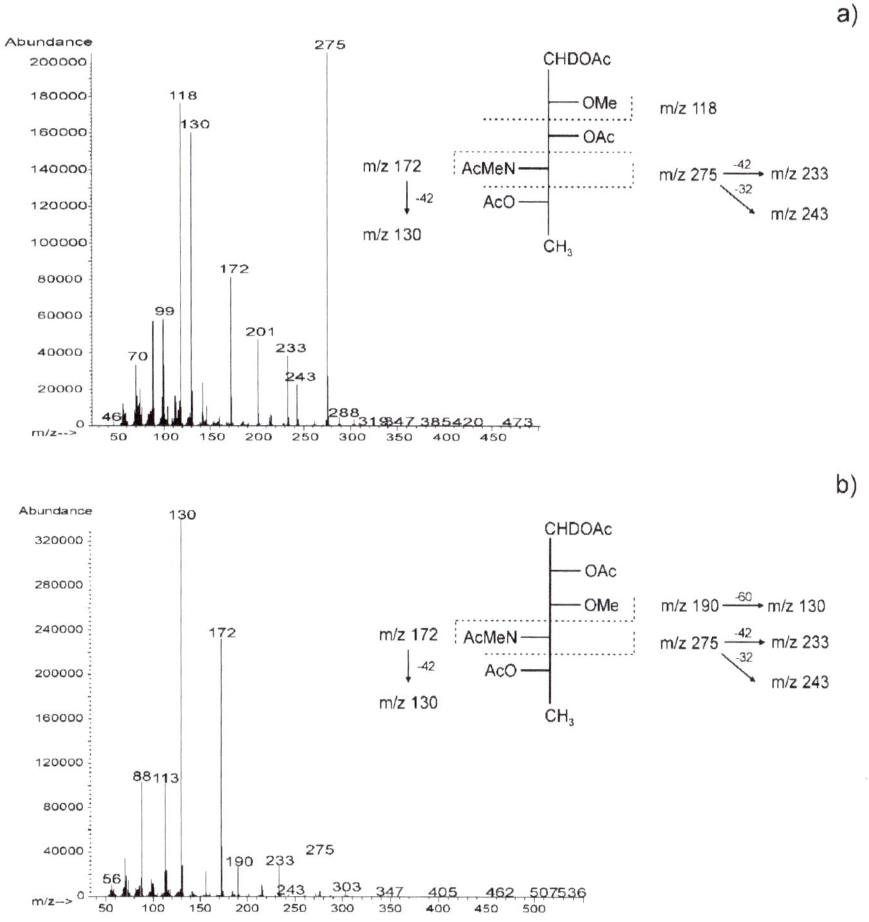

Figure 4. *EI* (electron impact) mass spectra and fragmentation pathways of 1,3,5-tri-*O*-acetyl-4,6-dideoxy-2-*O*-methyl-4-(*N*-methyl)acetamido)-mannitol-1-*d* (**a**) and 1,2,5-tri-*O*-acetyl-4,6-dideoxy-3-*O*-methyl-4-(*N*-methyl)acetamido)-mannitol-1-*d* (**b**) obtained from the O-PS of *A. veronii* bv. *sobria* strain K557. Diagnostic primary and secondary fragment ions are indicated. The mass difference 42, 60, or 32 indicates loss of chetene, acetic acid, or methanol, respectively.

The low-field region of the ^{1}H NMR spectrum of the O-polysaccharide (Figure 5) contained one major and three minor signals for anomeric protons at δ 5.18, 5.04, 5.00, and 4.97. The high-field region of the spectrum included signals for *N*-acetyl groups at δ 2.07, CH_3-C groups of 6-deoxy sugars in the range of δ 1.21–1.24. The ^{1}H and ^{13}C resonances of the O-PS of *A. veronii* bv. *sobria* K557 were assigned using 2D homonuclear ^{1}H,^{1}H DQF-COSY, TOCSY, NOESY, heteronuclear ^{1}H,^{13}C HSQC, and HMBC experiments. The ^{1}H and ^{13}C NMR data are collected in Table 3.

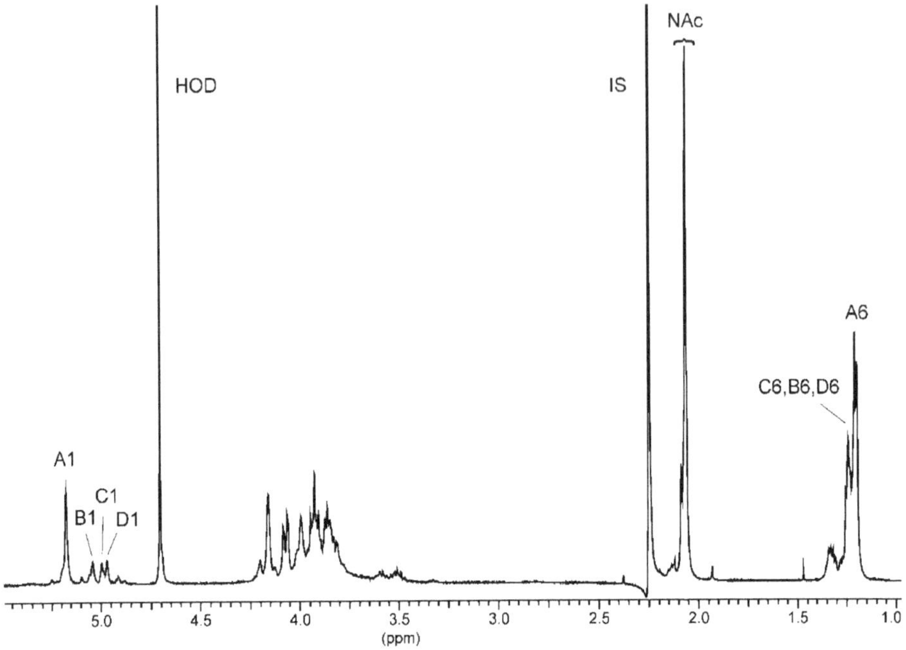

Figure 5. ^{1}H NMR spectrum of the O-PS of *A. veronii* bv. *sobria* strain K557. The spectrum was recorded in D_2O at 32 °C at 500 MHz. Capital letters and Arabic numerals refer to atoms in the sugar residues denoted as shown in Table 3. NAc—*N*-acetyl groups, IS—acetone as an internal standard (δ_H 2.225), asterisk—free acetic acid.

Table 3. ^{1}H (500 MHz) and ^{13}C NMR (125 MHz) data (δ, ppm) for the O-PS of *A. veronii* bv. *sobria* strain K557.

Sugar Residue		Chemical Shifts (δ, ppm)					
		H-1 C-1	H-2 C-2	H-3 C-3	H-4 C-4	H-5 C-5	H-6 C-6
→2)-α-L-Rhap4NAc-(1→	A	5.18 101.7	4.16 78.3	4.07 69.2	3.93 54.3	3.85 69.7	1.21 18.2
→2)-α-L-Rhap4NAc-(1→	B	5.04 102.1	3.81 79.8	4.01 69.2	3.87 54.3	3.91 69.6	1.22 18.2
→3)-α-L-RhapNAc-(1→	C	5.00 103.3	3.87 70.7	3.93 78.3	4.00 53.2	3.92 69.5	1.24 18.2
→3)-α-L-Rhap4NAc-(1→	D	4.97 103.5	4.20 70.2	3.99 78.4	3.94 53.2	3.92 69.5	1.24 18.2

Chemical shifts for NAc are δ_H 2.07 and δ_C 23.4 (CH_3) and 176.0 (CO).

The ^{1}H,^{1}H TOCSY, and DQF-COSY spectra revealed one major and three minor spin systems for monosaccharide residues, which were labelled **A–D** in the order of the decreasing chemical shifts of their anomeric protons. The high-field positions of H-6 (δ 1.21–1.24) and C-6 (δ 18.2) resonances and the values of vicinal coupling constants $^3J_{1,2}$ (~2 Hz), $^3J_{3,4}$ (8.5 Hz), $^3J_{4,5}$ (9 Hz), and $^3J_{5,6}$ (6 Hz) characteristic for *manno*-pyranose allowed assigning all the **A**, **B**, **C**, and **D** spin systems (H-1/C-1 cross-peaks at δ 5.18/101.7, 5.04/102.1, 5.0/103.3, and 4.97/103.5, respectively) to Rha4NAc residues. The latter conclusion was confirmed by the correlation signals at δ 3.93/54.3 (the major one), 3.87/54.3, 3.94/53.2, and 4.00/53.2 observed in the ^{1}H,^{13}C HSQC spectrum (Figure 6), which indicated that the O-PS contains *N*-acetamido sugar.

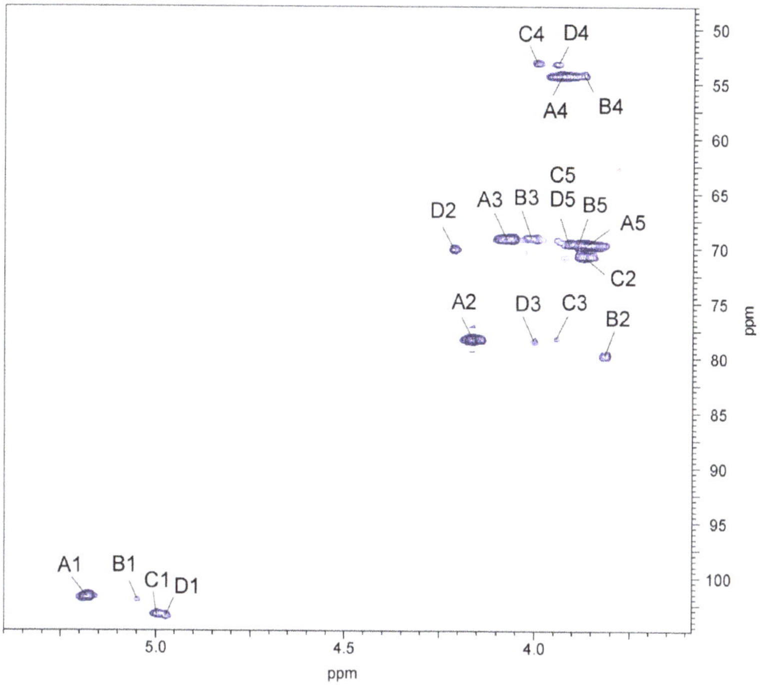

Figure 6. Part of a ^{1}H,^{13}C HSQC spectrum (500 x 125 MHz) of the O-PS of *A. veronii* bv. *sobria* strain K557. The spectrum was recorded at 32 °C in D_2O as a solvent. Capital letters and Arabic numerals refer to atoms in sugar residues denoted as shown in Table 3.

In the TOCSY spectrum, at the H-1 coordinate, the cross-peaks with H-2-H-5 were visible for Rha4NAc **A**, but only one cross-peak with H-2 for Rha4NAc **B**, and two ones with H-2 and H-3 for Rha4NAc **C** and **D**. In turn, starting from the H-2 proton signal, cross-peaks with H-3-H-6 were visible for all spin systems; however, some signals overlapped. The ^{1}H,^{1}H COSY spectrum allowed unambiguous differentiation between protons within the spin system **A** and only partly resolved the cross-peaks for the Rha4NAc **B**, **C** and **D**. The difficulties in the assignment of the H-3, H-4, and H-5 of Rha4NAc **B**, **C** and **D** were overcome in the ^{1}H,^{13}C HMBC and ^{1}H, ^{13}C HSQC experiments.

The ^{13}C NMR resonances of e.g., Rha4NAc **C** were assigned by the long-range H-6/C-4 and H-6/C-5 correlations at δ 1.24/53.2 and 1.24/69.5, and then the C-4/H-3 and C-4/H-2 correlations at δ 53.2/3.93 and 53.2/3.87, respectively, in the ^{1}H,^{13}C HMBC spectrum. In the ^{1}H,^{13}C HSQC spectrum, the cross-peak of the proton at the nitrogen-bearing carbon to the corresponding carbon at δ 4.00/53.2 was assigned to the H-4/C-4 correlation of Rha4NAc **C**. Moreover, in the ^{1}H,^{13}C HMBC spectrum, correlations of the anomeric proton with carbons C-2 and C-5 were found, and then the proton resonances were assigned from the ^{1}H,^{13}C HSQC spectrum. Similar long-range correlations were searched during identification of H-3, H-4 and H-5 proton signals of Rha4NAc **B** and **D**. The chemical shifts of the C-2 signal of Rha4NAc **B** and the C-3 signals of Rha4NAc **C** and **D** were confirmed after consideration of the methylation analysis data.

The α-configuration of all Rha4NAc residues was inferred from the relatively high-field position of the C-5 signals at δ 69.5–69.7 for the residues in the O-polysaccharides, compared with δ 68.6 and δ 72.4 for α-Rha4NAc and β-Rha4NAc, respectively [44,45].

The correlation signals between H-4 of all the Rha4NAc residues and carbonyl group signals at δ_C 176.0 and between the latter and methyl proton signals at δ_H 2.07, which were observed in the ^{1}H,^{13}C HMBC spectrum, confirmed that the residues building the O-PS were *N*-acetylated.

The positions of glycosylation were determined by downfield shifted signals of carbon atoms C-2 for residues Rha4NAc **A** (δ 78.3) and **B** (δ 79.8), and C-3 for Rha4NAc **C** (δ 78.3) and **D** (δ 78.4), compared with their positions in the spectra of the corresponding non-substituted monosaccharides [44–46].

In the NOESY spectrum (Figure 7), intraresidue H-1/H-2 and interresidue H-1/H-5 correlations at δ 5.18/4.16 and δ 5.18/3.85, respectively, typical of α-(1→2)-linked sugars with the *manno* configuration, indicated that the Rha4NAc **A** residues constitute the main O-polysaccharide (O-PS 1) being the homopolymer [47]. In turn, in the spectrum inter-residue NOE contacts were observed for protons of residues **B**→**C**, **C**→**D**, and **D**→**B**. The following correlations between the anomeric protons and protons at the linkage carbons: Rha*p*4NAc **B** H-1/Rha*p*4NAc **C** H-3 at δ 5.04/3.93; Rha*p*4NAc **C** H-1/Rha*p*4NAc **D** H-3 at δ 5.00/3.99; Rha*p*4NAc **D** H-1/Rha*p*4NAc **B** H-2 at δ 4.97/3.81 demonstrated that the other O-polysaccharide (O-PS 2) is a heteropolymer with a trisaccharide repeating unit (Figure 7, Table 3).

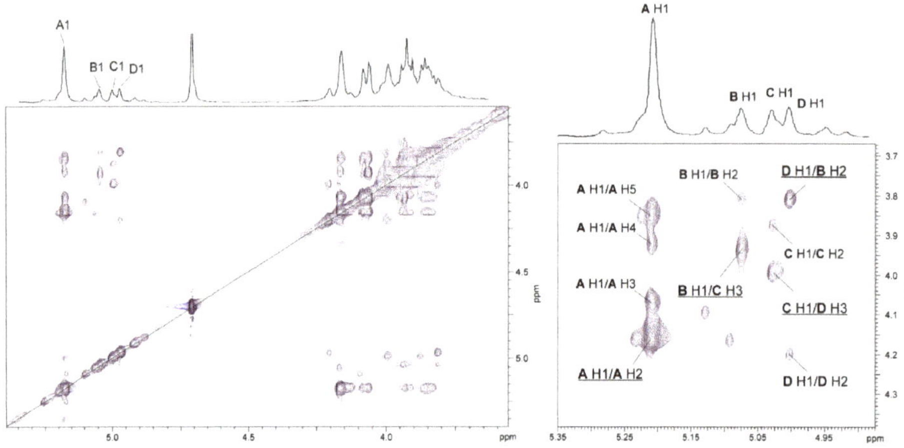

Figure 7. Parts of ^{1}H,^{1}H NOESY and ^{1}H NMR (insert) spectra of the O-PS of *A. veronii* bv. *sobria* strain K557. The map shows NOE contacts between anomeric protons and protons at the glycosidic linkages (underlined). Some other H/H correlations are depicted as well. Capital letters and Arabic numerals refer to atoms in the sugars denoted as shown in Table 3.

The sequence of monosaccharides in the repeating unit of O-PS 2 was confirmed, in the ^{1}H,^{13}C HMBC spectrum (Figure 8), by the following inter-glycosidic cross-peaks: **B** H-1/**C** C-3 at δ 5.04/78.3; **C** H-1/**D** H-3 at δ 5.00/78.4; **D** H-1/**B** H-2 at δ 4.97/79.8.

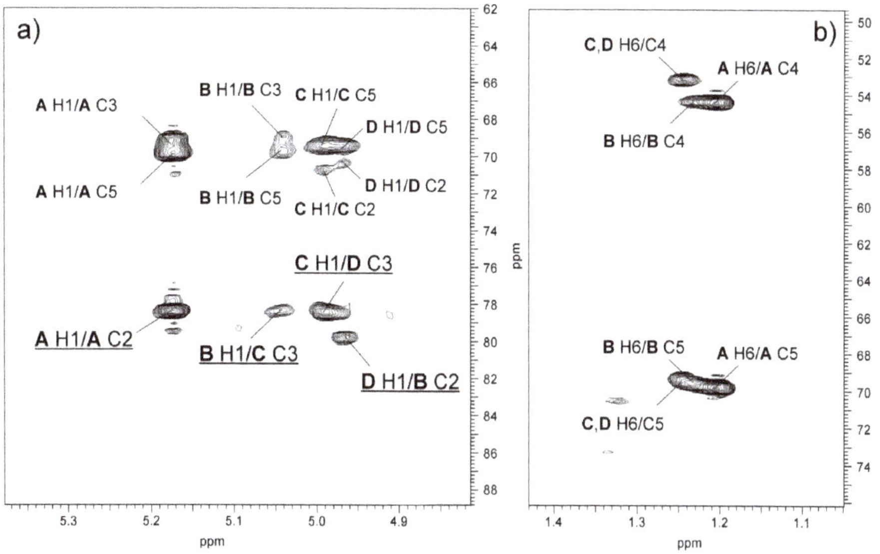

Figure 8. Regions of the $^{1}H,^{13}C$ HMBC spectrum of the O-PS of *A. veronii* bv. *sobria* strain K557. The maps show heteronuclear correlations for (**a**) anomeric protons, and (**b**) H-6 protons. Interresidue correlations between anomeric protons and carbons at the glycosidic linkages are underlined. Some other H/C correlations are depicted as well. Capital letters and Arabic numerals refer to protons or carbons in the sugar residues denoted as shown in Table 3.

In conclusion, the O-antigen of *A. veronii* bv. *sobria* K557, O6 consists of two structurally different O-polysaccharides, both containing only L-Rha4NAc residues. One of them (O-PS 1) is an α-(1→2)-linked homopolymer of L-Rha4NAc. The other polysaccharide, O-PS 2, is built up of trisaccharide repeating units composed of one α-(1→2)-, and two α-(1→3)-linked L-Rha*p*4NAc residues.

As judged by the comparison of the integral intensities of the H-1 proton signals, the content of O-PS 1 in the O-antigen of *A. veronii* bv. *sobria* K557 was predominant and was estimated at approximately 64%.

Based on the composition and methylation analyses as well as the NMR data, the following structures of the O-polysaccharides have been established:

O-PS 1 →2)-α-L-Rha*p*4NAc-(1→
A

O-PS 2 →2)-α-L-Rha*p*4NAc-(1→3)-α-L-Rha*p*4NAc-(1→3)-α-L-Rha*p*4NAc-(1→
B **C** **D**

The structure of the *A. veronii* bv. *sobria* K557 O-antigen is similar but not identical to the recently published O-PS of *Aeromonas hydrophila* JCM 3968, O6, which consisted of two structurally different O-polysaccharides as well (Figure 9). One of them (O-PS 1) was a heteropolymer built up of trisaccharide repeating units composed of one 4-substituted α-D-Gal*p*NAc and two α-(1→3)-linked L-Rha*p*4NAc residues. The other polysaccharide, O-PS 2, was an α-(1→2)-linked homopolymer of L-Rha4NAc [37].

The serological results (Western blotting and ELISA) allowed recognition of structural determinants that are most probably responsible for antibody binding (putative epitopes). The occurrence of common α-L-Rha*p*4NAc-(1→2)-α-L-Rha*p*4NAc and α-L-Rha*p*4NAc-(1→3)-α-L-Rha*p*4NAc disaccharides is sufficient for providing cross-reactivity between the *A. hydrophila* JCM 3968 O6 antiserum and the LPS of *A. veronii* bv. *sobria* K557 and a similar reaction in an opposite system.

On the other hand, an additional epitope (epitopes) most probably related to a 4-substituted α-D-GalpNAc residue or a →4)-α-D-GalpNAc-(1→3)-α-L-Rhap4NAc disaccharide fragment, which has not been found in the O-antigen studied here, determines the specificity of the *A. hydrophila* JCM 3968 O6 serotype. This putative epitope (epitopes) seems to play an important role in the immunospecificity of the reference O6 antiserum (Figure 9).

Based on the data obtained, we suggest the division of the O6 serogroup into two subgroups: serogroup O6a for strains that share α-L-Rhap4NAc-(1→2)-α-L-Rhap4NAc and α-L-Rhap4NAc-(1→3)-α-L-Rhap4NAc disaccharides and an additional →4)-α-D-GalpNAc-associated epitope, and the other serogroup O6b deprived of the latter structural fragment in the O-antigen, with strain *A. veronii* bv. *sobria* K557 as a representative of the new subgroup.

Aeromonas hydrophila JCM 3968, O6a (Dworaczek *et al.* 2019)

O-PS 1 →3)-α-L-Rhap4NAc-(1→**4)-α-D-GalpNAc**-(1→3)-α-L-Rhap4NAc-(1→

O-PS 2 →2)-α-L-Rhap4NAc-(1→

Aeromonas veronii bv. *sobria* K557, O6b (this work)

O-PS 1 →2)-α-L-Rhap4NAc-(1→

O-PS 2 →2)-α-L-Rhap4NAc-(1→3)-α-L-Rhap4NAc-(1→3)-α-L-Rhap4NAc-(1→

Figure 9. Structures of the O6a and O6b antigens repeating units. The additional sugar residue characteristic for serotype O6a and not defined in serotype O6b is shown in bold type.

3. Discussion

Almost every year, health disorders in freshwater fish are recorded on many farms in Poland. Although the development of a particular fish disease depends largely on climate conditions prevailing in a given zone and region, infections caused by *Aeromonas* spp. are the most common among bacterial fish diseases. As demonstrated recently by data collected during the last several years in Poland by Pękala-Safińska, health disorders caused by *Aeromonas* species were mostly observed in carp (*Cyprinus carpio* L.) and were usually manifested by skin lesions (MAI) in the form of ulceration as well as fish mortalities [48].

Previous studies performed on 558 isolates of mesophilic *Aeromonas* collected during 5 years in Polish culture facilities, among which 427 isolates were obtained from common carp and 121 from rainbow trout, revealed predominant occurrence of *A. veronii* bv. *sobria*, *A. sobria* and *A. salmonicida* species. In turn, *A. veroni* bv. *sobria A. bestiarum* and *A. salmonicida* were most frequently identified in both carp and trout samples; *A. veroni* bv. *sobria* was one of the dominant species only among carp isolates, and which is worth emphasizing, almost 80% of these isolates were classified as virulent [28,36]. Serological typing of all collected isolates using 44 antisera of the NIH scheme extended by 20 provisional serogroups for selected Polish isolates showed that O6 was one of the five most commonly identified serogroups, especially among carp isolates, and the other ones were O11, PGO1, O16, and O18. The report mentioned above also showed that, with the use of the antisera for serogroups from O45 to O96, there was little possibility for *Aeromonas* typing, since these groups occur rather rarely among fish isolates, especially in Polish aquacultures. In turn, when the 44 antisera of the NIH scheme were used, about 53% isolates were positively classified to appropriate somatic serotypes and this increased to about 88% when the NIH collection was extended by the sera for provisional serogroups of Polish origin [28,36]. Therefore, to obtain the most positive serotyping results, the postulate to include new provisional antisera against strains occurring in a given area and thus to extend the collection of antisera seems reasonable.

The inland aquaculture sector in Poland is mainly based on the culture of two species of freshwater fish - carp (49% of total production in 2015) and rainbow trout (38% of total production in 2015) [49]; thus, the lack of a commercially available vaccine dedicated to carp seems to be alarming. The most effective prevention of fish disease should involve immunoprophylaxis based on auto-vaccines chosen according to the needs of culture facilities and prepared from bacterial strains isolated from the fish or even the entire region. Auto-vaccines are confirmed to be highly effective against conditionally pathogenic microorganisms like *Aeromonas* sp., *Yersinia ruckeri*, and *Pseudomonas* sp. [50–53]. According to recently published studies, a vaccine containing whole cells and LPS of *Aeromonas* sp. seems to protect fish against MAS disease [54–56]. However, to avoid a failure of implemented prophylactic programs based on auto-vaccination, the serological and structural similarities and differences in O-chain polysaccharides in various serogroups and strains, which contribute to the immunospecificity of *Aeromonas*, should be carried out [57].

Here, the immunochemical investigation of LPS, especially the O-specific polysaccharide of *A. veronii* bv. *sobria* strain K557 were performed. SDS-PAGE and Western blotting revealed that the LPS-glycoforms of *A. veronii* bv. *sobria* K557 contained shorter O-chains than those of *A. hydrophila* JCM 3968; however, these molecules represented S-LPS species rather with the prevalence of the intermediate length O-antigen chains. In a recent study, Osawa et al. reported that the lengths of the *E. coli* O157 antigen could be modulated by *wzz* gene mutations and it has been shown that strains with long, intermediate, and short O-antigens vary in sensitivity to serum complement. The greater resistance of strains with intermediate and/or long length O-antigen chains to serum complement lysis than those with short O-chains is likely to have been optimized for pathogenesis during evolution [58].

Moreover, we established the structure of the O-antigen of *A. veronii* bv. *sobria* strain K557, and found that it consisted of two different O-polysaccharides. The serological studies indicated that the O-PS of *A. veronii* bv. *sobria* strain K557 is closely related but not identical to the O-antigen of *A. hydrophila* strain JCM 3968, which is a reference strain to *Aeromonas* serogroup O6. Another peculiar characteristic of both O-PSs is the presence of 4-amino-4,6-dideoxy-L-mannose (L-Rha4N, L-perosamine), quite an unusual amino sugar, for the first time identified as a compound building the O-antigen of *A. hydrophila* strain JCM 3968, O6 [37]. This is the second work that shows the occurrence of L-perosamine as a component of bacterial O-polysaccharides.

The chemical and mass spectrometry analyses of the phenol-soluble LPS of *A. veronii* bv. *sobria* strain K557 demonstrated the LPS glycoforms had hexaacylated lipid A species with a conserved architecture and a backbone composed of 1,4′-bisphosphorylated-β-(1→6)-linked-D-GlcN disaccharide. The residues of 3-hydroxytetradecanoate were predominant among fatty acids, similarly as previously reported for *A. hydrophila* [37]. However, some differences were found in the acylation profile of lipid A species, in comparison to those of *A. bestiarum* [59]. Amongst the ester-linked saturated fatty acids, not only dodecanoic (12:0) but also tetradecanoic (14:0) residues were found. The latter fatty acids were also detected in the lipid A of *A. veronii* strain Bs19, O16 [60].

The compositional analysis of the core oligosaccharide revealed two isomers of heptose (D,D-Hep and L,D-Hep), and MALDI-TOF MS confirmed that the core decasaccharide, with the following composition: $Hep_6Hex_2HexN_1Kdo_1P_1$, has a structure shared by the LPS core regions of the *A. hydrophila* and *A. bestiarum* species [37,59]. Interestingly, the mass spectrum of the core OS of *A. hydrophila* JCM 3968, which was isolated from R-LPS molecules after mild acid hydrolysis and separation by gel-permeation-chromatography, showed an ion at *m/z* 1856.59, which corresponded to the dephosphorylated core variant with the composition $Hep_6Hex_2HexN_1Kdo_{anh}$. On the other hand, the structure of the core OS in the SR-LPS molecules of *A. hydrophila* JCM 3968 was slightly different: $Hep_6Hex_1HexN_1HexNAc_1Kdo_1P_1$ [37]. In these species, the O-antigen was linked to the GalNAc residue, whereas in the rough R-LPS glycoforms, the galactose appeared to be a terminal outer core sugar, similarly as has been established for the core OS of the rough mutant strain of *A. hydrophila* AH-3, O34 [39].

In conclusion, the immunochemical studies of LPS, which is a glycolipid characterized by high heterogeneity amongst *Aeromonas* sp. bacteria, may facilitate selection of vaccine strains suitable for immunoprophylaxis of MAI/MAS diseases. The O-antigen of *A. veronii* bv. *sobria* strain K557, serotype O6, studied here is composed of two O-polysaccharides, both containing a unique sugar 4-amino-4,6-dideoxy-L-mannose (*N*-acetyl-L-perosamine, L-Rha*p*4NAc). The major O-polysaccharide (O-PS 1) is built up of an α1→2 linked of L-Rha*p*4NAc, whereas the other one, O-PS 2, has trisaccharide repeating units composed of α1→2 and α1→3 linked L-Rha*p*4NAc residues. The serological studies confirmed the structural analyses and showed that the O-antigens of *A. veronii* bv. *sobria* K557, i.e., the strain isolated from the common carp during an outbreak of MAI/MAS on a Polish fish farm, and *A. hydrophila* JCM 3968, represent the same *Aeromonas* serogroup O6 and are closely related but not identical. The recently studied O-antigen of *A. hydrophila* JCM 3968, O6 consisted of two structurally different O-polysaccharides. One of them was a heteropolymer built up of trisaccharide repeating units composed of one α-D-Gal*p*NAc and two α-(1→3)-linked L-Rha*p*4NAc residues. The other O-polysaccharide was an α-(1→2)-linked homopolymer of L-Rha4NAc. The consideration of the serological results in view of the known O-antigen structures enabled recognition of domains that could be responsible for antibody binding (putative epitopes). The occurrence of common α-L-Rha*p*4NAc-(1→2)-α-L-Rha*p*4NAc and α-L-Rha*p*4NAc-(1→3)-α-L-Rha*p*4NAc disaccharides is sufficient for providing cross-reactivity between *A. hydrophila* JCM 3968 O6 antiserum and the LPS of *A. veronii* bv. *sobria* 557, and *vice versa*. On the other hand, the additional epitope related to a 4-substituted α-D-Gal*p*NAc residue, which has not been found in the O-antigen studied here, determines the specificity of the *A. hydrophila* JCM 3968 O6-serotype. This putative epitope seems to play an important role in the immunospecificity of the reference O6 antiserum.

Therefore, based on the data obtained, we suggest division of the O6 serogroup into two subgroups. The serogroup O6a includes strains that share α-L-Rha*p*4NAc-(1→2)-α-L-Rha*p*4NAc and α-L-Rha*p*4NAc-(1→3)-α-L-Rha*p*4NAc disaccharides and an additional →4)-α-D-Gal*p*NAc-associated epitope. *A. veronii* bv. *sobria* strain K557 is a representative of the other subgroup O6b and its O-antigen is deprived of the latter structural fragment.

4. Materials and Methods

4.1. Bacterial Strain, Cultivation Conditions, and Isolation of LPS

Aeromonas veronii strain K557 was isolated from a common carp during an outbreak of motile aeromonad infection/motile aeromonad septicemia (MAI/MAS) on a Polish fish farm. The isolate was identified to the species level by restriction analysis of 16S rRNA gene amplified by polymerase chain reaction [28] and classified as *Aeromonas veronii* bv. *sobria* because of the positive reaction for arginine dihydrolase, and negative reactions for ornithine decarboxylase and aesculin hydrolysis [30,31]. Based on virulence-associated markers (hemolytic, gelatinolytic, and caseinolytic activities), strain K557 was classified as virulent for fish. For the LPS analysis, *A. veronii* bv. *sobria* K557 bacterium was cultivated with shaking (120 rpm) on tryptic soy broth (TSB) for 72 h at 28 °C. The cells were harvested by low-speed centrifugation (8000× *g*, 20 min). The recovered bacterial cell pellet was washed twice with 0.85% saline and once more with distilled water.

The bacterial cells (5 g dry mass) were digested with lysozyme, RNAse, and DNAse (24 h, 1 mg/g) and then with Proteinase K (36 h, 1 mg/g) in 50 mM phosphate buffer (pH 7.0) containing 5 mM $MgCl_2$. The suspension was dialyzed against distilled water and freeze-dried. The digested cells were extracted three times with aqueous 45% phenol at 68 °C, [38] and the separated layers were dialyzed against tap and distilled water. LPS species recovered from the phenol phase were purified by ultracentrifugation at 105,000× *g* and freeze-dried to give a yield of 3.9% of dry bacterial cell mass.

4.2. Degradation of LPS and Isolation of O-polysaccharide

The phenol-soluble S-LPS (110 mg) was hydrolyzed with aqueous 2.5% acetic acid at 100 °C for 3 h, and lipid A precipitate was removed by centrifugation. The supernatant was concentrated and then fractionated by GPC on a column (1.8 × 80 cm) of Sephadex G-50 fine (Pharmacia, Sweden) using 1% acetic acid as the eluent and monitoring with a differential refractometer (Knauer, Berlin, Germany). The yield of the O-PS fraction was 33% of the LPS mass subjected to hydrolysis.

4.3. Chemical Analyses

For neutral and amino sugar analysis, the LPS and O-PS samples were hydrolyzed with 2 M CF_3CO_2H (100 °C, 4 h) or 10 M HCl for 30 min at 80 °C, respectively, and reduced with $NaBD_4$; this was followed by acetylation with a 1:1 (*v*/*v*) mixture of acetic anhydride and pyridine (85 °C, 0.5 h).

To release acidic sugar, LPS was dephosphorylated with 48% aqueous hydrofluoric acid, HF (4 °C, 18 h) and dried under vacuum over sodium hydroxide [40]. Methanolysis was performed with 1 M MeOH/HCl (85 °C, 1 h), and the sample was extracted twice with hexane. The methanol layer was concentrated and the residue was dried and acetylated. The monosaccharides were identified as alditol and aminoalditol acetates [61] as well as acetylated methyl glycosides by GLC-MS.

For determination of the absolute configuration [42], the O-PS was subjected to 2-octanolysis (300 μL (*S*)-(+)-2-octanol or (*SR*)-(±)-2-octanol and 20 μL acetyl chloride, 100 °C, 3 h); the products were acetylated and analyzed by GLC-MS as above. A sample from the polysaccharide of *Citrobacter gillenii* O9a,9b [43] was used as the reference standard of D-Rha4N (D-perosamine).

Methylation of the O-PS (1.0 mg) was carried out with methyl iodide in dimethyl sulfoxide in the presence of powdered sodium hydroxide [62]. The products were recovered by extraction with chloroform/water (1:1, *v*/*v*), hydrolyzed with 10 M HCl for 30 min at 80 °C, *N*-acetylated, and reduced with $NaBD_4$. The partially methylated alditol acetates derivatives were analyzed by GLC-MS.

For fatty acid analysis, a sample of the lipid A (1 mg) was subjected to methanolysis in 2 M methanolic HCl (85 °C, 12 h). The resulting fatty acid methyl esters were extracted with hexane and converted to their *O*-trimethylsilyl (*O*-TMS) derivatives, as described [63,64]. The methanol layer containing methyl glycosides was dried and acetylated with a pyridine-acetic anhydride mixture. The fatty acid derivatives and acetylated methyl glycosides were analyzed by GLC-MS as above.

All the sample derivatives were analyzed on an Agilent Technologies 7890A gas chromatograph (Agilent Technologies, Wilmington, DE, USA) connected to a 5975C MSD detector (inert XL EI/CI, Agilent Technologies, Wilmington, DE, USA). The chromatograph was equipped with an HP-5MS capillary column (Agilent Technologies, 30 m × 0.25 mm, flow rate of 1 mL/min, He as carrier gas). The temperature program for all the derivatives was as follows: 150 °C for 5 min, then 150 to 310 °C at 5 °C/min, and the final temperature was maintained for 10 min.

4.4. NMR Spectroscopy

An O-PS sample was deuterium-exchanged by freeze-drying with D_2O and then examined in 99.98% D_2O using acetone as an internal standard (δ_H 2.225, δ_C 31.45). 1D and 2D NMR spectra were recorded at 32 °C on a 500 MHz NMR Varian Unity Inova instrument (Varian Associates, Palo Alto, CA, USA) using Varian software Vnmrj V. 4.2 rev. (Agilent Technologies, Santa Clara, CA, USA). The following homonuclear and heteronuclear shift-correlated two-dimensional experiments were conducted for signal assignments and determination of the sugar sequence: $^1H,^1H$ DQF-COSY, $^1H,^1H$ TOCSY, $^1H,^1H$ NOESY, $^1H,^{13}C$ HSQC, and $^1H,^{13}C$ HMBC. The mixing time was set to 100 and 200 ms in the TOCSY and NOESY experiments, respectively. The $^1H,^{13}C$ HSQC experiment with CRISIS based multiplicity editing was optimized for a coupling constant of 146 Hz. The heteronuclear multiple-bond correlation (HMBC) experiment was optimized for $J_{H,C}$ = 7 and 5 Hz, with 2-step low-pass filter 130 and 165 Hz to suppress one-bond correlations.

4.5. MALDI-TOF Mass Spectrometry (MS)

LPS was analyzed with matrix-assisted laser desorption/ionization time-of flight (MALDI-TOF) mass spectrometry (MS) using a Waters SYNAPT G2-*Si* HDMS instrument (Waters Corporation, Milford, MA, USA) equipped with a 1 kHz Nd:YAG laser system. Acquisition of the data was performed using MassLynx software version 4.1 SCN916 (Waters Corporation, Wilmslow, UK). Mass spectra were assigned with a multi-point external calibration using red phosphorous (Sigma-Aldrich, St. Louis, MO, USA) and recorded in the negative ion mode. An LPS sample (at a concentration of 10 μg/μL) was suspended in a water/methanol (1:1, *v*/*v*) solution containing 5 mM EDTA and then dissolved by ultrasonication. After desalting with some grains of cation exchange beads (Dowex 50WX8-200; Sigma-Aldrich, St. Louis, MO, USA), one microliter of the sample was transferred onto a well plate covered with a thin matrix film and allowed to dry at room temperature. The matrix solution was prepared from 2′,4′,6′-trihydroxyacetophenone (THAP) (200 mg/mL in methanol) mixed with nitrocellulose (15 mg/mL) suspended in 2-propanol/acetone (1:1, *v*/*v*) in proportion of 4:1 (*v*/*v*), according to the published method [65].

4.6. SDS-PAGE

LPS preparations were separated in 12.5% SDS-Tricine polyacrylamide electrophoresis gel and bands were visualized by silver staining after oxidation with periodate according to the published method [66].

4.7. Serological Studies

Polyclonal O-antisera against *A. hydrophila* JCM 3968, serogroup O6 and *A. veronii* bv. *sobria* strain K557 were obtained by immunization of New Zealand white rabbits with heat-inactivated bacteria according to the published procedure [28]. Rabbits were acclimatized in the animal facility of the National Veterinary Research Institute (Puławy, Poland), and all the experiments were performed according to the procedures approved by the local ethical committee (The Second Local Ethical Committee on Animal Testing in Lublin, the permission number 48/2012).

Western blots with rabbit antisera were performed after transferring SDS-PAGE-separated LPS profiles to Immobilon P (Millipore, St. Louis, MO, USA). The primary antibodies were detected using alkaline phosphatase-conjugated goat anti-rabbit antibodies (Sigma, St. Louis, MO, USA). Blots were developed with nitroblue tetrazolium and 5-bromo-4-chloro-3-indolylphosphate toluidine (Sigma) for 5 min, as described elsewhere [67].

The ELISA was performed as described previously [68] with some modifications, namely: 1–2 μg of the studied LPS per well were coated on flat-bottom Nunc-Immuno plates; the reaction was developed using rabbit-IgG specific peroxidase-conjugated goat antibodies (Jackson ImmunoResearch, West Grove, PA, USA); the final absorbance (A_{405}) was read with the help of a Multiskan Go microplate reader (Thermo Fisher Scientific USA, Vantaa, Finland).

Adsorption of antisera was carried out using wet masses of bacterial cells washed in PBS (phosphate-buffered saline). Bacterial mass (100 μL) was suspended in 1 mL of serum diluted 1:50 in PBS. After 0.5 h incubation on ice, the cells were removed by centrifugation and the process was repeated two or three more times [69].

Author Contributions: A.T.-S. and K.D. conceived and designed the experiments; A.T.-S. and K.D. funding acquisition; K.D. performed Western blotting studies; D.D. carried out ELISA experiments; A.P-S.. and K.D. performed serotyping with the agglutination test; K.D. and A.T.-S. performed chemical analyses; K.D. and A.T.-S. contributed to the interpretation of mass spectra and NMR data; All authors analyzed the data; A.T.-S. and K.D. original draft preparation; All authors have reviewed the paper.

Funding: This research was financially supported by the grant from the National Science Centre (Decision No. DEC-2011/03/B/NZ1/01203) and the Polish Ministry of Science and Higher Education research funds (BS-M-11-010-18-2-05 and BS-P-11-010-18-2-04).

Acknowledgments: The authors gratefully thank Andrzej Gamian for the gift of the O-polysaccharide of *Citrobacter gillenii* O9 (L. Hirszfeld Institute of Immunology and Experimental Therapy, Polish Academy of Sciences, Wrocław, Poland) and Paweł Sowiński for recording the NMR spectra (Intercollegiate NMR Laboratory, Department of Chemistry, Technical University of Gdańsk, Poland). The authors wish to acknowledge the assistance and technical support of Hubert Pietras during LPS isolation and purification.

Conflicts of Interest: The authors declare no conflict of interest.

References

1. Janda, J.M.; Duffy, P.S. Mesophilic aeromonads in human diseases: Current taxonomy, laboratory infection and infectious diseases spectrum. *Rev. Infect. Dis.* **1988**, *10*, 980–997. [CrossRef] [PubMed]
2. Janda, J.M. Recent advances in the study of the taxonomy, pathogenicity and infectious syndromes with the genus *Aeromonas*. *Clin. Microbiol. Rev.* **1991**, *4*, 397–410. [CrossRef] [PubMed]
3. Martin-Carnahan, A.; Joseph, S.W. Order XII. Aeromonadales ord. nov. In *Bergey's Manual of Systematic Bacteriology*, 2nd ed.; Brenner, D.J., Krieg, N.R., Staley, J.T., Eds.; Springer: New York, NY, USA, 2005; Volume 2, pp. 556–578.
4. Janda, J.M.; Abbott, S. The genus *Aeromonas*: Taxonomy, pathogenicity, and infection. *Clin. Microbiol. Rev.* **2010**, *23*, 35–73. [CrossRef] [PubMed]
5. Huys, G. The Family *Aeromonadaceae*. In *The Prokaryotes—Gammaproteobacteria*; Rosenberg, E., DeLong, E.F., Lory, S., Stackebrandt, E., Thompson, F., Eds.; Springer: Berlin/Heidelberg, Germany, 2014; pp. 27–57. [CrossRef]
6. Araujo, R.M.; Arribas, R.M.; Pares, R. Distribution of *Aeromonas* species in waters with different levels of pollution. *J. Appl. Bacteriol.* **1991**, *71*, 182–186. [CrossRef] [PubMed]
7. Cipriano, R.C.; Bullock, G.L. Furunculosis and other diseases caused by *Aeromonas salmonicida*. *Fish. Dis. Leafl.* **2001**, *66*, 1–32. [CrossRef]
8. Praveen, P.K.; Debnath, C.; Shekhar, S.; Dalai, N.; Ganguly, S. Incidence of *Aeromonas* spp. infection in fish and chicken meat and its related public health hazards: A review. *Vet. World* **2016**, *9*, 6–11. [CrossRef]
9. Ibrahem, M.D.; Mostafa, M.M.; Arab, R.M.H.; Rezk, M.A.; Elghobashy, H.; Fitzsimmons, K.; Diab, A.S. (Eds.) Prevalence of Aeromonas hydrophila infection in wild and cultured tilapia nilotica (O. niloticus) in Egypt. In Proceedings of the 8th International Symposium on Tilapia in Aquaculture, Cairo, Egypt, 12–14 October 2008; Volume 2, pp. 1257–1270.
10. Tomas, J.M. The Main *Aeromonas* Pathogenic Factors. *ISRN Microbiol.* **2012**, *2012*, 256261. [CrossRef]
11. Stratev, D.; Odeyemi, O.A. An overview of motile *Aeromonas* septicaemia management. *Aquacult. Int.* **2017**, *25*, 1095–1105. [CrossRef]
12. Hoel, S.; Vadstein, O.; Jakobsen, A.N. The significance of mesophilic *Aeromonas* spp. in minimally processed ready-to-eat seafood. *Microorganisms* **2019**, *7*, 91. [CrossRef]
13. Holmberg, S.D.; Schell, W.L.; Fanning, G.R.; Wachsmuth, I.K.; Blake, P.A.; Brenner, D.J.; Farmer, J.J. *Aeromonas* intestinal infections in the United States. *Ann. Int. Med.* **1986**, *105*, 683–689. [CrossRef]
14. Figueras, M.J. Clinical relevance of *Aeromonas sM503*. *Rev. Clin. Microbiol.* **2005**, *16*, 145–153. [CrossRef]
15. Mencacci, A.; Cenci, E.; Mazzolla, R.; Farinella, S.; D'Alo, F.; Vitali, M.; Bistoni, F. *Aeromonas veronii* biovar *veronii* septicaemia and acute suppurative cholangitis in a patient with hepatitis B. *J. Med. Microbiol.* **2003**, *52*, 727–730. [CrossRef] [PubMed]
16. Roberts, M.T.M.; Enoch, D.A.; Harris, K.A.; Karas, J.A. *Aeromonas veronii* biovar sobria bacteraemia with septic artritis confirmed by 16S rDNA PCR in an immunocompetent adult. *J. Clin. Microbiol.* **2006**, *55*, 241–243. [CrossRef]
17. Garduño, R.A.; Moore, A.R.; Oliver, G.; Lizama, A.L.; Garduño, E.; Kay, W.W. Host cell invasion and intracellular resistance by *Aeromonas salmonicida*: Role of the S-layer. *J. Clin. Microbiol.* **2000**, *46*, 660–668. [CrossRef]
18. Merino, S.; Rubires, X.; Aguillar, A.; Guillot, J.F.; Tomas, J.M. The role of the O-antigen lipopolysaccharide on the colonization in vivo of the germfree chicken gut by *Aeromonas hydrophila* serogroup O:34. *Microb. Pathog.* **1996**, *20*, 325–333. [CrossRef]
19. Rabaan, A.A.; Gryllos, I.; Tomas, J.M.; Shaw, J.G. Motility and polar flagellum are required for *Aeromonas caviae* adherence to HEp-2 cells. *Infect. Immun.* **2001**, *69*, 4257–4267. [CrossRef] [PubMed]

20. Raetz, C.R.H.; Whitfield, C. Lipopolysaccharide endotoxins. *Annu. Rev. Biochem.* **2002**, *71*, 635–700. [CrossRef]
21. Caroff, M.; Karibian, D. Structure of bacterial lipopolysaccharides. *Carbohydr. Res.* **2003**, *338*, 2431–2447. [CrossRef]
22. Whitfield, C.; Trent, M.S. Biosynthesis and export of bacterial lipopolysaccharides. *Annu. Rev. Biochem.* **2014**, *83*, 99–128. [CrossRef]
23. Maldonado, R.F.; Sá-Correia, I.; Valvano, M.A. Lipopolysaccharide modification in Gram-negative bacteria during chronic infection. *FEMS Microbiol. Rev.* **2016**, *40*, 480–493. [CrossRef]
24. Cao, H.; Wang, M.; Wang, Q.; Xu, T.; Du, Y.; Li, H. Identifying genetic diversity of O antigens in *Aeromonas hydrophila* for molecular serotype detection. *PLoS ONE* **2018**, *13*, e0203445. [CrossRef] [PubMed]
25. Sakazaki, R.; Shimada, T. *O*-Serogrouping for mesophilic *Aeromonas* strains. *Jpn. J. Med. Sci. Biol.* **1984**, *37*, 247–255. [CrossRef] [PubMed]
26. Thomas, L.V.; Gross, R.J.; Cheasty, T.; Rowe, B. Extended serogrouping scheme for motile, mesophilic *Aeromonas* species. *J. Clin. Microbiol.* **1990**, *28*, 980–984. [PubMed]
27. Kokka, R.P.; Vedros, N.A.; Janda, J.M. Electrophoretic analysis of the surface components of autoagglutinating surface array protein-positive and surface array protein-negative *Aeromonas hydrophila* and *Aeromonas sobria*. *J. Clin. Microbiol.* **1990**, *28*, 2240–2247. [PubMed]
28. Kozinska, A.; Pekala, A. Serotyping of *Aeromonas* species isolated from Polish fish farms in relation to species and virulence phenotype of the bacteria. *Bull. Vet. Inst. Pulawy* **2010**, *54*, 315–320.
29. Esteve, C.; Alcaide, E.; Canals, R.; Merino, S.; Blasco, D.; Figueras, M.J.; Tomas, J.M. Pathogenic *Aeromonas hydrophila* serogroup O:14 and O:81 strains with S-layer. *Appl. Environ. Microbiol.* **2004**, *70*, 5898–5904. [CrossRef] [PubMed]
30. Hickman-Brenner, F.W.; MacDonald, K.L.; Steigerwalt, A.G.; Fanning, G.R.; Brenner, D.J.; Farmer, J.J., III. *Aeromonas veronii*, a new ornithine decarboxylase-positive species that may cause diarrhea. *J. Clin. Microbiol.* **1987**, *25*, 900–906.
31. Altwegg, M.; Steigerwalt, A.G.; Altwegg-Bissig, R.; Luthy-Hottenstein, J.; Brenner, D.J. Biochemical identification of *Aeromonas* genospecies isolated from humans. *J. Clin. Microbiol.* **1990**, *28*, 258–264.
32. Rahman, M.; Colque-Navarro, P.; Kühn, I.; Huys, G.; Swings, J.; Möllby, R. Identification and characterization of pathogenic *Aeromonas veronii* bv. *sobria* associated with epizootic ulcerative syndrome in fish in Bangladesh. *Appl. Environ. Microbiol.* **2002**, *68*, 650–655. [CrossRef]
33. Cai, S.-H.; Wu, Z.-H.; Jian, J.-C.; Lu, Y.-S.; Tang, J.F. Characterization of pathogenic *Aeromonas veronii* bv. *veronii* associated with ulcerative syndrome from Chinese longsnout catfish (Leiocassis longirostris Günther). *Braz. J. Microbiol.* **2012**, *43*, 382–388. [CrossRef]
34. Turska-Szewczuk, A.; Lindner, B.; Pekala, A.; Palusinska-Szysz, M.; Choma, A.; Russa, R.; Holst, O. Structural analysis of the O-specific polysaccharide from the lipopolysaccharide of *Aeromonas veronii* bv. *sobria* strain K49. *Carbohydr. Res.* **2012**, *353*, 62–68. [CrossRef] [PubMed]
35. Turska-Szewczuk, A.; Pietras, H.; Duda, K.A.; Kozińska, A.; Pękala, A.; Holst, O. Structure of the O-specific polysaccharide from the lipopolysaccharide of *Aeromonas sobria* strain Pt312. *Carbohydr. Res.* **2015**, *403*, 142–148. [CrossRef] [PubMed]
36. Kozińska, A. Genotypic and Serological Analysis of Domestic Mesophilic Isolates *Aeromonas* sp. in Terms of Pathogenicity and the Type of Disease Symptoms Caused by Them in Fish. Habilitation Thesis, The National Veterinary Institute—The National Research Institute, Puławy, Poland, 2009.
37. Dworaczek, K.; Kurzylewska, M.; Karaś, M.A.; Janczarek, M.; Pękala-Safińska, A.; Turska-Szewczuk, A. A unique sugar L-perosamine (4-amino-4,6-dideoxy-L-mannose) is a compound building two O-chain polysaccharides in the lipopolysaccharide of *Aeromonas hydrophila* strain JCM 3968, serogroup O6. *Mar. Drugs* **2019**, *17*, 254. [CrossRef] [PubMed]
38. Westphal, O.; Jann, K. Bacterial lipopolysaccharide. Extraction with phenol-water and further applications of the procedure. *Meth. Carbohydr. Chem.* **1965**, *5*, 83–91.
39. Knirel, Y.A.; Vinogradov, E.; Jimenez, N.; Merino, S.; Tomas, J.M. Structural studies on the R-type lipopolysaccharide of *Aeromonas hydrophila*. *Carbohydr. Res.* **2004**, *339*, 787–793. [CrossRef] [PubMed]

40. Pieretti, G.; Corsaro, M.M.; Lanzetta, R.; Parrilli, M.; Nicolaus, B.; Gambacorta, A.; Lindner, B.; Holst, O. Structural characterization of the core region of the lipopolysaccharide from the haloalkaliphilic *Halomonas pantelleriensis*: Identification of the biological O-antigen repeating unit. *Eur. J. Org. Chem.* **2008**, *2008*, 721–728. [CrossRef]
41. Domon, B.; Costello, C.E. A systamatic nomenclature for carbohydrate fragmentations in FAB MS/MS spectra of glycoconjugates. *Glycoconj. J.* **1988**, *5*, 397–409. [CrossRef]
42. Leontein, K.; Lindberg, B.; Lönngren, J. Assignment of absolute configuration of sugars by GLC of their acetylated glycosides formed from chiral alcohols. *Carbohydr. Res.* **1978**, *62*, 359–362. [CrossRef]
43. Lipiński, T.; Zatonsky, G.V.; Kocharova, N.A.; Jaquinod, M.; Forest, E.; Shashkov, A.S.; Gamian, A.; Knirel, Y.A. Structures of two O-chain polysaccharides of *Citrobacter gillenii* O9a,9b lipopolysaccharide. A new homopolymer of 4-amino-4,6-dideoxy-D-mannose (perosamine). *Eur. J. Biochem.* **2002**, *269*, 93–99. [CrossRef]
44. Lipkind, G.M.; Shashkov, A.S.; Knirel, Y.A.; Vinogradov, E.V.; Kochetkov, N.K. A computer-assisted structural analysis of regular polysaccharides on the basis of 13C-n.m.r. data. *Carbohydr. Res.* **1988**, *175*, 59–75. [CrossRef]
45. Ovchinnikova, O.G.; Kocharova, N.A.; Katzenellenbogen, E.; Zatonsky, G.V.; Shashkov, A.S.; Knirel, Y.A.; Lipiński, T.; Gamian, A. Structures of two O-polysaccharides of the lipopolysaccharide of *Citrobacter youngae* PCM 1538 (serogroup O9). *Carbohydr. Res.* **2004**, *339*, 881–884. [CrossRef] [PubMed]
46. Jansson, P.E.; Kenne, L.; Widmalm, G. Computer-assisted structural analysis of polysaccharides with an extended version of CASPER using ^{1}H- and ^{13}C-NMR data. *Carbohydr. Res.* **1989**, *188*, 169–191. [CrossRef]
47. Knirel, Y.A.; Ovod, V.V.; Paramonov, N.A.; Krohn, K.J. Structural heterogeneity in the O polysaccharide of *Pseudomonas syringe* pv. coriandricola GSPB 2028 (NCPPB 3780, W-43). *Eur. J. Biochem.* **1998**, *258*, 716–721. [CrossRef] [PubMed]
48. Pękala-Safińska, A. Contemporary threats of bacterial infections in freshwater fish. *J. Vet. Res.* **2018**, *62*, 261–267. [CrossRef] [PubMed]
49. Lirski, A.; Myszkowski, L. Polish aquaculture in 2016 based on the analysis of questionnaire RRW-22. Part 1. *Komun. Ryb.* **2017**, *6*, 20–27.
50. Terech-Majewska, E. Improving disease prevention and treatment in controlled fish culture. *Arch. Pol. Fish.* **2016**, *24*, 115–165. [CrossRef]
51. Kozińska, A.; Pękala, A. Characteristics of disease spectrum in relation to species, serogroups, and adhesion ability of motile aeromonads in fish. *Sci. World J.* **2012**. [CrossRef] [PubMed]
52. Grudniewska, J.; Dobosz, S.; Terech-Majewska, E.; Zalewski, T.; Siwicki, A.K. Economic and health aspects of vaccinating against furunculosis and yersiniosis in rainbow trout culture. *Komun. Ryb.* **2010**, *1*, 18–21.
53. Siwicki, A.K.; Baranowski, P.; Dobosz, S.; Kuźmiński, H.; Grudniewska, J.; Kazun, K.; Głąbski, E.; Kazuń, B.; Terech-Majewska, E.; Trapkowska, S. Using new generation vaccines administered *per os* in granulate as prophylaxis against furunculosis and yersiniosis in salmonids. In *Current Challenges in Fish Disease Prevention and Treatment*; Siwicki, A.K., Ed.; Publisher IRS: Olsztyn, Poland, 2004; pp. 117–122.
54. Sukenda, S.; Romadhona, E.I.; Yuhana, M.; Pasaribu, W.; Hidayatullah, D. Efficacy of whole-cell and lipopolysaccharide vaccine of *Aeromonas hydrophila* on juvenile tilapia *Oreochromis niloticus* against motile aeromonad septicemia. *AACL Bioflux* **2018**, *11*, 1456–1466.
55. Dehghani, S.; Akhlaghi, M.; Dehghani, M. Efficacy of formalin-killed, heat-killed and lipopolysaccharide vaccines against motile aeromonads infection in rainbow trout (*Oncorhynchus mykiss*). *Glob. Veterin.* **2012**, *9*, 409–415. [CrossRef]
56. Fernandez, J.B.; Yambot, A.V.; Almeria, O. Vaccination of Nile tilapia (*Oreochromis niloticus*) using lipopolysaccharide (LPS) prepared from *Aeromonas hydrophila*. *Int. J. Fauna Biol. Stud.* **2014**, *1*, 1–3.
57. Pajdak, J.; Terech-Majewska, E.; Platt-Samoraj, A.; Schulz, P.; Siwicki, A.K.; Szweda, W. Characterization of pathogenic *Yersinia ruckeri* strains and their importance in rainbow trout immunoprophylaxis. *Med. Weter* **2017**, *73*, 579–584. [CrossRef]
58. Osawa, K.; Shigemura, K.; Iguchi, A.; Shirai, H.; Imayama, T.; Seto, K.; Raharjo, D.; Fujisawa, M.; Osawa, R.; Shirakawa, T. Modulation of the O-antigen chain length by the *wzz* gene in *Escherichia coli* O157 influences its sensitivities to serum complement. *Microbiol. Immunol.* **2013**, *57*, 616–623. [PubMed]

59. Turska-Szewczuk, A.; Lindner, B.; Komaniecka, I.; Kozinska, A.; Pekala, A.; Choma, A.; Holst, O. Structural and immunochemical studies of the lipopolysaccharide from the fish pathogen, *Aeromonas bestiarum* strain K296, serotype O18. *Mar. Drugs* **2013**, *11*, 1235–1255. [CrossRef] [PubMed]
60. Turska-Szewczuk, A.; Duda, K.A.; Schwudke, D.; Pekala, A.; Kozinska, A.; Holst, O. Structural studies of the lipopolysaccharide from the fish pathogen, *Aeromonas veronii* strain Bs19, serotype O16. *Mar. Drugs* **2014**, *12*, 1298–1316. [CrossRef] [PubMed]
61. Russa, R.; Urbanik-Sypniewska, T.; Lindström, K.; Mayer, H. Chemical characterization of two lipopolysaccharide species isolated from *Rhizobium loti* NZP2213. *Arch. Microbiol.* **1995**, *163*, 345–351. [CrossRef] [PubMed]
62. Ciucanu, I.; Kerek, F. A simple and rapid method for the permethylation of carbohydrates. *Carbohydr. Res.* **1984**, *131*, 209–217. [CrossRef]
63. Pieretti, G.; Corsaro, M.M.; Lanzetta, R.; Parrilli, M.; Vilches, S.; Merino, S.; Tomas, J.M. Structure of the core region from the lipopolysaccharide of *Plesiomonas shigelloides* strain 302-73 (serotype O1). *Eur. J. Org. Chem.* **2009**, *2009*, 1365–1371. [CrossRef]
64. Komaniecka, I.; Choma, A.; Lindner, B.; Holst, O. The structure of a novel lipid A from the lipopolysaccharide of *Bradyrhizobium elkanii* containing three mannose units in the backbone. *Chem. Eur. J.* **2010**, *16*, 2922–2929. [CrossRef]
65. Silipo, A.; Molinaro, A.; Sturiale, L.; Dow, J.M.; Erbs, G.; Lanzetta, R.; Newman, M.A.; Parrilli, M. The elicitation of plant innate immunity by lipooligosaccharide of *Xanthomonas campestris*. *J. Biol. Chem.* **2005**, *280*, 33660–33668. [CrossRef]
66. Tsai, C.M.; Frasch, C.E. A sensitive silver stain for detecting lipopolysaccharides in polyacrylamide gels. *Anal. Biochem.* **1982**, *119*, 115–119. [CrossRef]
67. Turska-Szewczuk, A.; Pietras, H.; Borucki, W.; Russa, R. Alteration of O-specific polysaccharide structure of symbiotically defective *Mesorhizobium loti* mutant 2213.1 derived from strain NZP2213. *Acta Biochim. Pol.* **2008**, *55*, 191–199. [PubMed]
68. Sidorczyk, Z.; Zych, K.; Toukach, F.V.; Arbatsky, N.P.; Zabłotni, A.; Shashkov, A.S.; Knirel, Y.A. Structure of the O-polysaccharide and classification of *Proteus mirabilis* G1 in *Proteus* serogroup O3. *Eur. J. Biochem.* **2002**, *269*, 1406–1412. [CrossRef] [PubMed]
69. Drzewiecka, D.; Arbatsky, N.P.; Shashkov, A.S.; Stączek, P.; Knirel, Y.A.; Sidorczyk, Z. Structure and serological properties of the O antigen of two clinical *Proteus mirabilis* strains classified into a new *Proteus* O77 serogroup. *FEMS Immunol. Med. Microbiol.* **2008**, *54*, 185–194. [CrossRef] [PubMed]

MDPI
St. Alban-Anlage 66
4052 Basel
Switzerland
Tel. +41 61 683 77 34
Fax +41 61 302 89 18
www.mdpi.com

Marine Drugs Editorial Office
E-mail: marinedrugs@mdpi.com
www.mdpi.com/journal/marinedrugs

www.ingramcontent.com/pod-product-compliance
Lightning Source LLC
LaVergne TN
LVHW071940080526
838202LV00064B/6645